Mycorrhizosphere and Pedogenesis

AF895982

Ajit Varma • Devendra K. Choudhary
Editors

Mycorrhizosphere and Pedogenesis

Springer

Editors
Ajit Varma
Amity Institute of Microbial Technology
Amity University
Noida, Uttar Pradesh, India

Devendra K. Choudhary
Amity Institute of Microbial Technology
Amity University
Noida, Uttar Pradesh, India

ISBN 978-981-13-6479-2 ISBN 978-981-13-6480-8 (eBook)
https://doi.org/10.1007/978-981-13-6480-8

© Springer Nature Singapore Pte Ltd. 2019
This work is subject to copyright. All rights are reserved by the Publisher, whether the whole or part of the material is concerned, specifically the rights of translation, reprinting, reuse of illustrations, recitation, broadcasting, reproduction on microfilms or in any other physical way, and transmission or information storage and retrieval, electronic adaptation, computer software, or by similar or dissimilar methodology now known or hereafter developed.
The use of general descriptive names, registered names, trademarks, service marks, etc. in this publication does not imply, even in the absence of a specific statement, that such names are exempt from the relevant protective laws and regulations and therefore free for general use.
The publisher, the authors, and the editors are safe to assume that the advice and information in this book are believed to be true and accurate at the date of publication. Neither the publisher nor the authors or the editors give a warranty, express or implied, with respect to the material contained herein or for any errors or omissions that may have been made. The publisher remains neutral with regard to jurisdictional claims in published maps and institutional affiliations.

This Springer imprint is published by the registered company Springer Nature Singapore Pte Ltd.
The registered company address is: 152 Beach Road, #21-01/04 Gateway East, Singapore 189721, Singapore

Preface

It is based on consensus that agriculture has a long history of research targeted on how to improve the efficacy of root symbionts, namely, rhizobia and mycorrhiza. A hopeful approach has been engaged to understand how natural selection regulates changes in mutualistic exchanges. An eloquent understanding of basic evolutionary processes can be employed to develop agricultural management practices that favor the most effective symbionts. It has been reported that mutually beneficial interactions between plant and associated rhizospheric microorganisms are ubiquitous which is important for ecosystem functioning. Reports observed that in rhizosphere, symbiotic nitrogen fixation has incurred through bacteria not in order, namely, *Rhizobium*, *Bradyrhizobium*, *Mesorhizobium*, *Sinorhizobium*, *Azorhizobium* spp., and several other so-called rhizobia. In addition, mycorrhizal fungi supply their host plants with mineral nutrients, namely, phosphorus (P) and other benefits.

Glimpses on microbial commune in the rhizosphere: What is the need of rhizobacterial assortment and commune investigation in rhizosphere? (i) Assists in measurement of soil health and biological displacement (ii) Provides commune fingerprinting (iii) Describes niche selectivity and ecological adaptation (iv) A parametric to analyze appropriate inoculants and transgenics in soil (v) Assists in differentiation of viable cells from nonviable.

The microbial commune in the rhizosphere has demonstrated to be a challenging task due to the vast diversity and the atrociousness of the population inhabiting the environment. An investigation has incurred through widespread perturbation of microbial community by changes in environmental conditions and soil management practices. In the present scenario, we should have interest in understanding the cooperative activities among microbial populations and how they affect agroecosystems when applied in agricultural soils.

Based on preceding reports, it has been reported that more than 80% of the terrestrial ecosystem is able to form mycorrhizal association wherein it involves bidirectional flow of nutrients and several other benign properties. In addition, mycorrhizal fungi are able to provide protection to the host plant against root and shoot pathogens.

Glimpses on mycorrhizal commune in the rhizosphere: An importance of mycorrhizal symbiosis! (i) Most of the land plants are mycorrhizal that appx. >80% (ii) Helps in bidirectional flow of nutrients in an ecosystem (iii) C- flow from plant to fungus (iv) ii Mineral flow from fungus to plant (v) An extension of hypha beyond nutrient depletion zone (vi) An extension into soil pores that helps in nutrients absorption (vii) Shows biocontrol mechanisms against root/shoot and soil pathogens (viii) Antibiosis through antibiotics (ix) Antibiosis through enzyme secretion (x) Induced systemic resistance (xi) Systemic acquired resistance.

Published reports reflect that several mycorrhizosphere bacteria also help in mycorrhiza formation wherein a variety of Gram-positive and Gram-negative strains are involved, the so-called mycorrhiza-helper bacteria (MHB). It was also shown subsequently that in natural agroecosystem, the occurrence of MHB along with diversity of AM fungi is considered as a key contributor to the diversity and productivity of plant community. The symphony of root-inhabiting AM community shows seasonal variation within individual host plants, and this can change with plant maturity. Various farming practices, namely, fertilizer input, cultivation, and fumigation, put forth deleterious effects on AM commune.

Therefore, in the present book, editors compiled research carried out on microbial occurrence and diversity of mycorrhiza, various tools to characterize them, and its impact on soil formation/health together with crop productivity.

Chapter 1 provides glimpses on the mycorrhizal fungi and their prominent role in nutrient transfer into host plants, presenting a view on the application of mycorrhiza for crop biofortification.

Chapter 2 focuses on the role of microorganisms in soil formation and the mechanisms for weathering process employed by such microflora, highlighting the current and advanced molecular approaches for studying soil microbial diversity.

Chapter 3 focuses on the role and significance of AM fungi in phytoremediation of hydrocarbon-contaminated sites. Additionally, metabolite formation during bioremediation of organic compounds is discussed. Furthermore, the factor affecting the bioremediation process is also summarized.

Chapter 4 focuses on crop rotation, soil processing, and other management factors that can affect the level and benefits of mycorrhizas. All field crops are included in the product rotation. To learn more about the benefits of mycorrhizas to field crops, more work should be done on product rotation.

Chapter 5 describes awareness about mycorrhiza utility among policy-makers and agriculturists which is a step toward sustainable agriculture, reforestation, and climate change-resilient farming and enhanced food security.

Chapter 6 represents a systematic review of the role of mycorrhiza in soil genesis using scientometric approach.

Chapter 7 highlights the concept of mycorrhizosphere, xenobiotic metabolism, molecular approaches for detoxifying the organic xenobiotics, and the role of mycorrhizosphere in stabilizing the environment in an eco-friendly way.

Chapter 8 represents definitions, descriptions, and histories of the important allied and/or corollary activities of soil morphology, survey, interpretation, and characterization.

Chapter 9 highlights the positive influence of microbial interactions on plant diseases and plant growth-promoting effect considering updated knowledge.

Chapter 10 describes nutrient development in soil which is carried out via biological transformation through action of microorganism. Without microbes, soil would be a virtually inert (lifeless) body, but with them, soil is truly a living, dynamic system. Microbes and the humus produced by them work as a glue to hold soil particles together in aggregates, hence improving soil tilth and decreasing soil depletion or erosion.

Chapter 11 focuses on the recent tools and techniques to study the mycorrhizosphere.

Chapter 12 focuses on the present scenario of pedosphere in terms of its structural composition, functions, and the interrelationship of the microflora and microfauna with the different layers of soil.

Chapter 13 focuses on the role of metagenomic analysis in exploring the AM fungi which are the most widespread symbionts in agroecosystems worldwide.

Chapter 14 highlights the mechanisms adapted by AM fungi for the biocontrol of soilborne phytopathogens.

Chapter 15 describes the role of various fungal species for biodegradation and transformation of environmental contaminants by enzymes and biomass.

Chapter 16 aims at dealing with the two processes together and thus has a comprehensive review literature on how this symbiosis drives pedogenesis and determines terrestrial biome of a particular ecosystem.

Chapter 17 focuses on term "phytoremediation" that has got more and more attention over the past decade. Due to the multifaceted applications of AM fungi, it has been widely used as a xenobiotic tool.

Chapter 18 focuses on the importance of mycorrhizal fungi which are nearly an indispensable part of the rhizosphere, because of their immense potential for bringing sustainability and stability in crop production.

Chapter 19 focuses on optimization of crop management practices, agriculture practices which increased proliferation, and diversity of mycorrhizal fungi which in turn increased agriculture production.

Chapter 20 emphasizes the exploration of metagenomics data over recent years, with special reference to extreme habitats that have given access to diverse and novel biocatalysts that may be of great value in mycorrhizosphere and pedogenesis.

Chapter 21 addresses the significance of mineral weathering by microbial interactions and the contribution of plant microbial communities on soil formation through nutrient cycling which further improves the soil functionality.

Chapter 22 deals with the role of microflora and microfauna in soil health and the various roles played by these two groups of organisms.

Chapter 23 focuses on various PCR-based and non-PCR-based molecular techniques that may be utilized to study the microbial diversity and structure within the mycorrhizosphere.

Noida, Uttar Pradesh, India Ajit Varma
Noida, Uttar Pradesh, India Devendra K. Choudhary

Contents

About the Editors

Ajit Varma completed his M.Sc. (1959) and Ph.D. (1964) degrees at Allahabad University, Allahabad, India. In the course of his professional career, he has also served as a Microbiologist (Assistant Professor), Indian Agricultural Research Institute (IARI), New Delhi (1963–1971), Senior Microbiologist (Associate Professor), IARI, New Delhi (1971–1974), Associate Professor, Jawaharlal Nehru University (JNU), New Delhi (1975–1984), and Professor, JNU, New Delhi (1985–2004). He has been a Visiting Professor and Visiting Research Scientist at the Technical University, Graz (Austria); University of Tuebingen, Tuebingen (Germany); Friedrich Schiller University, Jena (Germany); Philipps University, Marburg (Germany); Technical University, Munich (Germany); Kingston (Jamaica); Max Planck Visiting Professorship (Germany), Helmholtz Zentrum, Muenchen (Germany); Gutenberg University, Mainz (Germany); Consejo Superior de Investigaciones Científicas (CSIC), Madrid (Spain); University of Dundee (Scotland); University of Ljubljana (Slovenia); and International Centre For Genetic Engineering and Biotechnology (ICGEB) (Italy). His international awards/fellowships include the Commonwealth Fellowship (Australia); National Research Council (Canada); Alexander von Humboldt Foundation (Germany); National Science Foundation (USA); Indo-Czechoslovakia Exchange Programme (Prague); DAAD Fellowship (Germany); the Deutsches BMFT Programme, George-August University, Gottingen (Germany); and RAISA Fellowship. He was awarded a fellowship for Innovative Research in Biotechnology (Italy), Swiss Federal Research Fellowship (Switzerland), the BP Koirala Award (Nepal), and DFG-INSA Fellowship (Indo-Germany), as well as the FAMI Award, Association of Microbiologists of India, and Honorary Diploma, UMF, Cluj-Napoca, Romania. Dr. Varma has successfully completed major projects as PI sponsored by DBT, DST, DRDO, and ICAR. Besides, he has supervised more than 60 Ph.D. students and published over 300 research articles for national and international journals of repute, as well as several major review articles and chapters in books. He has published 90 books in the area of microbial technology, published by Academic Press, London; CRC Press, Florida, USA; IDRC, Canada; and Springer-Verlag, Germany. Dr. Varma has been the series editor for Springer-Verlag's series on soil biology, and has edited 50 volumes on soil biology. He was also nominated as Editor-in-Chief by IK International to make series of books on microbial and biotechnological research. Dr. Varma has been a member of the National Academy of Agriculture Sciences (FNAAS); International Society of Symbiosis, Boston, USA; Indian Science Congress Association; Executive Council, Amity University Uttar Pradesh; University Research Council, Amity University Uttar Pradesh; Academic Council, Amity University Rajasthan; ASSOCHAM Knowledge Millennium Council, ASSOCHAM Expert Committee on Agriculture and Food Processing; and ASSOCHAM Expert Committee on S&T and Innovation. He has vast experience in organizing national and international training workshops/symposia and congresses.

Devendra K. Choudhary has 16 years of research experience in microbiology and microbial biotechnology. He is an Associate Professor at Amity Institute of Microbial Technology (AIMT). Recently, he was associated with PDM and worked as Professor (Ad hoc) for 3 months. Before joining Amity University, he spent several years in Mody University of Science and Technology (MUST), Lakshmangarh, Rajasthan, preceded by Barkatullah University, Bhopal, as Lecturer cum Scientist. He pursued his Ph.D. in microbiology on July 11, 2005 from GB Pant University of Agriculture and Technology, Pantnagar, preceded by M.Sc. in microbiology from Maharshi Dayanand Saraswati (MDS) University, Ajmer, and qualified CSIR-UGC-NET in 2002. He is an active researcher and operated major projects with worth amount 10 million sponsored by DBT, DST, and SERB, New Delhi, India, as principal investigator and co-investigator. He has involved and contributed in NAAS policy paper number 36 in the year 2006. He has published more than 80 research/review articles along with several book chapters for reputed journals and edited books (Springer, Elsevier, and Taylor Francis/Wiley-Blackwell). He has supervised/supervising 11 research scholars for their doctoral program as supervisor and co-supervisor, wherein 7 have been awarded at MUST and four ongoing at AIMT. In addition, he is a recipient of NASI-membership 2016 along with Indian National Science Academy (INSA) visiting and summer research fellowship-2014, and his team received DR. RS Rana Memorial best research award 2013 sponsored by Association of Microbiologists of India. His team submitted bacteria to public domain and incurred two accession numbers from MTCC, IMTECH for submitted bacterial cultures (MTCC, 12057 & 12058) along with one MCC no 2607. Besides, his team submitted approximately 25 functional bacterial genes and 80 I6S rDNA sequences to National Center for Biotechnology Information (NCBI) with Accession Numbers. In recent times, he has edited seven books through Springer and filed three patents to India Patent Office, New Delhi.

Mycorrhizal Mediated Micronutrients Transportation in Food Based Plants: A Biofortification Strategy

1

Viabhav K. Upadhayay, Jyoti Singh, Amir Khan, Swati Lohani, and Ajay Veer Singh

1 Introduction

Micronutrients deficiency in cereals provides an impetus for fulfilling the goal of biofortification for the production of crops with increase content of micronutrients. Biofortification and standard fortification are two different terms, where biofortification consigns the nutrients aggregation inside plant cells while later involves use of additives with the foods. Inadequacy of micronutrients (zinc, iron, selenium, copper, manganese and vitamins) in both humans and plants is narrated as 'hidden hunger' (de Valenca et al. 2017), and bestows peril of mal-nutrition among world population. According to the global hunger index 2014, two billion people were reported to suffer from hidden hunger with aspect of micronutrients deficiency (von Grebmer et al. 2014). Besides promoting crop yields, modern agriculture is focused to produce nutritious safe food crops with enhanced micronutrients concentration in edible part of the plants. Intake of crop based food ultimately influences human health and consuming diet with deficiency of necessary micronutrients cause serious ailments in individuals. Various countries of the world primarily depend upon agriculture and production of major cereals such as maize, rice and wheat are notorious to provide about 30% of the calories to more than 4.5 billion people in 94 developing countries (Shiferaw et al. 2011). Effect of adverse environmental factors impedes the uptake of selected micronutrients from soil to plants, and hence lesser amount of micronutrients retards plant growth and development. Dousing of hidden hunger can be alleviate by direct (nutrition-specific) and indirect (nutrition-sensitive) interventions (Ruel and Alderman 2013). The focus of direct interventions relies on consumption behavior and includes dietary diversification, micronutrient supplementation, modification of food choices and fortification. Nutrition-sensitive

V. K. Upadhayay · J. Singh · A. Khan · S. Lohani · A. V. Singh (✉)
Department of Microbiology, College of Basic Sciences and Humanities, GBPUA&T, Pantnagar, Uttarakhand, India

© Springer Nature Singapore Pte Ltd. 2019

A. Varma, D. K. Choudhary (eds.), *Mycorrhizosphere and Pedogenesis*,
https://doi.org/10.1007/978-981-13-6480-8_1

interventions deal with the core determinants of malnutrition and include biofortification. Biofortification process deals with several strategies for augmenting bioavailability of essential nutrients in crops through genetic and agronomic approaches (Bouis et al. 2011). Genetic engineering and classical breeding are two phases of genetic biofortification. However, agronomic biofortification is carried out through application of micronutrient fertilizer to the soil or direct foliar application to the leaves of the crop (de Valenca et al. 2017). The approaches of agronomic and genetic biofortification to increase micronutrient concentration in crops for alleviating micronutrient malnutrition is considered to be lucrative and inapt in developing countries where rural population is most prevalent (Mayer et al. 2008). Therefore, to circumvent the dependency on agronomic and genetic biofortification, agriculturally important microorganisms can be used to confer biofortification of crops as a possible supplementary measure, which can furnish increased micronutrient concentrations in crop plants (Upadhayay et al. 2018; Bouis 2003). Mycorrhizal fungi are ubiquitous soil microorganisms that associate with the roots of almost all land plants and assist their hosts in increased nutrient uptake and the application of mycorrhiza could be an alternative and sustainable tool to augment micronutrients concentration in crops. Striking feature of mycorrhizal fungi is transferring macro elements such as P and N to the host plants. Besides these, the mycorrhizal fungal partner also provides micro nutrients such as Zn, Fe, Mn and Cu to the host plants. Inspite of that mycorrhizal fungi may increases the abiotic stress tolerance against drought, salinity, and heavy metal and biotic stress resistance from several root pathogens, and in return, the host plant transfer about 4–20% of its photosynthetic products to the mycorrhizal fungus (Wright et al. 1998). Intensive studies on mycorrhiza especially on arbuscular mycorrhizal fungi (AMF) have been performed for micro and macro nutrient acquisition from soil and their transportation in associated host plants. Since mycorrhizal fungi associated with crops enhance mineral nutrients in crop plants and elicit an idea to use mycorrhiza in crop biofortification with several elements besides phosphorus. Modern agriculture also requires the goal to fulfill the demand of enhanced production of crops in more sustainable way with increased concentration of essential micronutrients to combat against micronutrients related malnutrition. A number of crops are associated with mycorrhizal fungi and as a 'natural biofertilizers', the mycorrhizal fungal partner exert beneficial effect from feeding the host plant with micronutrients to provides protection from several adverse conditions.

2 Mycorrhiza: A Natural Phosphate Transporter for Co-Partner

Mycorrhizal fungi are an assorted group of diverse fungal taxa make association with the roots of more than 90% of all plant species and in return fungal partner attains up to 20% of plant-fixed carbon (Parniske 2008). Mutualistic relationship between soil fungi and the roots of vascular plants shows bidirectional pattern of nutrient transport, where plant supplies the fungi with photo-synthetically produced sugars, whereas the hyphae network improves the plant capability to absorb nutrients and water. Hence, such supportive system presents benevolence of advantageousness for both partners in terms of nutrient transfer. Four major mycorrhizal types are most commonly described on the basis of their structure and function i.e. arbuscular mycorrhiza (AM), ectomycorrhiza (EM), ericoid mycorrhiza and orchid mycorrhiza (van der Heijden et al. 2015). A number of mycorrhizal fungi could be classified as either ectomycorrhizal or arbuscular mycorrhizal where ectomycorrhizal fungi as typically members of the Ascomycota or Basidiomycota colonize root of trees, predominantly in forest areas, and create a complex network of mycelia adjacent the epidermal and outer cortical cells, known as a 'Hartig net' (Smith and Read 1997). On other hand arbuscular mycorrhizal fungi are all members of the phylum Glomeromycota and show specific nature as obligate biotrophic symbionts and differ from ectomycorrhizal fungi in terms of penetration of hyphae to cells of the inner root cortex to create specific branched structures called 'arbuscules'. Arbuscules form a particular nutrient transfer interface that is embedded with numerous plant and fungal transporters, assisting in nutrient transfer between the symbionts. Though the term 'mycorrhiza' was first used in 1885 which is derived from Greek words *mycos* ('fungus') and *rhiza* ('root') to illustrates mutualistic association between plants root and mycorrhizal fungi. In mycorrhiza development, the AM fungus undergoes numerous developmental phases. In the asymbiotic phase, germination of spores occurs and AM fungi exhibit restricted hyphal development in the absence of a host plant. But, in the presence of root exudates, they turn to the presymbiotic phase which is described by extensive hyphal branching. Mycorrhizal fungi and endophytic fungi both showed beneficial effects on plants, however, one feature that distinguishes mycorrhizal fungi from endophytic fungi is that mycorrhizal fungi do not survive in the absence of their suitable plant hosts because the majority of mycorrhizal fungi are obligate biotrophs. Phosphate (P) availability is second most imperative limiting factor for growth after nitrogen for the plants and as an essential macronutrient 'P' playing a central role in developmental and number of metabolic processes (Singh et al. 2010, 2011, 2013, 2018; Singh and Goel 2015). Plants require huge amount of soluble orthophosphate for growth and development, but unfortunately such forms of phosphorus in soil is inadequate because they react with available cations to form barely soluble calcium phosphates in alkaline soils or/and iron and aluminum phosphates in acid soils (Gyaneshwar et al. 2002; Taktek et al. 2016). Therefore, plants show inability to obtain enough P through the direct uptake pathway because of short P accessibility in the rhizosphere; and as a result, plants greatly rely on the mycorrhizal uptake pathway for P absorption. Mycorrhizal symbiosis is extensively dispersed in terrestrial ecosystems and inhabits a protected ecological niche via improving host plant nutrient uptake, particularly P uptake. It is well established that in mycorrhizal uptake pathway, P absorbed by external fungal hyphae is translocated to structures inside the roots and thus across the symbiotic interface to the plant cortical cells (Stonor et al. 2014). The improved uptake of P has been suggested to be due to certain aspects such as; (a) an enhancement in the absorbing surface with exploration of a high soil volume by the extramatrical mycelium, (b) the little hyphal diameter leading to an increased P-absorbing surface region and, compared to non-mycorrhizal roots, high P influx rates per surface unit, (c) the production of organic acids and phosphatases for releasing P from inorganic and organic complexes, respectively, and (d) formation of polyphosphates by mycorrhizal fungi and thus low internal P concentrations (Marschner and Dell 1994; Bücking and Heyser 2003). In mycorrhizal Pi (inorganic phosphate) uptake pathway" Pi is unloaded from fungal partner to photobiont in colonized root cortex cells where fungal hyphae create hyphal coils and arbuscules formed via repeated dichotomous branching of the fungal hypha. Here Pi departs the hypha into the peri-arbuscular space by an unknown mechanism where it is assimilated by the colonized cortex cell (Willmann et al. 2013). Moreover, mycorrhizal fungal partner during infection may influence the mineral nutrition of the host plant directly by improving plant growth through nutrient acquisition by the fungus, or indirectly by altering transpiration rates and the composition of rhizospheric microflora (Marschner and Dell 1994).

3 Essential Micronutrients and Related Problems with Their Deficiency

The supply of food with poor nutritious value to the poor communities of ever increasing world population is a major serious concern as most of the diet used is micronutrient deficient. According to the report of World Health Organization (WHO) (2002), hidden hunger with respect to micronutrients, especially Zn, Fe, I and Se affects half of the world population. Intake of micronutrient deficient diet engenders various health related risk in humans (Fig. 1.1). Regarding the significance of micronutrients in crop production, the modern agriculture has challenge that deals with the level of micronutrients in the major staple food crops as well as diets of humans and animals. Prevalence of micronutrient deficiency and their detrimental consequences on mortality, morbidity and disability result in extensive disease burden in low and middle income countries. Iron (Fe), a fourth most abundant and essential microelement on the earth crust, belongs to the 4th period and group VIII of the long form of periodic table (atomic number 26, and atomic weight of 55.845),

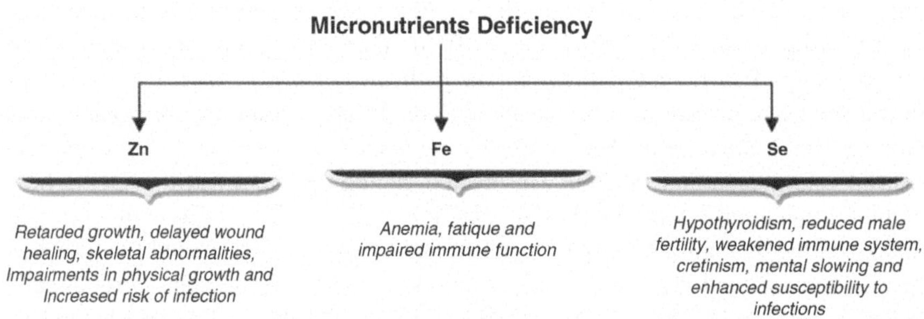

Fig. 1.1 Schematic representation of effects of deficiency of three major micronutrients (Zn, Fe and Se) in humans

which intercedes a range of cellular processes. Iron resides in both heme and non heme forms in plants and involved in energy production and other several redox reactions (Miller et al. 1995). Fe-containing proteins were determined for their various roles in cellular respiration, intermediary metabolism, oxygen transport, DNA stability and repair, and as well as photosynthesis in plants. However, Fe deficiency is a general phenomenon among the animal and plant kingdoms (López-Millán et al. 2013). Fe deficiency cause chlorosis in plants and considered as major limitation for crop yield, which ultimately affects human health via food-chain, particularly to those people whose diets mostly rely on plant resources.

Iron deficiency is the most important cause of nutritional anemia which is associated with impaired neurocognitive development and as well impaired immune factions in children (Murray-Kolb 2013). The Ministry of Health and Family Welfare, Government of India, in 2013, launched the National Iron Plus Initiative as a comprehensive strategy to combat the public health challenge of IDA, as iron deficiency contributes to more than 50% of anemia in India (MoHFW 2013). It is now well documented that iron deficiency has detrimental effects in patients with coronary artery disease, heart failure, and pulmonary hypertension, and possibly in patients undergoing cardiac surgery.

The micronutrient zinc (Zn) is vital for all organisms and required as a cofactor in more than 300 enzymes and plays important structural roles in several proteins, including transcription factors. Zinc deficiency is presently listed as a main risk factor for human health responsible for the development of various diseases (WHO 2002). Zinc deficiency associated with several health related problems such as retarded growth, delayed wound healing, skeletal abnormalities, diarrhea, increased abortion risk (Salgueiro et al. 2000), impairments of physical growth, increased risk of infections, and DNA damage and cancer development (Gibson 2006; Prasad 2007). Soils with inadequate zinc of various regions (China, India, Iran Pakistan and Turkey) are also critical reason for Zn deficiency in human beings of these particular regions (Cakmak et al. 1999; Hotz and Brown 2004). However, less solubility of Zn in soils rather than low total amount of Zn is the foremost reason for the common occurrence of Zn deficiency problem in crop plants. Another example of profound dietary nutrient is Selenium (Se), which essentially required for humans and animals. Selenium, a major constituent of selenocysteine at the active site of selenoproteins involved in a wide range of metabolic pathways, such as antioxidant defense, thyroid hormone metabolism and immune function (Rayman 2012). Deficiency of selenium is thought to affect 800 million people worldwide (Malagoli et al. 2015). Low intake of Se in the diet may associated with a number of health disorders, including heart diseases, hypothyroidism, reduced male fertility, weakened immune system and enhanced susceptibility to infections and cancer (Hatfield et al. 2014), oxidative stress-related conditions and epilepsy (Zeng and Combs 2008). Deficiency of Vitamin B6, Vitamin B12, Vitamin C, Vitamin E, folic acid, iron and zinc were also reported to cause DNA damage through adapting same strategies as radiation and various chemicals, and hence considered a major cause of cancer and other disabilities (Ames 2001).

4 Biofortification

To challenge the micronutrients malnutrition and maintaining elevated concentrations of major micronutrients, experts advocated for using different strategies for elemental biofortification of crop plants. Agronomic approaches, plant breeding and genetic engineering and application of soil microorganisms (plant growth promoting rhizobacteria and mycorrhiza) are key strategies which could be practiced for engendering substantial concentration of micronutrients (Zn, Fe, Se, Cu and Mn) in edible parts of plants. Each approach exerts its unique advances and also presents certain limitations.

4.1 Agronomic Biofortification

Production of food crops with increased content of iron, zinc and other micronutrients for feeding world population with improved human health is a major challenge. Various agricultural approaches to endeavored sustainable solutions to conquer the consequences of micronutrient malnutrition are urgently needed. The term "Agronomic biofortification" is used in perspective of application of nutrient-rich fertilizers to soil or on foliage to increase the micronutrients concentration in edible parts of the crops and consequently increase the intake of essential micronutrients by consumers. Application of Zn fertilizers to soil and in form of foliar application for improving Zn content in different parts of crops is important agronomic interventions for gaining prosperous Zn biofortification of food crops (Cakmak and Kutman 2017).

4.2 Genetic and Plant Breeding Approaches for Biofortification

Another approach for biofortification is "plant breeding" which has been practiced by farmers for hundreds of years. Conventional plant breeding denotes to the crossing of plants to yield descendants with traits of both parents. Since the last few decades, scientists at several research institutes/centers such as CGIAR, IRRI, CIMMYT, CIAT and IITA have collected data on the potential for breeding to increase the content of Fe, Zn, and provitamin A carotenoids considerably in edible parts of rice, wheat, maize, beans and cassava (Bouis 1996; Graham et al. 1999, 2001). Moreover, crops with higher levels of other micronutrients such as selenium have also been developed (CGIAR 2007). Rice varieties containing high proportions of Zn and Fe were crossed with high-yielding rice varieties to generate progeny with both high yield and elevated concentrations of micronutrients (Khush 2003). Biofortification carried out through genetic modification has two advantages compared with conventional plant breeding. First it takes less time to produce crops those expresses in a stable way the trait of interest (e.g. nutritional content). Second, it allows the transfer of particular genes or gene of interest

(Garcia-Casal et al. 2016). 'Golden Rice' is a paradigm of a genetically modified biofortified crop capable for synthesis of beta carotene.

4.3 Biofortification with Microorganisms

Microorganisms as invisible soil engineers maintain soil health, construct a hub for different biogeochemical cycles and facilitate complex mixtures of micronutrients into simpler form which are uptake by plants for their growth at various stages. Plant associated bacteria stimulate the growth of host plant by concerning certain ways such as increase mobility, uptake, and enrichment of nutrients in plants. Fixed forms of iron, phosphorus and zinc deposited in soil are unavailable to plants, so organic acids produced by soil microorganisms change unavailable form of these micronutrients to available form. Rampant use of chemical fertilizers for pleasing the demand of nutrients resulted in deterioration of soil lushness with altered soil microbial diversity that has led to reduce crop production. Therefore, application of microorganisms such as plant growth promoting bacteria (PGPB), cyanobacteria and fungi as bioinoculants can be a striking strategy for efficient and environment friendly alternatives to chemical fertilizers and pesticides. Plant growth promoting bacteria (PGPB) enhance crop production through various mechanisms namely biological nitrogen fixation, solubilization of insoluble minerals, production of phytohormones and biocontrol (Glick et al. 1999, Singh and Prasad 2014; Prasad et al. 2016, Yadav et al. 2016; Joshi et al. 2018). Mycorrhizal fungi make friendly association with plants and fulfill the need of nutrient requirements of host plant. As a bridge mycorrhiza allow crossing of macro elements (P, N) and as well microelements (Zn, Fe, Cu) from soil to plants. Mycorrhizal fungi and plant growth promoting rhizobacteria (PGPR) may exhibit variety of strategies for nutrient acquisition and exhibit plant growth through production of siderophore, organic acid, phytohormones, VOCs and controlling plant pathogens (Fig. 1.2).

Microorganisms certainly soil bacteria associated with plants facilitate better nutrient acquisition from soil to plants and this consequence led microorganisms for

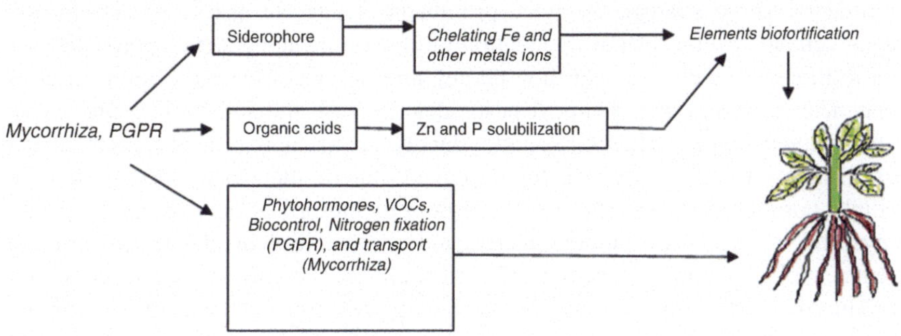

Fig. 1.2 Schematic diagram representing Mycorrhiza and PGPR mediated elemental biofortification with various plant growth promoting traits

using them as an ideal candidate for crop biofortification with Zn, Fe and selenium (Upadhayay et al. 2018). Variety of microorganisms including bacteria, cyanobacteria and fungi have been characterized to assist host plants growth and embeds increased concentration of nutrients in crops as achieved by genetic and agronomic biofortification. However, plant growth stimulating microorganisms are minutely investigated for biofortification strategies and they are needed to be launch after the category of agronomic and genetic approaches to develop proficient biofortification strategies for the staple crops.

5 Why Mycorrhiza in Biofortification of Food Based Plants/Crops?

At worldwide level micronutrient deficiency drastically affects health of women and children, that consequence into serious and wide spread negative health issues, especially in developing countries (WHO 2002). Unfortunately, populations of both of developed and developing countries devour cereals as main food components. And firmly dependence on cereal grains containing inadequate amount of micronutrients are an important cause for malnutrition among populations. Therefore, urgent need of natural fertilizers instead of chemical fertilizers is required to increase the concentration of important micronutrients in edible parts of the plants, which could be determined as eco-friendly way in sustainable agro-ecosystem. Application of soil microorganisms such as arbuscular mycorrhizal fungi (AMF or AM fungi) in form of natural fertilizers represent a key link between plants and soil mineral nutrients. Mycorrhiza make symbiotic association with most crop plants and directly associated with plant mineral nutrition, controlling over plant pathogens and increase drought tolerance of plants (Plenchette et al. 2005).

In present era, the productivity of crops increased through several scientific intervenes such as production of genetically modified plants, enhancement of micronutrient uptake efficiency of plant through plant breeding and agronomic approaches, protection of crops from pathogens by application of chemical pesticides and improvement in plant productivity and yield by application of chemical fertilizers. These strategies provide prominent results but also exert certain non approachable results such as application of agronomic and genetic approaches for biofortification is costlier and time taking, and on other side huge application of chemical fertilizers and chemical pesticides decrease the soil fertility and cause environmental issues. Therefore employing mycorrhiza may be a cost effective and environment friendly approach for improved mineral elements concentrations in edible portions of crops (Fig. 1.3). Moreover, biofortication strategy by using mycorrhiza can propose a more effective and sustainable element biofortification to reduce global human malnutrition (He and Nara 2007). Verities of siderophores are produced by mycorrhiza (ectomycorrhizal fungi, AM fungi and encoid mycorrhizal fungi) assisting as metal chelating agents especially for iron chelation. The widespread mycorrhizal mycelia extensively experienced soils substrates and obtain mineral elements including major (N, P, K) and micronutrients (Fe, Zn, Cu)

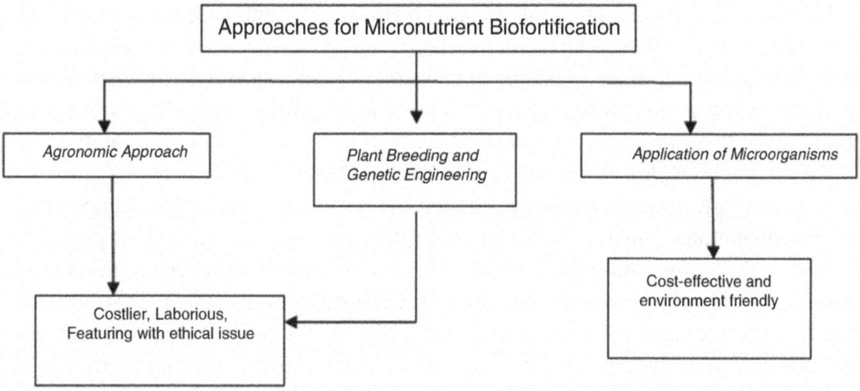

Fig. 1.3 Schematic flow chart representing different strategies for micronutrients biofortification

efficiently and effectively (Koide and Kabir 2000). Root colonization by vesicular arbuscular mycorrhizae increases spatial availability of zinc similarly to that of phosphorus. Mycorrhizal plants usually have higher zinc contents in the shoot dry matter and are less sensitive to zinc deficiency than non-mycorrhizal plants (Marschner 1993).

6 Mechanisms of Nutrients Transportation

Fungi utilize very miscellaneous substrates on the basis of their nutritional strategy by owing their filamentous organization. Mycorrhizal association strengthen plant growth by escalating nutrient uptake by means of several traits such as (a) by increases in absorbing surface area, (b) by mobilizing sparingly available nutrient sources, (c) by excreting chelating compounds and (d) by producing ectoenzymes. This association also protect root from various soil pathogens and thereby increase root growth and nutrient acquisition of the host root. However, plants associated with mycorrhiza can take up nutrients from the soil via two pathways:

I. **The plant pathway:** that embraces the direct uptake of nutrients from the soil by the root epidermis along with its root hairs. However, the nutrients uptake from the soil through the plant pathway is often limited by the low mobility of nutrients in the soil. For example the mobility of phosphate is subsequently low that its uptake leads rapidly to the development of depletion zones around the roots and limits the further phosphate uptake through the plant pathway to the low rate of diffusion (Schachtman et al. 1998).
II. **The mycorrhizal pathway:** that involves the uptake of nutrients via the extra radical mycelium (ERM) of the fungus and transport to the Hartig net in ecto-mycorrhiza (ECM) interactions or to the intraradicle mycelium (IRM) in arbuscular mycorrhizal (AM) interactions, and the uptake by the plant from the interfacial apoplast.

However, the involvement of the mycorrhizal or plant pathway toward total phosphorus uptake also depends on the fungal and plant species. *Glomus intraradices* has been known to suppress the expression of plant phosphorus transporters of the plant pathway the most, whereas *G. mosseae* had the least effect (Grunwald et al. 2009). In *Solanum lycopersicum*, *G. intraradices* shown approximately 100% of the plant's phosphorus via the mycorrhizal pathway, but the part of *Gigaspora rosea* to total phosphorus uptake was much lower (Smith et al. 2003). A high phosphorus uptake and transfer independent on the plant species was also reported by *Glomus caledonium*, although *Glomus invermaium* merely transferred momentous amounts of phosphorus to the host plant flax (Bucking et al. 2012). This indicates that the involvement of the mycorrhizal pathway to nutrient acquirement also depends on fungal specific effects lying on the plant pathway activity as well as on the competence with which mutual partners cooperate and exchange nutrients across the mycorrhizal interface.

6.1 Mechanism of Phosphorus Transport

According to Jones et al. (1998), the efficacy of mycorrhizal associated plants for inorganic phosphorus (Pi) uptake is 3.1–4.7 times higher than that of nonmycorrhizal plants. In soils, not sufficiently abounding with P, nutrient uptake by plants faraway exceeds the pace at which it diffuses into the root zone, resulting in zones of Pi depletion surrounding roots. AMF helps to conquer this problem by extending their external hyphae from root surfaces to areas of soil beyond the Pi depletion zone, thereby exploring a greater volume of the soil than is accessible to the unaided root (O'keefe and Sylvia 1991). Root hairs are the primary site for the Pi acquisition and in response to Pi scarcity both the density and length of root hairs increase to explore a larger volume of soil. P-deficient plants are characterized by increases in root/shoot ratio, root branching, root elongation, and root top soil exploration. The root hairs are commonly longer, while primary root growth is reduced (Lynch and Brown 2008; Vance 2010). Effective uptake phosphorus by mycorrhizal fungi is linked to:

(a) Polyphosphate formation in the hyphae while maintain the low orthophosphate concentration internally
(b) Emancipate of soluble phosphorus from the organic complexes by means of extracellular acid phosphatases and phytases production through hyphae in the soil.

Besides acidification of the rhizosphere, exudation of malate, citrate and oxalate greatly enhances Pi mobilization by chelation or ligand exchange (such as Al, Ca, Fe). Root induced acidification can decrease the rhizosphere pH by 2–3 units relative to the bulk soil (Marschner 1995). Phosphorus is generally taken up in the form of orthophosphate (Pi). It is conventionally known that orthophosphate transport into the root is arbitrated via mechanism of secondary transport system and dependent on P type H^+-ATPase activity in plasma membrane. The H^+-ATPase protein is present in the plant membrane in the region of arbuscule hyphae, which confirms the existence of nutrient transport activities at this root and fungus interface (Rausch et al. 2001). The transporters of orthophosphate has been cloned and revealed to be expressed in rhizodermis region (root epidermis including root hairs) of *Medicago truncatala* and *Solanum lycopersicum*. These transporters are involved in uptake of orthophosphate at the interface of root soil (Royzman et al. 1997). The encoded proteins are orthologous to GvPT, which expressed in extraradical hyphae of *Glomus versiforme* and were allocated to the Pht1 family of Pi transporters in plants (Field et al. 1996; Humbert et al. 2000). Although mRNA levels of Pht1 transporters are usually lower in mycorrhizal roots than in nonmycorrhizal plants (Leone et al. 2001), orthophosphate transporter's mRNA LePT1 was detected in arbuscule containing cells of mycorrhizal tomato, suggesting altered cellular localization of Pht1 transporters during fungal colonization in mycorrhizal roots. The molecular basis of the establishment and functioning of the arbuscular-mycorrhizal symbiosis is largely not understood. Hyphae of the ectomycorrhizal *Hebeloma cylindrosporum* have at least two high-affinity Pi transporters (HcPT1 and HcPT2) that are differentially expressed depending on the P availability and mycorrhizal status (Tarty et al. 2009).

6.2 Nitrogen

Ectomycorrhizal (ECM) fungi can uptake inorganic nitrogen sources very proficiently from soils, but their capability to utilize organic nitrogen sources, and to make these sources available for the host plant, is usually seen as a vital factor in the nitrogen nutrition of ECM plant species. Many ECM fungi can mobilize and utilize amides, amino acids such as alanine, glutamine and glutamate, which can represent a major nitrogen puddle, particularly in acid organic soils. The extraradical mycelium (ERM) of AM and ECM fungi can uptake the inorganic nitrogen sources ammonium or nitrate from the soil (Finlay et al. 1988). Ammonium ion is the most preferred nitrogen source for mycorrhizal fungi, because ammonium uptake is energetically more proficient than the uptake of nitrate (Toussaint et al. 2004). The expression of high affinity *AMT1* and *AMT2*, two ammonium transporters of the ECM fungus *H. cylindrosporum*, is regulated by the exogenous ammonium supply. The expression of both transporters is up regulated under low ammonium conditions, but down-regulated in response to an exogenous supply of ammonium. In addition to *AMT1* and *AMT2*, a low affinity ammonium transporter (*AMT3*) is expressed under non-limiting ammonium conditions, which enables the fungus to maintain a basal level of nitrogen uptake and assimilation also at high exogenous supply conditions (Javelle et al. 2003). *GintAMT1*, an ammonium transporter of the AM fungus *Glomus intraradices* seems to be mainly involved in the uptake of ammonium by the ERM under low ammonium availabilities. An exogenous supply of nitrate, stimulates the expression of a fungal nitrate transporter in the ERM of *G. intraradices*. After its uptake from the soil, nitrate is converted into ammonia via nitrite and nitrate reductases in AM and ECM fungi (Jin et al. 2005). Inorganic nitrogen is taken up by the fungal ERM and assimilated via nitrate reductase and the oxoglutarate aminotransferase GS–GOGAT cycle. It is then converted into arginine, which is translocated along the coenocytic fungal hyphae from the ERM into the IRM. Arginine is broken down in the IRM, releasing urea and ornithine, which are further broken down by the actions of urease and ornithine aminotransferase. Ammonia released from arginine breakdown passes to the host via ammonia channels (AMT). Amino acids from ornithine breakdown and ammonia assimilation in the IRM may be catabolized within the IRM or translocated to the ERM. (Bucking et al. 2012).

6.3 Iron

Fungal endophytes of ericaceous plants were the first group of mycorrhizal fungi investigated with regard to the structure of the main siderophores they release (Haselwandter et al. 1992). The principal siderophore of ericoid mycorrhizal fungi such as *Rhizoscyphus ericae* and *Oidiodendron griseum* was identified as the hexapeptide ferricrocin. Siderophore mediated iron uptake in mycorrhizal fungus is an energy-dependent process. Siderophores are part of a multiple module system for transporting ferric iron into a cell. Other components include a specific outer membrane receptor protein in the inner membrane are (a) Fec A protein complex, (b) Fep A protein complex and (c) TonB-ExbB-ExbD protein complex. Under iron deficiency bacteria synthesize siderophore and increase number of receptor molecules. Once the siderophore excreted outside from cell via membrane receptor it bind with iron complex and transport the iron inside the cell via Fec A and Fep A outer membrane receptor proteins. Afterwards, it transported to ABC-Transporter systems (from ATP binding cassette) (Davidson and Nikaido 1991), assembled of two proteins, one is permease, span the membrane and a second one which is capable in ATP hydrolysis to make available energy for transport (Fig. 1.4). Later siderophore iron complex release in cytoplasm with the help of membrane protein Ton B. In the cytoplasm of mycorrhiza, iron released from the complex by means of a mechanism which is still not clear: it may involve hydrolytic destruction of the siderophore molecule or the reduction of Fe^{3+} by a NADPH linked siderophore reductase. The resulting Fe^{2+} does not have a high affinity for siderophore and consequently dissociated from the complex.

In soil, plant roots normally coexist with bacteria and fungi which may produce siderophores capable of sequestering the accessible soluble iron and so interfere in the midst of plant growth and function (Singh et al. 2017). Plant root might be capable of taking up ferric complexes of siderophore and using these as sources of iron. To satisfy iron requirement of mycorrhiza in addition to plants have evolved specific potential mechanism to chelate insoluble iron by releasing siderophores. Fungal siderophores may arouse plant growth directly by means of increasing iron availability in rhizospheric region or indirectly competitively inhibiting plant pathogens growth with less efficient iron uptake system (Srivastava et al. 2013).

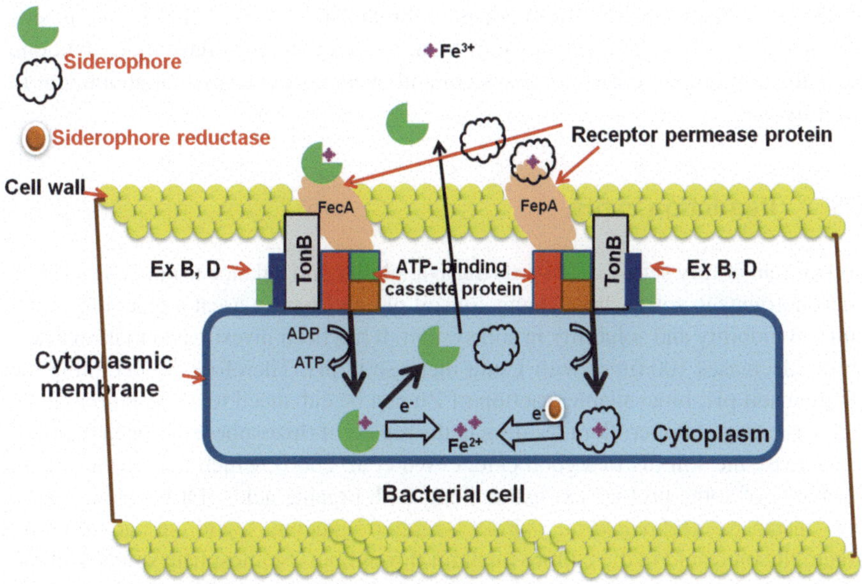

Fig. 1.4 Depiction of siderophore mediates iron transport. (Singh et al. 2017)

6.4 Zinc Solubilization

Fungi necessitate a variety of micronutrients for their growth and metabolic process. Among the nutrients, zinc is an element present in the enzyme system as co factor, metal activator of many enzymes, role in nutrition and physiology of both eukaryotic as well as prokaryotic organisms is widely studied, especially its importance for activity of many enzymes. Zinc deficiency in fungi along with bacteria is accompanied through impairment of the pigment formation (prodigiosin, chrisogenin, subtilin and melanin) (Saravanan et al. 2003). Moreover, there are sufficient reports indicating significant potential of these fungi in improving bioavailable fraction of Zn in plant rhizosphere and in plant tissues (Biari et al. 2008; Subramanian et al. 2009). While these fungi play a vital part in improvement of food quality. Consequently, they would be given prime importance in future while formulation strategies to ease Zn malnutrition in humans through food, particularly in developing countries where different diets is not obtainable to common people and they are not capable to afford food supplements. Among microbes, both fungi and bacteria have revealed incredible capability to improve Zn availability in the rhizosphere and augment Zn concentration in different plant regions as well (Subramanian et al. 2009). Rhizosphere microflora may cause mobilization or solubilization of Zn include:

1. Reduction in soil pH (Subramanian et al. 2009),
2. Chelation (Whiting et al. 2001)
3. Through improving root growth and root absorptive area (Burkert and Robson 1994).

These mechanisms differ from one microorganism to another. Several organisms may make use of single mechanism of them while others may have more than one mechanism to improve Zn in soil system and eventually improve Zn attainment in plant tissues.

6.5 Reduction in pH

Soil system is very sensitive towards the availability and balance in concentration of micronutrients in soil. A little change in soil pH may have a great impact on micronutrient mobility and solubility in soil system. It has been investigated that Zn availability decreases 100 times with 1 unit increase in pH. Therefore, by declining the alkaline soil pH, bioavailable fraction of Zn can be enhanced to a substantial level. A diverse range of microflora residing in the region of rhizosphere has been reported to decrease the soil pH to a good extent (Wu et al. 2006), which may occur due to discharge of some protons extrusion along with organic acids (Fasim et al. 2002). For instance, *Oidiodendron maius* secreted gluconic acid in addition to 2-ketogluconic acid in the culture medium throughout Zn phosphate solubilization. Furthermore, protons concentration was also found to be increase in the culture medium after incubation period (Di Simine et al. 1998). Similarly, Fasim et al. (2002) examine the Zn oxide as well as zinc phosphate solubilization accompanied through proton extrusion and 2-ketogluconic acid production. Martino et al. (2003) predicted that ericoid mycorrhizal fungi secrete organic acid for Zn solubilization from insoluble zinc oxide and zinc phosphate. An alteration in pH was recognized when *Beauveria caledonica* were used to solubilize zinc phosphate and ZnO (Fomina et al. 2004). Koide and Kabir (2000), proposed that mycorrhizal plants facilitate Zn availability by lowering the pH of soil by the release of some organic acids. Subramanian et al. (2009) also examine that bioavailability of Zn through bioinoculants and acid phosphatase activity in arbuscular mycorrhizae inoculated soil, which have decline the pH of rhizosphere and contributed the release of Zn from mineral fraction. Thus, pH decreases due to the release of organic acids and proton, facilitates Zn solubilization and its uptake by plants.

6.6 Zn Chelation

Zinc ions have high interaction with the soil constituent due to which its persistency is very low in soil. Due to low persistency there is high reactivity of Zn in soil system. However, bioavailability of Zn could also increase by means of Zn chelating compounds such as EDTA, potassium humate. These compounds are either man-made or synthesized and released by the roots of plant and arbuscular mycorrhizal (AM) fungus *Glomus caledonium*, Glomus spp. in the roots chelates the Zn and hence improve its bioavailability. The chelators of microorganisms are the metabolites, which create complexes with metal cations like Zn^{2+} (Tarkalson et al. 1998). Afterwards, These Zn chelators move towards the roots and release chelating ligand

Zn^{2+} at the surface of root, making them free to chelate other Zn ion from the soil. In some microorganisms, chelation has been recognized as prevailing phenomena to perk up bioavailability and uptake by plant roots.

6.7 Changes in Root Architecture

Zinc is immobile in soil and plant uptake zinc mainly by means of diffusion process. Because of poor native Zn bioavailability as well as less exogenous supply, zones of diminution are formed around roots. Consequently, for Zn uptake improvement it should be in close proximity of roots. It can be achieved also by application of more Zn or by getting better root growth along with surface area so that roots can uptake nutrients beyond the diminution zone. A rhizosphere microorganism particularly mycorrhizal fungus is extensively recognized for its impact on root architecture. Mycorrhizal plants uptake Zn over more distances, crossing the depletion zone. Jansa et al. (2003) examine that *Glomus intraradices* can uptake Zn from a distance of 50 mm from the maize roots. In the absence of Zn fertilization, rhizospheric bacteria and mycorrhizal fungus considerably augmented root length, root weight, spread root volume and Zn uptake in straw as well as in grain compared to the plants without any fungal inoculation and this increased the Zn concentration in the grain up to 4% (Subramanian et al. 2009)

7 Steps Lead to Mycorrhiza for Biofortification of Food Based Plants

7.1 Zn Mobilization

Production of food crops with substantial concentration of zinc is prime need to the modern world to conquer the effect of zinc malnutrition. Grains of cereal crops are used as a basic food stock in most countries of the world to feed a large numbers of peoples. Zn-enriched cereal grains, fruits and vegetables can potentially engender main health benefits; therefore, sufficient Zn content is notorious to improve crop productivity. Slow resupply of soil Zn and its low plant availability ultimately results in a forge which restricts crop plant growth, yield, and Zn concentration (Zn density, in human nutritional terms) in the edible portions of crop plants (Alloway 2008). As the mycorrhizal symbiosis is well-known to be a chief intercessor of plant P nutrition, the AM symbiosis is increasingly deliberated to be one of the key bestower to plant Zn nutrition (Thompson et al. 2013), and allow aggregation of zinc in edible parts of crops and thus presents suitable candidature for biofortification of plants. Increased Zn uptake occurs as a result of colonization of AMF on plant roots, which magnify the surface area by means of a hyphal network beyond the nutrient limiting zone of roots (Smith and Read 2008). Additionally, AMF assist in Zn acquirement from pores and patches of soil not reachable by plant roots (Bolan 1991). Quality of soils also provokes the problem of zinc deficiency in cereal crops. Generally, the problem of zinc deficiency is predictable in calcareous soils, sandy soils, peat soils, and soils accompanying with high content of phosphorus and silicon (Alloway 2008). Inoculation of local and foreign AMF also influences the concentration of micronutrients such as Fe and Zn in crop plants (Pellegrino and Bedini 2014). Indeed locally sourced AMF inoculation may be a right choice because of a better adaptation to the existing conditions and also because they could circumvent the ecological risks of the introduction of foreign AM fungal species (Schwartz et al. 2006). In case of chick pea, local AM fungal inoculation consequence into the increases in Zn concentration of about 16% in comparison with foreign AM fungal inoculation, respect with controls in field conditions (Pellegrino and Bedini 2014). Mycorrhizal plants maintain acidic pH in surroundings and facilitate solubilization of strongly bound Zn besides synergistic interaction with phosphorus (Subramanian et al. 2008). Somehow, synergistic interaction between zinc and phosphorus may assist in increased uptake of zinc through mycorrhizal hyphae, which consequently get remobilized into edible parts of crop plants. In a study AMF (*Glomus intraradices*) symbiosis enhances Zn level up to 15% in maize when subjected with varying level of Zn and P fertilizers and produced grains fortified with Zn with increased tryptophan content in open field conditions, and hence concluded the synergistic interaction between these two nutrients (Zn and P) for enhanced uptake of zinc (Subramanian et al. 2013). In a meta-analysis of 33 field studies, Pellegrino et al. (2015) described that AMF increased grain yield and Zn level in wheat. Different varieties of similar crop also vary in uptaking of Zn concentration with the same mycorrhizal association. As the AMF inoculation experienced higher accumulation of Zn concentration up to 101% and 75% in two new and old varieties of wheat (*Triticum turgidum* L. subsp. *durum* (Desf.) Husn.), respectively (Ercoli et al. 2017).

7.2 Iron

For most organisms including plants and animals, iron (Fe) is considered as an essential micronutrient. Fe dearth is one of the most common micronutrient deficiencies globally, and it has been estimated that due to Fe deficiency two billion people are being affected (Stoltzfus and Dreyfuss 1998) and 0.8 million deaths annually reported worldwide (WHO 2002). Moreover, Fe deficiency is ranked sixth among the threat for causing death and disability in developing countries with elevated mortality rates (WHO 2002). Then approaches are needed to alleviate Fe deficiency such as production of healthy food, supplementation and food fortification, but people from poor families, especially from developing countries cannot afford such strategies as these are cost effective. Biofortification with Fe in staples offers a cost-effective tool to rescue Fe deficiency in target populations worldwide. Extensive occurrence of AMF results in acidification of rhizospheric region due to secretion of organic acids and phenolic compounds, with the increases of soil Fe availability (White and Broadley 2009). Hyphal length of mycorrhiza may also involve in transport of Fe from soil to host plants. Hyphae mediated transport improved plant available Fe that may have propped up Fe nutrition of maize plants and biofortification of grains (Caris et al. 1998). In addition to the hyphal transport, mycorrhizal fungi produce Fe siderophores that may favour chelation and availability of Fe and other metals. Some fungal species were reported to produce more siderophore than bacteria (Milagres et al. 1999). Ericoid mycorrhizal fungi release ferricrocin or fusigen as the major siderophores. Ferricrocin was also found to be produced by the ectomycorrhizal fungi, *Cenococcum geophilum* and *Hebeloma crustuliniforme*. AM fungi are also reported to improve Fe-uptake rates of associated host plants (Lee and George 2005). Enzyme activities could be related for improved Fe nutrition in plants, and the important enzymes involved in iron nutrition including catalase and peroxidase were constantly higher in AMF inoculated than uninoculated plants despite of sterilized or natural soils (Subramanian et al. 2013).

7.3 Other Micronutrients

Selenium (Se) is considered as an essential micronutrient because of its antioxidant capacity and positive effects on human health (Cartes et al. 2005). The main source of Se for humans and animals is the soil–plant system, and a mineral imbalance can lead to Se-deficient food with consequences for human and animal nutrition (Govasmark and Salbu 2011). In general, soils around the world have low Se quantities. Therefore, agronomic Se fortification by using inorganic Se source is a current technology in order to maintain optimizes level of Se status in human diet for reducing disease risks. In soil, applied selenium is rapidly reduced to insoluble forms, and usually the crop nutrient use efficiency was less than 10% only. Selenium addition in commercial fertilizers may be a larger programme method that is too wasteful, as much of the Se used thereby will be lost for future utilization. Selenium content in plants is highly dependent on soil Se concentration. Thus, Se dietary intake varies greatly across the different regions of the world. In volcanic soils from southern Chile (Andisol), selenium (Se) can form stable complexes with clays and/or can be strongly adsorbed, resulting in low Se bioavailability to plants (Cartes et al. 2005; Mora et al. 2008). The microbial community associated to AMF and/or changes induced by the inoculated AMF on the structure of the microbial community present in the soil can also affect Se reduction processes and the uptake of this element by roots. Durán et al. (2013) observed a synergistic effect when co-inoculated several selenobacteria and the mycorrhizal fungus *Glomus claroideum* for enhancing Se levels in wheat grains. Besides the Fe, Zn and Se, other important micronutrients such as Mn and Cu are essential for plants and animals. Increasing the micronutrient density of staple crops or biofortification can play a vital role in improving human nutrition on a global scale (Rana et al. 2012). Mn deficiency is reported worldwide and differences in Mn efficiency among the crops are related to their ability to affect the solubility of Mn in the rhizosphere. The availability of Mn in the rhizosphere is affected by several factors including redox condition and pH, moisture, temperature and concentrations of other nutrients and heavy metal in soil solution. Tomato plant inoculated by AM fungi (*Rhizophagus-irregularis* and

Funneliformis mossae) was increased in appreciable Cu concentration in fruit with high antioxidants and carotenoids (Hart et al. 2015). Similarly, vesicular-arbuscular (VA) mycorrhiza (*Glomus mosseae*) augmented total Cu uptake up to 62% in white clover (*Trifolium repens* L.) (Li et al. 1991). Co-inoculation of AMF and *Pseudomonas* strain increased wheat yield with mineral nutrient concentrations of Cu, Fe, K, Mn and Zn (Mader et al. 2011).

8 Conclusion and Future Prospects

Mycorrhizal fungi were found to be capable to significantly augment zinc, iron and other micronutrients contents in several plants, which are globally required as major food crops containing less concentration of micronutrients. Thus mycorrhizal fungi establish their potent role for biofortification and lessen the dependency on costlier approaches such as agronomic intervention and genetic modification for enhancing micronutrients concentrations in edible parts of crops. Multiple beneficial traits of mycorrhizal fungi with various nutrients acquisitions processes made mycorrhiza as suitable candidate to formulate them for biofortification strategies to address the problem of hidden hunger.

References

Alloway, B. J. (2008). Micronutrients and crop production: An introduction. In B. J. Alloway (Ed.), *Micronutrient deficiencies in global crop production* (pp. 1–39). Dordrecht: Springer.

Ames, B. N. (2001). DNA damage from micronutrient deficiencies is likely to be a major cause of cancer. *Mutation Research/Fundamental and Molecular Mechanisms of Mutagenesis, 475*(1–2), 7–20. https://doi.org/10.1016/s0027-5107(01)00070-7.

Biari, A., Gholami, A., & Rahmani, H. (2008). Growth promotion and enhanced nutrient uptake of maize (*Zea mays* L.) by application of plant growth promoting rhizobacteria in arid region of Iran. *Journal of Biological Sciences, 8*(6), 1015–1020. https://doi.org/10.3923/jbs.2008.1015.1020.

Bolan, N. S. (1991). A critical review on the role of mycorrhizal fungi in the uptake of phosphorus by plants. *Plant and Soil, 134*(2), 189–207. https://doi.org/10.1007/bf00012037.

Bouis, H. (1996). Enrichment of food staples through plant breeding: A new strategy for fighting micronutrient malnutrition. *Nutrition Reviews, 54*(5), 131–137. https://doi.org/10.1111/j.1753-4887.1996.tb03915.x.

Bouis, H. E. (2003). Micronutrient fortification of plants through plant breeding: Can it improve nutrition in man at low cost? *Proceedings of the Nutrition Society, 62*(2), 403–411. https://doi.org/10.1079/pns2003262.

Bouis, H. E., Hotz, C., McClafferty, B., Meenakshi, J. V., & Pfeiffer, W. H. (2011). Biofortification: A new tool to reduce micronutrient malnutrition. *Food and Nutrition Bulletin, 32*(1_suppl1), S31–S40. https://doi.org/10.1177/15648265110321s105.

Bücking, H., & Heyser, W. (2003). Uptake and transfer of nutrients in ectomycorrhizal associations: Interactions between photosynthesis and phosphate nutrition. *Mycorrhiza, 13*(2), 59–68. https://doi.org/10.1007/s00572-002-0196-3.

Bücking, H., Liepold, E., & Ambilwade, P. (2012). The role of the mycorrhizal symbiosis in nutrient uptake of plants and the regulatory mechanisms underlying these transport processes. In N. K. Dhal & S. C. Sahu (Eds.), *Plant science* (pp. 107–539). Rijeka: Intech.

Burkert, B., & Robson, A. (1994). ^{65}Zn uptake in subterranean clover (*Trifolium subterraneum* L.) by three vesicular-arbuscular mycorrhizal fungi in a root-free sandy soil. *Soil Biology and Biochemistry, 26*(9), 1117–1124. https://doi.org/10.1016/0038-0717(94)90133-3.
Cakmak, I., & Kutman, U. B. (2017). Agronomic biofortification of cereals with zinc: A review. *European Journal of Soil Science, 69*(1), 172–180. https://doi.org/10.1111/ejss.12437.
Cakmak, I., Kalaycı, M., Ekiz, H., Braun, H., Kılınç, Y., & Yılmaz, A. (1999). Zinc deficiency as a practical problem in plant and human nutrition in Turkey: A NATO-science for stability project. *Field Crops Research, 60*(1–2), 175–188. https://doi.org/10.1016/s0378-4290(98)00139-7.
Caris, C., Hördt, W., Hawkins, H., Römheld, V., & George, E. (1998). Studies of iron transport by arbuscular mycorrhizal hyphae from soil to peanut and sorghum plants. *Mycorrhiza, 8*(1), 35–39. https://doi.org/10.1007/s005720050208.
Cartes, P., Gianfreda, L., & Mora, M. (2005). Uptake of selenium and its antioxidant activity in ryegrass when applied as selenate and selenite forms. *Plant and Soil, 276*(1–2), 359–367. https://doi.org/10.1007/s11104-005-5691-9.
CGIAR (Consultative Group on International Agricultural Research Science Council). (2007). *Report of the first external review of the HarvestPlus Challenge Program* (p. 2008). Rome: Science Council Secretariat.
Davidson, A. L., & Nikaido, H. (1991). Purification and characterization of the membrane-associated components of the maltose transport system from *Escherichia coli*. *Journal of Biological Chemistry, 266*(14), 8946–8951.
de Valença, A. W., Bake, A., Brouwer, I. D., & Giller, K. E. (2017). Agronomic biofortification of crops to fight hidden hunger in sub-Saharan Africa. *Global Food Security, 12*, 8–14. https://doi.org/10.1016/j.gfs.2016.12.001.
Di Simine, C. D., Sayer, J. A., & Gadd, G. M. (1998). Solubilization of zinc phosphate by a strain of *Pseudomonas fluorescens* isolated from a forest soil. *Biology and Fertility of Soils, 28*(1), 87–94. https://doi.org/10.1007/s003740050467.
Durán, P., Acuña, J., Jorquera, M., Azcón, R., Borie, F., Cornejo, P., & Mora, M. (2013). Enhanced selenium content in wheat grain by co-inoculation of selenobacteria and arbuscular mycorrhizal fungi: A preliminary study as a potential Se biofortification strategy. *Journal of Cereal Science, 57*(3), 275–280. https://doi.org/10.1016/j.jcs.2012.11.012.
Ercoli, L., Schüßler, A., Arduini, I., & Pellegrino, E. (2017). Strong increase of durum wheat iron and zinc content by field-inoculation with arbuscular mycorrhizal fungi at different soil nitrogen availabilities. *Plant and Soil, 419*(1–2), 153–167. https://doi.org/10.1007/s11104-017-3319-5.
Fasim, F., Ahmed, N., Parsons, R., & Gadd, G. M. (2002). Solubilization of zinc salts by a bacterium isolated from the air environment of a tannery. *FEMS Microbiology Letters, 213*(1), 1–6. https://doi.org/10.1111/j.1574-6968.2002.tb11277.x.
Field, S. J., Tsai, F. Y., Kuo, F., Zubiaga, A. M., Kaelin, W. G., Livingston, D. M., Okrin, S. H., & Greenberg, M. E. (1996). E2F-1 Functions in mice to promote apoptosis and suppress proliferation. *Cell, 85*(4), 549–561. https://doi.org/10.1016/s0092-8674(00)81255-6.
Finlay, R. D., EK, H., Odham, G., & Soderstrom, B. (1988). Mycelial uptake, translocation and assimilation of nitrogen from 15N-labelled ammonium by *Pinus sylvestris* plants infected with four different ectomycorrhizal fungi. *New Phytologist, 110*(1), 59–66. https://doi.org/10.1111/j.1469-8137.1988.tb00237.x.
Fomina, M., Alexander, I. J., Hillier, S., & Gadd, G. M. (2004). Zinc phosphate and pyromorphite solubilization by soil plant-symbiotic fungi. *Geomicrobiology Journal, 21*(5), 351–366. https://doi.org/10.1080/01490450490462066.
Garcia-Casal, M. N., Peña-Rosas, J. P., Pachón, H., De-Regil, L. M., Centeno Tablante, E., & Flores-Urrutia, M. C. (2016). Staple crops biofortified with increased micronutrient content: Effects on vitamin and mineral status, as well as health and cognitive function in the general population. *Cochrane Database of Systematic Reviews*. https://doi.org/10.1002/14651858.cd012311.
Gibson, R. S. (2006). Zinc: The missing link in combating micronutrient malnutrition in developing countries. *Proceedings of the Nutrition Society, 65*(01), 51–60. https://doi.org/10.1079/pns2005474.

Glick, B. R., Patten, C. L., Holguin, G., & Penrose, D. M. (1999). *Biochemical and genetic mechanisms used by plant growth promoting bacteria*. London: Imperial College Press. https://doi.org/10.1142/p130.

Govasmark, E., & Salbu, B. (2011). Translocation and re-translocation of selenium taken up from nutrient solution during vegetative growth in spring wheat. *Journal of the Science of Food and Agriculture, 91*(8), 1367–1372. https://doi.org/10.1002/jsfa.4387.

Graham, R., Senadhira, D., Beebe, S., Iglesias, C., & Monasterio, I. (1999). Breeding for micronutrient density in edible portions of staple food crops: Conventional approaches. *Field Crops Research, 60*(1–2), 57–80. https://doi.org/10.1016/s0378-4290(98)00133-6.

Graham, R. D., Welch, R. M., & Bouis, H. E. (2001). Addressing micronutrient malnutrition through enhancing the nutritional quality of staple foods: Principles, perspectives and knowledge gaps. *Advances in Agronomy, 70*, 77–142. https://doi.org/10.1016/s0065-2113(01)70004-1.

Grunwald, U., Guo, W., Fischer, K., Isayenkov, S., Ludwig-Müller, J., Hause, B., Yan, X., Küster, H., & Franken, P. (2009). Overlapping expression patterns and differential transcript levels of phosphate transporter genes in arbuscular mycorrhizal, Pi-fertilised and phytohormone-treated *Medicago truncatula* roots. *Planta, 229*(5), 1023–1034. https://doi.org/10.1007/s00425-008-0877-z.

Gyaneshwar, P., Kumar, G. N., Parekh, L. J., & Poole, P. S. (2002). Role of soil microorganisms in improving P nutrition of plants. In J. J. Adu-Gyamfi (Ed.), *Food security in nutrient-stressed environments: Exploiting plants' genetic capabilities* (Developments in plant and soil sciences, Vol. 95). Dordrecht: Springer. https://doi.org/10.1007/978-94-017-1570-6_15.

Hart, M., Ehret, D. L., Krumbein, A., Leung, C., Murch, S., Turi, C., & Franken, P. (2015). Inoculation with arbuscular mycorrhizal fungi improves the nutritional value of tomatoes. *Mycorrhiza, 25*(5), 359–376. https://doi.org/10.1007/s00572-014-0617-0.

Haselwandter, K., Dobernigg, B., Beck, W., Jung, G., Cansier, A., & Winkelmann, G. (1992). Isolation and identification of hydroxamate siderophores of ericoid mycorrhizal fungi. *Biometals, 5*(1), 51–56. https://doi.org/10.1007/bf01079698.

Hatfield, D. L., Tsuji, P. A., Carlson, B. A., & Gladyshev, V. N. (2014). Selenium and selenocysteine: Roles in cancer, health, and development. *Trends in Biochemical Sciences, 39*(3), 112–120. https://doi.org/10.1016/j.tibs.2013.12.007.

He, X., & Nara, K. (2007). Element biofortification: Can mycorrhizas potentially offer a more effective and sustainable pathway to curb human malnutrition? *Trends in Plant Science, 12*(8), 331–333. https://doi.org/10.1016/j.tplants.2007.06.008.

Hotz, C., & Brown, K. H. (2004). International Zinc Nutrition Consultative Group (IZiNCG) technical document no. 1. Assessment of the risk of zinc deficiency in populations and options for its control. *Food and Nutrition Bulletin, 25*, S94–S203.

Humbert, P. O., Rogers, C., Ganiatsas, S., Landsberg, R. L., Trimarchi, J. M., Dandapani, S., Brugnara, C., Erdman, S., Schrenzel, M., Bronson, R. T., & Lees, J. A. (2000). E2F4 is essential for normal erythrocyte maturation and neonatal viability. *Molecular Cell, 6*(2), 281–291. https://doi.org/10.1016/s1097-2765(00)00029-0.

Jansa, J., Mozafar, A., & Frossard, E. (2003). Long-distance transport of P and Zn through the hyphae of an arbuscular mycorrhizal fungus in symbiosis with maize. *Agronomie, 23*(5–6), 481–488. https://doi.org/10.1051/agro:2003013.

Javelle, A., Morel, M., Rodríguez-Pastrana, B., Botton, B., André, B., Marini, A. M., & Chalot, M. (2003). Molecular characterization, function and regulation of ammonium transporters (Amt) and ammonium-metabolizing enzymes (GS, NADP-GDH) in the ectomycorrhizal fungus *Hebeloma cylindrosporum*. *Molecular Microbiology, 47*(2), 411–430. https://doi.org/10.1046/j.1365-2958.2003.03303.x.

Jin, H., Pfeffer, P. E., Douds, D. D., Piotrowski, E., Lammers, P. J., & Shachar-Hill, Y. (2005). The uptake, metabolism, transport and transfer of nitrogen in an arbuscular mycorrhizal symbiosis. *New Phytologist, 168*(3), 687–696. https://doi.org/10.1111/j.1469-8137.2005.01536.x.

Jones, M. D., Durall, D. M., & Tinker, P. B. (1998). A comparison of arbuscular and ectomycorrhizal *Eucalyptus coccifera*: Growth response, phosphorus uptake efficiency and external hyphal production. *New Phytologist, 140*(1), 125–134. https://doi.org/10.1046/j.1469-8137.1998.00253.x.

Joshi, S., Singh, A. V., & Prasad, B. (2018). Enzymatic activity and plant growth promoting potential of endophytic bacteria isolated from *Ocimum sanctum* and *Aloe vera*. *International Journal of Current Microbiology and Applied Sciences, 7*(06), 2314–2326.

Khush, G. (2003). Productivity improvements in rice. *Nutrition Reviews, 61*(suppl_6), S114–S116. https://doi.org/10.1301/nr.2003.jun.s114-s116.

Koide, R. T., & Kabir, Z. (2000). Extraradical hyphae of the mycorrhizal fungus *Glomus intraradices* can hydrolyse organic phosphate. *New Phytologist, 148*(3), 511–517. https://doi.org/10.1046/j.1469-8137.2000.00776.x.

Lee, Y. J., & George, E. (2005). Contribution of mycorrhizal hyphae to the uptake of metal cations by cucumber plants at two levels of phosphorus supply. *Plant and Soil, 278*(1–2), 361–370. https://doi.org/10.1007/s11104-005-0373-1.

Leone, G., Sears, R., Huang, E., Rempel, R., Nuckolls, F., Park, C. H., Giangrande, P., Wu, L., Saavedra, H. I., Field, S. J., Thompson, M. A., Yang, H., Fujiwara, Y., Greenberg, M. E., Orkin, S., Smith, C., & Nevins, J. R. (2001). Myc requires distinct E2F activities to induce S phase and apoptosis. *Molecular Cell, 8*(1), 105–113. https://doi.org/10.1016/s1097-2765(01)00275-1.

Li, X. L., Marschner, H., & George, E. (1991). Acquisition of phosphorus and copper by VA-mycorrhizal hyphae and root-to-shoot transport in white clover. *Plant and Soil, 136*(1), 49–57. https://doi.org/10.1007/BF02465219.

López-Millán, A., Grusak, M. A., Abadía, A., & Abadía, J. (2013). Iron deficiency in plants: An insight from proteomic approaches. *Frontiers in Plant Science, 4*, 254. https://doi.org/10.3389/fpls.2013.00254.

Lynch, J. P., & Brown, K. M. (2008). Root strategies for phosphorus acquisition. In P. J. White & J. P. Hammond (Eds.), *The ecophysiology of plant–phosphorus interactions* (Vol. 7, pp. 83–116). Dordrecht: Springer.

Mäder, P., Kaiser, F., Adholeya, A., Singh, R., Uppal, H. S., Sharma, A. K., Srivastava, R., Sahai, V., Aragno, M., Wiemken, A., Johri, B. N., & Fried, P. M. (2011). Inoculation of root microorganisms for sustainable wheat–rice and wheat–black gram rotations in India. *Soil Biology and Biochemistry, 43*(3), 609–619.

Malagoli, M., Schiavon, M., Dall'Acqua, S., & Pilon-Smits, E. A. (2015). Effects of selenium biofortification on crop nutritional quality. *Frontiers in Plant Science, 6*, 280. https://doi.org/10.3389/fpls.2015.00280.

Marschner, H. (1993). Zinc uptake from soils. In A. D. Robson (Ed.), *Zinc in soils and plants* (Developments in plant and soil sciences, Vol. 55, pp. 59–77). Dordrecht: Springer.

Marschner, H. (1995). Functions of mineral nutrients. *Mineral Nutrition of Higher Plants*, 229–312. https://doi.org/10.1016/b978-012473542-2/50010-9.

Marschner, H., & Dell, B. (1994). Nutrient uptake in mycorrhizal symbiosis. *Plant and Soil, 159*(1), 89–102. https://doi.org/10.1007/bf00000098.

Martino, E., Perotto, S., Parsons, R., & Gadd, G. M. (2003). Solubilization of insoluble inorganic zinc compounds by ericoid mycorrhizal fungi derived from heavy metal polluted sites. *Soil Biology and Biochemistry, 35*(1), 133–141. https://doi.org/10.1016/s0038-0717(02)00247-x.

Mayer, J. E., Pfeiffer, W. H., & Beyer, P. (2008). Biofortified crops to alleviate micronutrient malnutrition. *Current Opinion in Plant Biology, 11*(2), 166–170. https://doi.org/10.1016/j.pbi.2008.01.007.

Milagres, A. M., Machuca, A., & Napoleão, D. (1999). Detection of siderophore production from several fungi and bacteria by a modification of chrome azurol S (CAS) agar plate assay. *Journal of Microbiological Methods, 37*(1), 1–6. https://doi.org/10.1016/s0167-7012(99)00028-7.

Miller, G. W., Huang, I. J., Welkie, G. W., & Pushnik, J. C. (1995). Function of iron in plants with special emphasis on chloroplasts and photosynthetic activity. In J. Abadía (Ed.), *Iron nutrition in soils and plants* (Developments in plant and soil sciences, Vol. 59, pp. 19–28). Dordrecht: Springer.

MoHFW. (2013). *Government of India. Guidelines for control of iron deficiency anemia*. Available at http://www.unicef.org/india/10._National_Iron_Plus_Initiative_Guidelines_for_Control_of_IDA.pdf. Accessed 29 Mar 2013.

Mora, M. L., Pinilla, L., Rosas, A., & Cartes, P. (2008). Selenium uptake and its influence on the antioxidative system of white clover as affected by lime and phosphorus fertilization. *Plant and Soil, 303*(1–2), 139–149. https://doi.org/10.1007/s11104-007-9494-z.

Murray-Kolb, L. E. (2013). Iron and brain functions. *Current Opinion in Clinical Nutrition and Metabolic Care, 16*(6), 703–707. https://doi.org/10.1097/mco.0b013e3283653ef8.

O'Keefe, D. M., & Sylvia, D. M. (1991). Mechanisms of vesicular-arbuscular mycorrhizal plant growth response. In D. K. Arora, B. Rai, K. G. Mukerji, & G. R. Knudsen (Eds.), *Handbook of applied mycology* (Soil and plants, Vol. 1, pp. 35–53). New York: Marcel Dekker, Inc.

Parniske, M. (2008). Arbuscular mycorrhiza: The mother of plant root endosymbioses. *Nature Reviews. Microbiology, 6*(10), 763–775. https://doi.org/10.1038/nrmicro1987.

Pellegrino, E., & Bedini, S. (2014). Enhancing ecosystem services in sustainable agriculture: Biofertilization and biofortification of chickpea (*Cicer arietinum* L.) by arbuscular mycorrhizal fungi. *Soil Biology and Biochemistry, 68*, 429–439. https://doi.org/10.1016/j.soilbio.2013.09.030.

Pellegrino, E., Öpik, M., Bonari, E., & Ercoli, L. (2015). Responses of wheat to arbuscular mycorrhizal fungi: A meta-analysis of field studies from 1975 to 2013. *Soil Biology and Biochemistry, 84*, 210–217. https://doi.org/10.1016/j.soilbio.2015.02.020.

Plenchette, C., Clermont-Dauphin, C., Meynard, J. M., & Fortin, J. A. (2005). Managing arbuscular mycorrhizal fungi in cropping systems. *Canadian Journal of Plant Science, 85*(1), 31–40. https://doi.org/10.4141/p03-159.

Prasad, A. S. (2007). Zinc: Mechanisms of host defense. *The Journal of Nutrition., 137*(5), 1345–1349. https://doi.org/10.1093/jn/137.5.1345.

Prasad, B., Kumar, A., Singh, A. V., & Kumar, A. (2016). Plant growth and seed yield attributes as influenced by bacterial isolates under glass house. *Progressive Research, 11*(IV), 2573–2576.

Rana, A., Saharan, B., Nain, L., Prasanna, R., & Shivay, Y. S. (2012). Enhancing micronutrient uptake and yield of wheat through bacterial PGPR consortia. *Soil Science and Plant Nutrition, 58*(5), 573–582. https://doi.org/10.1080/00380768.2012.716750.

Rausch, C., Daram, P., Brunner, S., Jansa, J., Laloi, M., Leggewie, G., Nikolaus Amrhein, N., & Bucher, M. (2001). A phosphate transporter expressed in arbuscule-containing cells in potato. *Nature, 414*, 462–470. https://doi.org/10.1038/35106601.

Rayman, M. P. (2012). Selenium and human health. *The Lancet., 379*(9822), 1256–1268. https://doi.org/10.1016/s0140-6736(11)61452-9.

Royzman, I., Whittaker, A. J., & Orr-Weaver, T. L. (1997). Mutations in Drosophila DP and E2F distinguish G1-S progression from an associated transcriptional program. *Genes & Development, 11*(15), 1999–2011. https://doi.org/10.1101/gad.11.15.1999.

Ruel, M. T., & Alderman, H. (2013). Nutrition-sensitive interventions and programmes: How can they help to accelerate progress in improving maternal and child nutrition? *The Lancet, 382*(9891), 536–551. https://doi.org/10.1016/s0140-6736(13)60843-0.

Salgueiro, M. J., Zubillaga, M., Lysionek, A., Cremaschi, G., Goldman, C. G., Caro, R., De Paoli, T., Hager, A., Weill, R., & Boccio, J. (2000). Zinc status and immune system relationship: A review. *Biological Trace Element Research, 76*(3), 193–205. https://doi.org/10.1385/btcr:76:3:193.

Saravanan, V. S., Subramoniam, S. R., & Raj, S. A. (2003). Assessing in vitro solubilization potential of different zinc solubilizing bacterial (zsb) isolates. *Brazilian Journal of Microbiology, 34*, 121–125.

Schachtman, D. P., Reid, R. J., & Ayling, S. (1998). Phosphorus uptake by plants: From soil to cell. *Plant Physiology, 116*(2), 447–453. https://doi.org/10.1104/pp.116.2.447.

Schwartz, M. W., Hoeksema, J. D., Gehring, C. A., Johnson, N. C., Klironomos, J. N., Abbott, L. K., & Pringle, A. (2006). The promise and the potential consequences of the global transport of mycorrhizal fungal inoculum. *Ecology Letters, 9*(5), 501–515. https://doi.org/10.1111/j.1461-0248.2006.00910.x.

Shiferaw, B., Prasanna, B. M., Hellin, J., & Bänziger, M. (2011). Crops that feed the world 6. Past successes and future challenges to the role played by maize in global food security. *Food Security, 3*(3), 307–327. https://doi.org/10.1007/s12571-011-0140-5.

Singh, A. V., & Goel, R. (2015). Plant growth promoting efficiency of *Chryseobacterium* sp. PSR10 on finger millet (*Eleusine coracana*). *Journal of Global Biosciences., 4*(6), 2569–2575.
Singh, A. V., & Prasad, B. (2014). Enhancement of plant growth, nodulation and seed yield through Plant Growth Promoting Rhizobacteria in Lentil (*Lens culinaris* Medik cv. VL125). *International Journal of Current Microbiology and Applied Sciences, 3*(6), 614–622.
Singh, A. V., Shah, S., & Prasad, B. (2010). Effect of phosphate solubilizing bacteria on plant growth promotion and nodulation in soybean (*Glycine max* (L.) Merr). *Journal of Hill Agriculture, 1*(1), 35–39.
Singh, A. V., Prasad, B., & Shah, S. (2011). Influence of phosphate solubilizing bacteria for enhancement of plant growth and seed yield in lentil. *Journal of Crop and Weed, 7*(1), 1–4.
Singh, A. V., Chandra, R., & Reeta, G. (2013). Phosphate solubilization by *Chryseobacterium* sp. and their combined effect with N and P fertilizers on plant growth promotion. *Archives of Agronomy and Soil Science, 59*(5), 641–651.
Singh, J., Singh, A. V., Prasad, B., & Shah, S. (2017). Sustainable agriculture strategies of wheat biofortification through microorganisms. In A. Kumar, A. Kumar, & B. Prasad (Eds.), *Wheat a premier food crop*. New Delhi: Kalyani Publishers.
Singh, A. V., Prasad, B., & Goel, R. (2018). Plant Growth Promoting Efficiency of Phosphate Solubilizing *Chryseobacterium* sp. PSR 10 with Different Doses of N and P Fertilizers on Lentil (*Lens culinaris* var. PL-5) Growth and Yield. *International Journal of Current Microbiology and Applied Sciences, 7*(05), 2280–2289.
Smith, S. E., & Read, D. J. (1997). *Mycorrhizal symbiosis* (2nd ed., ix + 605 pp). San Diego: Academic.
Smith, S. E., & Read, D. J. (2008). Growth and carbon economy of arbuscular mycorrhizal symbionts. In S. E. Smith & D. J. Read (Eds.), *Mycorrhizal symbiosis* (pp. 117–144). London: Academic. https://doi.org/10.1016/B978-012370526-6.50006-4.
Smith, S. E., Smith, F. A., & Jakobsen, I. (2003). Mycorrhizal fungi can dominate phosphate supply to plants irrespective of growth responses. *Plant Physiology, 133*(1), 16–20. https://doi.org/10.1104/pp.103.024380.
Srivastava, M. P., Tewari, R., & Sharma, N. (2013). Effect of different cultural variables on siderophores produced by *Trichoderma* spp. *International Journal of Advanced Research, 1*, 1–6.
Stoltzfus, R. J., & Dreyfuss, M. L. (1998). Micronutrient deficiency disorders. In *Guidelines for the use of iron supplements to prevent and treat iron deficiency anemia* (pp. 1–39). Washington, DC: ILSI Press.
Stonor, R. N., Smith, S. E., Manjarrez, M., Facelli, E., & Smith, F. A. (2014). Mycorrhizal responses in wheat: Shading decreases growth but does not lower the contribution of the fungal phosphate uptake pathway. *Mycorrhiza, 24*(6), 465–472. https://doi.org/10.1007/s00572-014-0556-9.
Subramanian, K. S., Bharathi, C., & Jegan, A. (2008). Response of maize to mycorrhizal colonization at varying levels of zinc and phosphorus. *Biology and Fertility of Soils, 45*(2), 133–144. https://doi.org/10.1007/s00374-008-0317-z.
Subramanian, K. S., Tenshia, V., Jayalakshmi, K., & Ramachandran, V. (2009). Biochemical changes and zinc fractions in arbuscular mycorrhizal fungus (*Glomus intraradices*) inoculated and uninoculated soils under differential zinc fertilization. *Applied Soil Ecology, 43*(1), 32–39. https://doi.org/10.1016/j.apsoil.2009.05.009.
Subramanian, K. S., Balakrishnan, N., & Senthil, N. (2013). Mycorrhizal symbiosis to increase the grain micronutrient content in maize. *Australian Journal of Crop Science, 7*(7), 900–910.
Taktek, S., St-Arnaud, M., Piché, Y., Fortin, J. A., & Antoun, H. (2016). Igneous phosphate rock solubilization by biofilm-forming mycorrhizobacteria and hyphobacteria associated with *Rhizoglomus irregulare* DAOM 197198. *Mycorrhiza, 27*(1), 13–22. https://doi.org/10.1007/s00572-016-0726-z.
Tarkalson, D. D., Jolley, V. D., Robbins, C. W., & Terry, R. E. (1998). Mycorrhizal colonization and nutrient uptake of dry bean in manure and compost manure treated subsoil and untreated topsoil and subsoil. *Journal of Plant Nutrition, 21*(9), 1867–1878. https://doi.org/10.1080/01904169809365529.

Tatry, M., El Kassis, E., Lambilliotte, R., Corratgé, C., Van Aarle, I., Amenc, L. K., Alary, R., Zimmermann, S., Sentenac, H., & Plassard, C. (2009). Two differentially regulated phosphate transporters from the symbiotic fungus *Hebeloma cylindrosporum* and phosphorus acquisition by ectomycorrhizal *Pinus pinaster*. *The Plant Journal., 57*(6), 1092–1102. https://doi.org/10.1111/j.1365-313x.2008.03749.x.

Thompson, J. P., Clewett, T. G., & Fiske, M. L. (2013). Field inoculation with arbuscular-mycorrhizal fungi overcomes phosphorus and zinc deficiencies of linseed (*Linum usitatissimum*) in a vertisol subject to long-fallow disorder. *Plant and Soil, 371*(1–2), 117–137. https://doi.org/10.1007/s11104-013-1679-z.

Toussaint, J., St-Arnaud, M., & Charest, C. (2004). Nitrogen transfer and assimilation between the arbuscular mycorrhizal fungus *Glomus intraradices* Schenck & Smith and Ri T-DNA roots of *Daucus carota* L. in an *in vitro* compartmented system. *Canadian Journal of Microbiology, 50*(4), 251–260. https://doi.org/10.1139/w04-009.

Upadhayay, V. K., Singh, A. V., & Pareek, N. (2018). An insight in decoding the multifarious and splendid role of microorganisms in crop biofortification. *International Journal of Current Microbiology and Applied Sciences, 7*(06), 2407–2418. https://doi.org/10.20546/ijcmas.2018.706.286.

van der Heijden, M. G. A., Martin, F. M., Selosse, M., & Sanders, I. R. (2015). Mycorrhizal ecology and evolution: The past, the present, and the future. *New Phytologist, 205*(4), 1406–1423. https://doi.org/10.1111/nph.13288.

Vance, C. P. (2010). Quantitative trait loci, epigenetics, sugars, and microRNAs: Quaternaries in phosphate acquisition and use. *Plant Physiology, 154*(2), 582–588. https://doi.org/10.1104/pp.110.161067.

von Grebmer, K., Saltzman, A., Birol, E., Wiesmann, D., Prasai, N., Yin, S., Yohannes, Y., Menon, P., Thompson, J., & Sonntag, A. (2014). *2014 Global hunger index: The challenge of hidden hunger*. Bonn/Washington, DC/Dublin: Welthungerhilfe/International Food Policy Research Institute/Concern Worldwide. https://doi.org/10.2499/9780896299580.

White, P. J., & Broadley, M. R. (2009). Biofortification of crops with seven mineral elements often lacking in human diets - iron, zinc, copper, calcium, magnesium, selenium and iodine. *New Phytologist, 182*(1), 49–84. https://doi.org/10.1111/j.1469-8137.2008.02738.x.

Whiting, S. N., De Souza, M. P., & Terry, N. (2001). Rhizosphere bacteria mobilize Zn for hyperaccumulation by *Thlaspica erulescens*. *Environmental Science & Technology, 35*(15), 3144–3150. https://doi.org/10.1021/es001938v.

WHO (World Health Organization). (2002). *The world health report 2002 – Reducing risks, promoting healthy life*. Geneva: World Health Organization.

Willmann, M., Gerlach, N., Buer, B., Polatajko, A., Nagy, R., Koebke, E., Jansa, J., Flisch, R., & Bucher, M. (2013). Mycorrhizal phosphate uptake pathway in maize: Vital for growth and cob development on nutrient poor agricultural and greenhouse soils. *Frontiers in Plant Science, 4*, 533. https://doi.org/10.3389/fpls.2013.00533.

Wright, D. P., Read, D. J., & Scholes, J. D. (1998). Mycorrhizal sink strength influences whole plant carbon balance of *Trifolium repens* L. *Plant, Cell and Environment, 21*(9), 881–891. https://doi.org/10.1046/j.1365-3040.1998.00351.x.

Wu, S., Cheung, K., Luo, Y., & Wong, M. (2006). Effects of inoculation of plant growth-promoting rhizobacteria on metal uptake by *Brassica juncea*. *Environmental Pollution, 140*(1), 124–135. https://doi.org/10.1016/j.envpol.2005.06.023.

Yadav, R., Singh, A. V., Kumar, M., & Yadav, S. (2016). Phytochemical analysis and plant growth promoting properties of endophytic fungi isolated from tulsi and Aloe vera. *International Journal of Agricultural and Statistical Sciences, 12*(1), 239–248.

Zeng, H., & Combs, G. F. (2008). Selenium as an anticancer nutrient: Roles in cell proliferation and tumor cell invasion. *The Journal of Nutritional Biochemistry., 19*(1), 1–7. https://doi.org/10.1016/j.jnutbio.2007.02.005.

Role of Microorganisms in Soil Genesis and Functions

2

N. Kaviya, Viabhav K. Upadhayay, Jyoti Singh, Amir Khan, Manisha Panwar, and Ajay Veer Singh

1 Introduction

Soil shows an important task in maintaining the ecosystem and provides supports to the plant roots and bestows them with essential minerals and nutrients, and protects plants from erosion and other destructive physical, biological and chemical activities. Soil have the profound habitat for different and varieties of living organisms such as insects and microorganisms. Microorganisms play foremost role in soil formation and soil ecology because they as 'natural soil engineers' regulate the flux of nutrients to plants and prop up nitrogen fixation, and ultimately promote detoxification of naturally occurring inorganic and organic pollutants in soil. Microorganisms associated with soil assist in liberation of essential nutrients from primary minerals, and thus released nutrients which are required essentially for both microorganisms and as well as for plants (Uroz et al. 2009). In soil, the phenomenon of weathering of rocks is a multifaceted interaction of three kinds of weathering processes (physical, chemical and biological). Plants, animals and microorganisms vigorously participate in the biogeochemical cycles which ultimately contribute to the process of pedogenesis through biological weathering (Gadd 2007). Microorganisms express imperative role in the weathering of rocks, and use released elements as nutrients (Calvaruso et al. 2006). They compel the important processes of mineral weathering, participate in process of soil structure formation and organic matter decomposition, and also play important role in nutrient cycling (Chorover et al. 2007; Feeney et al. 2006; Schimel 1995). Microorganisms, for instances bacteria, fungi, cyanobacteria and lichens have been considered as main entities for carrying out biological weathering of rocks (Gadd 2010). Hirsch et al. (1995) demonstrated that the byproduct of microbial metabolism in form of 'organic acids, produced by soil

N. Kaviya · V. K. Upadhayay · J. Singh · A. Khan · M. Panwar · A. V. Singh (✉)
Department of Microbiology, College of Basic Sciences and Humanities, Govind Ballabh Pant University of Agriculture and Technology, Pantnagar, Uttarakhand, India

© Springer Nature Singapore Pte Ltd. 2019

A. Varma, D. K. Choudhary (eds.), *Mycorrhizosphere and Pedogenesis*,
https://doi.org/10.1007/978-981-13-6480-8_2

microorganisms can dissolve rocks. Microorganisms show impact on the dissolving rate of minerals, as the microbial metabolites facilitate in leaching out some substances from rocks or minerals (Lian et al. 2008a, b). The scientific study on types of microorganisms opens several ways for knowing the mechanisms of various steps of soil formation through biological process. Moreover, several current approaches particularly molecular techniques have been developed to explore and identify the both culturable and unculturable microbial diversity in soil assist in soil formation. Present chapter is focused on depicting the role of microorganisms in soil formation and the mechanisms for weathering process employed by such microflora with highlighting the current and advanced molecular approaches for determining microbial diversity in soil.

2 Microbial Diversity and Soil Microorganisms (Eubacteria, Archaea, Fungi and Algae)

The earth is considered to be around 4.5 billion years old and the microbial diversity on earth has a much prolonged evolutionary history than plants/animals. The microbiological process of life spans about 3.8 billion years of organic evolution, shortly after the surface of earth was still very hot (Delong and Pace 2001). At that time there was a "reducing atmosphere" comprised of methane, carbon dioxide, ammonia, and hydrogen and was devoid of free oxygen. The earth atmosphere was shifted from anoxic to oxic states by oxygenic photosynthetic progenitors. The two earth's most plentiful cellular life forms, *Prochlorococcus* and *Synechoccocus*, filling the ocean to varying degrees from pole to pole, generating oxygen as a byproduct of sunlight-driven photosynthesis (Gilbert and Neufeld 2014). Plenty of adverse conditions also influence the survival of microorganisms. Some microorganisms produce spores, when they encounter environmental stresses such as high temperature, and such form of tough structure survive for longer periods and engender new vegetative cell when exposed to favorable conditions. Bacterium such as *Deinococcus radiodurans* possesses the capacity to survive under higher/lethal doses of radiation, i.e., 3000 times greater than the mortal dose for humans. However, numerous microorganisms possibly developed in the subsurface of landmasses or beneath the sea surface where they were protected to some extent from 'UV radiation'. On earth, soil contains most diverse habitats and for diverse assemblages of different types of soil microorganisms comes under the category of two well known groups such as 'prokaryotes' and 'eukaryotes' (Fig. 2.1). Interpretation of microbial community's dynamics is likely the most challenging task because of the surprisingly huge microbial diversity in soil, and the variable and complex matrix where soil microorganisms are fixed. Torsvik et al. (1990) evaluated that 1 g of soil contains 4000 different bacterial "genomic units" which were determined on the basis of DNA–DNA reassociation. Approximately 5000 bacterial species have been confirmed from the soil (Pace 1999), while Giller et al. (1997) estimated about 1,500,000 fungal species exist on earth. Whereas as macrofauna, 3000 species of earthworms (Lee 1985), 1,00,000 species of 'protozoa' and 500,000 species of 'nematodes'

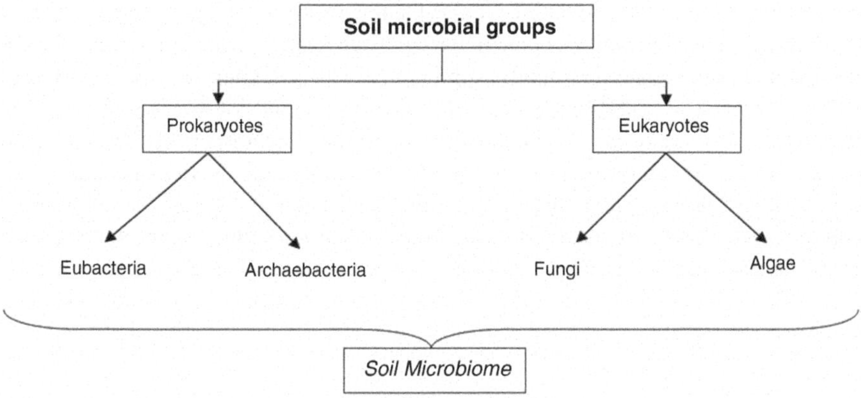

Fig. 2.1 Chief groups of microorganisms exist in soils

(Hawksworth and Mound 1991) have been reported in soil among soil flora and fauna, not to state the further invertebrate groups of the mesofauna (collembola, enchytraeids and mites) and macrofauna (ants, beetles, spiders and termites) (Giller et al. 1997). Microbial community composition may be one important control on soil processes. Soil microorganisms particularly both bacteria and fungi play very important roles in numerous biogeochemical cycles and as well help in cycling of different organic compounds. Soil microorganisms exert their effect on aboveground ecosystems as they contribute multifarious roles in plant health and nutrition, soil formation and soil fertility (O'Donnell et al. 2001). Prime factors considered to manage community composition are (Tiedje et al. 1999): (i) the key resources for growth i.e. fertility level e.g. various kinds of carbon or related compounds from plant litter, rhizosphere and invertebrates, key nutrients like N, P and K. (ii) Soil environment and its diverse characteristics (iii) Certain factors influencing organism dispersal, for instance soil structure, routes of dispersal, micro aggregate stability, and (iv) basis of population turnover such as nematode and protozoan grazing and controls on lytic enzymes. Plants also influence spatial distribution of soil bacteria and fungi. Anthropogenic activities such as intensive application of agricultural chemicals could negatively affect microbial diversity, and perhaps also present adverse effect on both above and below-ground functioning of ecosystem. Buckley and Schmidt (2001) reported higher amounts of 16S rRNA for all microbiological groups determined in uncultivated fields as compared to agricultural or cultivated fields, and this suggests a reduction in microbial activity in cultivated fields. Living organisms on earth are comprising three domains (i.e. Bacteria, Archaea, and Eucarya) and each domain containing two or more than two kingdoms. They come under the less complex cell constitutes called as prokaryotes (organisms without definite nucleus), consisting two groups of microorganisms such as eubacteria and archaebacteria. The eubacteria (also called as bacteria) comprises 'cyanobacteria', an important group formerly well-known as 'blue-green algae'. The cells are prokaryotic where the lipids present in membrane are primarily diacyl glycerol diesters, and the ribosomes having a eubacterial type of rRNA. Other important prokaryotic entities come under 'archaebacteria', where membrane lipids into cellular architecture are mainly isoprenoid glycerol diethers or diglycerol tetraethers. Archaea is subdivided into the two important known kingdoms, (1) Euryarchaeota, containing the methanogens and their phenotypically diverse relatives and (2) Crenarchaeota, which having the comparatively tight clustering of extreme thermophilic archaebacteria. On the other hand, the organisms with more complicated cellular structure possessing true nucleus are called eukaryotes (include plants, animal and fungi). The membrane of eukaryotic cells contains lipids mainly glycerol fatty acyl diesters and the eukaryotic kind of rRNA present in ribosome (Woese et al. 1990).

2.1 Eubacteria

Various kinds of soils have 'Eubacterial group' as the most dominant group of prokaryotic microorganisms. The number or population of bacteria depends on the depth of soil, and with the depth soil the microbial population decreases. Generally, 'horizon A' of a soil profile is rich in organic matters and holds high population of microorganisms than horizon B and C (SubbaRao 1997). Soil bacteria present various cellular forms such as cocci (spheres, 0.5 mm), bacilli (rods, 0.5–0.3 mm) or spirilli (spirals). However, bacilli are considered as most common in soil, while spirilli are found very rarely in natural environments (Baudoin et al. 2002). Soil bacteria further categorized in two important groups (a) 'autochthonous organisms' and (b) 'zymogenous organisms'. Autochthonous, also referred as indigenous populations, are more uniform and stable in soil, since they derive their nutrition from native soil organic or mineral matter (*Arthrobacter* and *Nocardia*). But, zymogenous bacteria need an external substrate or nutrients unlike autochthonous organisms, and their activity in soils is uneven and they often form resting structure 'propagules' (*Pseudomonas* and *Bacillus*). The population of zymogenous bacteria increases when specific nutrient substrates are added to the soil and declines gradually as the added substrate is exhausted (cellulose decomposing bacteria, nitrogen-utilizing bacteria, *Nitrosomonas* and *Nitrobacter*). The most predominant species of soil bacteria comes under three orders namely, *Pseudomonas*, *Eubacteria* and *Actinomycetes* (Benizri et al. 2001). The most common bacterial genera are *Achromobacter, Arthrobacter, Azosprillium, Bacillus, Clostridium, Corynebacterium, Flavobacterium, Micrococcus, Mycobacterium, Pseudomonas and Sarcina* (Lynch 1987). Although the bacterial population is also influenced by temperature and moisture, but certain bacteria have the capacity to survive under extreme climatic conditions. Bacteria can survive in area where the temperature is below freezing point such as arctic zones, and also can thrive in arid/desert soils, where temperatures go very high. They form a resting structure called 'spores' which assist in the survival of bacteria in most of adverse environmental conditions. Bacteria can be categorized into several groups on the basis of their temperature tolerance ability and these groups are assigned as mesophiles grow between

15–45 °C, psychrophiles grow below 10 °C and thermophiles grow between 45 and 65 °C. However, mesophilic bacteria encompass the immensity of soil bacteria (Barber and Lynch 1977). Other factors such as pH, farm practices, application of fertilizer and pesticide, and organic matter amendments also affect bacterial populations in soils. 'Autotrophic' and 'heterotrophic' bacteria are reside in a wide array of soils, where autotrophic bacteria such as purple and green bacteria build their own organic matter from CO_2 or inorganic carbon sources, while heterotrophic bacteria rely on pre-formed organic compounds or matter for their nutritional strength. 'Photoautotrophs' obtain their energy from sunlight that they grab and convert it into chemical energy with the help of bacteriochlorophyll pigment. Whereas, chemoautotrophs oxidize inorganic materials to obtain energy and parallely, they collect carbon from carbon dioxide (Tate et al. 1995). The cyanobacteria are gram-negative eubacteria identified by their capability to carry out oxygenic photosynthesis and having features common to bacteria and algae, and thus often named as "blue-green algae". Other than chlorophyll, phycocyanin is additional pigment which gives a special blue-green color to cyanobacteria. Soil exerts excellent habitat for cyanobacteria where important or dominant cyanobacteria belong to the genera *Anabaena*, *Aphanocapsa*, *Chroococcus*, *Cylindrospermum*, *Fischerella*, *Lyngbya*, *Microcoleus*, *Nostoc*, *Scytonema* and *Oscillatoria* (SubbaRao 1997; Benizri et al. 2002). Heterocyst is special structure in some cyanobacteria having imperative role in nitrogen fixation. The rice fields are an excellent habitat for the growth of some cyanobacteria where they are able to fix atmospheric nitrogen (Prescott et al. 1996).

2.2 Actinomycetes

Among soil biota, actinomycetes are also important soil microorganisms having adequate characteristics to classify them into a different group of prokaryotes. Indeed actinomycetes are grouped with other bacteria under class 'Schizomycetes', but are restricted to order 'Actinomycetales'. Actinomycetes exert certain resemblance to Fungi imperfecti or Deuteromycetes, which abundantly sporulates, and can be seen as distinct clumps or pellets in liquid cultures (Benson 1988). The abundant population of actinomycetes can be seen the presence of decaying organic matter. However, they are intolerant towards acidic conditions and the number of actinomycetes decreases below pH 5.0. Furthermore, waterlogged soil create unfavorable conditions for the growth of actinomycetes, but desert soils of both arid and semiarid regions maintain considerable populations, possibly due to spores resistance of towards desiccation. The proportion of actinomycetes in the entire microbial population enhances with the depth of soil. Therefore, actinomycetes can be isolated in ample numbers from C horizons of soil profiles. '*Streptomyces*' is the most common genus of actinomycetes while, *Nocardia* and *Micromonspora*, and in particular *Actinoplanes*, *Actinomyces* and *Streptosporangium*, are only encountered infrequently (Prescott et al. 1996; SubbaRao 1997). Temperature is another factor which influences the growth of actinomycetes, and temperatures ranged from 25 to 30 °C are much suitable. While in compost heap thermophilic actinomycetes (belong mostly to the genera *Thermoactinomyces* and *Streptomyces*) growing at 55 and 65 °C are most common and present in huge numbers.

2.3 Archaebacteria

'Archaebacteria' are primitive prokaryotes and are considered to be the first organisms to emerge on the earth. They have unique features for their survival in extreme hostile environments such as salt marshes and hot sulfur springs, where ordinary organisms cannot survive. They are phylogenetically very distant from 'eubacteria' and have typical characteristics. Moreover, these microorganisms are devoid of special cell-wall material 'peptidoglycan' a unique feature to describe and distinguish it from eubacteria, However, proteins and non-cellulosic polysaccharides takes part in formation of their cell wall structure. Branched chain lipids are act as main constituents of cell membranes that help archaebateria to tolerate extreme pH and temperatures. The rRNA component of archaea is also moderately dissimilar from those of other organisms (Huber et al. 2002). Archaebacterial group could be made-up of two subgroups referred as 'obligate' and 'facultative' anoxybionts, where obligate anoxybionts mainly includes methanogenic and halophilic species, those reside in habitat devoid of oxygen. However, facultative anoxybionts can be found in the presence and absence of oxygen (Kyrpides and Olsen 1999).

2.4 Fungi

Among the soil microorganisms fungi show the huge diversity and have distinctive features as they bear filamentous mycelium consisting of individual hyphae. Since fungi are heterotrophic in nature, but the type of organic materials exhibit a direct impact on fungal populations in soils. A unique trait of fungi emerges in acidic, neutral and alkaline soils, give them an advantage over population of actinomycetes and bacteria. Fungi are strict aerobic organisms and abundant fungal populations are present in arable soil. Moreover, fungi show a critical preference for different soil depths, and fungal species common at lower depths are rarely exist on the surface. Such distribution of fungal population is specified by the accessibility of organic substances/materials and by the ratio between oxygen and carbon dioxide in different depths of soil atmosphere. Fungi have been classified into 'phycomycetes', 'ascomycetes', 'basidiomycetes' and 'fungi imperfecti' or 'deuteromycetes'. Most common fungal isolates from soils are belong to the class 'Fungi Imperfecti' as they have very common trait to generate profuse asexual spores but not have sexual stages. The characterization of the members of these fungi is based on their septate mycelium and a special structure known as 'conidiophore', where this structure forms conidia or spores incessantly. However, other three classes of fungi show both sexual and asexual strategies for reproduction. Presence of non-septate mycelia is trait of phycomycetes and members of this group produce an imprecise number of specific spore cells known as 'sporangia'. In ascomycetes, species-specific number of meiotic spores is produced by the sporangium, while a higher degree of sporangium specialization known as 'basidia', is recognized in basidiomycetes. Fungi, specially 'ascomycetes' and 'basidimycetes', have the capacity to degrade complex organic compounds such as cellulose or lignin, but many members of these two fungi also live as root symbionts (mycorrhiza) and acquire photosynthetic product derived from plant partner in the form of simple sugar (Lynch and Hobbie 1988). '*Botrytis*', '*Chaetomium*', '*Aspergillus*', '*Cephalosporium*', '*Alternaria*', '*Cladosporium*', '*Cunninghamella*', '*Fusarium*', '*Gliocladium*', '*Monilia*', '*Rhizopus*', '*Mortierella*', '*Pillularia*', '*Pythium*', '*Rhizoctonia*', '*Mucor*', '*Scopulariopsis*', '*Trichoderma*', '*Zygorynchus*', '*Verticillium*', '*Penicillium*' and '*Trichothecium*' are the major genera of fungi present in soils (Hawksworth 1991a; SubbaRao 1997). Filamentous fungi show beneficial effect in soil as they participate in organic matter degradation and as well as assist in soil aggregation. '*Metarhizium*', '*Alternaria*', '*Dematium*', '*Aspergillus*', '*Gliocladium*' and '*Cladosporium*' are the some important genera of fungi having ability to produce certain substances related to humic substances in soil and hence assist in the maintenance of soil organic matter (Hawksworth 1991b).

2.5 Algae

In nature, the soil algae are ubiquitous when favorable conditions in form of moisture and sunlight are available. In soil system, the population of algae is not as abundant as bacteria and fungi, and structurally they may be unicellular (*Chlamydomonas*) or filamentous (*Ulothrix, Spirogyra*). As photoautotrophic organism algae utilize carbon dioxide from the environment and able to generate oxygen. Moreover, algae may also show the variety of their habitat, as they have also been found beneath the soil surface and at the location where sun light cannot be reached, besides the normal soil habitats. Though, such extreme locations express the lower population as compared to those of algae that dwell in normal habitat or soil surface (Metting 1988). '*Protosiphone*', '*Chlorella*', '*Chlrococcum*', '*Oedogonium*', '*Chlamydomonas*', and '*Chlorochytrium*' are the important genera of green algae inhabit in most soils (Lynch 1990).

3 Biosphere and Microorganisms

In soil there are variety of spheres exist which influence the microbial interaction and relevant microbial mediated biogeochemical processes. Soils can be partitioned into different spheres of influence such as the rhizosphere, the detritusphere, or the drilosphere (Nunan 2017). The detritusphere is a vastly and relevant microhabitat in soils where litter fermentation and humification layers over the soil surface have substantial root, saprophytic and mycorrhizal activity, and associated grazing fauna (Haynes 2014). It is also depicted as 'biogeochemical interface', where soil makes contact with fresh plant litter and act as a very important biochemical hot spot for activity of microorganisms and soil material cycling in soil (Kuzyakov 2010). Accordingly, the detritusphere is compatible to explicate the regulation of important soil functions. Bacteria are typically considered to be more active entities in carrying out the degradation of labile organic compounds and also involve in early stages of plant litter decomposition. On contrary, fungi are considered to be more important organism in the degradation of composite substances, and actively involved in later stages of litter decomposition process (Paterson et al. 2008). Rhizosphere is an important sphere of soil frontier and first time described by 'Lorenz Hiltner' in 1904, who was a German agronomist and plant physiologist by profession. 'Rhizosphere' is the region where soil volume interacts in direct way with plant roots where the products of rhizodeposition stimulate the activity and population of microorganisms, thus shifting the balance between mineralization and immobilization of Nitrogen (N) (Clarholm 1985). Additionally, rhizosphere is an interface between biota and geologic atmosphere, where roots show extreme physical pressures on adjacent soils. Rhizosphere is also the chemical milieu where several biogenic chemical reactions intermingle with minerals, and this root surrounding region presents the unique territory for a broad group of microorganisms. Therefore, rhizospheres are primarily significant for soil formation, and as well as participate in the formation of the most tremendously weathered soils of earth (Richter et al. 2007). "Drilosphere", is an another soil sphere consist of the following (a) an inner microenvironment of the earthworm's gut, (b) the exterior part of earthworm contacted with the soil, (c) 'surface' and 'belowground' casts and (d) 'burrows' and 'chambers' constructed by the earthworm, and all of which are considered as 'microhabitats' for a variety of microorganisms such as bacteria and fungi (Condron et al. 2010).

The soils of drilosphere are rich in P, N and humified organic substances in contrast to the nearby soils and such kind of soils (Giri et al. 2005) are also reported to have a huge proportion of the entire soil 'denitrifying' and 'nitrogen-fixing' bacteria (Wolters 1991). Both the microbial community and enzymatic activity in the drilosphere can be influenced by the input of labile carbon and energy (Lipiec et al. 2015) and, moreover fresh earthworm cast aggregates with an elevated carbon input stimulate the activity and development of microorganisms. On contrary, the microorganisms in structural site built by earthworm are an inevitable part of earthworms usual diet (Pizl and Novakova 2003) and consequently through the direct trophic effect such structural sites may influence both the microbial loads and activity (Andriuzzi et al. 2016). Hence, the enhanced functional microbial diversity and the better enzymatic activity in the majority of earthworm-influenced compartments 'drilosphere' make the soils less prone to degradation. Moreover, the biological and physical functions of the both 'drilosphere' and 'casts' influence numerous ecological processes at the local (burrow) and landscape scales, and help in improving and as well as conserving the soil quality (Lipiec et al. 2016).

4 Role of Microorganisms in Chemical Transformation

A process of conversion in which organic compounds transforms from one form to another form, it affects existence and toxicity of the compound and known as chemical transformation (Smitha et al. 2017). Soil dwelling bacteria and fungi play a major task in chemical transformation by means of their direct and indirect activities. This process is known as biotransformation. Microbes developed such mechanisms to acclimatize environmental changes. Biotransformation occurs by means of enzyme as well as non enzymatic way. A huge diversity of microorganisms are important because diverse microorganisms contains different enzymes and resides in various physiological pH, which leads to the wide range of biotransformation, these biotransformation are the major source of nutrient recycle in environment.

4.1 Phosphorus Transformation

'Phosphorus' is found in both organic and inorganic form and known as second most essential nutrient it participates in various cellular metabolic activities in plants and microorganisms. In phosphorus cycle, microorganisms bring many changes in phosphorus transformation such as alter inorganic phosphorus solubility, convert organic phosphorus in inorganic form by means of mineralization. Many bacteria such as *Pseudomonas* and *Bacillus* take part in inorganic phosphorus solubilization by secreting a wide range of organic acids (acetic acid, glycolic acid, formic acid and succinic acid) (Chalot et al. 2002). Many fungi such as *Fusarium* and *Penicillium* also secrets various organic acids that solubilizes insoluble form of phosphate (Sollins et al. 1981). Mineralization of organic phosphorus occurs by the action of phosphatases (Phytase, Nucleotidase, Sugar phosphatases, Nuclease and Phospholipases), hydrolyzes phosphorus-ester bond and releases orthophosphate. These phosphatase categorized into three groups according to their optimal pH, if enzymes having optimal pH '5', '7', and '9.5' then they can be categorized in 'acid', 'neutral' and 'alkaline phosphatases', respectively (Tabatabai 1982). Some fungi produce huge amount of 'acid', 'neutral' and 'alkaline phosphatases and take part in organic phosphorus mineralization (Bae and Barton 1989).

4.2 Nitrogen Transformation

Nitrogen availability for plants is the major area of concern to maintain sustainable ecosystem. Majorly nitrogen is present in three forms such as ammonium, nitrate and organic nitrogen. Soil microorganisms utilize this organic nitrogen in their metabolic activities and converts into ammonium and nitrate form by a process called nitrogen fixation. Nitrogen fixation occurs in symbiotic association by *Rhizobium*, *Bradyrhizobium* in association with leguminous plants as well as asymbiotically by aerobic bacteria such as *Azotobacter*, *Beijerinckia*, anaerobic bacteria *Clostridium*, organotrophic bacteria and free-living cyanobacteria. In nitrogen cycle, a diverse and huge group of bacteria are involved in ammonification. But small group of microorganisms i.e. chemoautotrophic bacteria (ammonium oxidizers and nitrite oxidizers) are able to convert ammonia into nitrate in nitrification process (Kaplan 1983). Nitrate and nitrite reduce into ammonia through bacteria such as *Mycobacterium* and into nitrogen by denitrifying bacteria such as *Pseudomonas*, *Bacillus*, *Thiobacillus* (Payne 1981). Fungi are versatile in chemo-heterotrophic metabolism by means of their specific enzyme (Cromack and Caldwell 1992), and metabolize various substances for nutrient and energy (Wainwright 1992). Ectotrophic mycorrhizal fungi participate in ammonification of organic nitrogen hence takes part in nitrogen transformation (Lakhanpal 2000).

5 Soil Genesis and Function of Microorganisms

The term 'soil' is referred to as "critical zone" of earth as it plays an imperative function in controlling environment and life sustainability on earth. The term 'mineral soil' is used for those soils derived from weathered rocks and minerals. Soil is an excellent habitat for various kinds of organisms where they contribute in the organic matter decomposition and subsequently formation of humus. As the plant leaves are fall onto the soil surface where several pertinent soil microorganisms can "attack" and decompose plant tissues. The organic matter derived from plants or leaves is utilized as rich energy source for microbial growth, and increasing their population in the soil. Soil microorganisms use easily degradable materials (simple sugars and carbohydrates) present in the plant parts and leave more resistant substances (such as fats and waxes) behind, which are not easily degradable. The material left at last is not simply decomposed, and such tough residues involves in the humus formation. Humus in soil acts as a gluing agent, and effectively holding primary soil particles together to generate secondary aggregates and thus, soil microorganisms and the humus assist in the soil formation and development.

5.1 Microorganisms in Rocks and Minerals

Weathering of rock is one of the most significant geochemical processes which taking place on the earth and results into the pedogenesis, maintenance of soil productivity, regulating atmospheric composition and global climate change. The weathering is the process of breakdown and disintegration of rocks and minerals which are commenced by certain physical agents and chemical processes, leading to the formation of regolith (parent material). A complex interaction of physical, chemical and biological activities participate in rock weathering processes. In physical weathering the rocks are disintegrated and are broken down to relatively smaller pieces, without creating any new substances. However, in the process of chemical weathering disintegration of rocks and minerals occurs by various chemical processes and it takes place mostly at the surface of rocks and minerals with vanishing of certain minerals and formation of secondary products. On other hand, in biological weathering the biological or living agents are accountable for both disintegration and breakdown of rocks and minerals. Plants, animals and microorganisms actively involved in the biogeochemical cycles which add to the pedogenesis through biological weathering (Gadd 2007). Microbial and plant root assisted biological weathering of rock plays a key task in maintaining and supply of various inorganic compounds/elements which essentially required as nutrients by plants (Chang and Li 1998). The roots of plants loosen the rock material and the process for crack formation commence. 'Root-pry' is an important phenomenon which occurs by big crack created through root expansion. Soil associated macro fauna including earthworm, snail and burrowing animals (such as rodents) also contribute in the process of biological weathering (Lian et al. 2008a, b). The weathering of major types of rocks through microorganisms results into the releasing of various types of elements, which can be required as nutrients (Calvaruso et al. 2006). Bacteria, cyanobacteria, fungi and lichens are the important microbial groups, and they all have been considered as agents for biological weathering of rocks. Rock surfaces, cracks and the pore spaces of sand, stone and granite are the unique territory for microorganisms where they sometimes form 'biofilms' that ultimately contribute to the disintegration of the rocks (De ta Torre et al. 1993; Puente et al. 2006). Hirsch et al. (1995) demonstrated that organic acids secreted by microorganisms helps in dissolving of rocks, and enhance the rate of rock disintegration. Rocks are the main resource of metals in the form of minerals and ores. Metal bio-reduction is considered to be an imperative feature for microbes to endure in such environmental conditions. Cations, anions and inorganic nutrients are needed for rock inhabiting microorganisms and plants. And these nutrients are released during the process of rock biodegradation (Chang and Li 1998). One important activity carried out by bacteria through rock weathering is the accessibility of trace element in the soil; and such bacteria improve the trace element uptake by plants. This is attained by microbial alteration of the absorptive properties of the roots such as improving and increasing the root length, surface area and the amount of root hairs indirectly or directly contributes to the translocation of trace elements by different processes. Microbial mediated mineral weathering process occurs by range of bio-deteriorating mechanisms that comprises uptake of elemental, redox reactions, production of acids, metabolites, chelating compounds and polymers (Welch and McPhail 2003) (Fig. 2.2). Bacterial communities participate in dissolving the primary rock-forming minerals to get essential nutrients and also perform as nucleation sites for the precipitation of secondary minerals. Plant roots in association with the microorganisms disrupt sheet silicates and hence expose new surface area where the further process of biochemical weathering occurs (April and Keller 1990). In general, the weathering process of silicates carried out by microbial communities is a key biological mechanism for nutrient requirements. The sequential steps of mineral weathering can also be influenced by a nutritional potential of minerals with the microbial organisms having capacity for the formation of beneficial minerals. Numerous environmental factors such as lighting, humidity, nutrients and rainfall amount influence the process of biological weathering. Moreover, microbial mediated degradation of rocks is dependent on other certain factors including location, climate and season.

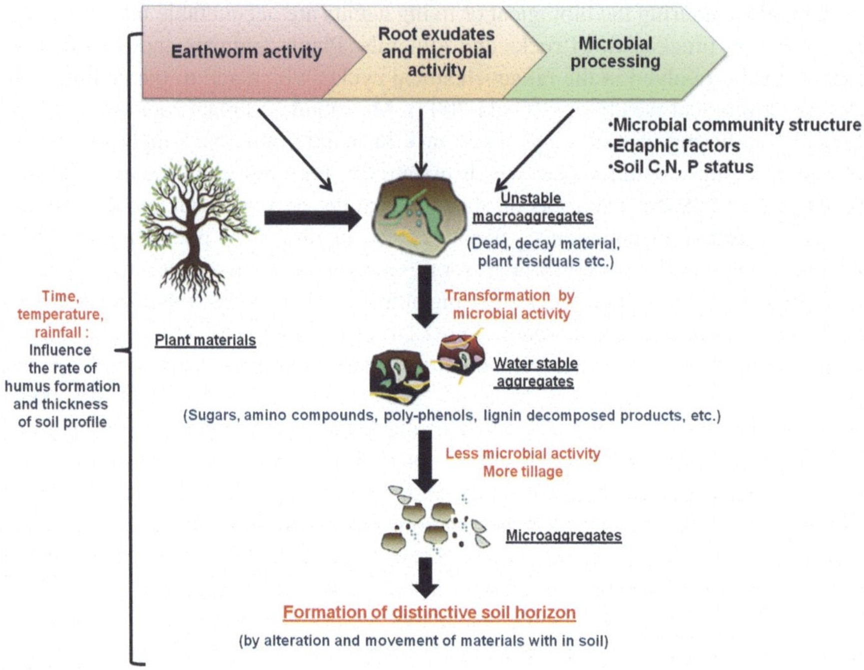

Fig. 2.2 Schematic representation of soil aggregates formation by biological and physical processes

Weathering process carried out by microorganisms can be either aerobic or anaerobic in nature, and may takes place in acidic, neutral and alkaline conditions (Berthelin 1983). Several factors such as wind, water and bird droppings can play a key role in the transfer of spores of both bacteria and fungi results into early colonization of rocks. Microorganisms may be entrained within the rocks through various processes including rainfall, snowmelt, and some aeolian transport (Cockell et al. 2009). Establishment and growth of microbial communities are also regulated by the elemental composition and physical properties of the rock (Gleeson et al. 2006). Lichens are the mutual association between fungus and a photosynthetic partner (either green algae or cyanobacteria), which are also helpful in weathering of rocks. Few lichen activities such as hypal penetration, contraction and extension of lichen thallus due to microclimatic wetting and drying helps in physical disintegration of rocks (Arino et al. 1997; Moses and Smith 1993). Chemical disruption of rocks is due to the production of respiratory CO_2, secretion of a variety of organic acids and salts which results in dissolving of minerals and formation of different biochemical compounds responsible for metal chelation. Weathering process of ‘sandstone basalt’, ‘granitic’ and ‘calcareous rocks’ through lichen, and the mode of action shown by lichen in rock weathering process are also well documented (Chen et al. 2000). Experimental works performed on hornblende granite in New Jersey (USA), depicted a three to fourfold enhancement in the rate of weathering of those rock surfaces which were covered by lichen as compared to non-covered/barren rock surfaces (Zambell et al. 2012). Chemical and mineral composition of rocks and 'micro-topography' influence the fungal rock colonization. 'Epilithic' and 'endolithic' fungal communities produce several compounds such as carbohydrate and osmolytes in response to 'desiccation' which extend the water retention period, and ultimately results into expanding the degree of chemical reactions responsible for silicate weathering. Moreover, mechanical strength of the hyphae is increased by melanin pigmentation, which is helpful to penetrate the crevices of an rock surface and also, provide shield from metal toxicity (Sterflinger and Krumbein 1997; Gadd 1993). Studies of Puente et al. (2004), revealed the fluorescent pseudomonads and *bacilli* were prominently responsible in the weathering process of limestone, igneous rocks and marble. Moreover, weathering of 'biotite' and 'anorthite' are potentially exhibited by bacteria associated with root and mycorrhiza (Balgoh et al. 2008). Bacteria also produced metal chelating organic compounds 'siderophores', assist in the weathering of iron and magnesium containing silicates (e.g. biotite) (Frey Klett et al. 2007). Bacteria and fungi presented a remarkable increase in the dissolution rates of apatite, feldspar, biotite, quartz and other related minerals (Barker et al. 1997).

5.2 Mineralization, Humification and Soil Aggregation

A process of progressive dismantling of organic material such as fertilizer into inorganic components is known as mineralization. Microorganisms are the major entities that take part in process of mineralization. The process of mineralization can also be influenced by numerous environmental factors such as oxygen availability, pH, temperature and water potential. Mineralization of organic fertilizers into nitrates governed by different microorganisms in three steps process such as aminizations, ammonification and nitrification. Aminization is referred to breakdown of huge and complex forms of proteins into short structural compounds such as amino acids, amines and amides. An organotrophic bacterium such as *Rhodospirillum rubrum* performs aminization by producing an enzyme 'proteases'. However, the process of formation of ammonia through organic compound is known as ammonification. The important bacterial groups that carry out this process include *Bacillus, Streptomyces* and *Proteus.* By mineralization process, microorganisms recycle nutrients in the soil and augments soil fertility and health (Buscot and Varma 2005) (Fig. 2.2). Secondly, 'humification' is a process governs by bacteria and fungi lead to the conversion of dead organic matter (leaves, twig) into humus. Humification influences diverse soil characteristics such as fertility, water availability, pore size and pH of soil. Decayed parts of plants, animal excreted substances containing organic compounds such as carbohydrates, protein, lignin and resin, which used by microbes as energy source and changed into humus. Fat, waxes and lignin are undegradable by many microorganisms but white rot fungi are able to metabolize it, which are the precursor for humus formation. Humus releases nitrogen containing compounds in soil and increases the soil fertility, also increases water holding capacity through increased porosity (De Macedo et al. 2002), which leads to the well soil structure and recycle nutrients in the environment. It enhances the cation exchange capacity of soil and thus increasing nutrients chelation activity (Szalay 1964). In the process of humification microorganisms produces mucilaginous substances, which increases soil adhering capability as well as allows better aeration (Huang et al. 2008). Therefore, microorganisms exhibit a key role in humification process and hence maintain structure, health and fertility of soil (Figs. 2.2 and 2.3).

Another important aspect is 'soil aggregation' where soil particles reside together in the form of stable clumps because of moist clay, gums, organic matter and fungal hyphae. Soil containing plenty of aggregates is called well-aggregated soil and is an important marker for soil health and environmental sustainability since soil aggregation stabilizes organic material, improve water holding capability, nutrients and air within soil micro-sites (Balesdent et al. 2000). A number of study exerted that microorganisms produce various sticky substances that can interact with soil particles (organic material, clay material and polyvalent metal) and leads to the development of soil aggregates (Davinic et al. 2012). Chenu (1993) reported

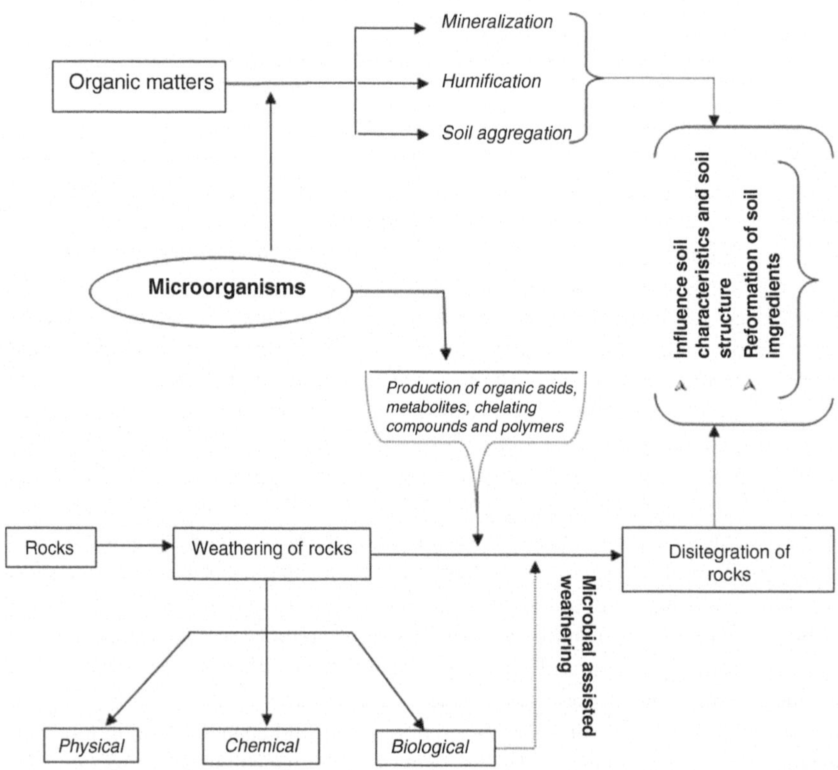

Fig. 2.3 Microbial mediated steps of organic matter decomposition and weathering of rocks for pedogenesis

increased water holding ability of soil aggregates due to addition of polysaccharides. Furthermore, bacteria have the ability to develop electrostatic charge, which holds small aggregates of soil together (Kallenbach et al. 2016). Fungi forms long hyphae which help in aggregation of soil by cross linking between soil particles and hence length of the hyphae is related to soil aggregation (Davinic et al. 2012). Moreover, arbuscular mycorrhizal fungi produce a glycoprotein 'glomalin' contain the trait of gluing the soil particle which leads to the phenomenon of soil aggregation (Chotte 2005).

5.3 Microbial Metabolism in Soil

Soil has been declared as porous medium comprising of organic materials, variety of minerals, water, gases and numerous communities or varieties of microorganisms. Diverse forms of microorganisms reside in soils present various kinds of cellular metabolism that ultimately influence the soil environment. Microbial metabolism is the process through which microorganisms uptake source of energy and nutrient, which are needed for their survival and reproduction process. Most of the soils broadly can be divided into two categories; (a) A mineral soil contains less than 20% organic carbon and, (b) an organic soil possesses at least this amount. The significance of organic matter within soils cannot be underrated. Soil organic matter (SOM) is an important source because it assists to retain the nutrients, holds the improved water capacity and maintains soil structure. Microbial mediated degradation of plant materials results into the evolution of CO_2 and the incorporation of the plant used carbon into additional microbial biomass. On contrary, a small fraction of the decayed plant material remnants in the soil medium as soil organic matter. Nitrogen (N) is another important element of the soil, but most of the soils are nitrogen deficient. Therefore, each year tons of nitrogen fertilizer is applied to agricultural field which consequently led to the soil highly polluted. Soil nitrogen is often considered in relation to the soil carbon content like the organic carbon to the nitrogen ratio (Nannipieri et al. 2017). The Nocardioforms, Coryneforms and the true filamentous bacteria such as Streptomycetes are important constituents of microbial community in soil (Madigan et al. 2011). These gram positive bacteria take part in the degradation of the hydrocarbons and additionally some members of these bacterial groups actively participate in degradation of pesticides and its residues (Quejigo et al. 2018). The filamentous actinomycetes, mainly of the genus *Streptomyces,* produce an odor causing compound called 'geosmin', which confers soils their characteristic odor. Supply of essential elements such as carbon (C), oxygen (O), hydrogen (H), nitrogen (N), phosphorus (P) and sulphur (S) from the soil are required for the growth of all living organisms. In soil the 'recycling' of these elements is a key process and also the basic measure to avoid exhaustion. Soil microorganisms take part in breaking down of dead organic matter and their conversion in to the forms which could further be utilized by other organisms. In this context, microbial enzymes act as key 'engines', and dominantly assist in driving the bio geochemical cycles (Falkowski et al. 2008). The carbon cycle globally is dominated because of the equilibrium between photosynthesis and respiration process. The shifting of carbon from the atmosphere to soil carried out by 'carbon fixing' autotrophic organisms, principally plants and as well 'photo' and 'chemoautotrophic, microorganisms that produce atmospheric carbon dioxide (CO_2) into organic matters. The fixed carbon is then reverse to the atmosphere through employing various pathways that account for the process of respiration of both types of microorganisms i.e., autotrophic and heterotrophic microorganisms. Numerous environmental factors including mineral nutrients, available water content, carbon and energy sources, ionic composition, pH, temperature, oxidation-reduction potential, spatial relationships, relation/interaction between microbiota and genetics of the microorganisms show their influence on the natural environment, activity and as well as population dynamics of microorganisms in soil (Nannipieri et al. 2017).

5.4 Microorganisms in Polluted Soil

Presence of toxic chemicals and other pollutants in the soil is in elevated concentrations is accountable for polluted soil, to be of hazard for plants, wildlife, humans and the soil itself. Because of the polluted soil, most of the arable land of world is turning to desert land and becoming non-arable at continually increasing rates, mainly in part affected from global warming and due to rampant application of agricultural fertilizers and pesticides. Soil pollution can exert a number of harmful effects on ecosystems and human, plants and animal health also. In bulk soils, some of the heavy metals severely affect the growth, cellular architecture and the metabolism of microorganisms, where heavy metal cause several functional disorder, which causes denaturation of protein or the disintegration of cell membranes (Lei-ta et al. 1995). However, the polluted soil become enrich with specific microorganisms which particularly degrade the pollutant and convert into non-pollutant or less harmful matter. Polluted soil is the specific site for those microorganisms who have adapted towards several environmental pollutants in form of heavy metal, pesticides and other chemicals. Therefore, polluted soil is a fine milieu for exploring the effect of environmental pollution on community structure of microorganisms, microbial biomass, and enzymatic activities of microorganisms (Kandeler et al. 2000). The environmental pollutants present in the soil, alters the soil composition and pH of the soil which directly influence the growth of specific microbiome which can adapt or survive in the conditions of heavy metal stress, salts and other extreme conditions due to the presence of pollutants. Depending on their concentrations, these pollutants can have devastating effects on ecosystems, as well as cause severe damage to humans and other animals. Microbial wealth of soil is generally considered synonymous to higher bio-availability of nutrients, and therefore soils are biologically more reactive to any management input(s). Microorganisms are the eminent natural agents for pursuing biodegradation reaction of several toxic compounds (such as pesticides, polycyclic aromatic hydrocarbons and munition wastes etc.) and remediate soils effectively. Numbers of bacterial classes such as gamma proteobacteria (gr. *Pseudomonas*, *Aerobacter*, *Acinetobacter*, *Moraxella*, *Plesiomonas*), betaproteobacteria (*Burkholderia*, *Neisseria*), alphaproteobacteria (*Sphingomonas*), *Actinobacteria* (*Micrococcus*), and *Flavobacteria* (*Flavobacterium*) are considered to be dynamic microbial bio-degraders (Mamta and Khursheed 2015). Several petroleum compounds show the health hazards to humans and other animals. Study of Mirdamadian et al. (2010) revealed that microorganisms namely *Bacillus* spp., *Rhodococcus* spp., *Pseudomonas* spp. and *Micrococcus* spp. were found to have capacity to degrade petroleum compounds.

6 Molecular Approach in Soil Microbiology

Soil presents the renowned hub for the range of microbial diversity and contained with various biospheres where microorganisms perform several metabolic activities and maintain the soil characteristics. To study the soil microorganism's dynamics, modern molecular approaches have been opened vast ways to explore the numerous diversity of microorganisms present in the environment, in which most of the organisms are non-cultivable due to its unknown growth requirements. Some advanced techniques are mentioned in Fig. 2.4 for characterizing the microbial diversity and

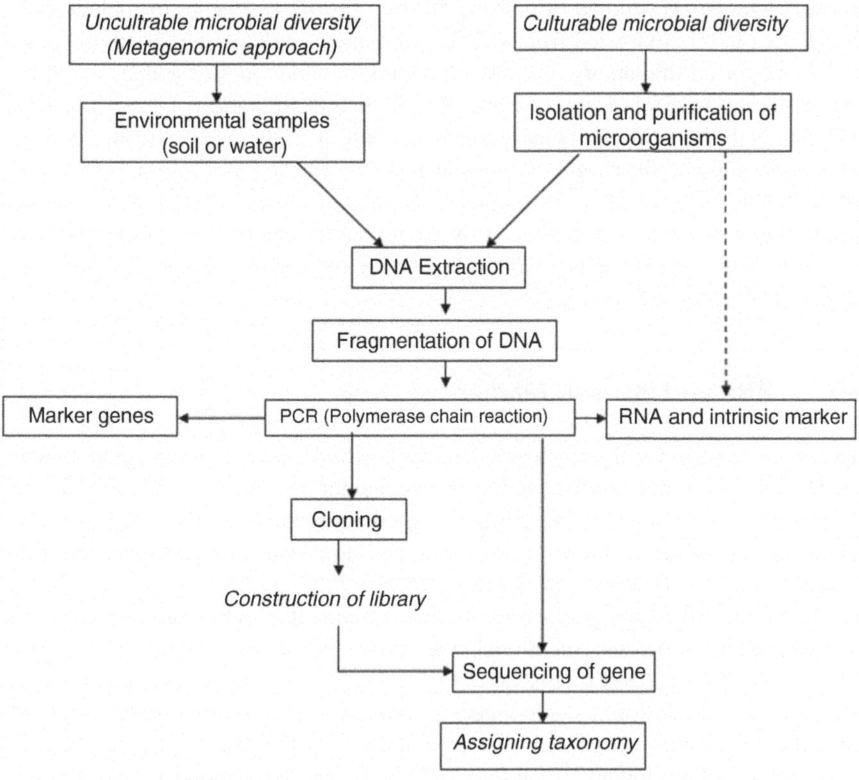

Fig. 2.4 Molecular techniques for determining culturable and unculturable microbial diversity

richness of culturable and unculturable microorganisms (Fig. 2.4). These microbial diversity take part in soil related functions such as decomposition of organic matters, nutrients recycling, production of organic acids and other related function lead to the soil genesis.

6.1 Marker Gene

In phylogenetics, group of orthologous genes can be used to define between taxonomic lineages as marker genes. Marker genes are genetic trait that can be detectable or a particular DNA segment that can be identify and track in the genome. Marker genes can serves as standard genes for another gene of interest called the target gene (Ren et al. 2016). These assets of the marker genes can either encoded by DNA and it can be detectable upon gene expression or be contained by a DNA itself. Generally two kinds of marker genes are presents i.e. intrinsic marker genes and recombinant marker genes. Intrinsic marker genes are naturally present within the genome of an organism such as *rrna, pmoA, narG, mcrA, chiA* etc. They are generally used for structural and functional diversity studies and examination of biofilm architectures (Jansson et al. 2000, Prosser 2002). The expression of intrinsic marker genes can be studied by detecting transcripts of these genes in nucleic acids, which are directly extracted from soil. Recombinant marker genes are those genes which are placed into an organism by utilization of techniques of genetic engineering. These are used for *in situ* gene transfer and their examination of growth survival and their activity in different environmental conditions. The fate of the microorganisms in the soil and the expression of selected activities in complex microbial communities can be studied using recombinant marker genes. Chromogenic marker genes, fluorescence marker genes, antibiotic resistance gene etc. are some examples of recombinant marker genes which directly or indirectly influence the different stages of soil genesis.

6.2 RNA and Intrinsic Marker

The genes encode for the ribose nucleic acids (RNA) especially its small subunit (SSU) is the important marker used in environmental microbiology. These SSU acts as intrinsic markers that are present in living microorganism, as they all possess the ribosome. Ribosome contains highly conserved sequences and participates in protein biosynthesis. However, presence of such highly conserved sequences in the RNA, certain part of the gene encoding this contains the highly variable segments that doesn't perform any functional role, rather can be act as intrinsic marker (Tourasse and Gouy 1997). This variability in the gene segment involved in tracing the biological evolution and phylogenetic relationship. The importance of rRNA encoding genes was recognized by Woese and Fox (1977) mainly for building the universal tree of phylogeny for all forms of life. In one of the pioneer study, fluorescent labeled rRNA targeted oligonucleotide was used for the detection of predominant genotype which is independent of cultivation by means on *in-situ* hybridization (Pace et al. 1986). The specificity of the different gene probe enables detection of different phylogenetic groups or species quantification without the desired cultivation. In addition, biofilms producing ability and symbiotic association tendency can easily be studied with this intrinsic marker. The SSU rRNA gene with an optimum size of 1540 nucleotide is more suitable for ecological study. In previously published literature symbiotic bacteria from marine ecology was successfully characterized by direct isolating and sequencing the 5S RNA from the larger subunit of prokaryotic RNA (Amann and Ludwig 2000). Thus all literature strongly supports the use of RNA as an intrinsic marker and can be best studied with polymerase chain reaction easily.

6.3 Cloning

Past the down to earth inquiries of the ways to improve vectors for library development and the ways to expand profitable existing libraries, a specialized inquiry that we find especially fascinating: what amount of the arrangement decent variety show in unique DNA separates is caught in built libraries, and what influences this? Despite the fact that less a worry for useful display, it is important to include the elements which impact library representativeness; explaining these elements may prompt improvement of effective procedures for getting maximum capacity of natural metagenomes. Already utilized shotgun sequencing to look at inclination in a fecal library of human (Lam and Charles 2015) likewise show the after effects of 16S rRNA quality sequencing to inspect predisposition in soil library of corn field (Cheng et al. 2014). Investigation at the phylum-level demonstrated that in spite of the fact that the fecal library contrasted significantly in the relative plenitude of phyla contrasted with its comparing extricate, the relative wealth of phyla in the soil library of corn field appeared to be like its concentrate. The abnormal state of host sullying could be because of special intensification of layout amid PCR in light of contrasts in DNA adaptation: however display in little amounts, straight DNA might be all the more effectively opened up finished supercoiled DNA or shut roundabout plasmid DNA (Chen et al. 2007). The way that specific taxa are under or over represented, won't not represent a boundary to screen, but rather it helpful to realize which arrangements are not prone to caught under libraries. An investigation which have contrasted shotgun sequences of unique examples relating with marine water metagenomic libraries (Temperton et al. 2009) have demonstrated AT-rich groupings are represented in libraries. The particular components influencing the "clonability" of DNA, and the systems that prompt DNA prohibition, still should be tentatively decided. The steadiness of outside DNA in *Escherichia coli* is affected by duplicate number of vector and, therefore, single-duplicate fosmids might be perfect as library spin, in spite of the fact that the achievement of some utilitarian screens might be subject to a higher quality measurement. Plasmid vectors those are not cos-based give an elective where cloning is significantly not so much troublesome as extensive section DNA require not be segregated and bundling and transduction are not required; the weaknesses, in any case, are that a littler embed measure implies that bigger operons won't be unblemished, and if the plasmid retain a big duplicate number—valid for customary cloning vectors—it may prompt more noteworthy embed unsteadiness and prohibition.

6.4 Metagenomics

The evolutionary history, functional and ecological biodiversity can be understand by genomic analysis of complex environmental samples instead of laboratory cultivation and/or isolation of individual specimens. Examples of environmental samples include soil, water, sediments, passively collected aquatic, terrestrial and benthic specimens, gut contents and faeces. More than conventional Sanger DNA-sequencing technology, the advanced technique of Sanger DNA-sequencing technology led to large-scale, broad-scope biosystematics projects with wide applications like barcode of life initiative (Hajibabaei et al. 2007). The primary purpose of identification of unknown specimens are to generate vast DNA libraries which can be employed by DNA barcoding which further enrolls standardized species-specific genomic regions of DNA. For example, across Animilia, the cytochrome c oxidase subunit I (COI) gene region is capable of discerning between closely related species (Hebert et al. 2003). Likewise, 16S ribosomal RNA (16S rRNA) sequences is generally used for identification of bacteria (Welch and Huse 2011). However for fungal studies, the internal transcribed spacer (ITS) region of the nuclear ribosomal DNA is employed (Nilsson et al. 2008). While, the regions of plastid DNA including maturase K (matK) and ribulose-bisphosphate carboxylase (rbcl) required for the Plant DNA barcoding (Burgess et al. 2011). For biodiversity analysis, a number of other marker genes have been employed at different phylogenetic depths or in taxonomic groups. Furthermore, for processing of complex environmental samples, the traditional DNA-sequencing method cannot be used because it can sequence specimens individually (Shokralla et al. 2012) and all samples often contains mixture of DNA of individuals present in soil. The conventional sequencing method is most efficient for the development of huge DNA barcode reference libraries but due to higher number of individuals from environmental sample is afar the scope of its capability. But by the use of next-generation sequencing (NGS) technologies, it is very much possible to recover DNA sequences from huge number of specimens from bulk samples of an environment which have the capability to read DNA, parallel from multiple templates; something that do effectively and with ever-lowering costs (Hajibabaei et al. 2007). These NGS technologies have huge potential and can generate parallel millions of sequencing reads (Esposito et al. 2016). This type of sequencing capacity generates numerous sequence reads from the fragmented library of a specific genome (i.e. genome sequencing) from a pool of cDNA library fragments generated through reverse transcription of RNA like RNA sequencing or transcriptome sequencing or from PCR amplified molecules like amplicon sequencing. These all generated sequences are without the need of a conventional method or vector-based cloning approach that is not easy to amplify and distinct DNA templates (Shokralla et al. 2012). NGS technologies based upon the diverse chemistry which includes base incorporation/detection tools and there are two main steps such as: 1st is amplicon library preparation and 2nd is diagnosis of the integrated nucleotides (Glenn 2011; Zhang et al. 2011). NGS technologies can be classified into two major groups in which one is PCR based and another one is non-PCR based. One of the groups are PCR-based technologies, includes four commercially available platforms: HiSeq 2000 (Illumina Inc., San Diego, CA, USA), AB SOLiD™ System (Life Technologies Corp., Carlsbad, CA, USA), Roche 454 Genome Sequencer (Roche Diagnostics Corp., Branford, CT, USA), and Ion Personal Genome Machine (Life Technologies, South San Francisco, CA, USA). The second group, called 'single molecule' sequencing (SMS) technologies, are non-PCR based and do not include an amplification step prior to sequencing. Two single-molecule sequencing systems have been recently announced: PacBio RS SMRT system (Pacific Biosciences, Menlo Park, CA, USA) and HeliScope (Helicos Bio-Sciences Corp., Cambridge, MA, USA) (Shokralla et al. 2012). In recent years, mass sequencing techniques of environmental samples has been utilized in full swing for ecology and biodiversity research. By the use of NGS technologies, analysis of environmental samples from various ecosystems like marine, terrestrial, freshwater, gut microbiota and soil can be performed (Buée et al. 2009). The numbers of studies solicit to reply the query that which type of flora and fauna are adjacent in environmental area of interest. By the use of sequence data processed by NGS techniques, researchers have been able to observe the modest changes in community structure and dynamics which may occur due to natural environmental fluctuations or anthropogenic sources (Fierer et al. 2007). The numerous studies have examined soil bacterial diversity by analyzing 16S rDNA amplicons (Singh et al. 2012; Rousk et al. 2010). Results of these studies suggest that agricultural management of soil may significantly influence the microbial diversity (Roesch et al. 2007). There are numbers of other studies which mainly focused on fungal diversity of soil in forest and agricultural system by examining ITS amplicons (Acosta-Martínez et al. 2008). The selected functional gene amplicons or total RNA serve as one of the alternate approach to target soil microbiota (Fierer et al. 2007; Leininger et al. 2006).

6.5 PCR

PCR (Polymerase chain reaction) based molecular approaches deliver a rapid and sensitive substitute to conventional culture techniques. PCR-based fingerprinting techniques comprised of three important steps: a) the nucleic acids extraction b) the rRNA/rDNA amplification c) using fingerprinting techniques analysis of PCR products (Agrawal et al. 2015). PCR-based 16S rDNA sequence profiles of an organism gives information regarding microbial diversity, identification and the phylogenetic relationship predictions (Pace 1997). So, various 16S rDNA-based PCR techniques such as DGGE (denaturing gradient gel electrophoresis), ARDRA (amplified ribosomal DNA restriction analysis), TGGE (temperature gradient gel electrophoresis), T-RFLPs (terminal restriction fragment length polymorphisms),

SSCPs (single-strand conformation polymorphisms) and RISA (ribosomal intergenic spacer analysis) can give deep information about community dynamics and composition of an ecosystem in terms of richness, evenness and can be utilized to compare diverse species exist in an environmental sample (Agrawal et al. 2015). Another PCR technique for instance 'qPCR' is relies on the real-time recognition of PCR product by reporter molecules, which flourish upon amplification as PCR product amplified in every amplification cycle. qPCR technique is rather distinctive among other methods of community analysis dynamics, in that it allows to produce comparative and fast quantitative estimation of the availability of exact phylogenetic groups of microbial populations present in soil of interest (Fierer et al. 2005).

7 Conclusion

Microorganisms are ubiquitous and present everywhere and even in extreme environments, and carry out various significant functions. Wide diversity of microbial organisms and their particular habitat present a suitable sphere where microorganisms play important role in different process such as recycling of elements and other activities. Pedogenesis is the process of soil formation from rock weathering and other aspects such as organic matter decomposition on earth surface. Besides soil degradation, the reformation of soil is an important criterion which maintains the soil ecosystem and sustains the life of approximately all living organisms. Microorganisms take part in weathering of soil via synthesis and secretion of numerous organic acids and decomposition\partial organic matters decomposition through multiple enzymatic and other processes for releasing the inorganic minerals and provoking soil structures formation. Soil is hub for microorganisms as 'natural engineers' which carry out the several important processes for maintenance of nutrient cycling and essential mineral (such as phosphorus and nitrogen) transformation for maintaining the integrity of nutrients in soils. Recycling of important elements in soil through microbial assisted activities promote the sustainability of earth inorganic and organic matters and thus maintains the soil structure and its multifarious properties. The decaying plant or animal residues/parts ultimately helps in formation of upper portion of soil and this unique consequence is met through the activities of soil microorganisms. Soil microbiomes also participate in humification, aggregation and stabilization of soils and hence promote the fertility of soils which directly related to improved plant growth and provoke various magnitudes of microbial related activities. Soil also presents the diversity of both culturable and unculturable microorganisms which are important in formation of new soils from recyclable materials, and the study of such microorganisms opens the doors for several scientists worldwide to explore and investigate the molecular techniques for identifying total microbiome of particular site or sphere.

References

Acosta Martínez, V., Dowd, S., Sun, Y., & Allen, V. (2008). Tag-encoded pyrosequencing analysis of bacterial diversity in a single soil type as affected by management and land use. *Soil Biology and Biochemistry, 40*, 2762–2770.

Agrawal, P. K., Agrawal, S., & Shrivastava, R. (2015). Modern molecular approaches for analyzing microbial diversity from mushroom compost ecosystem. *3 Biotech, 5*, 853–866.

Amann, R., & Ludwig, W. (2000). Ribosomal RNA-targeted nucleic acid probes for studies in microbial ecology. *FEMS Microbiology Reviews, 24*(5), 555–565.

Andriuzzi, W. S., Ngo, P. T., Geisen, S., Keith, A. M., Dumack, K., & Bolger, T. (2016). Organic matter composition and the protist and nematode communities around anecic earthworm burrows. *Biology and Fertility of Soils, 52*, 91–100.

April, R., & Keller, D. (1990). Mineralogy of the rhizosphere in forest soils of the eastern United States. *Biogeochemistry, 9*, 1–18.

Arino, X., Gomez-Bolea, A., & Saiz-Jimenez, C. (1997). Lichens on ancient mortars. *International Biodeterioration & Biodegradation, 40*, 217–224.

Bae, K. S., & Barton, L. L. (1989). Alkaline phosphates and other hydrolyases produced by *Cenococcum graniforme*, an ectomycorrhizal fungus. *Applied and Environmental Microbiology, 55*, 2511–2516.

Balesdent, J., Chenu, C., & Balabane, M. (2000). Relationship of soil organic matter dynamics to physical protection and tillage. *Soil and Tillage Research, 53*, 215–230.

Balogh-Brunstad, Z., Keller, C. K., Bormann, B. T., O'Brien, R., Wang, D., & Hawley, G. (2008). Chemical weathering and chemical denudation dynamics through ecosystem development and disturbance. *Global Biogeochemical Cycles, 22*, 1007.

Barber, D. A., & Lynch, J. M. (1977). Microbial growth in the rhizosphere. *Soil Biology and Biochemistry, 9*, 305–308.

Barker, W. W., Welch, S. A., & Banfield, J. F. (1997). Biogeochemical weathering of silicate minerals. In J. F. Banfield & K. H. Nealson (Eds.), *Geomicrobiology: Interactions between microbes and minerals* (Reviews in mineralogy) (Vol. 35, pp. 391–428). Chelsea: Mineralogical Society of America.

Baudoin, E., Benizri, E., & Guckert, A. (2002). Impact of growth stages on bacterial community structure along maize roots by metabolic and genetic fingerprinting. *Applied Soil Ecology, 19*, 135–145.

Benizri, E., Baudoin, E., & Guckert, A. (2001). Root colonization by inoculated plant growth promoting rhizobacteria. *Biocontrol Science and Technology, 11*(5), 557–574.

Benizri, E., Dedourge, O., Di Battista-Leboeuf, C., Nguyen, C. S., & Piutti, G. A. (2002). Effect of maize rhizodeposits on soil microbial community structure. *Applied Soil Ecology, 21*, 261–265.

Benson, D. R. (1988). The genus Frankia: Actinomycetes symbionts of plants. *Microbiological Sciences, 5*, 9–12.

Berthelin, J. (1983). *Microbial weathering processes. Microbial geochemistry* (pp. 223–262). London: Blackwell.

Buckley, D. H., & Schmidt, T. M. (2001). The structure of microbial communities in soil and the lasting impact of cultivation. *Microbial Ecology, 42*, 11–21.

Buée, M., Reich, M., Murat, C., Morin, E., Nilsson, R. H., Uroz, S., & Martin, F. (2009). 454 Pyrosequencing analyses of forest soils reveal an unexpectedly high fungal diversity. *New Phytologist, 184*(2), 449–456.

Burgess, K. S., Fazekas, A. J., Kesanakurti, P. R., Graham, S. W., Husband, B. C., Newmaster, S. G., Percy, D. M., Hajibabaei, M., & Barrett, S. C. H. (2011). Discriminating plant species in a local temperate flora using the rbcL+matK DNA barcode. *Methods in Ecology and Evolution, 2*, 333–340.

Buscot, F., & Varma, A. (2005). *Microorganisms in soils: Roles in genesis and functions* (pp. 159–171). New York: Berlin Heidelberg/Springer.

Calvaruso, C., Turpault, M. P., & Frey-Klett, P. (2006). Root-associated bacteria contribute to mineral weathering and to mineral nutrition in trees: A budgeting analysis. *Applied Environmental Microbiology, 72*, 1258–1266.

Chalot, M., Javelle, A., Blaudez, D., Lambilliote, R., Cooke, R., Sentenac, H., Wip, F. D., & Botton, B. (2002). An uptake on nutrient transport processes in ectomycorrhizas. *Plant and Soil, 244*, 165–175.

Chang, T. T., & Li, C. Y. (1998). Weathering of limestone, marble, and calcium phosphate by ectomycorrhizal fungi and associated microorganisms. *Taiwan Journal of Forest Science, 13*, 85–90.

Chen, J., Blume, H. P., & Beyer, L. (2000). Weathering of rocks induced by lichen colonization –a review. *Catena, 39*, 121–146.

Chen, J., Kadlubar, F. F., & Chen, J. Z. (2007). DNA supercoiling suppresses real-time PCR: A new approach to the quantification of mitochondrial DNA damage and repair. *Nucleic Acids Research, 35*, 1377–1388.

Cheng, J., Pinnell, L., Engel, K., Neufeld, J. D., & Charles, T. C. (2014). Versatile broad-host-range cosmids for construction of high quality metagenomic libraries. *Journal of Microbiological Methods, 99*, 27–34.

Chenu, C. (1993). Clay-or-sand polysaccharide associations as models for the interface between micro-organisms and soil: Water-related properties and microstructure. *Geoderma, 56*, 143–156.

Chorover, J., Kretzschmar, R., Garcia-Pichel, F., & Sparks, D. L. (2007). Soil biogeochemical processes within the critical zone. *Elements, 35*, 321–326.

Chotte, J. L. (2005). Importance of microorganisms for soil aggregation. In F. Buscot & A. Varma (Eds.), *Microorganisms in soils: Roles in genesis and functions* (pp. 107–119). Berlin: Springer.

Clarholm, M. (1985). Possible roles of roots, bacteria, protozoa and fungi in supplying nitrogen to plants. In A. H. Fitter, D. Atkinson, D. J. Read, & M. B. Usher (Eds.), *Ecological interactions in soil* (pp. 297–317). Oxford: Blackwell.

Cockell, C. S., Olsson, K., Herrera, A., & Meunier, A. (2009). Alteration textures in terrestrial volcanic glass and the associated bacterial community. *Geobiology, 7*, 50–65.

Condron, L., Strak, C., O'Callaghan, M., Clinton, P., & Huang, Z. (2010). The role of microbial communities in the formation and decomposition of soil organic matter. In G. Dixon & E. Tilson (Eds.), *Soil microbiology and sustainable crop production*. Dordrecht: Springer.

Cromack, K., & Caldwell, B. A. (1992). The role of fungi in litter decomposition and nutrient cycling. In G. C. Carroll & D. T. Wicklow (Eds.), *The fungal community, its organization and role in the ecosystem* (pp. 601–618). New York: Dekker.

Davinic, M., Fultz, L. M., Acosta-Martinez, V., Caldero´ n, F. J., Cox, S. B., Dowd, S. E., Allen, V. G., Zak, J. C., & Moore-Kucera, J. (2012). Pyrosequencing and mid-infrared spectroscopy reveal distinct aggregate stratification of soil bacterial communities and organic matter composition. *Soil Biology and Biochemistry, 46*, 63–72.

De Long, E. F., & Pace, N. R. (2001). Environmental diversity of bacteria and archaea. *Systematic Biology, 50*(4), 470–478.

De Macedo, J. R., Meneguelli, N. D., Ottoni Filho, T. B., & De Sousa Lima, J. A. (2002). Estimation of field capacity and moisture retention based on regression analysis involving chemical and physical properties in Alfisols and Ultisols of the state of Rio de Janeiro. *Communications in Soil Science and Plant Analysis, 33*(13–14), 2037–2055.

De ta Torre, M. A., Gome Z-Alarcon, G., & Palacios, J. M. (1993). In vitro biofilm formation by *Penidllium frequentans* strain on sandstone, granite, and limestone. *Applied Microbiology and Biotechnology, 40*, 408–415.

Esposito, A., Colantuono, C., Ruggieri, V., & Chiusano, M. L. (2016). Bioinformatics for agriculture in the next-generation sequencing era. *Chemical and Biological Technologies in Agriculture, 3*, 1–12.

Falkowski, P. G., Fenchel, T., & Delong, E. F. (2008). The microbial engines that drive Earth's biogeochemical cycles. *Science, 320*, 1034–1039.

Feeney, D. S., Crawford, J. W., Daniell, T., Hallett, P. D., Nunan, N., Ritz, K., Rivers, M., & Young, I. M. (2006). Three-dimensional microorganization of the soil–root–microbe system. *Microbial Ecology, 52*(1), 151–158.

Fierer, N., Jackson, J. A., Vilgalys, R., & Jackson, R. B. (2005). Assessment of soil microbial community structure by use of taxon-specific quantitative PCR assays. *Applied and Environmental Microbiology, 71*(7), 4117–4120.

Fierer, N., Breitbart, M., Nulton, J., Salamon, P., Lozupone, C., Jones, R., Robeson, M., Edwards, R. A., Felts, B., Rayhawk, S., Knight, R., Rohwer, F., & Jackson, R. B. (2007). Metagenomic and small-subunit rRNA analyses reveal the genetic diversity of bacteria, archaea, fungi, and viruses in soil. *Applied and Environmental Microbiology, 73*, 7059–7066.

Frey-Klett, J., Garbaye, J., & Tarkka, M. (2007). The mycorrhiza helper bacteria revisited. *New Phytologist, 176*, 22–36.

Gadd, G. M. (1993). Interactions of fungi with toxic metals. *New Phytologist, 124*, 25–60.

Gadd, G. M. (2007). Geomycology: Biogeochemical transformation of rocks, minerals, metals and radionuclides by fungi, bioweathering and bioremediation. *Mycological Research, 111*, 3–49.

Gadd, G. M. (2010). Metals, minerals and microbes: Geomicrobiology and bioremediation. *Microbiology, 156*, 609–643.

Gilbert, J. A., & Neufeld, J. D. (2014). Life in a world without microbes. *PLoS Biology, 12*(12), e1002020.

Giller, K. E., Beare, M. H., Lavelle, P., Izac, A. M. N., & Swift, M. J. (1997). Agricultural intensification, soil biodiversity and agroecosystem function. *Applied Soil Ecology, 6*, 3–16.

Giri, B., Giang, P. H., Kumari, R., Prasad, R., & Varma, A. (2005). Microbial diversity in soils. In A. Varma & F. Buscot (Eds.), *Microorganisms in soils: Roles in genesis and functions* (Soil biology, Vol. 3). Berlin, Heidelberg: Springer.

Gleeson, D. B., Kennedy, N. M., Clipson, N., Melville, K., Gadd, G. M., & McDermott, F. P. (2006). Mineralogical influences on bacterial community structure on a weathered pegmatitic granite. *Microbial Ecology, 51*, 526–534.

Glenn, T. C. (2011). Field guide to next-generation DNA sequencers. *Molecular Ecology Resources, 11*, 759–769.

Hajibabaei, M., Singer, G. A. C., Clare, E. L., & Hebert, P. D. N. (2007). Design and applicability of DNA arrays and DNA barcodes in biodiversity monitoring. *BMC Biology, 5*(1), 24.

Hawksworth, D. L. (1991a). *The biodiversity of microorganisms and invertebrates: Its role in sustainable agriculture* (p. 302). Melksham: CAB International/Redwood Press.

Hawksworth, D. L. (1991b). The fungal dimension of biodiversity: Magnitude, significance, and conservation. *Mycological Research, 95*, 641–655.

Hawksworth, D. L., & Mound. (1991). Biodiversity databases: The crucial significance of collections. In D. L. Hawksworth (Ed.), *The biodiversity of microorganisms and invertebrates: Its role in sustainable agriculture* (pp. 17–29). Wallingford: CAB International.

Haynes, R. J. (2014). Nature of the belowground ecosystem and its development during Pedogenesis. *Advances in Agronomy, 127*, 43–109.

Hebert, P. D. N., Ratnasingham, S., & de Waard, J. R. (2003). Barcoding animal life: Cytochrome c oxidase subunit 1 divergences among closely related species. *Proceedings of the Royal Society B: Biological Sciences, 270*, 96–99.

Hirsch, P., Eckhardt, F. E. W., & Palmer, R. J., Jr. (1995). Methods for the study of rock-inhabitating microorganisms—A mini review. *Journal of Microbiological Methods, 23*, 143–167.

Huang, D. L., Zeng, G. M., Feng, C. L., Hu, S., Jiang, X. Y., Tang, L., Su, F. F., Zhang, Y., Zeng, W., & Liu, H. L. (2008). Degradation of lead-contaminated lignocellulosic waste by *Phanerochaete chrysosporium* and the reduction of lead toxicity. *Environmental Science & Technology, 42*(13), 4946–4951.

Huber, H., Hohn, M. J., Rachel, R., Fuchs, T., Wimmer, V. C., & Stetter, K. O. (2002). A new phylum of archaea represented by a nanosized hyperthermophilic symbiont. *Nature, 417*, 63–67.

Jansson, J. K., van Elsas, J. D., & Bailey, M. J. (2000). *Tracking genetically engineered microorganisms*. Georgetown: Landes Bioscience.

Kallenbach, C. M., Frey, D. S., & Grandy, A. S. (2016). Direct evidence for microbial-derived soil organic matter formation and its ecophysiological controls. *Nature Communications, 7*, 13630.
Kandeler, E., Tscherko, D., Bruce, K. D., Stemmer, M., Hobbs, P. J., Bardgett, R. D., & Amelung, W. (2000). The structure and function of the soil microbial community in microhabitats of a heavy metal polluted soil. *Biology and Fertility of Soils, 32*, 390–400.
Kaplan, W. A. (1983). Nitrification. In E. J. Carpenter & D. G. Capone (Eds.), *Nitrogen in the marine environment* (pp. 139–190). New York: Academic.
Kuzyakov, Y. (2010). Priming effects: Interactions between living and dead organic matter. *Soil Biology and Biochemistry, 42*, 1363–1371.
Kyrpides, N. C., & Olsen, G. J. (1999). Archaeal and bacterial hyperthermophiles: Horizontal gene exchange or common ancestry. *Trends in Genetics, 15*, 298–299.
Lakhanpal, T. N. (2000). Ectomycorrhiza–an overview. In K. G. Mukerji, B. P. Chamola, & J. Singh (Eds.), *Mycorrhizal biology* (pp. 101–118). New York: Kluwer/Plenum.
Lam, K. N., & Charles, T. C. (2015). Strong spurious transcription likely contributes to DNA insert bias in typical metagenomic clone libraries. *Microbiome, 3*, 22.
Lee, K. E. (1985). *Earthworms their ecology and relationships with soils and land use*. Sydney: Academic.
Leininger, S., Urich, T., Schloter, M., Schwark, L., Qi, J., Nicol, G. W., Prosser, J. I., Schuster, S. C., & Schleper, C. (2006). Archaea predominate among ammonia-oxidizing prokaryotes in soils. *Nature, 442*, 806–809.
Leita, L., De Nobili, M., Muhlbachova, G., Mondini, C., Marchiol, L., & Zerbi, G. (1995). Bioavailability and effects of heavy metals on soil microbial biomass survival during laboratory incubation. *Biology and Fertility of Soils, 19*, 103–108.
Lian, B., Chen Ye, Z. H. U., Lijun, Y., & Ruidong. (2008a). Effect of microbial weathering on carbonate rocks. *Earth Science Frontiers, 15*(6), 90–99.
Lian, B., Wang, B., Pan, M., Liu, C., & Teng, H. (2008b). Microbial release of potassium from K-bearing minerals by thermophilic fungus *Aspergillus fumigatus*. *Geochimica et Cosmochimica Acta, 72*, 87–98.
Lipiec, J., Brzezińska, M., Turski, M., Szarlip, P., & Frąc, M. (2015). Wettability and biogeochemical properties of the drilosphere and casts of endogeic earthworms in pear orchard. *Soil and Tillage Research, 145*, 55–61.
Lipiec, J., Frąc, M., Brzezińska, M., Turski, M., & Oszust, K. (2016). Linking microbial enzymatic activities and functional diversity of soil around earthworm burrows and casts. *Frontiers in Microbiology, 7*, 1361.
Lynch, J. M. (1987). Microbial interactions in the rhizosphere. *Soil Microorganisms, 30*, 33–41.
Lynch, J. M. (1990). *The rhizosphere*. New York: Wiley.
Lynch, J. M., & Hobbie, J. B. (1988). *Microorganisms in action: Concepts and application in microbial ecology* (p. 36). Oxford: Blackwell.
Madigan, M. T., Martinko, J. M., Stahl, D. A., & David, P. (2011). *Clark Brock biology of microorganisms* (13th ed.). QR41.2.B77, 579—dc22.
Mamta, R. J. R., & Khursheed, A. W. (2015). Bioremediation of pesticides under the influence of bacteria and fungi Chapter 3. In *Handbook of research on uncovering new methods for ecosystem management through bioremediation* (pp. 51–72). Hershey: IGI Global.
Metting, B. (1988). Micro-algae in agriculture. In M. A. Borowitzka & L. A. Borowitzka (Eds.), *Microalgal biotechnology* (pp. 288–304). Cambridge: Cambridge University Press.
Mirdamadian, S. H., Emtiazi, G., Golabi, M. H., & Ghanavati, H. (2010). Biodegradation of petroleum and aromatic hydrocarbons by Bacteria isolated from petroleum-contaminated soil. *Journal of Petroleum & Environmental Biotechnology, 1*, 102.
Moses, C. A., & Smith, B. J. (1993). A note on the role of the lichen *Collema auriforma* in solution basin development on a carboniferous limestone substrate. *Earth Surface Processes and Landforms, 18*, 363–368.
Nannipieri, P., Ascher, J., Ceccherini, M. T., Landi, L., Pietramellara, G., & Renella, G. (2017). Microbial diversity and soil functions. *European Journal of Soil Science, 68*, 1–26.

Nilsson, R. H., Kristiansson, E., Ryberg, M., Hallenberg, N., & Larsson, K. H. (2008). Intraspecific ITS variability in the kingdom Fungi as expressed in the international sequence databases and ITS implications for molecular species identification. *Evolutionary Bioinformatics, 4*, 193–201.
Nunan, N. (2017). The microbial habitat in soil. Scale, heterogeneity and functional consequences. *Journal of Plant Nutrition and Soil Science, 180*, 425–429.
O'Donnell, A. G., Seasman, M., Macrae, A., Waite, I., & Davies, J. T. (2001). Plants and fertilizers as drivers of change in microbial community structure and function in soil. *Plant and Soil, 232*, 135–145.
Pace, N. R. (1997). A molecular view of microbial diversity and the biosphere. *Science, 276*(5313), 734–740.
Pace, N. R. (1999). Microbial ecology and diversity. *ASM News, 65*, 328–333.
Pace, N. R., Olsen, G. J., & Woese, C. R. (1986). Ribosomal RNA phylogeny and the primary lines of evolutionary descent. *Cell, 45*, 325–326.
Paterson, E., Osler, G., Dawson, L. A., Gebbing, T., Sim, A., & Ord, B. (2008). Labile and recalcitrant plant fractions are utilised by distinct microbial communities in soil: Independent of the presence of roots and mycorrhizal fungi. *Soil Biology and Biochemistry, 40*, 1103–1113.
Payne, J. W. (1981). *Denitrification*. New York: Wiley.
Pizl, V., & Novakova, A. (2003). Interactions between microfungi and *Eisenia andrei* (Oligochaeta) during cattle manure vermicomposting. *Pedobiologia, 47*, 895–899.
Prescott, L. M., Harley, J. P., & Klein, D. A. (1996). The diversity of the microbial world. In L. M. Prescott, J. P. Harley, & D. A. Klein (Eds.), *Microbiology*. Dubuque: WCB Publishers.
Prosser, J. I. (2002). Molecular and functional diversity in soil micro-organisms. *Plant and Soil, 244*, 9–17.
Puente, M. E., Bashan, Y., Li, C. Y., & Lebsky, V. K. (2004). Microbial populations and activities in the rhizoplane of rock-weathering desert plants I. Root colonization and weathering of igneous rocks. *Plant Biology, 6*, 629–642.
Puente, M. E., Rodriguez-Jaramillo, M. C., Li, C. Y., & Bashan, Y. (2006). Image analysis for quantification of bacterial rock weathering. *Journal of Microbiological Methods, 64*, 275–286.
Quejigo, J. R., Dominguez-Garay, A., Dorfler, A., Schroll, R., & Esteve-Nunez, A. (2018). Anodic shifting of the microbial community profile to enhance oxidative metabolism in soil. *Soil Biology and Biochemistry, 116*, 131–138.
Ren, R., Sun, Y., Zhao, Y., Geiser, D., Ma, H., & Zhou, X. (2016). Phylogenetic resolution of deep eukaryotic and fungal relationships using highly conserved low-copy nuclear genes. *Genome Biology and Evolution, 8*(9), 2683–2701.
Richter, D. D., Oh, N. H., Fimmen, R., & Jackson, J. (2007). The rhizosphere and soil formation. In *The rhizosphere: An ecological perspective* (pp. 179–200). Cambridge: Academic.
Roesch, L. F. W., Fulthorpe, R. R., Riva, A., Casella, G., Km, A., Kent, A. D., Daroub, S. H., Camargo, F. A. O., Farmerie, W. G., & Triplett, E. W. (2007). Pyrosequencing enumerates and contracts soil microbial diversity. *The ISME Journal, 1*, 283–290.
Rousk, J., Bååth, E., Brookes, P. C., Lauber, C. L., Lozupone, C., Caporaso, J. G., Knight, R., & Fierer, N. (2010). Soil bacterial and fungal communities across a pH gradient in an arable soil. *The ISME Journal, 4*, 1340–1351.
Schimel, J. (1995). Ecosystem consequences of microbial diversity and community structure. In F. S. Chapin & C. Korner (Eds.), *Arctic and alpine biodiversity: Patterns, causes and ecosystem consequences. Ecological studies* (Analysis and syntehsis, Vol. 113). Berlin, Heidelberg: Springer.
Shokralla, S., Spall, J. L., Gibson, J. F., & Hajibabaei, M. (2012). Next-generation sequencing technologies for environmental DNA research. *Molecular Ecology, 21*, 1794–1805.
Singh, A. V., Sharma, A., & Johri, B. N. (2012). Phylogenetic profiling of culturable bacteria associated with early phase of mushroom composting assessed by amplified rDNA restriction analysis. *Annals of Microbiology, 62*, 675–682.
Smitha, M. S., Singh, S., & Singh, R. (2017). Microbial biotransformation: A process for chemical alterations. *Journal of Bacteriology & Mycology: Open Access, 4*(2), 00085.

Sollins, P., Cromack, K., Jr., Li, C. Y., & Fogel, R. (1981). Role of low-molecular weight organic acids in the inorganic nutrition of fungi and higher plants. In G. C. Carroll & D. T. Wicklow (Eds.), *The fungal community, its organization and role in ecosystem*. New York: Dekker.

Sterflinger, K., & Krumbein, W. E. (1997). Dematiaceous fungi as a major agent for biopitting on Mediterranean marbles and limestones. *Geomicrobiology Journal, 14*, 219–222.

SubbaRao, N. S. (1997). *Soil microbiology*. Oxford: IBH Publ.

Szalay, A. (1964). Cation exchange properties of humic acids and their importance in the geochemical enrichment of UO_2^{++} and other cations. *Geochimica et Cosmochimica Acta, 28*, 1605–1614.

Tabatabai, M. A. (1982). Soil enzymes. In A. L. Page, M. Rh, & D. R. Keeney (Eds.), *Methods of soil analysis, part 2. Chemical and microbiological properties – Agronomy*. Madison: American Society of Agronomy, Soil Science Society of America.

Tate, K. R., Parshotam, A., & Ross, D. J. (1995). Soil carbon storage and turnover in temperate forests and grasslands – A New Zealand perspective. *Journal of Biogeography, 22*, 695–700.

Temperton, B., Field, D., Oliver, A., Tiwari, B., Mühling, M., & Joint, I. (2009). Bias in assessments of marine microbial biodiversity in fosmid libraries as evaluated by pyrosequencing. *The ISME Journal, 3*, 792–796.

Tiedje, J. M., Asuming-Brempong, S., Nüsslein, K., Marsh, T. L., & Flynn, S. J. (1999). Opening the black box of soil microbial diversity. *Applied Soil Ecology, 13*, 109–122.

Torsvik, V., Goksoyr, J., & Daae, F. L. (1990). High diversity in DNA of soil bacteria. *Applied Environmental and Microbiology, 56*, 782–787.

Tourasse, N. J., & Gouy, M. (1997). Evolutionary distances between nucleotide sequences based on the distribution of substitution rates among sites as estimated by parsimony. *Molecular Biology and Evolution, 14*(3), 287–298.

Uroz, S., Calvaruso, C., Marie-Pierre, T., & Frey-Klett, P. (2009). Mineral weathering by bacteria: Ecology, actors and mechanisms. *Trends in Microbiology, 17*, 378–387.

Wainwright, M. (1992). The impact of fungi on environmental biogeochemistry. In G. C. Carroll & D. T. Wicklow (Eds.), *The fungal community, its organization and role in the ecosystem* (pp. 601–618). New York: Dekker.

Welch, D. B. M., & Huse, S. M. (2011). Microbial diversity in the deep sea and the underexplored "Rare Biosphere". In *Handbook of molecular microbial ecology II: Metagenomics in different habitats* (pp. 243–252). Hoboken: Wiley-Blackwell. https://doi.org/10.1002/9781118010549.ch24.

Welch, S., & McPhail, D. (2003). Mobility of major and trace elements during biologically mediated weathering of granite. In I. C. Roach (Ed.), *CRC LEME Regional Regolith Symposia* (pp. 437–440). Millaa Millaa: CRC LEME.

Woese, C. R., & Fox, G. E. (1977). Phylogenetic structure of the prokaryotic domain: The primary kingdoms. *Proceedings of the National Academy of Sciences, 74*(11), 5088–5090.

Woese, C. R., Kandler, O., & Wheelis, M. L. (1990). Towards a natural system of organisms: Proposal for the domains archaea, bacteria, and eucarya. *Proceedings of the National Academy of Sciences, 87*, 4576–4579.

Wolters, V. (1991). Soil invertebrates – Effects on nutrient turnover and soil structure: A review. *Zeitschrift für Pflanzenernährung und Bodenkunde, 154*, 389–402.

Zambell, C. B., Adams, J. M., Gorring, M. L., & Schwartzman, D. W. (2012). Effect of lichen colonization on chemical weathering of hornblende granite as estimated by aqueous elemental flux. *Chemical Geology, 291*, 166–174.

Zhang, J., Chiodini, R., Badr, A., & Zhang, G. (2011). The impact of next generation sequencing on genomics. *Journal of Genetics and Genomics, 38*, 95–109.

Mycorrhizal Assisted Phytoremediation of Xenobiotics from Contaminated Soil

3

Awadhesh Kumar Shukla, Amit Kishore Singh, and Anjney Sharma

1 Introduction

Soil acts as a natural resource for human beings. It harbors a variety of microbes that directly or indirectly involved and play important roles in the plants-microbe interaction. Generally the plant-microbe possesses mutual interaction symbiotically in ubiquitous way and help in sustainable agriculture. Organic soil pollution has become a global concern due to rampant industrialization, sewage, oil spills accidents and oil processing etc. is the leading contributors of hydrocarbon in the biosphere (Gan et al. 2009; Małachowska-Jutsz et al. 2011; Rajtor and Piotrowska-Seget 2016). Contamination by xenobiotics poses huge threats to the soil quality, crop plants, food chain and ultimately creates health hazards to the human beings. Hence, remediation of soil is warranted in order to protect the environment from deterioration and improve yields of the crop plants for food quality that met the demand of increasing human population.

Till dates, there are various conventional physico-chemicals techniques have been developed and applied for the remediation of hydrocarbons. Conventional physico-chemical remediation methods are usually highly efficient, however, they are costly and possess potential to alter the soil structure, decrease soil microbial activity and consequently leads to the depletion of the nutrients essential for plant

All the authors have contributed equally

A. K. Shukla (✉)
Department of Botany, K. S. Saket P.G. College, Ayodhya, Uttar Pradesh, India

A. K. Singh
Kamla Nehru P.G. College, Raebareli, Uttar Pradesh, India

A. Sharma
ICAR-National Bureau of Agriculturally Important Microorganisms, Mau Nath Bhanjan, Uttar Pradesh, India

© Springer Nature Singapore Pte Ltd. 2019

A. Varma, D. K. Choudhary (eds.), *Mycorrhizosphere and Pedogenesis*,
https://doi.org/10.1007/978-981-13-6480-8_3

growth and yield (Khan et al. 2000; Gan et al. 2009). In recent past the researchers are paying their attention on the use of biological means for detoxification of the environment. The most promising biological method proposed for clean-up of contaminated environment is phytoremediation. This technology involves the plants and their associated microorganisms for removal of toxic organic compounds from the environmental components contaminated with it. It has been established that phytoremediation is one of the most cost effective, eco-friendly and promising technology for the removal of pollutants from the soils (Gao et al. 2011). Effective degradation of pollutants in the soil is achieved due to the plant-stimulated microbial degradation in the rhizosphere (Joner et al. 2001). It has been demonstrated that involvement of the catabolic potential of both, microorganisms and plants is the most effective approaches for decontamination of pollutants from the environmental components like soil. The earlier researchers have also pointed out that the surface area adjoining the root, soil contact and microbial activity of rhizosphere are the major drawback in the phytoremediation process. However, these limitations are theoretically overcome by the mycorrhizal associations with the plants.

Arbuscular mycorrhizal fungi (AMF) show symbiotic association with higher plants which is an integral part of terrestrial ecosystems. It is reported that the exploitation of mycorrhizal fungi offers a potential advantage in bioremediation and phytoremediaiton due to that they get the direct supply of carbon source from their host in order to support growth into contaminated environment (Finlay 2008). AMF hyphae have potential to create a extensive underground network of mycelium that are directly connected through plant roots, soil and adjoining microflora (Parniske 2008; van der Heijden and Horton 2009). It is established that the surface area created by fungal hyphae is approximately 100 fold greater the area covered by the root system while, the length can be several orders of magnitude larger than that of the plant root, and hence the fungal hyphae occupy larger area of soil than plant roots (Khan et al. 2000). Such an extensive network of fungal mycelium helps to release nutrients and organic contaminants from soil particles thus facilitates nutrients and water uptake by plants (Leyval et al. 2002; Rabie 2005).

Khan et al. (2000) and Liao et al. (2003) have suggested through their studies that AM have shown positive effects on potential stabilization of plant and ability to detoxify the hydrocarbons in the contaminated soils. This chapter highlights the diversity of arbuscular mycorrhiza and their potential exploitation in phytoremediation of organic and inorganic pollutants from the contaminated soils.

2 Mechanisms of Biodegradation and its Metabolite

The application of AMF may improve the phytoremediation efficacy via plant growth, enhance the biodegradative activity of roots and rhizospheric microorganisms and also promotes the adsorption and bioaccumulation of hydrocarbons by roots (Monika Rajtor and Zofia Piotrowska-Seget 2016). Further they suggested that dissolved organic carbon (DOC) released by the AMF hyphae stimulates the development and increase the enzymatic activity of hydrocarbon-degrading microorganisms. Mycorrhizal colonization has potential to alter the root exudation pattern and enhance the enzymatic activity of oxidoreductases that directly involved in oxidative degradation of aromatic hydrocarbons. Besides this they also protect the plants against oxidative stress, elevate the lipid content, increase volume of the root system in order to create large area for absorption and consequently contribute to enhance the absorption of hydrophobic hydrocarbons. Figure 3.1 depictes the process of phytoremediation of xenobiotic compounds using mycorrhiza. From this figure it is seen that mycoorrhiza played important roles in reduction of trace element translocation, alleviate stress, prevent soil erosion and also produces chelating agents for the capturing of heavy metal ions in the vicinity of the soil. Mycorrhiza also helps in the phytostablization via metal uptake through hyphae, spore, glomalin production and adsorption of the metal.

Binet et al. (2001) reported that the anthraquinone was identified as a metabolite of anthracene through GC-MS. They observed that the concentration of anthracene was found to be in large amount in the soil planted with rycgrass than in unplanted controls. Similarly the concentration of anthracene was found to be significantly smaller in the mycorrhizal associated plant than nonmycorrhizal plant. Another study demonstrated the presence of atrazine metabolites in the roots of *Zea mays* grown in pots contaminated with deethylatrazine and deisopropylatrazine. Furthermore, they observed that AMF colonization enhanced the metabolization of atrazine (Huang et al. 2007; Lenoir et al. 2016). It is confirmed that in the

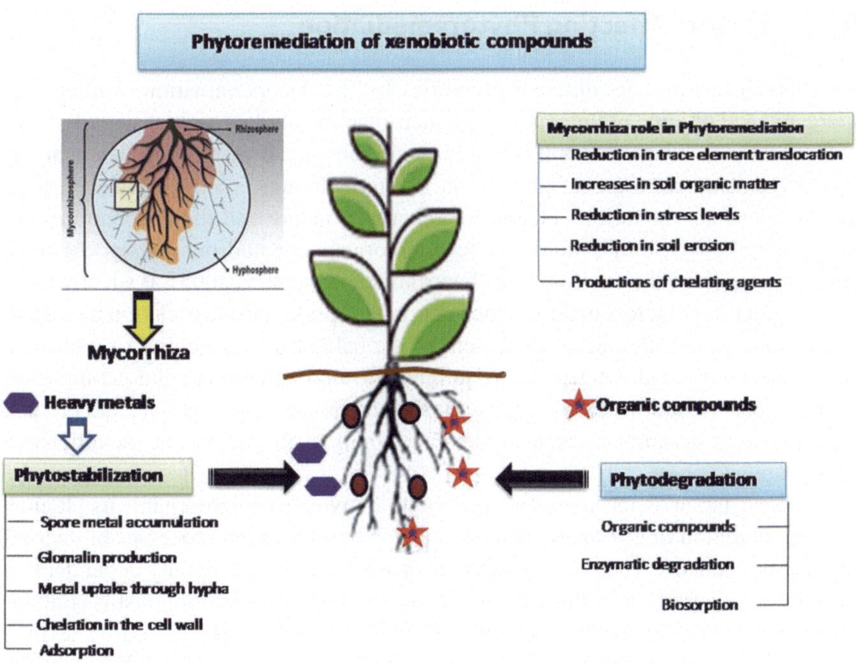

Fig. 3.1 Mechanism of phytoremediation of xenobiotic compounds using mycorrhiza

hyphosphere and mycorhizosphere zones high enzymatic activities such as dehydrogenase, catalase, dioxygenase etc. were observed (Rabie 2005; Corgie et al. 2006; Huang et al. 2009).

3 Biodegradation of Polyaromatic Hydrocarbons

Hydrocarbons affect the root colonization and rhizosphere microorganisms. Phytoremediation of soils contaminated with pyrene and phenanthrene in the presence of arbuscular mycorrhiza with host plant *Medicago sativa* have been studied in detail by Gao et al. (2011). They experimentally proved that more than 88.1% and 98.6% of pyrene and phenanthrene were degraded after incubation for70 days at initial concentrations of 74 and 103 mg/kg, respectively. Later on Aranda et al. (2013) studied the effects of PAH on the mycorrhizal associated with *Dacus carota* roots. They observed the increase in dry weight of mycorrhizal roots in the absence of PAH. They determined experimentally that in the presence of phenanthrene and dibenzothiophene at concentration of 60 μM the root biomass of mycorrhiza got reduced upto 60% of initial concentration. Further increase in the concentration from 60 to 120 μM the biomass drastically decreased to 80–92%, respectively in the presence of phenanthrene and dibenzothiophene. Details of some studies related to mycorrhiza associated phytoremediation are summarized in Table 3.1.

4 Factors Affecting Phytoremediation

Microorganisms possess inherent properties for the decontamination of numerous inorganic and organic pollutants having their own metabolic machinery and potential capacity to adapt into inhospitable environments. It is well proven that the microorganisms are the best players on site remediation process. However, their efficacy depends on various factors like chemical nature and the concentration of pollutants, their ease of availability to microorganisms, and the physicochemical characterization of the environment (Fantroussi and Agathos 2005). Basically there are two important factors that influences the rate of pollutants degradation by microorganisms are; (1) the microbes present in the habitat have potential to withstand with pollution load, nutrients requirements and environmental conditions. Furthermore, the abiotic factors like temperature, moisture content, pH and organic matter content, aeration, nutrient content and soil type also determine the efficiency of phytoremediation.

A biotic factor determines the metabolic activity of microorganisms. It may include inhibition of enzymatic activities and the proliferation processes of degrading microorganisms. The rate of hydrocarbon degradation is generally dependent on the concentration of the pollutants and the total number of microorganisms containing enzymes for decontamination. The development of huge amount of hyphae may also obstruct the nutrient translocation and aeration for mycorrhiza during bioremediation.

Table 3.1 A summary of mycorrhizal mediated phytoremediation of pollutants

No.	Pollutants	Fungus	Inferences	References
1.	Anthracene	*Glomus mosseae*, *Glomus intraradices*	Inoculation of fungal strain with jute (*Corchorus capsularis*) enhanced anthracene removal and also improved the plant growth	Cheung et al. (2008)
2.	Phenanthrene, pyrene	*Glomus mosseae* *G. etunicatum*	High rate of hydrocarbon degradation were observed in the presence of inoculation with mycorrhiza	Gao et al. (2011)
3.	Phenanthrene and pyrene	*Glomus mosseae*	High degradation rates were observed in rhizospheric zone compared to near rhizosphere and bulk soil	Wu et al. (2011)
4.	Phenanthrene and pyrene	*Glomus mosseae* and bacterium Acinetobacter sp.	Significant removal of PAH was observed bacteria fungi and rye grass	Yu et al. (2011)
5.	Phenanthrene, pyrene, dibenz(a,h)-anthracene	*Glomus intraradices*	Removal of hydrocarbons were found to be dominant and their accumulation was negligible	Zhou et al. (2013)
6.	Anthracene, phenanthrene and dibenzothiophene	*Rhizophagus custos*	The presence of anthracene have shown root growth of mycorrhiza while phenanthrene and dibenzothiophene inhibited the development of mycorrhizal roots	Aranda et al. (2013)
7.	Polycyclic aromatic hydrocarbons (PAHs)	*Glomus caledoniun*	PAHs degradation was found to be highest in combination with *Glomus caledoniun*, *Festuca arundinacea* and earthworm	Lu and Lu (2015)

5 Conclusions

It has been investigated that soils contaminated with pollutants possess very limited diversity of indigenous AM fungi hence, it is essential to enrich and isolate microorganisms from the contaminated environment and their potential could be exploited for decontamination of the hydrocarbons. Emphasis should be given on the selection of plant types and indigenous AMF strain that would be better choice to enhance the phytoremediation process. Special attention should be given on the interaction between plants roots and mycorrhiza colonization so that enhance the production of enzymes responsible for bioremediation of hydrocarbon. AMF mediated phytoremediation highlights a great potential for the remediation of hydrocarbon polluted soil, additionally, more comprehensive experiments are required in exhaustive way to optimize the methods and overcome its limitations. In order to fully elucidate the influence of hydrocarbons on mycorrhizal interactions, there is need to investigate the molecular characterization of microorganisms in response to hydrocarbons.

However, in future more concerted efforts are required in order to fully understand the mechanisms tracing of degradation pathway.

References

Aranda, E., Martín Scervino, J., Godoy, P., Reina, R., Antonio Ocampo, J., Regina-Michaela, W., & García-Romera, I. (2013). Role of arbuscular mycorrhizal fungus *Rhizophagus custos* in the dissipation of PAHs under root-organ culture conditions. *Environmental Pollution, 181*, 182–189.

Binet, P., Portal, J. M., & Leyval, C. (2001). Application of GC-MS to the study of anthracene disappearance in the rhizosphere of ryegrass. *Organic Geochemistry, 32*, 217–222.

Cheung, K. C., Zhang, J. Y., Deng, H. H., Ou, Y. K., Leung, H. M., Wu, S. C., & Wong, M. H. (2008). Interaction of higher plant (jute), electrofused bacteria and mycorrhiza on anthracene biodegradation. *Bioresource Technology, 99*, 2148–2155.

Corgie, S. C., Fons, F., Beguiristain, T., & Leyval, C. (2006). Biodegradation of phenanthrene, spatial distribution of bacteria populations and dioxygenase expression in the mycorrhizosphere of *Lolium perenne* inoculated with *Glomus mosseae*. *Mycorrhiza, 16*, 207–212.

El Fantroussi, S., & Agathos, S. N. (2005). Is bioaugmentation a feasible strategy for pollutant removal and site remediation? *Current Opinion in Microbiology, 8*, 268–275.

Finlay, R. D. (2008). Ecological aspects of mycorrhizal symbiosis: With special emphasis on the functional diversity of interactions involving the extraradical mycelium. *Journal of Experimental Botany, 59*, 1115–1126.

Gan, S., Lau, E. V., & Ng, H. K. (2009). Remediation of soils contaminated with polycyclic aromatic hydrocarbons (PAH). *Journal of Hazardous Materials, 172*, 532–549.

Gao, Y., Li, Q., Ling, W., & Zhu, X. (2011). Arbuscular mycorrhizal phytoremediation of soils contaminated with phenanthrene and pyrene. *Journal of Hazardous Materials, 185*, 703–709.

Huang, H., Zhang, S., Shan, X. Q., Chen, B. D., Zhu, Y. G., & Bell, J. N. (2007). Effect of arbuscular mycorrhizal fungus (*Glomus caledonium*) on the accumulation and metabolism of atrazine in maize (*Zea mays* L.) and atrazine dissipation in soil. *Environmental Pollution, 146*, 452–457.

Huang, H. L., Zhang, S., Wu, N., Luo, L., & Christie, P. (2009). Influence of *Glomus etunicatum*/*Zea mays* mycorrhiza on atrazine degradation, soil phosphatase and dehydrogenase activities, and soil microbial community structure. *Soil Biology and Biochemistry, 41*, 726–734.

Joner, E. J., Johnsen, A., Loibner, A. P., Szolar, O. H. J., Portal, J. M., & Leyval, C. (2001). Rhizosphere effects on microbial community structure and dissipation and toxicity of polycyclic aromatic hydrocarbons (PAHs) in spiked soil. *Environmental Science & Technology, 35*, 2773–2777.

Khan, A. G., Kuek, C., Chaudhry, T. M., Khoo, C. S., & Hayes, W. J. (2000). Role of plants, mycorrhizae and phytochelators in heavy metal contaminated land remediation. *Chemosphere, 41*, 197–207.

Lenoir, I., Lounes-Hadjsahraoui, A., & Fontaine, J. (2016). Arbuscular mycorrhizal fungal-assisted phytoremediation of soil contaminated with persistent organic pollutants: A review. *European Journal of Soil Science, 67*, 624–640.

Leyval, C., Joner, E. J., Del Val, C., & Haselwandter, K. (2002). Potential of arbuscular mycorrhizal fungi for bioremediation. In S. Gianinazzi, H. Schüepp, J. M. Barea, & K. Haselwandter (Eds.), *Mycorrhizal technology in agriculture: From genes to bioproducts* (pp. 175–186). Birkhauser: Springer.

Liao, J. P., Lin, X. G., Cao, Z. H., Shi, Y. Q., & Wong, M. H. (2003). Interactions between arbuscular mycorrhizae and heavy metals under sand culture experiment. *Chemosphere, 50*, 847–853.

Lu, Y. F., & Lu, M. (2015). Remediation of PAH-contaminated soil by the combination of tall fescue, arbuscular mycorrhizal fungus and epigeic earthworms. *Journal of Hazardous Materials, 285*, 535–541.

Małachowska-Jutsz, A., Rudek, J., & Janosz, W. (2011). The effect of ribwort (*Plantago lanceolata*) and its mycorrhizas on the growth of microflora in soil contaminated with used engine oil. *Archives of Environmental Protection, 37*, 99–113.

Parniske, M. (2008). Arbuscular mycorrhiza: The mother of plant root endosymbioses. *Nature Reviews. Microbiology, 6*, 763–775.

Rabie, G. H. (2005). Role of arbuscular mycorrhizal fungi in phytoremediation of soil rhizosphere spiked with poly aromatic hydrocarbons. *Mycobiology, 33*, 41–50.

Rajtor, M., & Piotrowska-Seget, Z. (2016). Prospects for arbuscular mycorrhizal fungi (AMF) to assist in phytoremediation of soil hydrocarbon contaminants. *Chemosphere, 162*, 105–116.

van der Heijden, M. G., & Horton, T. R. (2009). Socialism in soil? The importance of mycorrhizal fungal networks for facilitation in natural ecosystems. *Journal of Ecology, 97*, 1139–1150.

Wu, F. Y., Yu, X. Z., Wu, S. C., Lin, X. G., & Wong, M. H. (2011). Phenanthrene and pyrene uptake by arbuscular mycorrhizal maize and their dissipation in soil. *Journal of Hazardous Materials, 187*, 341–347.

Yu, X. Z., Wu, S. C., Wu, F. Y., & Wong, M. H. (2011). Enhanced dissipation of PAHs from soil using mycorrhizal ryegrass and PAH-degrading bacteria. *Journal of Hazardous Materials, 186*, 1206–1217.

Zhou, X., Zhou, J., Xiang, X., Cebron, A., Beguiristain, T., & Leyval, C. (2013). Impact of four plant species and arbuscular mycorrhizal (AM) fungi on polycyclic aromatic hydrocarbon (PAH) dissipation in spiked soil. *Polish Journal of Environmental Studies, 22*, 1239–1245.

Relationship Between Field Crops and Mycorrhiza

4

Demet Altındal and Nüket Altındal

Headings

1. The symbiotic relationship between mycorrhiza and the roots of plants contributes significantly to plant nutrition and growth.
2. Mycorrhiza allows the nutrients and water to be taken by the plant by entering the places where the roots cannot reach in the soil.
3. Some field crops such as wheat, barley, maize, soybean and peanuts are more producable with mycorrhiza application.
4. In order to reduce excessive fertilization in agricultural areas, symbiosis between nitrogen-binding bacteria and legumes should be encouraged.

1 Introduction

Mycorrhiza (Palta et al. 2010), constituted by the words of "mykes" and "rhiza" which means fungi and root in Greek, is a symbiotic life based on the mutual benefit relationship between mycorrhiza fungi and plant roots during the absorption of soil water and minerals. This term was used in 1885 by a German forest pathologist named A.B. Frank to describe the fungus-tree partnership. Today, more than 95% of mycorrhizal fungi belong to different species are known to live in a symbiotic way by specializing in different species of plants belonging to different families (Ortaş 1997). As a result of this symbiotic life, the plant gives mycorrhiza carbon compound carbohydrates as a source of energy, whereas mycorrhiza fungi colonize

D. Altındal (✉)
Fethiye Ali Sıtkı Mefharet Koçman Vocational School, Department of Crop and Animal Production, Organic Farming Programme, Muğla Sıtkı Koçman University, Muğla, Turkey

N. Altındal
Sivaslı Vocational School, Department of Crop and Animal Production, Programme of Medical and Aromatic Plants, Uşak University, Uşak, Turkey

© Springer Nature Singapore Pte Ltd. 2019

A. Varma, D. K. Choudhary (eds.), *Mycorrhizosphere and Pedogenesis*,
https://doi.org/10.1007/978-981-13-6480-8_4

as a real extension of the plant root system, but do not create disease. It ensures that the plant roots enter the deep soil and that the plant nutrients and the water can be picked up by the plant (Smith and Read 2008). The symbiotic relationship between the host plant and mycorrhiza maintains the viability of plants in addition to the nutrient cycle in the ecosystem (Jeffries and Dodd 1991; Ortaş 1997). Because some plant species has very weak and thick root structure, they cannot benefit from soil in sufficient level. Therefore, this plant roots cannot provide enough contact with the soil and the absorption of nutrients in the plants cannot performed in adequate level. Especially for such plants, mycorrhizas are thought to be of much greater importance.

Plant root secretions contribute to the formation of hyphes by providing germination of the mycorrhizal fungi in the soil. These hyphes are directed towards host plant roots, making infection. Thus, the connection between the root and the hyphes is established and a relationship is formed based on the mutual benefit of both organisms. Therefore, mycorrhizal fungi settle on the plant root surface, stem tissues, cell and intercellular spaces, continuing their lives at plant roots.

Mycorrhiza spores occur as ectomorphisms in forest trees and some fruit trees and infection forms in root (Sieverding 1991; Smith and Read 2008) and occur as endomycorrhizas in many cultural plants and fruit trees due to morphological and physiological structures (Marschner 1995).

In the form of ectomycorrhizal life, the fungus develops by branching between the root cells without entering the root cells, but emerging from the root surface and wrapping it like a felt.

Arbuscular Mycorrhiza formation, especially in endomycorrhizal forms of life, is important in terms of the contribution for the plant development. Endomycorrhiza develop both in the space between cells and intracellular spaces in fungicortex, forming oval-looking lipid 'vesicles'. These structures store and transport nutrients (Marschner 1995). In addition, numerous branched, thin-walled structures called "Arbusculel" are formed at host plant roots, being highly resistant, allowing transportation between plant and Arbuscular fungi. Since most plant species are infected with arbuscular type mycorrhiza, these mycorrhizas are called arbuscule mycorrhiza (AM).

In this useful symbiotic life in nature, mycorrhizal fungi interact with these beneficial microorganisms in the soil and dry matter production increases by 25% and plant growth is encouraged in a positive way (Sharma et al. 1992).

Effects of mycorrhizal fungi on plant growth are as follows:

- It increases the intake of minerals. P, Zn, Ca, Cu, Fe, Mg and Mn intake (Bieleski 1973) also increases the intake of minerals in mobile status such as N, K.
- It helps water intake along with the nutrients. Therefore, intensive fertilization and irrigation is not required to increase the use of nutrients.
- It provides the synthesis of organic compounds such as nitrogen, vitamin and amino acids that increase plant growth.
- He establishes amensal relationships with plant pathogens based on antibiotic production

- It neutralizes toxic compounds that are harmful to plants, protects the plant against biotic and abiotic stresses.

So far, the scientific research on mycorrhiza has been generally conducted under greenhouse and controlled conditions. In field conditions, there is a large number of uncontrollable factors due to the natural environment, in this way an effective mycorrhiza infection and development cannot be provided.

The symbiotic relationship between mycorrhiza and the roots of plants contributes significantly to plant nutrition and growth. This symbiotic life has been observed to increase productivity in various field crops such as wheat, maize, sorghum, alfalfa, potatoes, canola, chickpeas, black-eyed peas and cotton. It has been reported that alfalfa and white clover performs a good root infection in the mycorrhiza environment (Raj et al. 1981).

The white clover was infected with *Rhizoglomus intraradices* and *Paraglomus occultum* mycorrhiza, and it was found that applications promote plant growth, root development and chlorophyll production (Lu and Wu 2017).

Erlita and Hariani (2017) stated that the use of microorganisms such as mycorrhiza is an important parameter in improving the soil, and mycorrhizal fungi increased plant height, leaf number and area, root volume and yield of maize.

In a soybean study, mycorrhizal application increased the number of branches, 100 seed weight, biomass dry weight and yield per hectare (Muktiyanta et al. 2018).

2 The Effect of Mycorrhizas on Biotic and Abiotic Stress Factors

The positive effect of mycorrhizal fungi on plant yield depends on many factors. It increases growth of plants and resistance against the stress, especially in poor soil environments. Mycorrhizas help not only plants but also leads to the less pollution by pesticides and mineral fertilizers. It also helps plants to get water. Mycorrhizal infections of water-resistant and non-resistant maize and wheat species regulate the water intake of the plant, leading significant differences. In addition, plant roots are protected against pathogens and stress factors such as heavy metal toxicity, salinity, thus increases the plant's resistance.

Mycorrhiza fungi has an important role in the resistance of *Medicago sativa* plant grafted with *Rhizophagus irregularis* fungi (AM) in arsenic soils (Li et al. 2018). In addition, mycorrhiza fungi can increase the resistance of the plant to drought by means of plant physiology and morphology, such as hypes and root growth and capillary root formation (Davies et al. 1992).

In a study investigating the effects of mycorrhizas on sorghum plant development, photosynthesis and stoma conductivity in dry conditions; the plant yield, photosynthesis rate and stoma conductivity increased significantly and it was determined that fertilization of sorghum seedlings (*Sorghum bicolor* (L.) Moench cv NK-367) with a vesicular arbuscular mycorrhizal (VAM) fungi (*Glomus intraradices* (Schenck and Smith)) increased plant yield in dry conditions (Ibrahim et al. 1990).

In a study investigating the effect of salt applications (0, 100 Mg, Na and Cl/kg) and zinc applications (0, 25, 50 Mg and Zn/kg) in environments with and without mycorrhiza on maize development and phosphorus and zinc intake; the mycorrhiza grafting was found to increase the phosphorus and zinc content significantly in wet and dry weight compared to the applications without mycorrhiza. Salt application in the research decreased plant height and wet weight, while increased phosphorus intake. Depending on the applications of Zinc, the plant height, wet weight and dry weight, phosphorus and zinc content also increased (Sönmez et al. 2013).)

In the study, *Azospirillum brasilense* and vesicular arbuscular mycorrhiza fungi (*Glomus fasciculus*) were applied to wheat seeds, it was determined that both applications under water stress increased the leaf area, total chlorophyll and nitrate reductase activity, and maximum biomass production and grain yield were obtained (Panwar 1993).

In a study, the effect of arbuscular mycorrhizal (AM) fungus (*Glomus intraradices* (Schenck and Smith)) on the metabolic changes in tropical maize in dry conditions was investigated and it was reported that there was a significant amount of sugar and protein in ammonium thippeno sequia CO cultivars, especially sensitive to drought, and grafting may be of physiological importance to help plants to resist drought (Subramanian and Charest 1995).

In a study, researchers demonstrated that the low root zone temperature reduced soybean [*Glycine max* (L.) Merr.] nodulation and nitrogen fixation. The optimum root zone temperature for mycorrhizal infection of *Glomus versiforme* is between 21–22 °C. Mycorrhizal colonization increases up to flowering, and nodule formation (number of nodules) decreases at lower root zone temperatures (Zhang et al. 1995).

In a study on the yield of VAM fungus-grafted sorghum plants in dry conditions, it was determined that the development, photosynthesis rate and stomatal resistance of grafted plants in medium dry conditions were higher, and also VAM fungus in dry conditions increased the yield of sorghum plants (Ortaş et al. 1996).

In a study, the nutritional content and leaf water status of alfalfa (*Medicago sativa* L. cv Aragón) plants which were grafted with mycorrhizal fungus and / or Rhizobium in dry conditions were investigated, the mycorrhizal grafting was found to cause an increase in alfalfa nutrient content during drought that affects leaf-water relations (Goicoechea et al. 1997).

Zhu et al. (2010*)* investigated the growth characteristics, lipid peroxidation, activity of antioxidant enzymes in leaves and roots of maize (*Zea mays* L.) plant under the hot stress grafted with *Glomus etunicatum* Arbuscular mycorrhizal (AM) fungus. It was reported that AM fungus could alleviate the damage caused by temperature stress on maize plants by reducing maize lipid peroxidation and membrane permeability and increasing antioxidant enzyme activity, and it promoted plant biomass and supported plant growth.

In a study conducted on wheat; it was found that arbuscular mycorrhizal fungi (AMF), which infect plants and trigger tolerance to aluminum (Al) stress, reduce Al phytotoxicity in acidic soil (Aguilera et al. 2018).

3 Effect of Mycorrhizas Against Plant Diseases

Mycorrhizal fungi help plants in defense against soilborne diseases, form a physical barrier against pathogens, produce a number of useful antibiotics, and play a role in direct defense against fusarium and other pathogens.

There are many studies that suggest that arbuscular mycorrhizal fungus (AMF) inhibits, improves or does not affect the development of the disease. In these studies, the result is that the mycorrhizal fungus can be used for biological struggle against especially soil-borne pathogens. This complex mechanism must be well understood to establish an effective symbiont-pathogen relationship (Smith and Read 2008).

Mycorrhiza plants have been shown to be more resistant to diseases, but mycorrhiza cannot eliminate the disease, reduce the symptoms and severity of the disease, so increase the resistance of the plant. In a study on wilting disease of cotton, *G. fasciculatum* mycorrhiza species was observed to increase the wilting (Menge et al. 1980). In addition, these endomicorizas protect the plant against microorganisms in the parts of rhizosphere close to the root (Sieverding 1991).

Afek et al. (1990) found that *Pythium ultimum* was significantly suppressed in cotton plants with *G intraradices* in previously fumigated soils.

In a study conducted on sensitive and tolerated cotton varieties for Verticlium wilting, the application of *G mosseae* in tolerant Carmen cotton variety, *G mosseae* and *G margarita* variety increased the plant height, the wet and dry weight of the plant in sensitive Sayar 314. In addition, root colonization varied between 19–60% in fungal applications of Sayar 314 (Demir et al. 2010).

In a study on cotton, following Verticlium wilt, the effect of *G. fasciculatum* and soil phosphorus was investigated. The plant growth increased in the high dose of P, the AMF formation was inhibited and Verticillium wilting increased. But at a low dose of P, the plants grafted with *G. fasciculatum* were not affected by *V. dahliae* (Davis et al. 1979).

In maize plants grafted with Glomus sp., R. solani and R. solani + Glomus sp., the proportion of diseased plants originating from R. solani was 67%, the rate of Glomus + R. solani was 33% in sterile conditions, the diseased plant originating from R. solani alone in field conditions was 33%, Glomus + R. Solani was also 8% (Ilag et al. 1987).

G. mossea mycorrhizal fungus species was effective in reducing the infections of Fusarium vasinfectum being another important cause of cotton wilt (Hu and Gui 1991).

The effects of mycorrhizal fungi of *Glomus mosseae*, *Glomus versiforme* and *Sclerocystis sinusa* on Verticillium dahliae of cotton were investigated and the tolerance of plants to disease was found to increase. *Glomus versiforme* is proven to be the most effective against the disease. It also increased plant growth parameters, such as the number of flowers of the *G. versiforme* species, and thus leading an increase in the number of seeds (Liu 1995).

4 Effect of Mycorrhiza on the Intake of Plant Nutrients

Nitrogen fixation is made by mycorrhizal fungi. Thus, the biological structure of the soil is supported to be efficient. Bacteria that make N fixation such as mycorrhiza rhizobium are effective in forming nodules. It is also known to increase the intake of elements such as P, Zn, CA, Cu, Fe, Mg and Mn by plants and increase the intake of nutrients in the soil by 60 times thanks to its hyphes (Bieleski. 1973). In addition, mycorrhiza has little effect on the intake of nutrients such as N, K, which are mobile within the plant.

In this study, the effects of vesiculular-arbuscular mycorrhizal fungi (VAM) on the production of beans and summer wheat dry matter and the intake of phosphorus, zinc, copper, iron and manganese were investigated; the researchers found that VAM increased the production of dry matter in all plant growth conditions, and P, Zn, CU and Fe in beans, and P and Zn in wheat (Kucey and Janzen 1987).

In a study, the effects of arbuscular Mycorrhizal fungi on nitrogen concentration in berseem clover were investigated; it was determined that mycorrhiza fungi had significant effect on nitrogen concentration and increased it by 30% and 40.3%, respectively (Aram and Golchin 2013).

Arbuscular mycorrhizal (AM) fungi provide a significant contribution to the activity of N-binding soil organisms in nitrogen (N) intake from complex organic sources, promoting plant growth (Bukovská et al. 2018).

Pacovsky (1989) grafted Brady rhizobium japonicus variety and vesicular arbuscular mycorrhizal (VAM) fungus (*Glomus fasciculatum*) to soybean [*Glycine max* (L.) Merr. cv]. It has been determined that the quality and quantity of carbohydrates, proteins and amino acids are dependent on the physiological changes caused by infection from the N2-fixing bacteria or endomiorizal fungi.

Just one application of intensive P to soil can cause increased yield for several years, but it is stated that it can cause problems in Zn intake. In a study on wheat (*Triticum aestivum* L.), increasing P levels in the soil through P application significantly decreased the total Zn content of the aboveground plant parts and plant total Zn. Again, high dose P application significantly reduced the vesicular-arbuscular mycorrhizal colonization (VAM) in plant roots (Singh et al. 1986).

Although it is accepted by many researchers that mycorrhizal functions depend on the intake of phosphorus by plants, it is emphasized that it can also change depending on the plant genotypes. Researches were shown that when the plant was well infected with mycorrhiza fungi, the phosphorus intake by the plant genotypes was altered through root infection, spore production or root secretions (Smith et al. 1992).

In a study, white clover (*Trifolium repens* L.) were grown in limestone in separate compartments for mycorrhiza (Glomus mosseae [Nicol. & Gerd.]) development and root development. Phosphorus (P) was administered to the mycorrhizal compartment in doses of 0, 20 and 50 mg kg^{-1}. In mycorrhiza plants, root and shoot dry weight increased. Increased P source showed a P increase in the roots, whereas mycorrhizal hyphae development was not significantly affected by the P level change. Increased P levels and mycorrhizal hyphae enhanced total P intake from 34% to 90% (Li et al. 1991).

Phosphorus is difficult to obtain from the soil by plants, many plants creates a symbiotic relationship with soil fungi to deal with this problem. In greenhouse studies on the intake of mycorrhiza nutrients, it was determined that the absolute necessary plant nutrients, which are slowly taken by plants in the soil, increase the intake of phosphorus, especially. If there is too much P in the soil, mycorrhiza fungi become ineffective, and some other nutrients in the soil besides P cannot be taken (Robson et al. 1993).

Phosphorus intake is increased by 80% by plants with mycorrhizal infection in plant roots within soil with insufficient phosphorus. For example, phosphorus intake regions in maize plant without mycorrhiza were associated with root length (Marschner 1995).

Loit et al. (2018) reported that there was an urgent need to develop new approaches to improve sustainable agriculture, for this purpose, investigated the extent of colonization of arbuscular mycorrhiza (AM) fungi in the potato roots under traditional and organic farming systems. In the study, lower AM fungal colonization occurred in potato roots grown in traditional farming systems compared to organic farming systems, AM fungus colonization in potato root was lower in high-P-containing soils, and fertilizer application in organic fields positively affected AM fungus colonization.

Some field crops, such as barley, maize, soybean and peanut, which are mycorrhiza-dependent plants, have increased the effect with the application of mycorrhiza. Endomycorrhizae are important contributors to plant development by transporting many nutrients, especially P and Zn, from places where root cannot reach by establishing symbiotic life in many cultivated plants such as wheat, maize, soybean, tobacco, sugarcane (Marschner 1995).

In soils that are inadequate in terms of Zn; the intake of this mineral increased when maize and wheat were infected with VAM (Faber et al. 1991)

Using two different mycorrhiza species (*Glomus intradices*, *Glomus clarium*) in two breeding environments, the effects of different compost applications on the development of white clover, nutrient intake and mycorrhiza infection were investigated (Yüksel 2006). In this study, *G. intradices* was found to be the most effective mycorrhiza species. All mycorrhiza species had a high degree of mycorrhizal root infection according to the control (*G. clarium* % 36,7, *G. intradices* 37%). The highest % P content (0.27%) was detected in the *G. clarium* mycorrhizae while the low % P content was measured in the control plants. Again, in general, Zn contents were generally higher in plants grafted with *G. intradices* than grafted with *G. clarium*.

In order to determine the effect of mycorrhiza and iron (Fe) application on the bioavailability of zinc (Zn) in sorghum-sudangrass plant with and without mycorrhiza, Fe and Zn were applied at certain doses in the greenhouse soil. After all, it was revealed that the mycorrhiza did not affect Zn in the soil, concentration of Fe had not an effect on the Zn bioavailability in the soil and further studies are needed to determine the effects of Fe and mycorrhiza on the Zn bioavailability (İnal and Sönmez 2011).

Subramanian et al. (2011) reported that maize plants grafted with mycorrhizal fungi had higher P and Zn concentrations than ungrafted maize plants and that mycorrhizal symbiosis in zinc deficiency is important in increasing the activities of antioxidant enzymes and nutritional status of the host maize plant.

In a study investigating the nutrient intake, yield and some quality features of cotton plants by zinc applications (0, 25, 50, 75 kg ha-1) and mycorrhiza (*Glomus mosseae*) grafting, Mycorrhiza grafting increased N, P, K, Zn and Cu in cotton leaves; zinc applications increased N, P, K, Zn, and also the fiber durability reached the highest value at the highest Zn dose (Ceylan et al. 2016)

Mycorrhiza colonization in plants are controlled as physiological and/or genetic. Fungal signal in plants supports the formation and progression of AMF infection in cells. A similar number of AMF species were determined in the root samples of plants in the soil with the DNA amplification method. The molecular understanding of AMF relationship in plants is very important to choose a correct species. A spore bank should be established to represent the AMF taxon which is important in the field crops that are cultivated in large areas.

Plant-arbuscular mycorrhizal fungi (AMF) interaction can vary according to plant varieties and thus affect yield. In a study, AMF grafting was done in wild black-eyed pea and cultured varieties, and more nodules at the root occurred in modern cultures than wild ones, and P and N intake increased (Oruru et al. 2018).

In order to cope with the problem that may arise as a result of climate change in the future, a large number of plants including AM fungi have been identified and have been found to increase the development and productivity of many oil seed plants (soybean, peanut, sesame etc. both legume and legume plants) (Sharma et al. 2014).

Plant roots contain symbiotic arbuscular mycorrhizal (AM) fungi that can affect plant growth and health, and the effect of microbial interactions on the roots is depend on the genetic characteristics of the host plant. Researchers have revealed that AM symbiosis increases biomass in many varieties of chickpeas (Bazghaleh et al. 2018)

Crop rotation, soil processing and other management factors can affect the level and benefits of mycorrhizas. All field crops are included in the product rotation. To learn more about the benefits of mycorrhizas to field crops, more work should be done on product rotation. Typically, legume plants are used in agriculture to reduce soil pathogens and increase the yield of subsequent grain crops. The grafting of mycorrhiza in maize plant after canola increased plant yield and nutrient intake compared to those without mycorrhiza (Akpınar 2011).

Bakhshandeh et al. (2017) cultivated two wheat genotypes (IAW2013 and 249), after canola or chickpea, with different nitrogen (N) and phosphorus (P) fertilizer application (0–100 kgN ha^{-1} and 0–20 kg P ha^{-1}). In this study, although product rotation did not affect the present soil of N and P, the AMF colonization in wheat was 60% higher in cultivation after chickpea than after canola. Wheat yield after chickpea increased for IAW2013 genotype, and both genotypes were positively correlated with AMF colonization. N and P fertilization reduced AMF colonization.

Deng et al. (2017) investigated plant growth and root colonization of arbuscular mycorrhizal fungi (AMF) in the wheat-maize rotation study, which is a field plants in calcareous soils. In the research, wheat and especially maize had relatively high levels of AM fungi colonization, and AMF had an effective role for both plants.

It is known that the pasture legumes grown in the field can reduce the need for P fertilization and that the roots of these plants are usually colonized by arbuscular mycorrhizal fungi (AMF). Recent research suggests that a second community of root colonialist fungi forming Arbusculae may exist. It is also known that these root colonies fungi may be affected by the presence of different communities and these effects may vary among plant species (Jeffery et al. 2018).

5 Conclusion

As the effects of climate change and human beings in nature increase, the global diversity and function of arbuscular mycorrhizal fungi diminishes. Importance should be given to microbial symbiosis to increase the sustainability of global agricultural systems. So far, applications such as natural symbioses, classical agricultural methods, over-fertilization, eutrophication reduce the microbial diversity of the soil. Symbiosis between nitrogen-binding bacteria and legumes should be encouraged to reduce N application in agricultural areas. Many legumes are important among field crops for making the soil productive, for obtaining efficient products, for supplying the need for food in human and animal nutrition. In this context, as a result of agricultural applications such as product rotation in nature, mycorrhiza becomes active. Although mycorrhiza grafting is low-cost in small-scale production such as onions or strawberries, it is possible to manage the mycorrhizas naturally occurring in large agricultural areas by including field plants to rotation.

References

Afek, U., Menge, J. A., & Johnson, E. L. V. (1990). Effect of *Pythium ultimum* and metalaxyl treatments on root length and mikorizal colonization of cotton, onion and pepper. *Plant Disease, 74*, 1177–1120.

Aguilera, P., Larsenb, J., Boriea, F., Berríosa, D., Tapiaa, C., & Cornejoa, P. (2018). New evidences on the contribution of arbuscular mycorrhizal fungi inducing Al tolerance in wheat. *Rhizosphere, 5*, 43–50.

Akpınar, Ç. (2011). *The effects of mycorrhizae inoculation on yield and nutrient uptake of canola following by maize*. Çukurova University Institute of Natural and Applied Sciences Department of Soil Science and Plant Nutrition, PhD thesis, 185 page.

Aram, H., & Golchin, A. (2013). The effects of arbuscular mycorrhizal fungi on nitrogen concentration of berseem clover in contaminated soil with cadmium. *Journal of Chemical Health Risks, 3*(2), 53–58.

Bakhshandeh, S., Corneo, P. E., Mariotte, P., Kertesz, M. A., & Dijkstra, F. A. (2017). Effect of crop rotation on mycorrhizal colonization and wheat yield under different fertilizer treatments. *Agriculture, Ecosystems and Environment, 247*, 130–136.

Bazghaleh, N., Hamel, C., Gan, Y., Tar'an, B., & Knight, J. D. (2018). genotypic variation in the response of chickpea to arbuscular mycorrhizal fungi and non-mycorrhizal fungal endophytes. *Canadian Journal of Microbiology, 64*(4), 265–275.

Bieleski, R. L. (1973). Phosphate pools, phosphate avability. *Annual Review of Plant Physiology, 24*, 225–252.

Bukovská, P., Bonkowski, M., Konvalinková, T., Beskid, O., Hujslová, M., Püschel, D., Řezáčová, V., Gutiérrez-Núñez, M. S., Gryndler, M., & Jansa, J. (2018). Utilization of organic nitrogen by arbuscular mycorrhizal fungi-is there a specific role for protists and ammonia oxidizers? *Mycorrhiza, 28*(3), 269–283.

Ceylan, Ş., Mordoğan, N., & Çakıcı, H. (2016). Effect of zinc and mycorrhizae application on nutrient content yield and quality in cotton. *Ege Journal of Agricultural Research (EJAR), 53*(2), 117–123.

Davies, F. T., Potter, J. R., & Linderman, R. G. (1992). Mycorrhiza and repeated drought exposure affect drought resistance and exraradical hyphae development of paper plants independent of plant size and nutrient content. *Journal of Plant Physiology, 139*, 289–294.

Davis, R. M., Menge, J. A., & Erwin, D. C. (1979). Influence of Glomus fasciculatum and soil phosphorus on Andrticillium wilt of cotton. *Phytopathology, 69*, 453–456.

Demir, S., Bars Orak, A., & Demirer Durak, E. (2010). Arbuscular mycorrhizal fungus(AMF) dependency of some cotton cultivars ındicating different response against Verticillium wilt. *Yuzuncu Yil University Journal of Agricultural Sciences, 20*(3), 201–207.

Deng, Y., Feng, G., Chen, X., & Zou, C. (2017). Arbuscular mycorrhizal fungal colonization is considerable at optimal Olsen-P levels for maximized yields in an intensive wheat-maize cropping system. *Field Crops Research, 209*, 1–9.

Erlita, D., & Hariani, F. (2017). Pemberian mikoriza dan pupuk organik terhadap pertumbuhan dan produksi tanaman jagung (Zea Mays). *Agrium: Jurnal Ilmu Pertanian, 20*(3), 268–272.

Faber, B. A., Zaroski, R. J., Munns, D. A., & Shackal, K. (1991). A method of measuring hyphal nutrient and water uptake in mycorrhizal plants. *Canadian Journal of Botany, 69*, 87–94.

Goicoechea, N., Antolin, M. C., & Sanchez-Diaz, M. (1997). Influence of arbuscular mycorrhizae and rhizobium on nutrient content and water relations in drought stressed alfalfa. *Plant and Soil, 192*(2), 261–268.

Hu, Z. J., & Gui, X. D. (1991). Pretransplant inoculation with VA mycorrhizal fungi and Fusarium blight of cotton. *Soil Biology and Biochemistry, 23*, 201–203.

Ibrahim, M. A., Campbell, W. F., Rupp, L. A., & Allen, E. B. (1990). Effects of mycorrhizae on sorghum growth, photosynthesis and stomatal conductance under drought conditions. *Journal Arid Soil Research and Rehabilitation, 4*(2), 99–107.

Ilag, L. L., Rosales, A. M., & Mew, T. W. (1987, May 3–8). Effect of Glomus sp. on Rhizoctonia infection in selected crops. In *Proceedings of the 7th North American Conference on mycorrhizae* (Vol. 201), Gainessville, FL USA.

İnal, S., & Sönmez, O. (2011). The effects of mycorrhiza and different iron doses applications on the zinc toxicity. *Journal of Agriculture Faculty HR.U, 15*(2), 1–11.

Jeffery, R. P., Simpson, R. J., Lambers, H., Orchard, S., Kidd, D. R., Haling, R. E., & Ryan, M. H. (2018). Contrasting communities of arbuscule-forming root symbionts change external critical phosphorus requirements of some annual pasture legumes. *Applied Soil Ecology, 126*, 88–97.

Jeffries, P., & Dodd, J. C. (1991). The use of mycorrhizal inoculants in forestry and agriculture. In D. K. Arora, B. Raj, K. G. Mukerji, & G. R. Knudsen (Eds.), *Handbook of applied mycology* (pp. 155–185). New York: Marcel Dekker.

Kucey, R. M. N., & Janzen, H. H. (1987). Effect of VAM and reduced nutrient availability on growth and phosphorus and micronutrient uptake of wheat and field beans under greenhouse conditions. *Plant and Soil, 104*, 71–78.

Li, X.-L., Marschner, H., & George, E. (1991). Acquisition of phosphorus and copper by VA-mycorrhizal hyphae and root-to-shoot transport in white clover. *Plant and Soil, 136*, 49–57.

Li, J., Sun, Y., Jiang, X., Chen, B., & Zhang, X. (2018). Arbuscular mycorrhizal fungi alleviate arsenic toxicity to *Medicago sativa* by influencing arsenic speciation and partitioning. *Ecotoxicology and Environmental Safety, 15*(157), 235–243.

Liu, R. J. (1995). Effect of vesicular-arbuscular mycorrhizal fungi on Verticillium wilt of cotton. *Mycorrhiza, 5*(4), 293–297.

Loit, K., Soonvald, L., Kukk, M., Astover, A., Runno-Paurson, E., Kaart, T., & Öpik, M. (2018). The indigenous arbuscular mycorrhizal fungal colonisation potential in potato roots is affected by agricultural treatments. *Agronomy Research, 16*(2), 510 522.

Lu, L. H., & Wu, Q. S. (2017). Mycorrhizas promote plant growth, root morphology and chlorophyll production in white clover. *Biotechnology, 16*, 34–39.

Marschner, H. (1995). *Mineral nutrition of high plants. Second edition* (Vol. 37, pp. 155–159). London: Elsevier.

Menge, J. A., Jarrel, W., Labanauskas, W. M., Ojala, J. C., Huszar, C. E., Johnson, L. V., & Sibert, D. (1980). Predicting mycorrhizal dependency of troyer citrange on Glomus fasciculatus in California citrus soils. *Soil Science Society of American Journal, 46*(4), 762–768.

Muktiyanta, M. N. A., Samanhudi, S., Yunus, A., Pujiasmanto, B., & Minardi, S. (2018). Effectiveness of cow manure and mycorrhiza on the growth of soybean. *IOP Conference Series: Earth and Environmental Science, 142*(1), 012065.

Ortaş, İ. (1997). What is Mycorrhiza?. *Journal of TUBİTAK, 351*, Ankara.

Ortaş, İ., Harris, P. J., & Rowell, D. L. (1996). Enhanced uptake of phosphorus by mycorrhizal sorghum plants as influenced by forms of nitrogen. *Plant and Soil, 184*, 255–264.

Oruru, M. B., Njeru, E. M., Pasquet, R., & Runo, S. (2018). Response of a wild-type and modern cowpea cultivars to arbuscular mycorrhizal inoculation in sterilized and non-sterilized soil. *Journal of Plant Nutrition, 41*(1), 90–101.

Pacovsky, R. S. (1989). Carbohydrate, protein and amino acid status of Glycine-Glomus-Bradyrhizobium symbioses. *Physiologia Plantarum, 75*(3), 346–354.

Palta, Ş., Demir, S., Şengönül, K., Kara, Ö., & Şensoy, H. (2010). Arbuscular mycorrhizal fungi (AMF), their relationships with plants and soil, range rehabilitation. *Journal of Bartin Faculty of Forestry, 12*(18), 87–98.

Panwar, J. D. S. (1993). Response of VAM and Azospirillum inoculation to water, status and grain yield in wheat under, water stress condition. *Indian Journal of Plant Physiology, 36*, 41–43.

Raj, J., Bagyaraj, D. J., & Manjunath, A. (1981). Influence of soil inoculation with vesicular-arbuscular mycorrhiza and a phosphate-dissolving bacterium on plant growth and ^{32}P-uptake. *Soil Biology and Biochemistry, 13*(2), 105–108.

Robson, A., Abbott, L. K., & Schweiger, P. F. (1993). Benefits of VA mycorrhizas in agricultural and horticultural production. In *9th North American conference on mycorrhizae (abstracts)* (Vol. 51, pp. 231–239), Guelph, Ontario, Canada.

Sharma, A. K., Johri, B. N., & Gianinazzi, S. (1992). Vesicular arbuscular mycorrhizae in relation to plant disease. *World Journal of Microbiology and Biotechnology, 8*(6), 559–563.

Sharma, M. P., Sharma, S. S., Prasad, R. D., Pal, K. K., & Dey, R. (2014). Chapter: Application of arbuscular mycorrhizal fungi in production of annual oilseed crops. In Z. M. Solaiman, L. K. Abbot, & A. Varma (Eds.), *Mycorrhizal fungi: Use in sustainable agriculture and land restoration* (Soil biology) (Vol. 41, pp. 119–148). Berlin/Heidelberg: Springer.

Sieverding, E. (1991). *Vesicular-arbusculer mycorrhiza management in tropical agrosystems.* Technical Cooperation, Eschborn, Federal Republic of Germany, 371 pp.

Singh, J. P., Karamanos, R. E., & Stewart, J. W. B. (1986). Phosphorus-induced zinc deficiency in wheat on residual phosphorus plots. *Agronomy Journal, 78*(4), 668–675.

Smith, S. E., & Read, D. J. (2008). *Mycorrhizal symbiosis. Third edition (Hardcover).* New York: Academic Press is an Imprint of Elsevier. 800 p.

Smith, S. E., Robson, A. D., & Abbott, L. K. (1992). The involvement of mycorrhizas in assessment of genetically dependent efficiency of nutrient uptake and use. *Plant and Soil, 146*(1–2), 169–179.

Sönmez, F., Çığ, F., Erman, M., & Tüfenkçi, Ş. (2013). Effects of zinc, salt and mycorrhiza applications on the development and the phosphorus and zinc uptake of maize. *Yuzuncu Yil University Journal of Agricultural Sciences, 23*(1), 1–9.

Subramanian, K. S., & Charest, C. (1995). Influence of arbuscular mycorrhizae on the metabolism of maize under drought stress. *Mycorrhiza, 5*(4), 273–278.

Subramanian, K. S., Tenshia, J. S. V., Jayalakshmi, K., & Ramachandran, V. (2011). Antioxidant enzyme activities in arbuscular mycorrhizal (Glomus intraradices) fungus inoculated and non-inoculated maize plants under zinc deficiency. *Indian Journal of Microbiology, 51*(1), 37–43.

Yüksel, A. (2006). *The effects of different compost processes on two different growing environments on clover and onion plant's growing plant nutrient elements uptake and mycorrizae infection quality*. Msc thesis, Department of Soil Science Institue of Naturaland Applied Science University of Çukurova, page 101.

Zhang, F., Hamel, C., Kianmehr, H., & Smith, D. L. (1995). Root-zone temperature and soybean [*Glycine max*. (L.) merr.] vesicular-arbuscular mycorrhizae: Development and interactions with the nitrogen fixing symbiosis. *Environmental and Experimental Botany, 35*(3), 287–298.

Zhu, X., Song, F., & Xu, H. (2010). Influence of arbuscular mycorrhiza on lipid peroxidation and antioxidant enzyme activity of maize plants under temperature stress. *Mycorrhiza, 20*(5), 325–332.

Exploring the Role of Mycorrhizae as Soil Ecosystem Engineer

5

Antra Chatterjee, Shbbir R. Khan, and Huma Vaseem

1 Introduction

Environmental degradation in consequences of different ramped anthropogenic activities including industrialization has been worldwide highlighted. Although heavy metals are already present in the soil but indiscriminate activities such as mining, metal smelting, transportation activities, unsustainable agriculture practices and application of fertilizers etc. increase the concentration of these heavy metals in soil which harshly affects the physio-chemical and biological properties of the soil and cause the deterioration of soil quality. According to UN report (2000), nearly 2 billion hectares of land are affected by anthropogenic activities.

Soil is a one of the basic components of the terrestrial ecosystem which needs proper preservation. Soil is most vulnerable to anthropogenic activities and revealed a significant impact on ecosystem function, affecting both flora and fauna. Soil may be defined as a natural body consist of different horizon of minerals (Bakshi and Verma 2011) which serves as a natural medium to support the life on earth. Deterioration of soil health is a matter of concern for human being, as soil plays a pivotal role in delivering the ecosystem services as well as controlling terrestrial ecosystems through nutrient recycling (Jenny 1980; Bardgett 2005). Land degradation due to different unsustainable agriculture practices and mining activities etc., are among the emerging problems in all over the world.

Agriculture is the largest interface between humans and environment. In today's scenario, sustainable agriculture practises impart major challenge for agriculturist and farmers (Robertson and Swinton 2005). Climate change and shifting of climate

A. Chatterjee
Department of Botany, Institute of Science, Banaras Hindu University, Varanasi, India

S. R. Khan
Department of Zoology, Institute of Science, Banaras Hindu University, Varanasi, India

H. Vaseem (✉)
Department of Zoology, Faculty of Life Science, Aligarh Muslim University, Aligarh, India

© Springer Nature Singapore Pte Ltd. 2019

A. Varma, D. K. Choudhary (eds.), *Mycorrhizosphere and Pedogenesis*,
https://doi.org/10.1007/978-981-13-6480-8_5

showed potential impacts on crop productivity (Tubiello et al. 2002; Challinor et al. 2007). Along with that in agricultural practices such as application of indiscriminate chemical fertilizers (nitrogen, phosphorus, and potassium), enhance the crop productivity but simultaneously reduce the microbial diversity and soil enzymatic activities (Islam et al. 2009, 2010) which further reduces the soil quality. With the increase in awareness about the negative effects of chemicals, fertilizers and climate change on plants and human health, most of the agriculturists, agronomist are focussing towards crop yield with sustainable agriculture practices and mycorrhizal association may prove to be a milestone in the field of sustainable agriculture and crop protection against climate change. Therefore, we need to develop crop management strategies that optimise soil fertility, biological diversity and crop robustness (Altieri 1995). Microbial activities, respiration and metabolic quotient (qCO_2) enhance the quality and fertility of soil (Masto et al. 2006, 2007) as well as crop productivity.

Different stress factors such as heavy metal pollution, sewage sludge, application of chemicals and pesticides undermines the soil quality as well as exert toxic effects on the food webs and the structure of biotic communities (Boddington and Dodd 2000; Salminen et al. 2001; Aktar et al. 2009; Angelovičová et al. 2014), as these heavy metals are toxic in a very low concentration. Moreover they persist in the ecosystem for long period and extraction of these toxic substances from ecosystem is extremely difficult (Gisbert et al. 2000; Susarla et al. 2002). Thus such contaminated soil poses a serious risk to environmental health and is a great concern for ecologist and environmentalists (Li et al. 2014; Pandey et al. 2016). Various techniques and strategies such as encapsulation, solidification, stabilization, electrokinetics, vitrification, vapour extraction, soil washing and flushing have been used to clean up heavy metals but unfortunately they are very expensive and many of them do not support plant growth (Marques et al. 2009; Hashim et al. 2011). On the other side by virtue of economical and environmental friendly techniques, applications of these microorganisms are being encouraged by ecologists in the field of soil heavy metal remediation, reforestation and rehabilitation of degraded land.

Present chapter discusses different approaches of mycorrhizae in sustainable agricultural practices including better nutrient cycling, improving crop yield and remediation of toxic heavy metals from soil. In addition, the importance of mycorrhizal fungi in creating an environment for the development of plants in different successional stages of reclaimed land and their potential role as ecological engineering in the restoration of degraded land has also been elucidated. To accelerate ecological rehabilitation and restoration processes extensive restoration projects are required to be implemented. In recent years, ecosystem restoration has emerged as one of the central theme of global environmental policies (Aradottir and Hagen 2013; Jacobs et al. 2015). Most importantly, in (2010) UN Convention on Biological Diversity targeted to restore at least 15% of the world's degraded ecosystems 2011–2020 (CBD 2010).

2 Microorganism as Ecological Engineers

On earth microbes are diverse and most abundant organism, ubiquitously present (Chaudhry et al. 2012), simultaneously interact with the abiotic and biotic components of the ecosystem and perform a complex task for the ecosystem functioning. These microorganism exchange signals and coordinates with the other factors of an ecosystem and work as an ecological engineer. Ecological engineers are the organism that modify, maintain and create habitats by directly or indirectly regulating the biotic and abiotic available resources to other species (Jones et al. 1994). Enhancement and maintenance of soil health is very significant for ecosystem productivity and development (Tilman et al. 1997; Ash et al. 2010). Soil microorganism such as bacteria and fungi play a crucial role in biogeochemical cycling, maintenance of soil quality through decomposition of plant and animal organic matter for plant growth and soil structure and fertility (Wall and Virginia 1999; Jeffries et al. 2002; Chaudhry et al. 2012). Soil microorganisms are paramount for the recycling of micronutrients and thereby enhancing soil quality, the interaction of soil bacteria and their symbiotic relation with the plants as well (Brockwell et al. 1995). Such microbial communities are well known indicators of soil heavy metal pollution (McGrath et al. 1995; Djukic and Mandic 2006; Chen et al. 2014). Hence, it is essential to know the key role of soil-inhabiting organisms which serves as ecosystem engineers and supporting ecosystem services (Daily et al. 1997; Millennium Ecosystem Assessment 2005).

Microbes have gained attention due to their roles in bioremediation of contaminated soil and improving plant growth (Glick 2010; Khan et al. 2013). According to Weyens et al. (2010) microbes enhance the immune power of the plants to overcome toxic effects of metals which in turn accelerate the plant growth and phytoremediation activity. Many biological approaches have been implemented for the remediation of heavy metal polluted soils (Shu et al. 2002; Sekabira et al. 2011; Huang et al. 2013; Khan et al. 2013; Ul Hassan et al. 2017). Many scientific reports suggested that microorganism such as bacteria (Nath et al. 2012; Poornima et al. 2014) and fungal species (Andrade et al. 2010; Medina et al. 2010) have a potential to remediate heavy metal from contaminated soil and play an important role in rehabilitation of degraded land (Tscherko et al. 2004; Šourková et al. 2005). Some researchers suggested that bacterial and fungal inoculants along with the organic components of soil could be a potential option to be utilized in crop integrated nutrient management strategy of degraded soils (Medina et al. 2010; Chaer et al. 2011).

3 Mycorrhiza: Classification and Importance

Mycorrhizas are among the land marks in the evolutionary history of life forms on earth. Remy et al. (1994), reported a 400 million year old vesicular arbuscular mycorrhizae of which arbuscules were morphologically similar to those of living arbuscular mycorrhizae and suggested that nutrient transport mutualism may have been in existence when plants occupied the land. Mycorrhizas are symbiotic association between fungi and roots of most of the vascular plants. Soil fungi belong to Basidiomycetes, Ascomycetes and Zygomyectes form non-pathogenic symbiotic union, colonize the cortical tissues of roots of vascular plants in the duration of active plant growth both in natural environment and in cultivation (Miller and Jastrow 1994; Smith and Read 1997a, b). Such symbiotic association benefits the host plant with mineral nutrients from the soil as well as providing stress tolerance by manipulating water relations and pathogen resistance whereas photosynthetically derived products of host plant used by fungi as carbon source. Except for the waterlogged regions fungi involved in mycorrhizal association are ubiquitously found in temperate, tropical and arctic regions (Smith and Read 1997a, b). A wide range of host plants have been found with mycorrhizal association includes economically important crops as well as economically important forest trees belong to different families of angiosperms, gymnosperms as well as many pteridophytes and bryophytes. Because of their beneficial traits like providing tolerance against abiotic stresses such as drought (Abdelmoneim et al. 2014), heavy metal (Bano and Ashfaq 2013) and salinity (Porcel et al. 2012) as well as protecting from pathogen attack (Lewandowski et al. 2013) to host plant, glomalin (glycoprotein) synthesis and release in soil by them resulted in improved soil structure and elevated organic matter content hence improving soil fertility and preventing soil erosion (Singh et al. 2013). Mycorrhizal technology can be utilized in forestry as well as agricultural lands for improved nutrient consumption, more efficient land use and better crop productivity. Traditional classification of mycorrhizas includes two forms, ectotrophic and endotrophic on the basis of location of the fungal hyphae in respect of the root tissues of the host plant. In ectotrophic mycorrhizas fungal hyphae are present outside the root whereas in endotrophic mycorrhizas they are present inside the host root tissue. Isaac (1992) classified mycorrhiza on the basis of fungal associates and structural characteristics of mycorrhizas at maturity. Smith and Read (2008a, b) divided mycorrhiza into seven categories of equal rank viz., Arbuscular mycorrhiza which are endomycorrhizal association in which both vesicles and arbuscles are developed together and found in 90% of land plants including bryophytes, pteridophytes, gymnosperm (except Pinaceae) and most of angiosperms; Ectomycorrhiza which causes extensive branching and growth of roots and modification of branching pattern and occur in 3% of all seed plants in forests of temperate regions; Ericoid mycorrhiza are found in the different members of Ericaceae which grow in acidic soil with lower content of phosphorous and nitrogen; ectendomychorrhiza are intermediate in form between ecto- and endomycorrhizae; Arbutoid mycorrhiza form intercellular hartig net, usually restricted to the outer layer of root cells and found in few members of Ericaceae and pyrolaceae; monotropoid mycorrhiza are restricted to Monotropaceae family and the fungi does not penetrate the plant cell walls, carbon source is nearby plants of *Monotropa* and comes via their common mycorrhizal partner; Orchid mycorrhiza provide nutrition for the seeds to germinate in orchids. Diversity of Arbuscular mycorrhiza in structural features have been thoroughly studied (Widden 1996; Imhof 1997, 1999, 2003, 2007; Dickson 2004; Dominguez and Sersic 2004), whereas the study on division of mycorrhizae between ecto-, ectendo-, and arbutoid mycorrhizas is comparatively less (Brundrett 2004; Smith and Read 2008a, b). Imhof (2009), suggested hierarchical classification on the basis of structural similarities and emphasizes ecto-related mycorrhizas, arbuscular mycorrhizas and orchid mycorrhizas have independently developed mycorrhizal groups in a hierarchical system. Most mycorrhizal fungi are not host specific and simultaneously colonize a large number of plants (Van der Heijden and Horton 2009). According to study in natural ecosystems, plants fulfil 80% of nitrogen and up to 90% of phosphorus requirement from mycorrhizal fungi (Van der Heijden et al. 2008). Previous study revealed that parts of grasslands, savanna, boreal, temperate & tropical forests ecosystem are highly dominant and extensively colonized to mycorrhizal fungal network (Read 1991; Van der Heijden and Horton 2009). In some of the cases it has been recorded that mycorrhizal plants acquire nutrients from their fungal networks rather than soil (Van der Heijden et al. 1998; Hartnett and Wilson 1999) and influence the temporal stability of a plant community (Yang et al. 2014). Mycorrhizal symbiotic relation with plants root and their importance in plant growth in sustainable agriculture are recently been highlighted (Smith and Read 1997a, b; Jeffries et al. 2003). Mycorrhizal organisms are a combo tool for the sustainable management of agricultural ecosystems, with particular importance in plant health, soil fertility and as a bio-protector, as well as in producing compounds which directly stimulate plant growth, such as vitamins or plant hormones (Azcón-Aguilar and Barea 1996; Barea 1997, 2000). Mycorrhizal fungi are key interest to agriculturist because they influence the plant productivity and plant diversity, as they can easily connect plants, below ground via a hyphal network and allowing the movement of resources among coexisting plants (Van der Heijden et al. 2015).

4 Contribution of Mycorrhizal Fungi in Sustainable Agriculture Practices

Global warming and climate change has triggered the alteration of weather, which shifts the climate in different region of the earth, increase the chances of drought as well as crop vulnerability to infection and pest infection (Rosenzweig et al. 2001), along with that different anthropogenic activities discussed above cause the decline of soil quality, these all may lead to increase the problem of food demand and food insecurity. The microbial inoculants in the rhizosphere is affected by mycorrhizal co-inoculation thus these inoculation improved the establishment of both inoculated and indigenous phosphate-solubilizing rhizobacteria (Toro et al. 1997; Andrade et al. 1998; Ravnskov et al. 1999; Barea et al. 2002). According to study approximately 10–100 m mycorrhizal mycelium can be found per cm root, almost all tropical crops constitute mycorrhizal association, not all but almost most of them are strongly responsive to arbuscular mycorrhizas (McGonigle and Miller 1999; Cardoso and Kuyper 2006). Only few families and genera of plants with some exception do not form arbuscular mycorrhizas association these includes; Brassicaceae, Caryophyllaceae, Cyperaceae, Juncaceae, Chenopodiaceae, and

Amaranthaceae (Cardoso and Kuyper 2006). In recent years many researchers reviewed the role of mycorrhizae in enhancement of soil quality and sustainable agriculture (Barea et al. 2002; Gianinazzi et al. 2002; Jeffries et al. 2002; Ryan and Graham 2002; Harrier and Watson 2003). Arbuscular mycorrhizae formed a symbiotic relationship with most of the terrestrial plants root and successfully able to established a connecting link with the 80% root of the plants families and help to improve plant growth through uptake of nutrients from soil (Fitter et al. 2000; Gianinazzi et al. 2010) and account for 5–50% of the biomass of soil microbes (Olsson et al. 1999). Hyphae of AM fungi may be upto 54–900 kg ha^{-1} and the by-products produced by them may account upto 3000 kg (Lovelock et al. 2004; Zhu and Miller 2003). AM played important role in nutrient uptake and represent important carbon sinks to plants (Rillig 2004a, b). AM regulates plant diversity and influence primary production as well as actively participate in protection against root pathogens, improve plant-water relations including drought tolerance as well as host plant growth. (Newsham et al. 1995; Augé 2001; Rillig 2004a). The AM fungi extraradical mycelium highly influence the soil structure by holding the soil particles together and soil microbial community activities (Miller and Jastrow 2000; Rillig 2004a, b) and stabilize soil aggregates. Due to the long persistence time in the soil, hyphae of AM fungi is more important because AM fungi produces glomalin a specific soil-protein, but unfortunately biochemical nature of this protein is still not very clear and glomalin is quantified by measuring glomalin related soil-protein (GRSP), this protein has a longer residence time in soil than hyphae and present in soil in large amount, glomalin is stably glue hyphae to soil, helps in the formation of a 'sticky' string-bag of hyphae providing long term contribution to soil stability and aggregation (Wright and Upadhyaya 1998; Jastrow and Miller 1997; Rillig et al. 2002; Rillig 2004a). According to early report approximately 3.2% of total soil C and 5% of soil N in rain forest soil was in the form of glomalin and up to 5% of soil C and 4% of soil N stocks were derived from glomalin, whereas hyphae and glomalin together form approximately 15% of soil organic C in a grassland ecosystem (Rilling et al. 2001; Lovelock et al. 2004). Moreover, glomalin production increases carbon flow to soil simultaneously impact soil aggregation process (Cardoso and Kuyper 2006). Thus Glomalin production by AM fungi may helpful in the mitigation of soil erosion and management of cropping systems to enhance soil stability (Cardoso and Kuyper 2006). Although, how glomalin production enhances AM fungi fitness is still not known. Deficiency of phosphorous in soil in acidic soil retards plant growth. It is usually caused by adsorption of $H_2PO_4^-$ to Al and Fe (hydr)oxides which convert large amount of P into unavailable form of P to plants (Cardoso and Kuyper 2006). Mycorrhiza benefits plant growth as it can easily mine P from soil, thus AM fungi may be considered as a "biofertlizer" as well as capable of improving other macronutrients such as N and K, successfully enhance uptake of ammonium (NH_4^+) in plants, which is less mobile than nitrate (NO_3^-) in acidic soil as well as highly helpfull in arid conditions (Cardoso and Kuyper 2006; Lehmann et al. 2001; Lekberg and Koide 2005). AM fungal hyphae is more viable to penetrate decomposing organic material than plant root. Therefore by capturing simple organic nitrogen compounds AM fungi is very usefull in N cycling (Hawkins et al. 2000; Hodge et al. 2001). In this regard, recent work also reported that plants seem to be better than fungi (Hodge 2001). Many report suggested that AM also contribute in C recycling through redistribution of recently fixed C through the soil (Drigo et al. 2010; Nottingham et al. 2013; Fernandez et al. 2016). An early report suggested that when crops such as sunflower, sorghum, chickpea etc., are grown in long periods of bare fallow, exhibited deficient P and Zn and poor growth, may suffered long fallow disease and it is caused due to low AM colonisation association (Thompson 1991). However, Daeia et al. (2009) also concluded that under salinity wheat root colonized by AM *Glomus etunicatum* and *G. mosseae* have significant effect and relatively increase the nutrient uptake, act as a barrier for Na^+ and Cl^- adsorption by plant. Combined application of bacterial inoculants and AM fungi are more effective inoculation treatments and may be an important bio-resource for efficient bio-inoculants development for *Vigna radiata* productivity (Yasmeen et al. 2012). Moreover, Zaidi and Khan (2005) suggested that mixture of inoculation of N_2-fixing microbes, phosphate-solubilising bacteria along with AM fungi improved plant vitality and nutrient uptake and dramatically increase wheat crop yield. Another study revealed that when P-solubilising bacteria, AMF and *Azotobacter* applied together in sunflower farming, increased plant height and total chlorophyll content as well as significantly enhanced crop yield (Patra et al. 2013). AM fungi are also helpfull to enhance the nodulation and nitrogen fixation in legume plants which leads to higher P uptake through the AM hyphae rather than from soil (Valdenegro et al. 2001; Mortimer et al. 2008) but the enhancement of nodulation depends on the specification of the species of AM fungi (Valdenegro et al. 2001).

In addition to the role of mycorrhiza on plant growth, mycorrhizal symbionts could positively act as a defence line against pathogen attack (Dong and Zhang 2006). Thus, management of mycorrhizosphere exerted a significant effect on agro-ecosystem function and crop productivity through nutrient enrichment for plants, as Van der Heijden et al. (1998) suggested that below ground diversity of AM fungi is one of the major contributing factors for plant diversity and ecosystem variations.

5 Attributes of Mycorrhizal Fungi Towards Heavy Metal Remediation and Restoration of Degraded Land

In recent decade, increase in soil quality and soil health by the ecosystem engineers has received widespread attention (Jouquet et al. 2006, 2007; Holec and Frouz 2006; Khan et al. 2017). Mycorrhizal fungi have a strong bonding with the host plant root, thus easily facilitate the uptake of relatively immobile nutrients (such as P, N, Zn, Mg etc.,) which are present in low concentration in the soil (Kaur et al. 2014). For the reclamation of degraded land such as mine spoils, it is essential to create a stabilizing plant cover composed of herbs and shrubs. Previous report suggested that degraded land with heavy metal pollution are devoid of mycorrhizal symbiosis, showed poor root-shoot growth and biomass Such findings indicates that lack of mycorrhizal symbiosis hindered the re-vegetation process of degraded land (Pawlowska et al. 1996). It is documented that the mycorrhizal fungi provide resistance to soil against heavy metal pollution (Newsham et al. 1995) Interaction between plant and rizospheric microbes enhances biomass production and provides tolerance against heavy metal pollution to plants and these mycorrhizal fungi actively participated in the phytoremediation processes and produces high levels of root and shoot biomass as well as protect soil from erosion one of the crucial part of phytoremediation (Zak and Parkinson 1982; Leyval et al. 1997; Glick 2003; Belimov et al. 2005), as well as played an active role in the restoration of degraded land (Lebeau et al. 2008). Earlier research also revealed that mycorrhiza formation provides an eliminating barrier to plants against soil heavy metal pollution, which increases the plant tolerance to stress and improve plants nutrition (Turnau et al. 1993; Weissenhorn et al. 1995a, b; Gamalero et al. 2009). Thus, arbuscular mycorrhizal fungi (AMF), played an important role in soil restoration because of their high colonization and symbiotic ability with most of the terrestrial plants as well as high adaptability to variety of climatic conditions, improve plant nutrition and one of the very reliable source in the project of reclamation and revegetation of degraded and polluted soil (Turnau et al. 2005, 2008; Smith and Read 2008a, b; Bothe et al. 2010). Ecological role in phytoremediation and reclamation of AMF is limited and not very clear (Galli et al. 1994; Haselwandter et al. 1994; Pawlowska et al. 1996).

Mycorrhizal networks are believed to be a 'superorganisms' (Clements 1936). Restoration ecologists are paying attention towards fungi because of their key involvement in soil structural dynamics in paedogenesis as well as in constituting an important part of soil biomass (Ritz and Young 2004; Meena et al. 2014; Rashid et al. 2016) (Fig. 5.1).

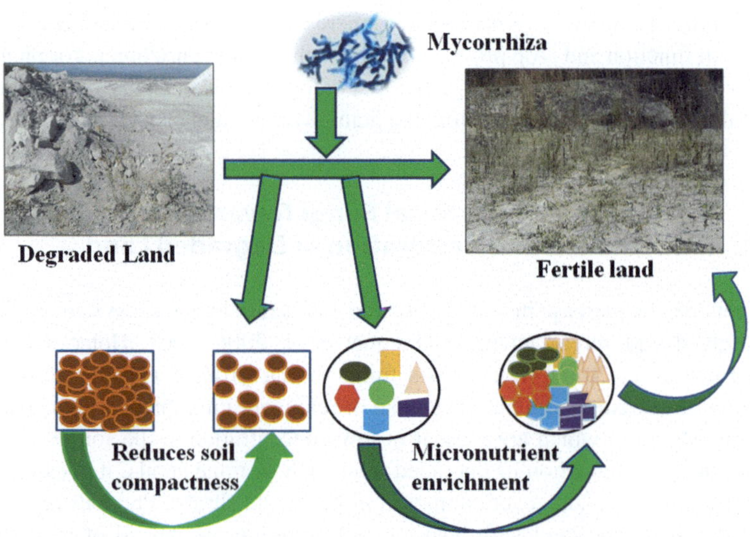

Fig. 5.1 Figure

When heavy metals enter the soil environment, they persist in that environment for a long time, because these elements are non degradable therefore remediation of heavy metals contaminated soil is one of the most challenging task. Traditional methods are generally very expensive and risky due to the generation of hazardous by-products (Malekzadeh et al. 2011). Fortunately phytoremediation is one of the most emerging alternative approaches available for the decontamination of soils through accumulation and detoxification of soil contaminants. Many researchers suggested that success of phytoremediation not only depends on the plants but also on the interaction of the plant roots with rhizospheric microbes along with the speciation and concentrations of heavy metals in the soil (Vivas et al. 2003; Khan 2005).

In this chapter we are emphasizing mainly on Arbuscular Mycorrhizal Fungi (AMF) and Ectomycorrhizal Fungi (EMF) because they have been typically responded in some way to polluted and drought condition (Khan 2005; McCormick et al. 2006, 2009; Medina et al. 2010). AMF provide growth and development to the plant in the polluted degraded soil such as mine sites, sewage-sludge and thereby improving soil quality (Marx 1975, 1980; Vivas et al. 2003; Yao et al. 2003; Gaur and Adholeya 2004). AMF generate links between soil and roots, thus consequently contribute in phytoremediation of heavy metals. Introduction of an AMF inoculum in degraded and metal contaminated site could be more economical and user friendly for re-vegetation of reclaimed sites by influencing available soil heavy metals (Leyval et al. 1997; Gaur and Adholeya 2004). Several report suggested AMF are beneficial for plant under heavy metal stress (Gildon and Tinker 1981; Weissenhorn et al. 1993; Griffioen et al. 1994), many researcher isolated heavy metal tolerant strains of AMF such as *Glomus aggregatum, G. constrictum, G. fasciculatum* from Zn contaminated mine site in Kansas, USA, Tamil Nadu in India and Netherlands (Dueck et al. 1986; Ietswaart et al. 1992; Sambandan et al. 1992; Shetty et al. 1994). Early report revealed that corn grown in metal contaminated soil showed significantly higher biomass and lower Cd, Cu and Zn concentrations when exhibited with AMF than those grown without AMF. Bioavailability of heavy metals did not affects AM colonization in corn, cultivated in Pb-Zn polluted soil (Turnau et al. 1993; Weissenhorn et al. 1995a, b). AM fungi have a good potential of heavy metal accumulation in contaminated soil. External mycelia of AMF occupy wider volume of soil by spreading beyond the root exploration zone thus can easily access to greater volume of heavy metals. A report suggested that species of *Glomus mosseae and G. versiforme* accumulate over 1200 $mgkg^{-1}$ and 600 $mgkg^{-1}$ of Zn respectively (Khan et al. 2000; Chen et al. 2001; Malcova et al. 2003). AMF also play an important role in protection of plants from arsenic and cadmium contamination. Inoculation of AMF indirectly enhances plant nutrition, thereby increasing growth, which further causes the diluting effects of these heavy metals (Chen et al. 2007). A report from southern Poland suggested that plants associated with colonized AMF at calamine spoil enriched in Cd, Pb and Zn (Orlowska et al. 2002). These fungi constitute a biological barrier against the uptake of heavy metal from root to shoots thus reduce the excess uptake of heavy metals such as Mn, Zn, and Cd (Heggo et al. 1990; Joner et al. 2000). Furthermore, a report suggested that when maize, grass and shrub (*Lygedum spartum* and *Anthyllis cytisoides*) grown in contaminated sites accumulated lower concentration of heavy metal when grown in AMF colonized soil as compared to non AMF inoculated soil. Similar results were also drawn by different researchers in different species of plants (Kaldorf et al. 1999; Tonin et al. 2001). Arbuscular mycorrhizal fungi consequently contribute to phytoremediation by influencing heavy metals availability by enhancing plants tolerance ability (Andrade et al. 2010; Ul Hassan et al. 2017; Marques et al. 2009). Thus mechanisms exerted by AMF to decrease soil and plant heavy metal stress through immobilization, adsorption to chitin in the cell walls changes in rhizosphere pH, and the regulation of gene expression under stress conditions (Joner et al. 2000; Li and Christie 2001; Christie et al. 2004; Wang et al. 2007; Upadhyaya et al. 2010) exclusively stimulate phytoextraction. Interactions of mycorrhizal AM fungi and plants are highly beneficial which enhance plants biomass production and provide tolerance to heavy metal pollution and may considered as an important and easy technique of phytoremediation. Thus AMF facilitate the process of phytoremediation in contaminated soils, furthermore many studies suggested that indigenous AMF which easily colonized degraded/polluted sites is highly helpful to restore metal contaminated soils (reviewed by Leyval et al. 1997; Khan 2004; Gamalero et al. 2009). Although AMF is beneficial in phytoremediation, early report suggested that AMF differ in their susceptibility and tolerance to heavy metals, thus the effectiveness of the phytoremediation by AMF also depends on the types and level of heavy metals in soil, symbiotic relation with plants as well as plants growth (Weissenhorn et al. 1995a; del Val et al. 1999; Wang et al. 2007). The ectomycorrhiza (EM) formed an association between ectomycorrhizal fungi and the root tips of many plant species (Smith and Read 2008a, b). Functional EM fungal hyphae cover the root tips and form a hyphal mantle inside the root, these hyphae penetrate the forest soil deep to access nutrients and in exchange provide abundant nourishment to plants in the form of water and micronutrients, which are unavailable to plant roots, while in reward these fungi obtained stored carbohydrates from the host plants (Druebert et al. 2009; Kaiser et al. 2010; Nehls et al. 2010; Pena et al. 2010).

6 Mycorrhiza Involvement in Rehabilitation of Degraded Land

During re-vegetation process roots of different plants are colonized by many different EMF and AMF, these form a common mycorrhizal network through fungal hyphae between tress/plants (Khan 2004; Martin et al. 2007; Martin and Nehls 2009; Rooney et al. 2009). Because of this common mycorrhizal network, water and nutrients between different individual trees are easily distributed, which may be highly beneficial for the stability and fitness of forest ecosystems against adverse degraded environments (Bingham and Simard 2011, 2012; Selosse et al. 2006; Van der Heijden and Horton 2009) which may also increase the re-vegetation of reclaimed land. There is worldwide agreement that global climate change may cause more frequent drought in different parts of the world (Dai 2011; Trenberth et al. 2013) and it is well established that polluted and degraded land such as mine sites are devoid of organic matter, total carbon/nitrogen, moisture, create xeric conditions and may be vulnerable to drought like condition (Singh et al. 2002; Frouz et al. 2006; Sheoran et al. 2010; Chaubey et al. 2012). Previously it is demonstrated that mycorrhizal fungi can improve the ability of their host plants to resist drought (Augé 2001; Lehto and Zwiazek 2011). In drought affected area colonization of both AM and EM fungi promotes drought tolerance in host plants, but in some cases it is reported that AM is more potent to facilitate plant resistance against drought as well as it is more predominant in arid region (Querejeta et al. 2009; Swaty et al. 2016). However, overall mycorrhizal fungi not only have potential against heavy metal pollution but also capable to promote drought tolerance to the plants by developing a larger positive effects on plant–water relationships in drought vulnerable area (Augé 2001; Yao et al. 2003; Gaur and Adholeya 2004; Lehto and Zwiazek 2011; Augé et al. 2015) as well as influence plant community structure and plants succession in degraded land, and may successfully assist phytostabilization of heavy metal contaminated land, thus these mycorrhizal fungi may prove to be highly beneficial in the restoration and reclamation process of degraded land.

7 Conclusion and Future Perspective

The omnipresence of mycorrhiza and its relations with plant's root and soil in terrestrial ecosystem make them a key functional group of soil biota. Study of mycorrhizal fungi and its symbiotic association with plants in the field is quite complex, because of their synergistic and antagonistic effects depends upon the identity of mycorrhiza and its colonization with the organism they are in symbiosis with. The effects of mycorrhiza are highly dependent on the soil type, its compactness, pH, management practices, plant species, soil-borne microorganisms, and type of metal pollutants etc. It is notable for the ecologists that all mycorrhizal associations are not beneficial because interactions between mycorrhiza and other microorganisms can be either detrimental or favourable to plants. However, such association with different types of soil micro-organisms like nitrogen fixing bacteria, P-solubilizing bacteria and soil fungi together with their antagonist behaviour towards plant pathogen may prove to be highly beneficial in the field of sustainable agriculture practices. Because of the ability of wide-spread mycelial network to penetrate deeper section in soil in search of water and nutrients they can be beneficial in the reclamation of degraded and polluted sites. Till date role of mycorrhiza in soil aggregation is not duly encouraged by the restoration ecologists. The excellent potential of phytoremediation by mycorrhizal plant is highly beneficial in the remediation of contaminated sites. However, knowledge of glomalin and its mechanism is less understood in the field of plant and soil ecology, and therefore further research on glomalin and its mechanism may prove to be a breakthrough in the field of restoration of degraded and reclaimed land. Extensive research is required for the understanding of beneficial mycorrhizal associations with different plant species under different types of environmental conditions at different level of land degradation. Thus application of mycorrhizal fungi can be immensely useful in: (a) restoration projects for the reclamation of degraded land (b) bringing new degraded barren land in crop cultivation thereby increasing net crop production (c) promoting low cost chemical free climate change resilient natural farming can prove to be highly beneficial for the developing countries like India.

Acknowledgement Huma Vaseem is thankful to UGC for Start-Up grant, Antra Chatterjee is thankful to CSIR-UGC for senior research fellow and Shbbir R. Khan is thankful to MANF-UGC.

References

Abdelmoneim, T. S., Tarek, A. A. M., Almaghrabi, O. A., Hassan, S. A., & Ismail, A. (2014). Increasing plant tolerance to drought stress by inoculation with arbuscular mycorrhizal fungi. *Life Science Journal, 11*(1), 10–17.

Aktar, W., Sengupta, D., & Chowdhury, A. (2009). Impact of pesticides use in agriculture: Their benefits and hazards. *Interdisc Toxicol, 2*(1), 1–12. https://doi.org/10.2478/v10102-009-0001-7.

Altieri, M. A. (1995). *Agroecology: The science of sustainable agriculture*. Boulder: Westview Press.

Andrade, G., Linderman, R. G., & Bethlenfalvay, G. J. (1998). Bacterial associations with the mycorrhizosphere and hyphosphere of the arbuscular mycorrhizal fungus Glomus mosseae. *Plant and Soil, 202*, 79–87.

Andrade, S. A. L., Gratão, P. L., Azevedo, R. A., et al. (2010). Biochemical and physiological changes in jack bean under mycorrhizal symbiosis growing in soil with increasing Cu concentrations. *Environmental and Experimental Botany, 68*, 198–207. https://doi.org/10.1016/j.envexpbot.2009.11.009.

Angelovičová, L., Lodenius, M., Tulisalo, E., & Fazekašová, D. (2014). Effect of heavy metals on soil enzyme activity at different field conditions in middle Spis mining area (Slovakia). *Bulletin of Environmental Contamination and Toxicology, 93*(6), 670–675.

Aradottir, A. L., & Hagen, D. (2013). Ecological restoration: approaches and impacts on vegetation, soils and society. *Advances in Agronomy, 120*, 173–222. https://doi.org/10.1016/b978-0-12-407686-0.00003-8.

Ash, N., Blanco, H., Brown, C., et al. (2010). *Ecosystems and human well-being*. Washington, DC: Island Press.

Augé, R. M. (2001). Water relations, drought and vesicular-arbuscular mycorrhizal symbiosis. *Mycorrhiza, 11*, 3–42.

Augé, R. M., Toler, H. D., & Saxton, A. M. (2015). Arbuscular mycorrhizal symbiosis alters stomatal conductance of host plants more under drought than under amply watered conditions: A meta-analysis. *Mycorrhiza, 25*, 13–24.

Azcón-Aguilar, C., & Barea, J. M. (1996). Arbuscular mycorrhizas and biological control of soilborne plant pathogens: An overview of the mechanisms involved. *Mycorrhiza, 6*, 457–464.

Bakshi, M., & Varma, A. (2011). Soil enzyme: The state-of-art. In G. Shukla & A. Varma (Eds.), *Soil enzymology, soil biology* (Vol. 22, pp. 1–24). Berlin/Heidelberg: Springer.

Bano, S. A., & Ashfaq, D. (2013). Role of mycorrhiza to reduce heavy metal stress. *Natural Science, 5*(12), 16–20.

Bardgett, R. (2005). *The biology of soil. A community and ecosystems approach* (242 pages). Oxford: Oxford University Press.

Barea, J. M. (1997). Mycorrhiza/bacteria interactions on plant growth promotion. In A. Ogoshi, L. Kobayashi, Y. Homma, F. Kodama, N. Kondon, & S. Akino (Eds.), *Plant growth-promoting rhizobacteria, present status and future prospects* (pp. 150–158). Paris: OECD.

Barea, J. M. (2000). Rhizosphere and mycorrhiza of field crops. In J. P. Toutant, E. Balazs, E. Galante, J. M. Lynch, J. S. Schepers, D. Werner, & P. A. Werry (Eds.), *Biological resource management: connecting science and policy* (pp. 110–125). Berlin/Heidelberg/New York: (OECD) INRA Editions/Springer.

Barea, J. M., Azcon, R., & Azcon-Aguilar, C. (2002). Mycorrhizosphere interactions to improve plant fitness and soil quality. *Antonie Van Leeuwenhoek, 81*, 343–351.

Belimov, A. A., Hontzeas, N., Safronova, V. I., et al. (2005). Cadmium-tolerant plant growth-promoting bacteria associated with the roots of Indian mustard (*Brassica juncea* L. Czern.). *Soil Biology and Biochemistry, 37*, 241–250.

Bingham, M. A., & Simard, S. W. (2011). Do mycorrhizal network benefits to survival and growth of interior Douglas-fir seedlings increase with soil moisture stress? *Ecology and Evolution*. https://doi.org/10.1002/ece3.24.

Bingham, M. A., & Simard, S. W. (2012). Ectomycorrhizal networks of *Pseudotsuga menziesii* var. *glauca* trees facilitate establishment of conspecific seedlings under drought. *Ecosystems, 15*, 188–199.

Boddington, C. L., & Dodd, J. C. (2000). The effect of agricultural practices on the development of indigenous arbuscular mycorrhizal fungi. I. Field studies in an Indonesian ultisol. *Plant and Soil, 218*, 137–144.

Bothe, H., Turnau, K., & Regvar, M. (2010). The potential role of arbuscular mycorrhizal fungi in protecting endangered plants and habitats. *Mycorrhiza, 20*, 445–458.

Brockwell, J., Bottomley, P. J., & Thies, J. E. (1995). Manipulation of rhizobia microflora for improving legume productivity and soil fertility: A critical assessment. *Plant and Soil, 174*, 143–180.

Brundrett, M. C. (2004). Diversity and classification of mycorrhizal associations. *Biological Reviews of the Cambridge Philosophical Society, 79*, 473–495.

Cardoso, I. M., & Kuyper, T. W. (2006). Mycorrhizas and tropical soil fertility. *Agricultue, Ecosystems and Environment, 116*, 72–84.

CBD. (2010). *Aichi biodiversity targets*. Available at: http://www.cbd.int/sp/targets/

Chaer, G. M., Resende, A. S., Campello, E. F. C., et al. (2011). Nitrogen-fixing legume tree species for the reclamation of severely degraded lands in Brazil. *Tree Physiology, 31*, 139–149.

Challinor, A., Wheeler, T., & Garforth, C. (2007). Assessing the vulnerability of food crop systems in Africa to climate change. *Climatic Change, 83*, 381–399.

Chaubey, O. P., Bohre, P., & Singhal, P. K. (2012). Impact of bioreclamation of coal mine spoil on nutritional and microbial characteristics—A case study. *International Journal of Bio-Science and Bio-Technology, 4*, 69–79.

Chaudhry, V., Rehman, A., Mishra, A., Chauhan, P., & Nautiyal, C. (2012). Changes in bacterial community structure of agricultural land due to long-term organic and chemical amendments. *Microbial Ecology, 64*, 450–460.

Chen, B., Christie, P., & Li, L. (2001). A modified glass bead compartment cultivation system for studies on nutrient and trace metal uptake by arbuscular mycorrhiza. *Chemosphere, 42*, 185–192.

Chen, B. D., Xiao, X. Y., Zhu, Y. G., et al. (2007). The arbuscular mycorrhizal fungus *Glomus mosseae* gives contradictory effects on phosphorus and arsenic acquisition by *Medicago sativa* Linn. *The Science of the Total Environment, 379*, 226–234.

Chen, J., He, F., Zhang, X., Sun, X., Zheng, J., & Zheng, J. (2014). Heavy metal pollution decreases microbial abundance: Diversity and activity within particle-size fractions of a paddy soil. *FEMS Microbiology Ecology, 87*, 164–181.

Christie, P., Li, X. L., & Chen, B. D. (2004). Arbuscular mycorrhiza can depress translocation of zinc to shoots of host plants in soils moderately polluted with zinc. *Plant and Soil, 261*, 209–217.

Clements, F. E. (1936). Nature and structure of the climax. *Journal of Ecology, 24*, 252–284.

Daeia, G., Ardekania, M. R., Rejalic, F., et al. (2009). Alleviation of salinity stress on wheat yield, yield components and nutrient uptake using arbuscular mycorrhizal fungi under field conditions. *Journal of Plant Physiology, 166*, 617–625.

Dai, A. (2011). Drought under global warming: A review. *Wiley Interdisciplinary Reviews: Climate Change, 2*, 45–65.

Daily, G. C., Alexander, S., Ehrlich, P., et al. (1997). Ecosystem services: Benefits supplied to human societies by natural ecosystems. *Issues in Ecology, 2*, 1–16.
del Val, C., Barea, J. M., & Azcòn-Aguilar, C. (1999). Assessing the tolerance to heavy metals of arbuscular mycorrhizal fungi isolated from sewage sludge-contaminated soils. *Applied Soil Ecology, 11*, 261–269.
Dickson, S. (2004). The Arum-Paris continuum of mycorrhizal symbioses. *The New Phytologist, 163*, 187–200.
Djukic, D., & Mandic, L. (2006). Microorganisms as indicators of soil pollution with heavy metals. *Acta Agriculturae Serbica, 11*, 45–55.
Dominguez, L. S., & Sersic, A. (2004). The southernmost mycoheterotrophic plant, *Arachnitis uniflora*: Root morphology and anatomy. *Mycologia, 96*, 1143–1151.
Dong, L. Q., & Zhang, K. Q. (2006). Microbial control of plant-parasitic nematodes: A five-party interaction. *Plant and Soil, 288*, 31–45.
Drigo, B., Pijl, A. S., Duyts, H., et al. (2010). Shifting carbon flow from roots into associated microbial communities in response to elevated atmospheric CO2. *Proceedings of the National Academy of Sciences of the United States of America, 107*, 10938–10942.
Druebert, C., Lang, C., Valtanen, K., et al. (2009). Beech carbon productivity as driver of ectomycorrhizal abundance and diversity. *Plant, Cell & Environment, 32*, 992–1003.
Dueck, T. A., Visser, P., Ernst, W. H. O., et al. (1986). Vesicular-arbuscular mycorrhizae decrease zinc-toxicity to grasses growing in zinc-polluted soil. *Soil Biology and Biochemistry, 18*, 331–333.
Fernandez, C. W., Langley, J. A., Chapman, S., et al. (2016). The decomposition of ectomycorrhizal fungal necromass. *Soil Biology and Biochemistry, 93*, 38–49.
Fitter, A. H., Heinemeyer, A., & Staddon, P. L. (2000). The impact of elevated CO2 and global climate change on arbuscular mycorrhizas: A mycocentric approach. *The New Phytologist, 147*, 179–187.
Frouz, J., Elhottová, D., Kuráž, V., et al. (2006). Effect of soil macrofauna on other soil biota and soil. Formation in reclaimed and unreclaimed post mining sites: Result of field microcosm experiment. *Applied Soil Ecology, 33*, 308–320.
Galli, U., Schüepp, H., & Brunold, C. (1994). Heavy metal binding by mycorrhizal fungi. *Physiologia Plantarum, 92*, 364–368.
Gamalero, E., Lingua, G., Berta, G., et al. (2009). Beneficial role of plant growth promoting bacteria and arbuscular mycorrhizal fungi on plant responses to heavy metal stress. *Canadian Journal of Microbiology, 55*, 501–514.
Gaur, A., & Adholeya, A. (2004). Prospects of arbuscular mycorrhizal fungi in phytoremediation of heavy metal contaminated soils. *Current Science, 86*, 528–534.
Gianinazzi, S., Schüepp, H., Barea, J. M., et al. (2002). *Mycorrhizal technology in agriculture—From genes to bioproducts*. Basel: Birkha¨user Verlag.
Gianinazzi, S., Gollotte, A., Binet, M. N., et al. (2010). Agroecology: The key role of arbuscular mycorrhizas in ecosystem services. *Mycorrhiza*. https://doi.org/10.1007/s00572-010-0333-3.
Gildon, A., & Tinker, P. B. (1981). A heavy metal tolerant strain of a mycorrhizal fungus. *Transactions of the British Mycological Society, 77*, 648–649.
Gisbert, C., Ros, R., De, H. A., et al. (2000). A plant genetically modified that accumulates Pb is especially promising for phytoremediation. *Biochemical and Biophysical Research Communications, 303*, 440–445.
Glick, B. (2003). Phytoremediation: Synergistic use of plants and bacteria to clean up the environment. *Biotechnology Advances, 21*, 383–393.
Glick, B. R. (2010). Using soil bacteria to facilitate phytoremediation. *Biotechnology Advances, 28*, 367–374.
Griffioen, W. A. J., Ietswaart, J. H., & Ernst, W. H. O. (1994). Mycorrhizal infection of an Agrostis capillaris population on a copper contaminated soil. *Plant and Soil, 158*, 83–89.
Harrier, L. A., & Watson, C. A. (2003). The role of arbuscular mycorrhizal fungi in sustainable cropping systems. *Advances in Agronomy, 79*, 185–225.

Hartnett, D. C., & Wilson, W. T. (1999). Mycorrhizae influence plant community structure and diversity in tall grass prairie. *Ecology, 80*, 1187–1195.
Haselwandter, K., Leyval, C., & Sanders, F. E. (1994). Impact of arbuscular mycorrhizal fungi on plant uptake of heavy metals and radionuclides from soil. In S. Gianinazzi & H. Schüepp (Eds.), *Impact of arbuscular mycorrhizas on sustainable agriculture and natural ecosystems* (pp. 179–189). Basel: Birkhäuser.
Hashim, M. A., Mukhopadhyay, S., Sahu, J. N., et al. (2011). Remediation technologies for heavy metal contaminated groundwater. *Journal of Environmental Management, 92*, 2355–2388.
Hawkins, H. J., Johansen, A., & George, E. (2000). Uptake and transport of organic and inorganic nitrogen by arbuscular mycorrhizal fungi. *Plant and Soil, 226*, 275–285.
Heggo, A., Angle, J. S., & Chaney, R. L. (1990). Effects of vesicular Arbuscular mycorrhizal fungi on heavy-metal uptake by soybeans. *Soil Biology and Biochemistry, 22*, 865–869.
Hodge, A. (2001). Arbuscular mycorrhizal fungi influence decomposition of, but not plant nutrient capture from, glycine patches in soil. *The New Phytologist, 151*, 725–734.
Hodge, A., Campbell, C. D., & Fitter, H. (2001). An arbuscular mycorrhizal fungus accelerates decomposition and acquires nitrogen directly from organic material. *Nature, 413*, 297–299.
Holec, M., & Frouz, J. (2006). The effect of two ant species Lasius niger and Lasius flavus on soil properties in two contrasting habitats. *European Journal of Soil Biology, 42*, 213–217.
Huang, L., Xie, J., Lv, B., et al. (2013). Optimization of nutrient component for diesel oil degradation by *Acinetobacter beijerinckii* ZRS. *Marine Pollution Bulletin, 76*, 325–332. https://doi.org/10.1016/jmarpolbul.2013.03.037.
Ietswaart, J. H., Griffioen, W. A. J., & Ernst, W. H. O. (1992). Seasonality of VAM infection in three populations of *Agrostis capillaries* (Gramineae) on soil with or without heavy metal enrichment. *Plant and Soil, 139*, 67–73.
Imhof, S. (1997). Root anatomy and mycotrophy of the achlorophyllous *Voyria tenella* Hooker (Gentianaceae). *Botanica Acta: Journal of the German Botanical Society, 110*, 298–305.
Imhof, S. (1999). Anatomy and mycotrophy of the achlorophyllous *Afrothismia winkleri* (Engl.) Schltr. (Burmanniaceae). *The New Phytologist, 144*, 533–540.
Imhof, S. (2003). A dorsiventral mycorrhizal root in the achlorophyllous *Sciaphila polygyna* (Triuridaceae). *Mycorrhiza, 13*, 327–332.
Imhof, S. (2007). Specialized mycorrhizal colonization pattern in achlorophyllous *Epirixanthes* spp. (Polygalaceae). *Plant Biology, 9*, 786–792.
Imhof, S. (2009). Arbuscular, ecto-related, orchid mycorrhizas-three independent structural lineages towards mycoheterotrophy: Implications for classification? *Mycorrhiza, 19*, 357–363.
Isaac, S. (1992). *Fungal-plant interactions*. Cambridge: Chapman & Hall.
Islam, M. R., Trivedi, P., Palaniappan, P., et al. (2009). Evaluating the effect of fertilizer application on soil microbial community structure in rice based cropping system using fatty acid methyl esters (FAME) analysis. *World Journal of Microbiology and Biotechnology, 25*, 1115.
Islam, M. R., Trivedi, P., Madhaiyan, M., et al. (2010). Isolation, enumeration, and characterization of diazotrophic bacteria from paddy soil sample under long-term fertilizer management experiment. *Biology and Fertility of Soils, 46*, 261.
Jacobs, D. F., Oliet, J. A., Aronson, J., et al. (2015). Restoring forests: What constitutes success in the twenty-first century? *New Forest, 46*, 601–614. https://doi.org/10.1007/s11056-015-9513-5.
Jastrow, J. D., & Miller, R. M. (1997). Soil aggregate stabilization and carbon sequestration: Feedbacks through organomineral associations. In R. Lal, J. M. Kimble, R. F. Follett, & B. A. Stewart (Eds.), *Soil processes and the carbon cycle* (pp. 207–223). Boca Raton: CRC Press.
Jeffries, P., Gianinazzi, S., Perotto, S., et al. (2002). The contribution of arbuscular mycorrhizal fungi in sustainable maintenance of plant health and soil fertility. *Biology and Fertility of Soils, 37*, 1–16.
Jeffries, P., Gianinazzi, S., Perotto, S., et al. (2003). The contribution of arbuscular mycorrhizal fungi in sustainable maintenance of plant health and soil fertility. *Biology and Fertility of Soils, 37*, 1–16.
Jenny, H. (1980). *The soil resource: Origin and behaviour* (Ecological studies, Vol. 37, 377 pages). New York: Springer.

Joner, E. J., Leyval, C., & Briones, R. (2000). Metal binding capacity of arbuscular mycorrhizal mycelium. *Biology and Fertility of Soils, 226*, 227–234.

Jones, C. G., Lawton, J. H., & Shachak, M. (1994). Organisms as ecosystem engineers. *Oikos, 69*, 373–386.

Jouquet, P., Dauber, J., Lagerlof, J., Lavelle, P., et al. (2006). Soil invertebrates as ecosystem engineers: Intended and accidental effects on soil and feedback loops. *Applied Soil Ecology, 32*, 153–164.

Jouquet, P., Bernard-Reversat, F., Bottinelli, N., et al. (2007). Influence of changes in land use and earthworm activities on carbon and nitrogen dynamics in a steepland ecosystem in Northern Vietnam. *Biology and Fertility of Soils, 44*, 69–77.

Kaiser, C., Koranda, M., Kitzler, B., et al. (2010). Belowground carbon allocation by trees drives seasonal patterns of extracellular enzyme activities by altering microbial community composition in a beech forest soil. *The New Phytologist, 187*, 843–858.

Kaldorf, M., Kuhn, A. J., Schroder, W. H., et al. (1999). Selective element deposits in maize colonized by a heavy metal tolerance conferring arbuscular mycorrhizal fungus. *Journal of Plant Physiology, 154*, 718–728.

Kaur, R., Singh, A., & Kang, J. S. (2014). Influence on different types of mycorrhizal fungi on crop productivity. *Current Agriculture Research Journal, 2*, 51–54.

Khan, A. G. (2004). *Co-inoculum of vesicular-arbuscular mycorrhizal fungi, mycorrhiza-helper-bacteria, and plant growth promoting rhizobacteria for phytoremediation of heavy metal contaminated soils*. Proceedings of the V international conference on environmental geochemistry in the tropics, Haiko, Hainan, China. Institute Soil Science, Chinese Academy of Science, Nanjing, PR China. March 21–26, p. 68.

Khan, A. G. (2005). Mycorrhizas and phytoremediation. In N. Willey (Ed.), *Methods in biotechnology: Phytoremediation: Methods and reviews*. Totowa: Humana Press Inc.

Khan, A. G., Kuek, C., Chaudhry, T. M., et al. (2000). Role of plants, mycorrhizae and phytochelators in heavy metal contaminated land remediation. *Chemosphere, 41*, 197e207.

Khan, S., Afzal, M., Iqbal, S., et al. (2013). Plant-bacteria partnerships for the remediation of hydrocarbon contaminated soils. *Chemosphere, 90*, 1317–1332.

Khan, S. R., Singh, S. K., & Rastogi, N. (2017). Heavy metal accumulation and ecosystem engineering by two common mine site-nesting ant species: Implications for pollution-level assessment and bioremediation of coal mine soil. *Environmental Monitoring and Assessment, 189*, 195. https://doi.org/10.1007/s10661-017-5865-y.

Lebeau, T., Braud, A., & Jezequel, K. (2008). Performance of bioaugmentation assisted phytoextraction applied to metal contaminated soils: A review. *Environmental Pollution, 153*, 497–522.

Lehmann, J., Cravo, M. S., Mace^do, J. L. V., et al. (2001). Phosphorus management for perennial crops in central Amazonian upland soils. *Plant and Soil, 237*, 309–319.

Lehto, T., & Zwiazek, J. J. (2011). Ectomycorrhizas and water relations of trees: A review. *Mycorrhiza, 21*, 71–90.

Lekberg, Y., & Koide, R. T. (2005). Is plant performance limited by abundance of arbuscular mycorrhizal fungi? A meta-analysis of studies published between 1988 and 2003. *The New Phytologist, 168*, 189–204.

Lewandowski, T. L., Dunfield, K. E., & Antunes, P. M. (2013). Isolate identity determines plant tolerance to pathogen attack in assembled mycorrhizal communities. *PLoS One, 8*(4), e61329.

Leyval, C., Turnau, K., & Haselwandter, K. (1997). Effect of heavy metal pollution on mycorrhizal colonization and function: Physiological, ecological and applied aspects. *Mycorrhiza, 7*, 139–153.

Li, X. L., & Christie, P. (2001). Changes in soil solution Zn and pH and uptake of Zn by arbuscular mycorrhizal red clover in Zn contaminated soil. *Chemosphere, 42*, 201–207.

Li, Z., Ma, Z., Van der Kuijp, T. J., et al. (2014). A review of soil heavy metal pollution from mines in China: Pollution and health risk assessment. *Science of the Total Environment, 468*, 843–853.

Lovelock, C. E., Wright, S. F., Clark, D. A., et al. (2004). Soil stocks of glomalin produced by arbuscular mycorrhizal fungi across a tropical rain forest landscape. *Journal of Ecology, 92*, 278–287.

Malcova, R., Vosátka, M., & Gryndler, M. (2003). Effects of inoculation with *Glomus intraradices* on lead uptake by *Zea mays* L. and *Agrostis capillaris* L. *Applied Soil Ecology, 2003*(23), 55–67.

Malekzadeh, E., Alikhani, A. H., Savaghebi-Fioozabadi, R. G., et al. (2011). Influence of arbuscular mycorrhizal fungi and an improving growth bacterium on Cd uptake and maize growth in Cd-polluted soils. *Spanish Journal of Agricultural Research, 9*, 1213–1223.

Marques, A. P. G. C., Rangel, A. O. S. S., & Castro, P. M. L. (2009). Remediation of heavy metal contaminated soils: Phytoremediation as a potentially promising clean-up technology. *Critical Reviews in Environmental Science and Technology, 39*(8), 622–654.

Martin, F., & Nehls, U. (2009). Harnessing ectomycorrhizal genomics for ecological insights. *Current Opinion in Plant Biology, 12*, 508–515.

Martin, F., Kohler, A., & Duplessis, S. (2007). Living in harmony in the wood underground: Ectomycorrhizal genomics. *Current Opinion in Plant Biology, 10*, 204–210.

Marx, D. H. (1975). Mycorrhizae and the establishment of trees on stripmined land. *The Ohio Journal of Science, 75*, 88–297.

Marx, D. H. (1980). Role of mycorrhizae in forestation of surface mines. In: Proceedings of Trees for Reclamation. 27–28 October, 1980, Lexington, KY, pp. 109–116.

Masto, R. E., Chhonkar, P. K., Singh, D., et al. (2006). Changes in soil biological and biochemical characteristics in a long-term field trial on a sub-tropical Inceptisol. *Soil Biology and Biochemistry, 38*, 1577–1158.

Masto, R. E., Chhonkar, P. K., Singh, D., et al. (2007). Soil quality response to long-term nutrient and crop management on a semiarid Inceptisol. *Agriculture, Ecosystems and Environment, 118*, 130–142.

McCormick, M. K., Whigham, D. F., Sloan, D., et al. (2006). Orchid–fungus fidelity: A marriage meant to last? *Ecology, 87*, 903–911.

McCormick, M. K., Whigham, D. F., O'Neill, J. P., et al. (2009). Abundance and distribution of Corallorhiza odontorhiza reflect variations in climate and ectomycorrhizae. *Ecological Monographs, 79*, 619–635.

McGonigle, T. P., & Miller, M. H. (1999). Winter survival of extraradical hyphae and spores of arbuscular mycorrhizal fungi in the field. *Applied Soil Ecology, 12*, 41–50.

McGrath, S. P., Chaudri, A. M., & Giller, K. E. (1995). Long-term effects of metal in sewage sludge on soils, microorganisms and plants. *Journal of Industrial Microbiology, 14*, 94–104.

Medina, A., Roldán, A., & Azcón, R. (2010). The effectiveness of arbuscular-mycorrhizal fungi and *Aspergillus niger* or *Phanerochaete chrysosporium* treated organic amendments from olive residues upon plant growth in a semi-arid degraded soil. *Journal of Environmental Management, 91*, 2553. https://doi.org/10.1016/j.jenvman.2010.07.008.

Meena, V. S., Maurya, B. R., & Verma, J. P. (2014). Does a rhizospheric microorganism enhance K^+ availability in agricultural soils? *Microbiological Research, 169*, 337–347.

Millennium Ecosystem Assessment. (2005). Ecosystems and human well-being. In *Millennium ecosystem assessment*. Washington, DC: Island Press.

Miller, R. M., & Jastrow, J. D. (1994). Vesicular-arbuscular mycorrhizae and biogeochemical cycling. In F. L. Pfleger & R. G. Linderman (Eds.), *Mycorrhizae and plant health* (pp. 189–212). St. Paul: APS Press, The American Phytopathlogical Society.

Miller, R. M., & Jastrow, J. D. (2000). Mycorrhizal fungi influence soil structure. In Y. Kapulnik & D. D. Douds (Eds.), *Arbuscular mycorrhizas: Physiology and function* (pp. 3–18). Dordrecht: Kluwer Academic.

Mortimer, P. E., Pe'rez-Ferna'ndez, M. A., & Valentine, A. J. (2008). The role of arbuscular mycorrhizal colonization in the carbon and nutrient economy of the tripartite symbiosis with nodulated *Phaseolus vulgaris*. *Soil Biology and Biochemistry, 40*, 1019–1027.

Nath, S., Deb, B., & Sharma, I. (2012). Isolation and characterization of cadmium and lead resistant bacteria. *Global Advanced Research Journal of Microbiology, 1*(11), 194–198.

Nehls, U., Gohringer, F., Wittulsky, S., et al. (2010). Fungal carbohydrate support in the ectomycorrhizal symbiosis: A review. *Plant Biology, 12*, 292–301.

Newsham, K. K., Fitter, A. H., & Watkinson, A. R. (1995). Arbuscular mycorrhiza protect an annual grass from root pathogenic fungi in the field. *Journal of Ecology, 83*, 991–1000.

Nottingham, A. T., Turner, B. L., Winter, K., et al. (2013). Root and arbuscular mycorrhizal mycelial interactions with soil microorganisms in lowland tropical forest. *FEMS Microbiology Ecology, 85*, 37–50.

Olsson, P. A., Thingstrup, I., Jakobsen, I., et al. (1999). Estimation of the biomass of arbuscular mycorrhizal fungi in a linseed field. *Soil Biology and Biochemistry, 31*, 1879–1887.

Orlowska, E., Zubek, S., Jurkiewicz, A., et al. (2002). Influence of restoration on arbuscular mycorrhiza of *Biscutella laevigata* L. (Brassicaceae) and *Plantago lanceolata* L. (Plantaginaceae) from calamine spoil mounds. *Mycorrhiza, 12*, 153–160.

Pandey, B., Agrawal, M., & Singh, S. (2016). Ecological risk assessment of soil contamination by trace elements around coal mining area. *Journal of Soils and Sediments, 16*, 159–168.

Patra, P., Pati, B. K., Ghosh, G. K., et al. (2013). Effect of bio-fertilizers and Sulphur on growth, yield, and oil content of hybrid sunflower (*Helianthus annuus* L.) in a typical lateritic soil. *2*, 603. https://doi.org/10.4172/scientificreports.603.

Pawlowska, T. B., Blaszkowski, J., & Rühling, A. (1996). The mycorrhizal status of plants colonizing a calamine spoil mound in southern Poland. *Mycorrhiza, 6*, 499–505.

Pena, R., Offermann, C., Simon, J., et al. (2010). Girdling affects ectomycorrhizal fungal (EMF) diversity and reveals functional differences in EMF community composition in a beech forest. *Applied and Environmental Microbiology, 76*, 1831–1841.

Poornima, M., Kumar, R. S., & Thomas, P. D. (2014). Isolation and molecular characterization of bacterial Strains from tannery effluent and reduction of chromium. *International Journal of Current Microbiology and Applied Sciences, 3*, 530–538.

Porcel, R., Aroca, R., & Ruiz-Lozano, J. M. (2012). Salinity stress alleviation using arbuscular mycorrhizal fungi: A review. *Agronomy for Sustainable Development, 31*(1), 181–200.

Querejeta, J. I., Egerton-Warburton, L. M., & Allen, M. F. (2009). Topographic position modulates the mycorrhizal response of oak trees to interannual rainfall variability. *Ecology, 90*, 649–662.

Rashid, M. I., Mujawar, L. H., Shahzad, T., et al. (2016). Bacteria and fungi can contribute to nutrients bioavailability and aggregate formation in degraded soils. *Microbiolgical Research, 183*, 26–41.

Ravnskov, S., Nybroe, O., & Jakobsen, I. (1999). Influence of an Arbuscular mycorrhizal fungus on *Pseudomonas fluorescens* DF57 in rhizosphere and hyphosphere soil. *The New Phytologist, 142*, 113–122.

Read, D. J. (1991). Mycorrhizas in ecosystems. *Experientia, 47*, 376–391.

Remy, W., Taylor, T. N., Hass, H., & Kerp, H. (1994). 4-hundred million year old vesicular-arbuscular mycorrhizae. *PNAS USA, 91*, 11841–11843.

Rillig, M. C. (2004a). Arbuscular mycorrhizae and terrestrial ecosystemprocesses. *Ecology Letters, 7*, 740–754.

Rillig, M. C. (2004b). Arbuscular mycorrhizae, glomalin and soil aggregation. *Canadian Journal of Soil Science, 84*(4), 355–363.

Rillig, M. C., Wright, S. F., Nichols, K. A., et al. (2001). Large contribution of arbuscular mycorrhizal fungi to soil carbon pools in tropical forest soils. *Plant and Soil, 233*, 167–177.

Rillig, M. C., Wright, S. F., Nichols, K. A., Schmid, W. F., & Torn, M. S. (2002). The role of arbuscular mycorrhizal fungi and glomalin in soil aggregation: Comparing effects of five plant species. *Plant and Soil, 238*, 325–333.

Ritz, K., & Young, I. M. (2004). Interaction between soil structure and fungi. *Mycologist, 18*, 52–59.

Robertson, G. P., & Swinton, S. M. (2005). Reconciling agricultural productivity and environmental integrity: A grand challenge for agriculture. *Frontiers in Ecology and the Environment, 3*, 38–46.

Rooney, D. C., Killham, K., Bending, G. D., et al. (2009). Mycorrhizas and biomass crops: Opportunities for future sustainable development. *Trends in Plant Science, 14*, 542–549.

Rosenzweig, C., Iglesias, A., Yang, X. B., et al. (2001). Climate change and extreme weather events: Implications for food production, plant diseases, and pests. *Global Change and Human Health, 2*, 90–104.

Ryan, M. H., & Graham, J. H. (2002). Is there a role for arbuscular mycorrhizal fungi in production agriculture? *Plant and Soil, 244*, 263–271.

Salminen, J., Anh, B. T., & van Gestel, C. A. M. (2001). Indirect effects of zinc on soli microbes via a keystone enchytraeid species. *Environmental Toxicology and Chemistry, 20*, 1167–1174.

Sambandan, K., Kannan, K., & Raman, N. (1992). Distribution of vesicular-arbuscular mycorrhizal fungi in heavy metal polluted soils of Tamil Nadu, India. *Journal of Environmental Biology, 13*, 159–167.

Sekabira, K., Oryem-Origa, H., Mutumba, G., et al. (2011). Heavy metal phytoremediation by *Commelina benghalensis* (L.) and *Cynodon dactylon* (L.) growing in urban stream sediments. *International Journal of Plant Physiology and Biochemistry, 3*, 133–142.

Selosse, M. A., Richard, R., XSimar, H., et al. (2006). Mycorrhizal networks: des liaisons dangereuses. *Trends in Ecology & Evolution, 21*, 621–628.

Sheoran, V., Sheoran, A. S., & Poonia, P. (2010). Soil reclamation of abandoned mine land by revegetation: A review. *International Journal of Soil, Sediment and Water, 3*, 1–20.

Shetty, K. G., Hetrick, B. A. D., Figge, D. A. H., et al. (1994). Effects of mycorrhizae and other soil microbes on revegetation of heavy metal contaminated mine spoil. *Environmental Pollution, 86*, 181–188.

Shu, W. S., Yeb, Z. H., Lana, C. Y., et al. (2002). Lead, zinc and copper accumulation and tolerance in populations of Paspalum distichum and Cynodon dactylon. *Environmental Pollution, 120*, 445–453.

Singh, A. N., Raghubanshi, A. S., & Singh, J. S. (2002). Plantations as a tool for mine spoil restoration. *Current Science, 82*, 1436–1441.

Singh, P. K., Singh, M., & Tripathi, B. N. (2013). Glomalin: an arbuscular mycorrhizal fungal soil protein. *Protoplasma, 250*(3), 663–669.

Smith, S. E., & Read, D. J. (1997a). *Mycorrhizal symbiosis*. New York: Academic.

Smith, S. E., & Read, D. J. (1997b). *Mycorrhizal symbiosis*. London: Academic.

Smith, S. E., & Read, D. J. (2008a). *Mycorrhizal symbiosis*. San Diego: Academic.

Smith, S. E., & Read, D. J. (2008b). *Mycorrhizal symbiosis* (3rd ed.). London: Academic.

Šourková, M., Frouz, J., & Šantrůčková, H. (2005). Accumulation of carbon, nitrogen and phosphorus during soil formation on alder spoil heaps after brown-coal mining, near Sokolov (Czech Republic). *Geoderma, 124*, 203–214.

Susarla, S., Medina, V. F., & McCutcheon, S. C. (2002). Phytoremediation: An ecological solution to organic chemical contamination. *Ecological Engineering, 18*, 647–658.

Swaty, R., Michael, H. M., Deckert, R., & Gehring, C. A. (2016). Mapping the potential mycorrhizal associations of the conterminous United States of America. *Fungal Ecology*, 1–9.

Thompson, J. P. (1991). Improving the mycorrhizal condition of the soil through cultural practices and effects on growth and phosphorus uptake in plants. In C. Johansen, K. K. Lee, & K. L. Sahrawat (Eds.), *Phosphorus nutrition of grain legumes in the semi-arid tropics* (pp. 117–137). Andhra Pradesh: ICRISAT.

Tilman, D., Lehman, C. L., & Thomson, K. T. (1997). Plant diversity and ecosystem productivity: Theoretical considerations. *Proceedings of the National Academy of Sciences of the United States of America, 94*, 1857–1861.

Tonin, C., Vandenkoornhuyse, P., Joner, E. J., et al. (2001). Assessment of arbuscular mycorrhizal fungi diversity in the rhizosphere of *Viola calaminaria* and effect of these fungi on heavy metal uptake by clover. *Mycorrhiza, 10*, 161–168.

Toro, M., Azcón, R., & Barea, J. M. (1997). Improvement of arbuscular mycorrhizal development by inoculation with phosphate solubilizing rhizobacteria to improve rock phosphate bioavailability (^{32}P) and nutrient cycling. *Applied and Environmental Microbiology, 63*, 4408–4412.

Trenberth, K. E., Dai, A., van der Schrier, G., et al. (2013). Global warming and changes in drought. *Nature Climate Change, 4*, 17–22.

Tscherko, D., Hammesfahr, U., Marx, M. C., et al. (2004). Shifts in rhizosphere microbial communities and enzyme activity of Poa alpina across an alpine chronosequence. *Soil Biology and Biochemistry, 36*, 1685–1698.

Tubiello, F. N., Rosenzweig, C., Goldberg, R. A., et al. (2002). Effects of climate change on us crop production: Simulation results using two different GCM scenarios. Part I: Wheat, potato, maize, and citrus. *Climate Research, 20*, 259–270.

Turnau, K., Kottke, I., & Oberwinkler, F. (1993). Element localization in mycorrhizal roots of *Pteridium aquilinum* (L.) Kuhn collected from experimental plots treated with cadmium dust. *The New Phytologist, 123*, 313–324.

Turnau, K., Jurkiewicz, A., & Lingua, G. (2005). Role of arbuscular mycorrhiza and associated microorganisms in phytoremediation of heavy metal polluted sites. In M. N. V. Prasad, K. S. Sajwan, & R. Naidu (Eds.), *Trace elements in the environment* (pp. 235–252). Boca Raton: CRC Press.

Turnau, K., Anielska, T., & Ryszka, P. (2008). Establishment of arbuscular mycorrhizal plants originating from xerothermic grasslands on heavy metal rich industrial wastes-new solution for waste revegetation. *Plant and Soil, 305*, 267–280.

Ul Hassan, Z., Ali, S., Rizwan, M., et al. (2017). *Role of bioremediation agents (bacteria, fungi, and algae) in alleviating heavy metal toxicity* (Probiotics in agroecosystem, pp. 517–537). Singapore: Springer.

UN. (2000). UN Secretary General's report A/504/2000 Chapter C. "Defending the Soil".

Upadhyaya, H., Panda, S. K., Bhattacharjee, M. K., et al. (2010). Role of arbuscular mycorrhiza in heavy metal tolerance in plants: Prospects for phytoremediation. *Journal of Phytology, 2*, 16–27.

Valdenegro, M., Barea, J. M., & Azcòn, R. (2001). Influence of arbuscular mycorrhizal fungi, *Rhizobium meliloti* strains and PGPR inoculation on the growth of *Medicago arborea* used as model legume for revegetation and biological reactivation in a semi-arid Mediterranean area. *Plant Growth Regulation, 34*, 233–240.

Van der Heijden, M. G. A., & Horton, T. R. (2009). Socialism in soil? The importance of mycorrhizal fungal networks for facilitation in natural ecosystems. *Journal of Ecology, 97*, 1139–1150.

Van der Heijden, M. G. A., Klironomos, J. N., Ursic, M., et al. (1998). Mycorrhizal fungal diversity determines plant biodiversity, ecosystem variability and productivity. *Nature, 396*, 69–72.

Van der Heijden, M. G. A., Bardgett, R. D., & van Straalen, N. M. (2008). The unseen majority: Soil microbes as drivers of plant diversity and productivity in terrestrial ecosystems. *Ecology Letters, 11*, 296–310.

Van der Heijden, M. G. A., Martin, F. M., Selosse, M. A., et al. (2015). Mycorrhizal ecology and evolution: The past, the present, and the future. *New Phytologist, 205*, 1406–1423.

Vivas, A., Vörös, A., Biro, B., et al. (2003). Beneficial effects of indigenous Cd tolerant and Cd-sensitive *Glomus mosseae* associated with a Cd-adapted strain of *Brevibacillus* sp. in improving plant tolerance to Cd contamination. *Applied Soil Ecology, 24*, 177–186.

Wall, D. H., & Virginia, R. A. (1999). Controls on soil biodiversity: Insights from extreme environments. *Applied Soil Ecology, 13*, 137–150.

Wang, F. Y., Lin, X. G., & Yin, R. (2007). Role of microbial inoculation and chitosan in phytoextraction of Cu, Zn, Pb and Cd by *Elsholtzia splendens* – a field case. *Environmental Pollution, 147*, 248–255.

Weissenhorn, I., Leyval, C., & Berthelin, J. (1993). Cd-tolerant arbuscular mycorrhizal (AM) fungi from heavy metal-polluted soils. *Plant and Soil, 157*, 247–256.

Weissenhorn, I., Leyval, C., & Berthelin, J. (1995a). Bioavailability of heavy metals and abundance of arbuscular mycorrhiza in a soil polluted by atmospheric deposition from a smelter. *Biology and Fertility of Soils, 19*, 22–28.

Weissenhorn, I., Leyval, C., Belgy, G., et al. (1995b). Arbuscular mycorrhizal contribution to heavy metal uptake by maize (*Zea mays* L.) in pot culture with contaminated soil. *Mycorrhiza, 5*, 245–251.

Weyens, N., Truyens, S., Dupae, J., et al. (2010). Potential of the TCE-Degrading endophyte *Pseudomonas putida* W619-TCE to improve plant growth and reduce TCE phytotoxicity and evapotranspiration in poplar cuttings. S. *Environmental Pollution, 158*, 2915–2919.

Widden, P. (1996). The morphology of vesicular-arbuscular mycorrhizae in *Clintonia borealis* and *Medeola virginiana*. *Canadian Journal of Botany, 74*, 679–685.

Wright, S. F., & Upadhyaya, A. (1998). A survey of soils for aggregate stability and glomalin, a glycoprotein produced by hyphae of arbuscular mycorrhizal fungi. *Plant and Soil, 198*, 97–107.

Yang, G., Liu, N., Lu, W., et al. (2014). The interaction between arbuscular mycorrhizal fungi and soil phosphorus availability influences plant community productivity and ecosystem stability. *Journal of Ecology, 102*, 1072–1082.

Yao, Q., Li, X., Weidang, A., et al. (2003). Bi-directional transfer of phosphorus between red clover and perennial ryegrass via arbuscular mycorrhizal hyphal links. *European Journal of Soil Biology, 39*, 47–54.

Yasmeen, T., Hameed, S., Tariq, M., et al. (2012). Vigna radiata root associated mycorrhizae and their helping bacteria for improving crop productivity. *Pakistan Journal of Botany, 44*, 87–94.

Zaidi, A., & Khan, M. S. (2005). Interactive effect of Rhizotrophic microorganisms on growth, yield and nutrient uptake of wheat. *Journal of Plant Nutrition, 28*, 2079–2092.

Zak, J. C., & Parkinson, D. (1982). Initial vesicular-arbuscular mycorrhizal development of the slender wheatgrass on two amended mine soils. *Canadian Journal of Botany, 60*, 2241–2248.

Zhu, Y. G., & Miller, R. M. (2003). Carbon cycling by arbuscular mycorrhizal fungi in soil—Plant systems. *Trends in Plant Science, 8*, 407–409.

A Systematic Review on the Role of Mycorrhiza in Soil Genesis Using Scientometrics Analysis

6

Geetanjali Baruah and Jagajjit Sahu

1 Introduction

Mycorrhiza is a beneficial relationship between higher plant roots and a fungus colonizing those roots internally or on the root surfaces by which both the associations draw benefits, fungi provides the necessary minerals and water from the soil and the plant provides the photosynthesized food to its associated fungus. Although roots carry out the functions related to water and nutrient uptake, there are certain limitations with respect to various factors like distance covered, the nutrient form of minerals. The term Mycorrhiza, a Greek word, can be broken into two parts i.e. 'mykes' means fungus and 'rhiza' translates to root. Plants which grow in soils or environments with the abundant presence of water and nutrients may not have mycorrhiza which means not all higher plants need such mutualism. Mycorrhiza is of two types based on their site of colonization namely ectomycorrhiza and endomycorrhiza. Ectomycorrhiza is generally found in association with woody plants (generally 5–10%) acknowledged by the presence of a condensed hyphal sheath called mantle around the root surface. In contrast, endomycorrhiza has their presence in a vast category of plant species which includes crops to greenhouse plants. This type of association is based on the penetration of the cortical cells and the formation of arbuscules and vesicles by the fungi. Endomycorrhiza is further subdivided into specific types: *Arbuscular mycorrhiza* (AM), *Ericaceous mycorrhizae*, *Arbutoid mycorrhiza*, and *Orchidaceous mycorrhizae*.

G. Baruah
Department of Agricultural Biotechnology, Assam Agricultural University, Jorhat, Assam, India

J. Sahu (✉)
Department of Mycology and Plant Pathology, Institute of Agricultural Sciences, Banaras Hindu University, Varanasi, India

© Springer Nature Singapore Pte Ltd. 2019

A. Varma, D. K. Choudhary (eds.), *Mycorrhizosphere and Pedogenesis*,
https://doi.org/10.1007/978-981-13-6480-8_6

Plants being sessile organisms will always have a dependency on beneficial microbes with special reference to mycorrhizal associations for daily uptake of minerals and water for photosynthesis process. Moreover, mycorrhizal hyphae improve the growth of the host plant, helps survive interplant competition, converts different nutrient forms to the ones that can easily be absorbed by plants, provides biotic stress resistance, confers drought resistance and offers better scope of water acquisition by extending the branches of the roots in a manner which could have been impossible without the associated fungi (Jansa et al. 2013). Therefore, being mutually beneficial mycorrhizal associations have a unique and very important role in the nutrient cycling of soil together with the capability of remediating pollutants into useful minerals thereby bearing significance in environmental health as well.

Soil health, also termed as soil quality, if defined can be framed as the continued capacity of the soil to sustain plants, animals, and humans in the ecosystem which is why it represents a very important aspect of maintaining soil health to make sustainable agricultural production for the future generations. When proficient and eco-friendly ways to help sustain the soil health are explored the role of Mycorrhiza cannot be denied. The term "mycorrhizosphere", defining the rhizosphere surrounding and influenced by Mycorrhiza (Linderman 1988) has a role in stabilizing the soil matrix via mycorrhizal-induced soil aggregation. The penetration of roots and fungal hyphae into soil micro pores infers to the proliferation of the mycorrhizosphere which has roles in the enhancement of the soil's aggregation qualities and thereby improves the water retention and nutrient-holding capacity of the soil (Audet 2014).

Hence, mycorrhizosphere along with its role in soil health bears significant research interest as this is the region around a mycorrhizal fungus of interest, in which nutrients are released from the fungal activity to increase its population and activities. Being such an important topic and the availability of a number of reports in this aspect drove us to conduct a scientometric analysis to look deeper into the works done and the approaches used in the past.

Though scientometrics is not a new field, it is certainly an emerging one because of the huge growth of literature data and efficient data analytics approaches. Scientists of this era are preferring this approach over the manual methods of reviewing the available reports. The explosion of information in the form of electronic documents have made it difficult for selection of proper and most related reports of interest. Scientometrics not only provides a systematic review but also too many important information which is associated with statistical parameters. Databases such as PubMed, Scopus and WoS stores the bibliometric information for a number of journals (Falagas et al. 2008). PubMed is a more common literature database which keeps a record of most of the publications in the field of life science from MEDLINE. The Elsevier's abstract and citation database, commonly known as Scopus came into the picture in 2004 which stores publication information for journals in life sciences, social sciences, and physical sciences. WoS is maintained by Clarivate Analytics (United States) and provides access to the reports from all fields of science indexed by Thomson Reuter's Science Citation Index. WoS is the most widely used indexing system and trusted by a maximum number of scientists as an authentic source of publication data source. In this chapter, we provide a review of the key works done related to mycorrhiza in soil genesis with the support of results from scientometric analysis using WoS data. The entire analysis was performed using R, the most used data analysis tool of today.

2 Scientometric Analysis

We used R scripts and Bioconductor packages such as bibliometrix to perform the scientometric analysis (Aria and Cuccurullo 2017). R is an open-source statistical package and full-fledged programming tool. The analysis provided with so much of information related to authors, affiliation, annual growth, collaboration, etc. Mycorrhiza was used as a keyword and searched on WoS server and then screened to select 1477 records based on their classification under Soil science. The bibliometric raw data was collected in the bibtex format with "bib" extension. This was directly submitted to an in-house R wrapper script to perform scientometrics. The output showed published reports from 1989 to 2018 in 47 journals with an average of 42.23 publications per year. The number of publications ranged from a lowest of 15 in the year 1989 to highest of 72 in 2013 (Fig. 6.1). In 2018, there are 33 publications and a simple assumption is that it will be difficult to be the highest as only 3 months left for the year 2018 to be over and the current number is quite less than for the year 2013. The publication trend observed is not of a proper growth, rather a mixed one. The year 2012 consecutive to 2013 contains the second highest number of reports.

We concentrated mostly on topical classification to observe the research areas in which all the reports have been distributed. The reports selected for the analysis

Fig. 6.1 Bar graph for number of publications used in this study for scientometric analysis (from the year 1989 to August 2018). The number of publications have also been mentioned on individual bars

have already gone through the first stage of screening which is availability under the WoS classification of soil science. These records were again classified using WoS sub-classification system in R and a total of 7 categories; AGRICULTURE & MULTIDISCIPLINARY, AGRONOMY, APPLIED & ANALYTICAL CHEMISTRY, ECOLOGY, ENVIRONMENTAL SCIENCES, PLANT SCIENCES, and WATER RESOURCES & MULTIDISCIPLINARY. A Venn diagram was constructed to depict all the categories and a number of reports belonging to the individual and more than one categories (Fig. 6.2). As there were a number of papers which did not fall into the above categories but certainly belong to the broader classification group SOIL SCIENCE, we kept it as the universal set. The universal set contained the total number of reports selected at the beginning which is 1477, among which 676 records fall into no subcategories. The highest number of papers was found in the sub-category, PLANT SCIENCE with 657 whereas WATER RESOURCES & MULTIDISCIPLINARY contained only 5 records which is the lowest. We considered the most related sub-categories that are AGRICULTURE & MULTIDISCIPLINARY, ECOLOGY, and ENVIRONMENTAL SCIENCES for further manual curation and review. The total number of publications selected was

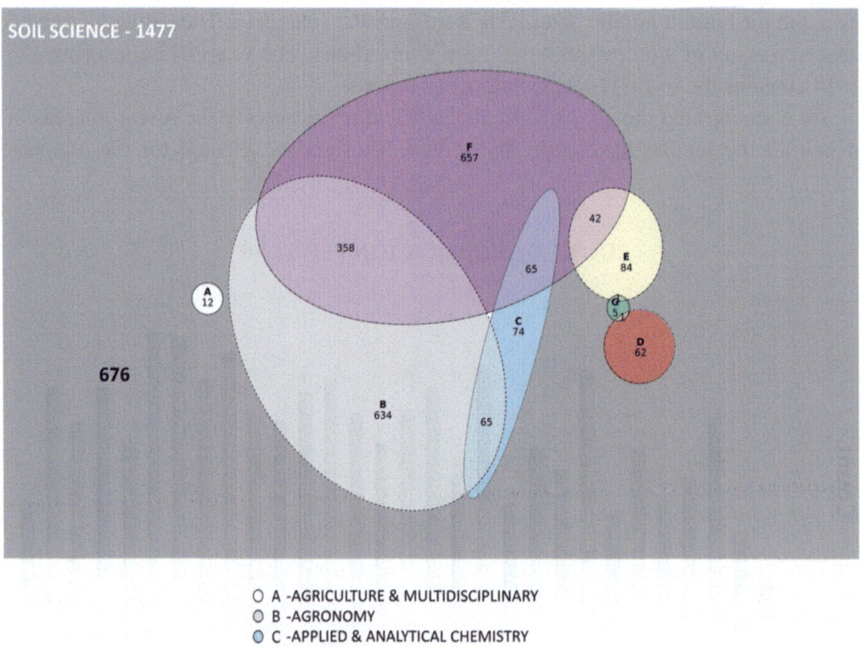

Fig. 6.2 A Venn diagram representing all the 7 sub-categories inside the universal set SOIL SCIENCE. The legends for the sets have been mentioned on the bottom of the graph along with the color of the respective ellipse and single letter representation

158 distributed as 12, 62 and 84, across the above categories respectively. Interestingly we did not find any common records for these categories. These records were checked manually and the key publications have been included in this chapter.

3 Key Works in Selected Categories

3.1 Agriculture & Multidisciplinary

In relation to agriculture, soil quality or the role of soil types is an integral part of it. Mycorrhizosphere, on the other hand is part and parcel of the soil ecology as approximately 80% of crop's roots are inhabited by the beneficial fungi associations. Therefore, in a category which we termed as AGRICULTURE & MULTIDISCIPLINARY, an overview of the research scenario through scientometric analysis has been given below.

Many aspects are taken into consideration with respect to research scopes to decipher the role and contribution of mycorrhiza in agriculture whether be it positive or negative. Mycorrhiza has the capability to help survive its hosts from adverse conditions, also change in soil texture, soil health, different crop rotation strategies and nature of fertilizers used has impacts on mycorrhizal colonization efficiency and proliferation. In one of the study in mid-nineties to see if sewage sludge-treated soil has any role in colonization of Mycorrhiza in different plants revealed that an increased heavy metal concentration in sewage has the negative impact on beneficial fungal infectivity in the roots of the plants grown in such soils (Loth and Hofner 1995). In two categories of no-tillage citrus orchards in Southern China; one covered with natural grass and the other with herbicide treatment, a notable variation in spore density, rhizospheric microbe populations and enzyme activities, hyphal length density, catalase activity and phosphatase activity and percentage of root length with arbuscules (RLA) has been seen in case of the combination of natural grass cover/no-tillage. Also, analysis showed a positive correlation between the hyphal length density and organic matter deposited in the orchard soil (Wang et al. 2011).

In traditional methods of Agriculture, Crop rotation is one of the practices of growing different types of crops in the same farming land in a sequenced manner, to maintain the soil fertility and to prevent the deposition of only one set of nutrients. It also helps in the reduction of soil erosion and increases crop yield. The effects of wheat and lentil pre-cropping on *Zea mays* L. showed better growth of that maize plants in case of lentil pre-cropping than the maize plants that had been cultivated after wheat cultivation in an experiment conducted in Southeastern Turkey. At the time of harvest; determination of plant dry weight, root length, P and Zn concentrations in plant tissues along with the extent of root colonization by arbuscular mycorrhiza were done, based on which the performance of maize was judged. Together with the crop rotation strategy, they have also employed a gradually increased level (0, 20, 40 and 80 mg kg^{-1}) of P fertilizers and in both the cases a collinear increase in P in plant tissues have been observed. They did not see any significant difference in mycorrhizal colonization in both the cases of pre-cropping though (Almaca and Ortas 2010).

Root and soil samples of dominant halophytes (*Artemisia santonicum*, *Aster tripolium*, *Festuca pseudovina*, *Lepidium crassifolium*, *Plantago maritima* and *Puccinellia limosa*) from four different locations of Hungary with saline soils were examined to decode the relation between soil salinity and endomycorrhizal colonization. As a result, a lack of symbiotic relationship in saline soil has been observed i.e. at increasing salt concentrations, mycorrhizal growth ceases. Another interesting finding showed a positive correlation between the mycorrhiza colonization and the plenty of oligotrophic bacteria known to of stable (k-strategist) group (Fuzy et al. 2010).

Wildfire is very well known to have roles in changing soil texture and nutrient composition. In unmanaged soils with fire history in Northern Portugal was evaluated to have a perception for development of strategic restoration programs by studying the mycorrhizal colonization in plants grown there. They found that the mycorrhizal propagules that survived fire were not sufficient for effective root colonization as in situ inoculum but correlations between soil nutrients and mycorrhizal parameters were found in this study explaining some role of mycorrhizal associations in fire-affected soil (Dias et al. 2010).

Mycorrhizal fungi have been proven to be useful in many plant categories including ornamental horticultural crops. In an attempt to test the effect of the application of Mycorrhiza on the growth of two cultivars of lisianthus (*Eustoma grandiflorum*) resulted in improved plant growth and flexibility towards abiotic and biotic stresses which helped to increase such application in diverse ornamental crops (Meir et al. 2010). Mycorrhiza application in the horticultural production of cucumber in the Eastern Mediterranean region of Turkey has been tested and found that it prominently increased the survival of the cucumber seedling, fruit yield, shoot concentrations of minerals P and Zn. Indigenous mycorrhiza inoculum was successful in colonizing plant roots and resulted in better survival and yield (Ortas 2010).

As mentioned earlier, mycorrhizal association confers resistance to crops with such mutualism against biotic as well as abiotic factors in the environment. The evidence of antagonistic effect imposed on soil-borne pathogens by root colonization with AM fungi by favoring the establishment of rhizobacteria over the pathogens opens a new chapter in the achievement of biotic stress resistance (Lioussanne 2010).

In an attempt to characterize the natural diversity among AM population by molecular techniques and taxonomic identification, from mediterranean sand dune ecosystems the seven species i.e. *Scutellospora persica*, *Glomus ambisporum*, *Glomus diaphanum*, *Glomus clarum*, *Glomus intraradices*, *Glomus microaggregatum* and *Gigaspora margarita* were identified (Koske and Walker 1986; Smith and Schenck 1985; Morton and Walker 1984; Nicolson and Schenck 1979; Schenck and Smith 1982; Koske and Gemma 1986; Becker and Hall 1976). Out of which the most abundant spores were of Glomus in extracted soil samples. Through molecular investigation, the most abundant fungi forming AM in the roots were found to be of the Gigasporaceae group followed by fungi of Glomus group A and Glomus group B. (Camprubi et al. 2010).

3.2 Ecology

As the general awareness regarding the disturbed ecological balance, a sense of responsibility is also increasing. Mycorrhizae are of global presence in terrestrial ecosystems, so, the role of this symbiotic association in plant population dynamics, formation of community structures and ecosystem operational activities has been grabbing the attention in the field of ecology. Over past decades, the thought process has been shifted to inquiries of their functional relevance in a broader ecological context from basic mycorrhizal biology.

The role of mycorrhiza in soil health is also contributed by its role towards the efflux of carbon-dioxide (CO2) in soil respiration (Pandey et al. 2010). The total global soil carbon pool is around three times the carbon present in the atmospheric pool and 4.5 times of the biotic pool (Gruber et al. 2004). In an effort to decode the influences of autotrophic and heterotrophic components on soil CO2 efflux resulted in the information that the contribution by rhizospheric respiration was found to be 36% ± 21%, the involvement by mycorrhizal respiration was 9% ± 9% while the contribution by heterotrophic respiration was 55% ± 21% on average (Papp et al. 2018).

Cover cropping is a famous way to enhance soil health in agriculture. There is an interconnected relationship among cover crops, soil health and soil microbial community has been found recently. It shows that AM fungi to be plentiful in oat and cereal rye cover crops, which means depending on the cover crop used particular microbiota will be seen in soil ecology (Finney et al. 2017). In an intense study in the fields of Argentina where soybean in rapeseed cultivation is mostly used for crop rotation, its effect on mycorrhizal colonization of soybean roots, nodulation and overall growth of the crop was assayed. As plants of the family Brassicaceae (ex. Rapeseed) do not have the association with AM fungi, an interesting finding correlating to the fact showed overall 30% decrease in the AMF soybean root colonization (Valetti et al. 2016). The role of arbuscular and ectomycorrhizal fungal inoculation in improving soil aggregate stability in revegetation strategies to get rid of soil erosion was studied using selected plant and mycorrhiza combination applied for a period of 9–10 months. Out of which the combination of *T. glauca* and *P. microcarpus* showed a significant increase in aggregate stability. Also, the average increase in all the other factors like aggregate stability, soil organic carbon, both above- and below-ground dry biomass, and root length density showed positive results. Therefore, the application of innate plant species with wisely chosen suitable mycorrhizal fungi presents a promising technique to support in the revegetation of barren lands (Demenois et al. 2017).

The role of arbuscular mycorrhiza *Rhizophagus intraradices* (Ri) and endophyte *Piriformospora indica* (Pi) in finger millet is compared under drought stress, it was found that although individual inoculation with both fungi have the capability to improve the growth and development of finger millet in drought-affected soils, cumulative inoculation shows a much-folded increase in drought tolerance via a stronger antioxidant defense system resulting in higher chlorophyll content, and an enriched osmoregulatory network (Tyagi et al. 2017). AM fungus colonization enriched plant growth and development in Citrus under drought stress by indirectly affecting the soil moisture retention through the effect of glomalin, a glycoprotein produced plentifully on hyphae and spores of AM, on soil water-stable wholes, keeping in mind the direct mineral nutritional (Wu et al. 2008). The saline soil has been credited for lowering of mycorrhizal activity around plant roots. When the saline soil of Urmia lake of northern Iran was examined it was proven that moderate salinity promotes vesicle formation in *Medicago sativa*. In contrast, arbuscle formation in *Allium cepa* and salinity was found to be negatively correlated signifying less symbiotic activity in saline soils (Barin et al. 2013).

With the increasing awareness generated regarding ecological imbalance is providing the opportunity to human to hypothesize various strategies to maintain the natural balance. As a result, the usage of indigenous and local tree species to reverse the deforested land scenario has been increased. It simultaneously increased the concern regarding its effect on the microbial community of the lands of interest. In a study conducted in West Africa, the forest systems made with anthropological effort showed an increase in the legume-nodulating rhizobia as well as increased phosphorus and nitrogen amount in the soil. But in case of the AM fungal groups, their number of taxa significantly went down in the man-made forests when compared with the deforested land pointing out the possible negative impact of using indigenous and native tree species for such purpose (Sene et al. 2012). Combined application of AM fungi has been proven to be better than individual application of AMs. In a greenhouse experiment involving four AM fungi taxa namely *Glomus claroideum*, *Glomus geosporum*, *Glomus intraradices* and *Glomus mosseae* when applied singly as well as a mixture showed a different level of effects on growth and biomass production of four grassland species proving mixed application as a better option (Zaller et al. 2011). Experiments to figure out the cumulative effect of mycorrhiza and earthworms in the growth and development of maize crops and phytoremediation of cd-polluted soils were conducted. It brought conclusion crediting the reason for much enhanced tolerance to Cd toxicity of maize to the contribution of both individual and interactive action of mycorrhiza and earthworms which also resulted in improved plant growth and Phosphorus nutrition, and constraining the Cd transfer to biomass (Aghababaei et al. 2014).

Although the importance AM associations with plant roots has been emphasized every time, it is a well-known fact that for a sustainable agricultural strategy integrated approach including inoculation of AM fungi, earthworms, rhizobacteria, well planned crop rotation results in a higher accumulation of nitrogen, better yield, proper accumulation of biomass in the above ground parts of the crop of interest can be achieved conferring the positive impact of agroecosystem biodiversity (Zarea et al. 2009).

The positive impact of mycorrhizal association against biotic stress was proved by antagonistic effect of soil inhabiting AM fungi and *Pseudomonas fluorescens* on Root-rot of *Phaseolus vulgaris* caused by *Rhizoctonia solani* in the northern India. Combined application of both these microbial group resulted in better disease resistance as well as a significant increase in yield although individual microbial category too showed almost similar enhancements but to a lesser extent. On the other hand, mixed inoculations with mustard oil cake were recommended as the best solutions for root-rot management (Neeraj and Singh 2011).

AM fungi are found almost in more than 80% of crops and hence is often said as the ubiquitous inhabitants of soil ecology. The role of AM in the nutrient cycles of P and C is also well recognized. The role of environmental factors affecting the community and efficiency of AM has now becoming an important aspect as mycorrhizal infectivity assay prominently could function as a comprehensive and worldwide indicator of soil health. Experiments to determine the effect of the environmental components in physico-chemical, biological, and geographic parameters of soils showed a negative correlation between the soil P and the AM establishment in the plants (Jansa et al. 2009).

3.3 Environmental Sciences

The benefits of knowing the role of mycorrhizal associations in improving the polluted lands and in help combating the host plants from abiotic stresses, especially drought, imposed due to rapid changes in the environment as a result of various anthropogenic activities provides infinite chances for the formulation of such mitigation strategies. Some of the approaches of the implication of mycorrhiza for the environmental benefit are discussed below.

Heavy metal (HM) deposition in soil is a very important aspect of environmental pollution which eventually results in plant tissue toxicity due to HM accumulation in the plant cells. Mycorrhiza, being a lucrative option for sustainable soil restoration, has been in use for reducing the concentration of Cd, Co, Cr, Cu, Ni, Pb, and Zn. The role of HMs in remediation of contaminated soil is tested further with mycorrhized as well as non-mycorrhized *Helichrysum italicum*. Results showed to have the role of mycorrhized ones in minimizing the phyto-toxicity caused by HMs with a predicted exclusion mechanism provided by the mycorrhiza with improvement in plant health as well as better soil quality (Brunetti et al. 2018).

Mycorrhizal effect on solubilizing of soil P increasing its availability to the plants are very well recognized but the impact of long-term usage of P fertilizers have not grabbed much attention from the researchers. In a study conducted in Uruguay, where excessive use of P fertilizers in farmlands is reported due to naturally low P availability in soil. However, no adverse effect in AM fungi biodiversity has been reported due to the escalation in available P in the rhizosphere (Garcia et al. 2017).

Application of swine slurry is a popular fertilizer in agricultural lands of many countries. But, its effect on soil nutrition as well as on AM fungi is suspected to be negative. In an experiment designed to determine the exact roleplay of long-term swine slurry application has been proven to be harmful against AM colonization without much disturbance created towards soil health and thus crop productivity (Balota et al. 2016).

There are reports saying most of the turfgrass species are the inhabitants of AM fungi, but no detailed facts are known about it. When individual and mixed inoculation of AMs *Glomus intraradices* Schenck & Smith, *Glomus etunicatum* Becker & Gerdemann, and *Glomus deserticola* Trappe & John, on two turfgrasses, *Poa pratensis* L. and *Festuca arundinacea* Schreb., it resulted in grander biomass production, yield and better nutrient uptake (especially P) to that of non-treated plants (Elhindi et al. 2018).

During adversity in the surrounding, crops get affected in many layers. In a greenhouse investigation on *Capsicum annuum* L. plants, concerning effect of saline soil and availability of P on growth parameters, mineral uptake and trickling of ions, chlorophyll content, level of soluble sugar and proline and alkaline phosphatase activity in presence or absence of AM fungi; interesting outcomes are found. The results showed that plants with mycorrhizal associations are better survivors of adverse environments with better growth rates and higher membrane integrity under salt stress than the ones with the absence of mycorrhiza (Beltrano et al. 2013). Under drought stress, inoculation of AM fungi (e.g. *Glomus intraradices*) in the seedlings of sorghum showed enhanced drought tolerance providing growth, increased photosynthesis thereby increasing yield, and stomatal conductance which were not observed in un-colonized plants. (Ibrahim et al. 1990).

As an added advantage of the application of mycorrhizal fungi is for its role played in preventing wind mediated soil erosion. Wind induced soil loss decreased significantly with increasing percentage of root colonization by mycorrhizal fungi in two plant species *Lolium perenne* and *Anthyllis vulneraria* with an increased biomass production in the above ground parts (Burri et al. 2013). Shukla and team while studying the distribution of mycorrhiza and inoculation capacity across soil depth with a simultaneous change in soil pH and soil humidity in *Ocimum sanctum* and *Withania somnifera* found that it is dependent on the soil depth. The rationale cause was stated as due to the presence of less number of roots fewer mycorrhiza is found. Also, other factors like variation in soil pH and soil moisture through soil depths has some role to play in the mycorrhizal community (Shukla et al. 2013).

Soil respiration (Rs) is the second largest carbon flux in most of the ecosystems accounting for 60–90% of the total ecosystem respiration where mycorrhizal fungi play critical roles in its regulation as almost 90% of plant species are found to be forming mycorrhizal associations (Longdoz et al. 2000; Zhu and Miller 2003; Smith and Read 2008). The change in environmental factors (temperature sensitivity) throughout the year and biotic factors like leaf area index (LAI) have roles in the regulation of Rs Different mycorrhiza showed different coping strategy with respect to change in temperature profile as well as the change in precipitation rates and thus found to be variable across different mycorrhizal associations (Shi et al. 2012).

The positive role of different cropping strategies is never denied in agriculture. The effect of a cover crop farming system involving white clover living mulch is when studied for its role in the proliferation of Mycorrhiza; it has been found to increase the P uptake of the crop of interest by encouraging arbuscular mycorrhizal establishment by native fungi (Deguchi et al. 2012).

With the increased demands of the livestock industry, farmers are relying on chemical fertilizers for the fulfillment of the yield and fodder demands. Under draught stress in presence of a sewage sludge pollutant, the role of AM fungi symbiosis has been studied to decipher the mechanisms behind the tolerance provided by this association to the host plant. AM fungi are found to play a role in overcoming the harsh effects of draught stress by improving water relation and in role of the enzymes involved which was not found in case of non-mycorrhizal plants by causing decreased oxidative damage to rhizosphere (Khalvati et al. 2010). As per the global need to increase the production of food, there is an emergent need to make use of the barren lands. Heavy metal contaminated sites have very less capability to support plant growth due to their poor soil texture, low water-holding capacity, lack of nutrients and essential minerals. In order to make such lands capable of supporting plant life, application of mycorrhizal propagules due to their ability to provide host plants with heavy metal tolerance and avoidance to hydro stress emerges as a beneficial and cheap solution (Medina and Azcon 2010).

The role of volcanic ash-derived soils is well recognized in improving the economies of many countries especially in Asia, Africa, and America. In Chile, half of the arable soils have origins from volcanic ashes. This study showed the prominent role of AM fungi in the storage of higher amounts of organic carbon and provides a basis for further studies regarding the management and application of the most suitable AM fungal propagules (Borie et al. 2010). Re-establishment of wastelands into agriculturally beneficial land area requires much effort. In a survey carried out in the SE coast of Spain with respect to seasonal variations, the harsh effects of drastic changes soil environment affect mycorrhizal population minimizing its colonization efficiency and reduction in spore numbers. But the survival of mycorrhizal propagules was seen proposing the development of adaptation levels to such stress, making inoculation of propagules in barren lands a promising solution (Diaz and Honrubia 1994).

Mycorrhiza is also applicable in biotransformation of toxic chemicals as it is suspected to play a role in providing bio methylation processes mechanisms to convert toxic metals into bio available forms would favor the detoxification of As, especially at the interface of plant roots and rhizospheric soil seen in sunflower plants with arsenic contamination (Ultra et al. 2007).

4 Conclusion

Scientometrics, a powerful data mining approach has been used in several studies in the last 5 years. Our study also used the same technique to reveal important classifications which have been discussed elaborately. Use of R has only enhanced the approach which can be reproduced easily. With the increasing demand for food supply, a sense of responsibility for fulfilling the growing need has emerged. Although, there have been many schemes proposed keeping in mind about the current scenario which involves various fields from conventional agricultural approaches such as plant breeding to a cutting-edge approach of application of biotechnological tools; application of mycorrhizal fungi still stands its ground well. This chapter provides a basic idea about timeline of works done on Mycorrhiza with reference to Pedogenesis. From the application of AM fungi in infertile deforested lands to improve the soil nutrition value; diverse remediation strategies involving mycorrhiza such as conversion of heavy metals into bioavailable forms till reducing the phytotoxicity levels in crops due to HMs it plays diverse roles. Also, native mycorrhizal colonies together with earthworm and indigenous bacterial colonies help in increasing the crop yield, plant growth and development by providing better mineral nutrition. The research works based on mycorrhizal application for better crop quality, and improved soil health are very much diverse and this makes mycorrhizal application in all the fields a very useful and easy way for any such improvement.

Acknowledgements We offer our sincere regards to Department of Mycology and Plant Pathology, Institute of Agricultural Sciences, Banaras Hindu University, Varanasi, India for the workspace to conduct the scientometric analysis. We also acknowledge WoS for the raw bibliometric data.

References

Aghababaei, F., Raiesi, F., & Hosseinpur, A. (2014). The significant contribution of mycorrhizal fungi and earthworms to maize protection and phytoremediation in cd-polluted soils. *Pedobiologia, 57*, 223–233.

Almaca, A., & Ortas, I. (2010). Growth response of maize plants (Zea mays L.) to wheat and lentil pre-cropping and to indigenous mycorrhizae in field soil. *Spanish Journal of Agricultural Research, 8*, S131–S136.

Aria, M., & Cuccurullo, C. (2017). Bibliometrix: An R-tool for comprehensive science mapping analysis. *Journal of Informetrics, 11*, 959–975.

Audet, P. (2014). Arbuscular Mycorrhizal Fungi and metal phytoremediation: Ecophysiological complementarity in relation to environmental stress. In *Emerging technologies and management of crop stress tolerance* (Vol. 2). San Diego: Academic. https://doi.org/10.1016/B978-0-12-800875-1.00006-5.

Balota, E. L., Machineski, O., Honda, C., Yada, I. F. U., Barbosa, G. M. C., Nakatani, A. S., & Coyne, M. S. (2016). Response of arbuscular mycorrhizal fungi in different soil tillage systems to long-term swine slurry application. *Land Degradation & Development, 27*, 1141–1150.

Barin, M., Aliasgharzad, N., Olsson, P. A., Rasouli-Sadaghiani, M. H., & Moghddam, M. (2013). Abundance of arbuscular mycorrhizal fungi in relation to soil salinity around Lake Urmia in northern Iran analyzed by use of lipid biomarkers and microscopy. *Pedobiologia, 56*, 225–232.

Becker, W. N., & Hall, I. R. (1976). Gigaspora margarita, a new species in the Endogonaceae. *Mycotaxon, 4*, 155–160.

Beltrano, J., Ruscitti, M., Arango, M. C., & Ronco, M. (2013). Effects of arbuscular mycorrhiza inoculation on plant growth, biological and physiological parameters and mineral nutrition in pepper grown under different salinity and p levels. *Journal of Soil Science and Plant Nutrition, 13*, 123–141.

Borie, F., Rubio, R., Morales, A., Curaqueo, G., & Cornejo, P. (2010). Arbuscular mycorrhizae in agricultural and forest ecosystems in Chile. *Journal of Soil Science and Plant Nutrition, 10*, 185–206.

Brunetti, G., Ruta, C., Traversa, A., d'Ambruoso, G., Tarraf, W., De Mastro, F., De Mastro, G., & Cocozza, C. (2018). Remediation of a heavy metals contaminated soil using mycorrhized and non-mycorrhized Helichrysum italicum (roth) don. *Land Degradation & Development, 29*, 91–104.

Burri, K., Gromke, C., & Graf, F. (2013). Mycorrhizal fungi protect the soil from wind erosion: A wind tunnel study. *Land Degradation & Development, 24*, 385–392.

Camprubi, A., Calvet, C., Cabot, P., Pitet, M., & Estaun, V. (2010). Arbuscular mycorrhizal fungi associated with psammophilic vegetation in Mediterranean coastal sand dunes. *Spanish Journal of Agricultural Research, 8*, S96–S102.

Deguchi, S., Uozumi, S., Touno, E., Kaneko, M., & Tawaraya, K. (2012). Arbuscular mycorrhizal colonization increases phosphorus uptake and growth of corn in a white clover living mulch system. *Soil Science and Plant Nutrition, 58*, 169–172.

Demenois, J., Rey, F., Stokes, A., & Carriconde, F. (2017). Does arbuscular and ectomycorrhizal fungal inoculation improve soil aggregate stability? A case study on three tropical species growing in ultramafic ferralsols. *Pedobiologia, 64*, 8–14.

Dias, J. M., Oliveira, R. S., Franco, A. R., Ritz, K., Nunan, N., & Castro, P. M. L. (2010). Assessment of mycorrhizal colonisation and soil nutrients in unmanaged fire-impacted soils from two target restoration sites. *Spanish Journal of Agricultural Research, 8*, S86–S95.

Diaz, G., & Honrubia, M. (1994). A mycorrhizal survey of plants growing on mine wastes in Southeast Spain. *Arid Soil Research and Rehabilitation, 8*, 59–68.

Elhindi, K., Al-Suhaibani, N., El-Hendawy, S., & Al-Mana, F. (2018). Effects of arbuscular mycorrhizal fungi on the growth of two turfgrasses grown under greenhouse conditions. *Soil Science and Plant Nutrition, 64*, 238–243.

Falagas, M. E., Pitsouni, E. I., Malietzis, G. A., & Pappas, G. (2008). Comparison of Pub Med, Scopus, Web of Science, and Google Scholar: Strengths and weaknesses. *The FASEB Journal, 22*, 338–342.

Finney, D. M., Buyer, J. S., & Kaye, J. P. (2017). Living cover crops have immediate impacts on soil microbial community structure and function. *Journal of Soil and Water Conservation, 72*, 361–373.

Fuzy, A., Biro, B., & Toth, T. (2010). Short communication. Effect of saline soil parameters on endomycorrhizal colonisation of dominant halophytes in four Hungarian sites. *Spanish Journal of Agricultural Research, 8*, S144–S148.

Garcia, S., Pezzani, F., & Rodriguez-Blanco, A. (2017). Long-term phosphorus fertilization effects on arbuscular mycorrhizal fungal diversity in Uruguayan grasses. *Journal of Soil Science and Plant Nutrition, 17*, 1013–1027.

Gruber, N., Friedlingstein, P., Field, C. B., Valentini, R., Heimann, M., Richey, J. E., Lankao, P. R., Schulze, E. D., & Chen, C. T. A. (2004). The vulnerability of the carbon cycle in the 21st century: An assessment of carbon-climate-human interactions. In C. B. Field & M. R. Raupach (Eds.), *SCOPE 62: The global carbon cycle*. Washington, DC/Covelo/London: Island Press.

Ibrahim, M. A., Campbell, W. F., Rupp, L. A., & Allen, E. B. (1990). Effects of mycorrhizae on sorghum growth, photosynthesis, and stomatal conductance under drought conditions. *Arid Soil Research and Rehabilitation, 4*, 99–107.

Jansa, J., Oberholzer, H. R., & Egli, S. (2009). Environmental determinants of the arbuscular mycorrhizal fungal infectivity of Swiss agricultural soils. *European Journal of Soil Biology, 45*, 400–408.

Jansa, J., Bukovská, P., & Gryndler, M. (2013). Mycorrhizal hyphae as ecological niche for highly specialized hypersymbionts – or just soil free-riders? *Frontiers in Plant Science, 16*, 4–134.

Khalvati, M., Bartha, B., Dupigny, A., & Schroeder, P. (2010). Arbuscular mycorrhizal association is beneficial for growth and detoxification of xenobiotics of barley under drought stress. *Journal of Soils and Sediments, 10*, 54–64.

Koske, R. E., & Gemma, J. N. (1986). Glomus microaggregatum, a new species in the Endogonaceae. *Mycotaxon, 26*, 125–132.

Koske, R. E., & Walker, C. (1986). Glomus globiferum: A new species of Endogonaceae with a hyphal peridium. *Mycotaxon, 26*, 133–142.

Linderman, R. G. (1988). mycorrhizal interactions with the rhizosphere microflora: The mycorrhizosphere effect. Supervisory research plant pathologist, Agricultural Research Service, United States Department of Agriculture, Horticultural Crops Research Laboratory. Corvallis. OR. *Phytopathology, 78*, 366–371.

Lioussanne, L. (2010). The role of the arbuscular mycorrhiza-associated rhizobacteria in the biocontrol of soilborne phytopathogens. *Spanish Journal of Agricultural Research, 8*, S51–S61.

Longdoz, B., Yernaux, M., & Aubinet, M. (2000). Soil CO2 efflux measurements in a mixed forest: Impact of chamber distances, spatial variability and seasonal evolution. *Global Change Biology, 6*, 907–917.

Loth, F. G., & Hofner, W. (1995). Influence of sewage sludge treated soils on the infectivity of VA-mycorrhizal fungi isolates in different plants. *Agribiological Research-Zeitschrift Fur Agrarbiologie Agrikulturchemie Okologie, 48*, 269–281.

Medina, A., & Azcon, R. (2010). Effectiveness of the application of arbuscular mycorrhiza fungi and organic amendments to improve soil quality and plant performance under stress conditions. *Journal of Soil Science and Plant Nutrition, 10*, 354–372.

Meir, D., Pivonia, S., Levita, R., Dori, I., Ganot, L., Meir, S., Salim, S., Resnick, N., Wininger, S., Shlomo, E., & Koltai, H. (2010). Application of mycorrhizae to ornamental horticultural crops: Lisianthus (Eustoma grandiflorum) as a test case. *Spanish Journal of Agricultural Research, 8*, S5–S10.

Morton, J. B., & Walker, C. (1984). Glomus diaphanum: A new species in the Endogonaceae common to West Virginia. *Mycotaxon, 21*, 431–440.

Neeraj, N., & Singh, K. (2011). Organic amendments to soil inoculated arbuscular mycorrhizal fungi and Pseudomonas fluorescens treatments reduce the development of root-rot disease and enhance the yield of Phaseolus vulgaris L. *European Journal of Soil Biology, 47*, 288–295.

Nicolson, T. N., & Schenck, N. C. (1979). Endogonaceous mycorrhizal. Endophytes in Florida. *Mycologia, 71*, 178–198.

Ortas, I. (2010). Effect of mycorrhiza application on plant growth and nutrient uptake in cucumber production under field conditions. *Spanish Journal of Agricultural Research, 8*, S116–S122.

Pandey, R. R., Sharma, G., Singh, T. B., & Tripathi, S. K. (2010). Factors influencing soil CO2 efflux in a northeastern Indian oak forest and plantation. *African Journal of Plant Science, 4*, 280–289.

Papp, M., Foti, S., Nagy, Z., Pinter, K., Posta, K., Fekete, S., Csintalan, Z., & Balogh, J. (2018). Rhizospheric, mycorrhizal and heterotrophic respiration in dry grasslands. *European Journal of Soil Biology, 85*, 43–52.

Schenck, N. C., & Smith, G. S. (1982). Additional new and unreported species of mycorrhizal fungi (Endogonaceae) from Florida. *Mycologia, 74*, 77–92.

Sene, G., Samba-Mbaye, R., Thiao, M., Khasa, D., Kane, A., Manga, A., Mbaye, M. S., & Sylla, S. N. (2012). The abundance and diversity of legume-nodulating rhizobia and arbuscular mycorrhizal fungal communities in soil samples from deforested and man-made forest systems in a semiarid Sahel region in Senegal. *European Journal of Soil Biology, 52*, 30–40.

Shi, Z., Wang, F., & Liu, Y. (2012). Response of soil respiration under different mycorrhizal strategies to precipitation and temperature. *Journal of Soil Science and Plant Nutrition, 12*, 411–420.

Shukla, A., Vyas, D., & Jha, A. (2013). Soil depth: An overriding factor for distribution of arbuscular mycorrhizal fungi. *Journal of Soil Science and Plant Nutrition, 13*, 23–33.

Smith, S. E., & Read, D. J. (2008). *Mycorrhizal symbiosis* (787 p). Amsterdam: Academic.

Smith, G. S., & Schenck, N. C. (1985). Two new dimorphic species in the Endogonaceae: Glomus ambisporum and Glomus heterosporum. *Mycologia, 4*, 566–574.

Tyagi, J., Varma, A., & Pudake, R. N. (2017). Evaluation of comparative effects of arbuscular mycorrhiza (Rhizophagus intraradices) and endophyte (Piriformospora indica) association with finger millet (Eleusine coracana) under drought stress. *European Journal of Soil Biology, 81*, 1–10.

Ultra, V. U. Y., Tanaka, S., Sakurai, K., & Iwasaki, K. (2007). Arbuscular mycorrhizal fungus (Glomus aggregatum) influences biotransformation of arsenic in the rhizosphere of sunflower (Helianthus annuus L.). *Soil Science and Plant Nutrition, 53*, 499–508.

Valetti, L., Iriarte, L., & Fabra, A. (2016). Effect of previous cropping of rapeseed (Brassica napus L.) on soybean (Glycine max) root mycorrhization, nodulation, and plant growth. *European Journal of Soil Biology, 76*, 103–106.

Walker, C., & Sanders, F. E. (1986). Taxonomic concepts in the Endogonaceae: III. The separation of Scutellospora gen. nov. from Gigaspora Gerd. & Trappe. *Mycotaxon, 27*, 169–182.

Wang, P., Zhang, J. J., Xia, R. X., Shu, B., Wang, M. Y., Wu, Q. S., & Dong, T. (2011). Arbuscular mycorrhiza, rhizospheric microbe populations and soil enzyme activities in citrus orchards under two types of no-tillage soil management. *Spanish Journal of Agricultural Research, 9*, 1307–1318.

Wu, Q. S., Xia, R. X., & Zou, Y. N. (2008). Improved soil structure and citrus growth after inoculation with three arbuscular mycorrhizal fungi under drought stress. *European Journal of Soil Biology, 44*, 122–128.

Zaller, J. G., Frank, T., & Drapela, T. (2011). Soil sand content can alter effects of different taxa of mycorrhizal fungi on plant biomass production of grassland species. *European Journal of Soil Biology, 47*, 175–181.

Zarea, M. J., Ghalavand, A., Goltapeh, E. M., Rejali, F., & Zamaniyan, M. (2009). Effects of mixed cropping, earthworms (Pheretima sp.), and arbuscular mycorrhizal fungi (Glomus mosseae) on plant yield, mycorrhizal colonization rate, soil microbial biomass, and nitrogenase activity of free-living rhizosphere bacteria. *Pedobiologia, 52*, 223–235.

Zhu, Y. G., & Miller, R. M. (2003). Carbon cycling by arbuscular mycorrhizal fungi in soil-plant systems. *Trends in Plant Science, 8*, 407–409.

Influence of Xenobiotics on the Mycorrhizosphere

7

R. Koshila Ravi, S. Anusuya, M. Balachandar, S. Yuvarani, K. Nagaraj, and T. Muthukumar

1 Introduction

In agriculture, various compounds or substances that do not occur naturally such as pollutants or other organic substances and toxic chemicals are introduced through the application of agrochemicals into the farmlands. These unnatural substances in the natural environment are known as xenobiotics. Xenobiotic substances are of major concern as they are not only hazardous to the environment but also affect the human health (Reiger et al. 2002; Varsha et al. 2011; Visioli 2015). Some of the xenobiotics commonly encountered in day-to-day life include pesticides, phenols, plastics, hydrocarbons, fuels, and polyaromatic compounds (Bulucea et al. 2012; Serra et al. 2013). Pesticides are the frequent xenobiotic compounds occurring in soil due to their extensive use in the agriculture. The soil absorbs the chemicals present in the pesticides and fixes them temporarily (Sonon and Schwab 2004). In due course, these substances tend to accumulate in the environment leading to bioaccumulation and biomagnification (Maurya and Malik 2016). Therefore, it is immensely important to detoxify, degrade or to eliminate such substance that is toxic to biological systems in an efficient and potential way through biological processes (Díaz 2004; Gonzalez et al. 2010; Furukawa 2018).

Biodegradation is one of the biological and eco-friendly methods that help in the elimination of xenobiotic compounds through living organisms, particularly, microorganisms (Singh 2008). Xenobiotics could resist biodegradation or undergoes partial degradation (Romi Singh 2017). Some of the xenobiotic chemicals persist in the soil for a prolonged period, while others that are even biodegradable when remain mobile in the soil can be toxic causing detrimental effects like polluting the

R. Koshila Ravi · S. Anusuya · M. Balachandar · S. Yuvarani · K. Nagaraj
T. Muthukumar (✉)
Root and Soil Biology Laboratory, Department of Botany, Bharathiar University, Coimbatore, Tamil Nadu, India

© Springer Nature Singapore Pte Ltd. 2019

A. Varma, D. K. Choudhary (eds.), *Mycorrhizosphere and Pedogenesis*,
https://doi.org/10.1007/978-981-13-6480-8_7

groundwater (Singh and Walker 2006). Biodegradation comprises of three processes, namely, mineralization, biotransformation and co-metabolism. Mineralization involves break down of organic chemicals into inorganic substances resulting in water, carbon dioxide and ammonia as byproducts. The organic compounds undergo structural variation during the process of biotransformation. In co-metabolism, non-growth substrate is transformed in the presence of growth substrate (Harms et al. 2011; Arora et al. 2012; Edwards and Kjellerup 2013).

The microorganisms involved in the degradation of the naturally occurring harmful substance include bacteria and fungi (Peng et al. 2008; Leja and Lewandowicz 2010; Harms et al. 2011). These microbes have the ability to degrade and detoxify the chemical toxic substances that is referred as microbial infallibility (Alexander 1965). Microbial degradation involves the removal of toxic contaminants from the soil in an eco-friendly approach (Silambarasan and Abraham 2013). Mycoremediation, a part of the microbial degradation is the process involving fungi have an essential role in the degradation of both organic and inorganic pollutants prevalent in an environment (Bhandari 2018). Some of the microorganisms when continuously exposed to xenobiotics tend to generate the capacity to degrade them. This is due to the modification of active site of the enzymes that show higher affinity to xenobiotics which results in their degradation through the development of new enzymatic pathways (Karpouzas and Singh 2006; Gianfreda and Rao 2008). Most of the filamentous fungi are resistant to environmental stresses as they secrete large amounts of extracellular enzymes during their colonization of the soil. This leads to higher bioremediation of the xenobiotics especially pesticides (Mangan et al. 2010). White rot fungi are most widely used for decomposing and degrading toxic xenobiotic compounds (Gao et al. 2010; Marco-Urrea and Reddy 2012).

Mycorrhiza denotes a symbiotic association between soil fungi and plant roots. About 80% of terrestrial plants are capable of associating with mycorrhizal fungi (Smith and Read 2008). Mycorrhizal fungi help plants in the acquisition of water, phosphorous (P) and other essential nutrients and the colonizing fungi, in turn, acquire carbon from their host plant (Smith and Smith 2011). Mycorrhizae are broadly classified into two major groups as ectomycorrhiza and endomycorrhiza based on the fungal penetration of the root cortical cells. Ectomycorrhizae (ECM) forms a dense mycelium sheath around the roots and intercellular hyphae that penetrate the root, but the fungal hyphae never enter the cortical cells. Endomycorrhzae is characterized by the presence of external hyphae and the intraradical hyphae that invades widely the root cortical cells. Endomycorrhiza is further divided into arbuscular mycorrhiza, orchid and ericoid mycorrhiza (ERM) (Roth-Bejerano et al. 2014). Arbuscular mycorrhizal fungi (AMF) are widespread and the most common mycorrhizal type (Smith and Read 2008). The AMF produces specialized structures called arbuscules that act as the transit point for the movement of nutrients from the soil to plants and the storage structure vesicles (Smith and Read 2008). Arbuscular mycorrhizal fungi benefits their host plants by enhancing the plant growth in low fertility soils and also improve plant's water relation. Further, AMF influence their host plant physiology by making them less susceptible to pathogens, salinity, soil pollution, drought and other environmental stresses (Gianinazzi et al. 2010).

Mycorrhizal fungi are of great importance in encouraging the degradation of organic contaminants in the soil (Khalvati et al. 2010). As mycorrhizal fungal-strains are known to have better resistance to toxic elements, they could enhance and assist in the process of phytoremediation (Turnau et al. 2002). Mycorrhizal fungi also extend their hyphae deep into the contaminated soil to degrade the pollutants called rhizodegradation (Fang et al. 2001; Husaini et al. 2008). Despite providing water and nutrients to plants, mycorrhiza also aids in the improvement of the soil structure. They also act as filters by blocking the harmful compounds within their mycelium which in turn leads to a reduction in the toxicity levels in the plants (Dubey and Fulekar 2011). Several persistent organic compounds like atrazine, polycyclic aromatic hydrocarbons (PAHs) are known to be degraded in the mycorrhizosphere either directly or indirectly (Joner and Leyval 2001; Volante et al. 2005; Huang et al. 2005; Sainz et al. 2006). Successful rhizodegradation of organic pollutants such as pesticides (Hsu and Bartha 1979; Joner and Leyval 2009) and chlorinated organic compounds (Anderson et al. 1993; Sicliliano and Germide 1999) has been demonstrated. *Morchella conica* and *Tylospcno fibrilnsa* represent a typical mycorhizons that are known to degrade organic pollutants naturally (Bennet et al. 2002).

In this chapter, we highlight the rhizospheric effect, mycorrhizosphere, the role of AMF in the degradation of pollutants, molecular mechanisms and pathways involved in degradation of xenobiotics. In addition, we also discuss the degradation of organic pollutant by AMF even under stress conditions.

2 Rhizosphere

The rhizosphere is the region of the soil that surrounds plant roots. This region and is characterized by the activity of roots like the production of exudates and soil microbes that are associated with it (Mendes et al. 2013). Rhizosphere encompasses many microorganisms, enables their diversity and activity and other interactions constituting a hot spot for many organisms (Hinsinger et al. 2009). Rhizosphere processes involve microbial colonization, physical, chemical and biological changes brought about by the plants through translocation of water and minerals and in turn releasing carbon dioxide nutrients and a wide range of chemical substances (Philippot et al. 2013). The chemical exudates that are released into the rhizosphere region by the root cells are known as rhizodeposits and the process is termed as rhizodeposition (Hinsinger et al. 2005). Root exudates include organic acids, polysaccharides, secondary metabolites, amino acids, organic carbon and nitrogen (N) that are released by plants into the surrounding soil (Dennis et al. 2010; Baetz and Martinoia 2014). One of the important phenomena the rhizosphere effect is the exudation of carbon by the germinating seeds or plant roots that penetrates into the soil and promotes microbial proliferation and their activity (Farrar et al. 2003). Many studies have reported the rhizosphere effects on biotransformation for different compounds such as enhancing the tolerance of plants to phytotoxic elements in the soil owing to the ability of plants to induce micro-organisms that detoxify the xenobiotics (Dubey and Fulekar 2013; Agrawal and Dixit 2015).

In the rhizosphere, the interactions between the microbe and plants may be positive, neutral or negative depending upon the particular host plant and the microbe involved in the prevailing environmental conditions (Raaijmakers et al. 2009; Bais et al. 2006; Gianfreda 2015). The positive interactions are favoured by mycorrhizal fungi, plant growth promoting rhizobacteria and other beneficial microbes; and the negative interaction involves association of some pathogenic microbes, parasitic plants and other harmful invertebrates (Raaijmakers et al. 2009).

Remediation of soil organic pollutants is one of the important features of rhizosphere interactions (Crowley et al. 1996; Fan et al. 2008). The rhizosphere microbes are considered as one of the main contributors to the degradation of organic compounds (Kuiper et al. 2004). In the degradation process of organic pollutants, phytoremediation involves rhizodegradation and phytodegradation. Rhizodegradation involves microbial degradation whereas; phytodegradation involves degradation of compounds that are taken up by the plants (Flathman and Lanza 1998). The increased microbial activity in the rhizosphere region leads to higher degradation of organic compounds in the rhizosphere when compared to the surrounding bulk soil (Cunningham et al. 1996). The root exudates containing organic substances provide N and carbon sources to soil microbes that have the ability to degrade organic pollutants (Kaimi et al. 2006). Some of the plants suitable for rhizoremediation include different grass varieties and legumes such as alfalfa (*Medicago sativa* L.) (Lee et al. 2008; Zhang et al. 2013). The remediation of xenobiotics is discussed further in the upcoming sections.

3 Hyphosphere

The zone enclosing or surrounding the fungal hyphae are referred to as hyphosphere (Johansson et al. 2004). Hyphosphere comprises of the enormous amount of AMF hyphae (Marschner 1995). Apart from AMF hyphae, ectomycorrhizal fungi also produce fungal mycelium in the soil (Van Elsas et al. 2008). The presence of high rates of AMF contributes to increased soil fungal biomass (Leake et al. 2004). It is familiar that AMF associate with plant roots (Smith and Read 2008). The surface of AMF hyphae are also colonized by the different group of bacteria that depend on the exudates produced by the hyphae for their endurance (Scheublin et al. 2010).

Arbuscular mycorrhizal fungi release plant photosynthates into the hyphosphere zone; it enhances the nutrients availability and promotes the degradation of organic compounds by the soil microbes (Jansa et al. 2013). The chief hyphal exudates are glucose, acetate, oligosaccharides etc., that are absorbed by the associating bacteria as carbon sources resulting in the modification of bacterial composition in the hyphosphere region (Toljander et al. 2007). As AMF are not capable of secreting extracellular enzymes, the process of maintaining the microorganisms could help in the degradation process of organic compounds present in the soil (Smith and Smith 2011). The bacteria associated with the mycorrhizal fungal hyphae in the hyphosphere of maize plants has been shown to increase the mineralization of phytate (Wang et al. 2013). The association between AMF hyphae of *Rhizophagus irregularis* (Blaszk., Wubet, Renker & Buscot) C. Walker & A. Schüßler and phosphate solubilizing bacteria, *Pseudomonas alcaligenes* Monias helps in the mineralization of phytate in the root-free soil zone (Zhang et al. 2014).

4 Mycorrhizosphere: The Extension of Rhizosphere

The microbial communities inhabiting the soil include bacteria, fungi, actinomycetes, algae and protozoa. Apart from these soil microorganisms, mycorrhizal fungi are considered as a major soil microbiota that colonizes the roots of the majority of the vascular plants. The exudates that are released by the mycorrhizal hyphae influence the activities of microbes in the rhizosphere region (Duponnois et al. 2006). The regions surrounded by the mycorrhizal fungus that facilitate the microbial activities in the rhizosphere through the release of nutrients and exudates are termed as mycorrhizosphere (Rambelli 1973). Mycorrhizosphere comprises to two compartments, rhizosphere that includes the soil surrounding the root system; second, the soil zone influenced by the fungal hyphae known as the hyphosphere region (Marschner 1995; Johansson et al. 2004).

The microbial communities prevailing in the mycorrhizosphere region have a pivotal role in agriculture, as the application of microbes in the crop productivity have reduced the dependence on inorganic fertilizers, pesticides, herbicides and insecticides that in turn gets accumulated in the soil for prolonged period causing detrimental effect to soil and the degradation of these organic chemicals becomes a difficult task (Adesemoye et al. 2009; Savci 2012). The mycorrhizal fungi decrease the detrimental effects of soil pollutants. The fungal structures produced by AMF within plant roots enlarge the surface area for exchange of metabolites between the plant and the fungus (Auge 2001). The AMF produce extraradical hyphae that directly interrelate with the soil helps in the transfer of P and other nutrients to the plant (Smith and Read 2008). The AMF hyphae have a major role in stabilizing the soil structures through soil aggregation (Rillig et al. 2002; Borie et al. 2008). The extraradical hyphae modify the host plant physiology either directly or indirectly by interacting with other soil microbes in the mycorrhizosphere (Johansson et al. 2004).

5 Influence of Xenobiotics on AMF

In the field, AMF is exposed to a wide range of xenobiotics in the form of fertilizers and biocides. The interaction between the AMF, pesticides and the host plant is very complex under natural conditions (Giovannetti et al. 2006). The response of AMF to xenobiotics like biocides varies widely. Biocides can have inhibitory (Zocco et al. 2011; Channabasava et al. 2015), neutral (Schweiger and Jakobsen 1998) or stimulatory (Spokes et al. 1981; Channabasava et al. 2015) effect on AMF. Karpouzas et al. (2014) studied the influence of the herbicide nicosulfuron on mycorrhization and AMF community structure in maize (*Zea mays* L.). The results of the study indicated that the influence of the herbicide on AMF in pot experiment could vary substantially from those of the field conditions. In the pot experiment, mycorrhizal colonization, AMF richness and plant biomass significantly declined on exposure to nicosulfuron at ×100 and ×1000 dose rates. Contrarily, nicosulfuron had no effects on AMF even at these high application rates under field conditions. Similarly, the widely used herbicide glyphosate affected the root colonization by AMF and reduced the viability of the spores by 5.8–7.7-folds (Druillea et al. 2013a). The influence of glyphosate is important as spores are the chief perennating propagule of AMF where plant growth is seasonal or in areas where the soil is left barren between two cropping seasons. Later the same authors (Druillea et al. 2013b) showed that the influence of glyphosate on AMF may be due to the direct or indirect effect of the herbicide on AMF. Glyphosate can affect AMF directly by influencing the germination of the spores and the hyphal spread in the soil. The indirect effect the herbicide on AMF propagule formation results from the reduced carbon availability to the fungi due to the interference of glyphosate on the host physiology (Druillea et al. 2013b). Though the effects of pesticides on AMF vary with toxicity and application rates, some recent studies do indicate the influence of the host plant in the response of AMF to pesticides. In a greenhouse study application of different concentrations of the organophosphate insecticide phoxim inhibited root colonization by *Rhizophagus intraradices* (N.C. Schenck & G.S. Sm.) C. Walker & A. Schüßler and *Funneliformis mosseae* (T.H. Nicolson & Gerd.) C. Walker & A. Schüßler in green onion (*Allium fistulosum* L.) and not in carrot (*Daucus carota* L.) (Wang et al. 2011b). In a subsequent study Wang et al. (2011a) showed that colonization of green onion roots by *Glomus caledonium* (T.H. Nicolson & Gerd.) Trappe & Gerd., was higher compared to *Acaulospora mellea* Spain & N. C. Schenck., in phoxim contaminated soils.

Pesticides not only affect the total root length colonized by mycorrhizal fungi but also affect the different AMF within roots variedly. For instance, development of different fungal structures by *R. intraradices* in Ri T-DNA-transformed roots of chicory (*Cichorium intybus* L.), was differently affected by the two fungicides fenhexamid and fenpropimorph that are known to inhibit sterol biosynthesis. The presence of fenpropimorph in the growing medium reduced the percentage of root length containing intraradical hyphae and vesicles and totally inhibited the development of arbuscules. In contrast, the presence of fenhexamid in the growing medium failed to impart any significant influence on the development of the AMF structures. Fenpropimorph also affected the development of the extraradical hyphae and sporulation of *R. intraradices* to a greater extent than fenhexamid (Campagnac et al. 2010).

The sensitivity to xenobiotics appears to vary with and within AMF groups. For example, AMF belonging to the *Glomus* group generally are more sensitive to the presence of high concentrations of nicosulfuron in the soil (Karpouzas et al. 2014). Contrarily, taxa belonging to Paraglomeraceae and *Claroideoglomus etunicatum* (W.N. Becker & Gerd.) C. Walker & A. Schüßler (=*Glomus etunicatum*) occurs in maize roots even at high concentrations of the herbicide nicosulfuron suggesting a tolerance to xenobiotic-induced stress. Indeed, *C. etunicatum* is more tolerant to various natural or synthetic pesticides than the other *Glomus* species (Schreiner and Bethlenfalvay 1997; Ipsilantis et al. 2012). Similarly, *F. mosseae* has a global distribution as it adapts well to different soil conditions (Lenoir et al. 2016). In spite of its worldwide distribution, *F. mosseae* is sensitive to the presence of pesticides in the soil. Giovannetti et al. (2006) tested the toxic effects of 14 commonly used agricultural pesticides on spore germination and hyphal development of *F. mosseae* under axenic conditions. The results of the study indicated that the spore germination, as well as the development of the fungal mycelium, was inhibited by most of the pesticides tested. Some of the pesticides were toxic to the fungal processes even at concentrations much lower than those recommended for field application (Giovannetti et al. 2006). Polycyclic aromatic compounds and fungicides can reduce the number of branched absorbing structures in the extraradical mycelium of the AMF (Aranda et al. 2013).

Introduction of pesticides dictate changes in the nature of the interaction between the different microorganisms residing in the soil due their contrasting toxicities for these organisms. AMF establishes a close relationship with the soil microorganisms in addition to their associated host plant. Some of these microorganisms play an important role in facilitating the colonization of the host roots and growth of the AMF. The mycorrhizal association also affects the microbial populations in the plant rhizosphere in xenobiotic contaminated soils. Alfalfa plants dual inoculated with the AM fungus *G. caledonium* and *Rhizobium* and grown on polychlorinated biphenyls contaminated field soils harboured 31% more bacteria, 12% more fungi and 21% more biphenyl degrading bacteria in their rhizosphere than uninoculated plants (Teng et al. 2010). This clearly shows that AMF could support a wide range of soil microorganisms in pesticide contaminated soils.

6 Rhizospheric Microbial Community in Degradation

The plant microbial community in the rhizosphere can degrade various contaminations that are of concern to the environment and human health as well. The remediation of toxic chemicals needs an understanding of factors that direct the process of biotransformation in the rhizosphere. The increased microbial population in the rhizosphere than in other regions of the soil is one of the important factors for degradation of xenobiotic compounds by microorganisms (Anderson and Coats 1995). The degradation rate of xenobiotic compounds depends upon the plant species, as each plant differ in the production of primary and secondary metabolites and in interaction with other organisms in the rhizosphere (Shann and Boyle 1994). The degradation of organic xenobiotics was owed to rhizodeposits due to their ability to increase the bioavailability as they act as xenobiotic structural analogues in the rhizosphere (Shaw and Burns 2007). The percentage of culturable microorganisms residing in the rhizosphere range between 2 and 7% of the total microorganism present in the soil, thus contributing to higher secretion of root exudates that helps in the microbial degradation when compared to the root-free soil with >1% culturable microbial populations (Zelenev et al. 2005).

The root exudates also enhance the survival of microorganisms even in polluted or contaminated soils ((Rohrbacher and St-Arnaud 2016). Degradation by rhizosphere microbial communities in hydrocarbon-contaminated soils revealed unique improvements with specific plant species as well. For instance, the microbial degradation of hydrocarbon was higher in the rhizosphere of ryegrass (*Lolium perenne* L.) when they were grown in petroleum-hydrocarbon polluted soil (Corgié et al. 2004). The microbes may also neutralize or degrade the polluted soil by the direct break down of toxic substances within their tissues (Doty et al. 2007). In another study, He et al. (2005) compared the degradation rates of pentachlorophenol in the rhizosphere and far rhizosphere regions and reported the higher degradation of pentachlorophenol by microbes inhabiting the rhizosphere of ryegrass when compared to the other regions. The higher urease and phosphatase activity in the rhizosphere of ryegrass was cited as the possible reason for increased pentachlorophenol degradation (He et al. 2005).

Some of the bacterial communities residing in the rhizosphere are potential degraders of xenobiotic compounds. The bacteria involved in the removal or degradation of phenanetherene are *Pseudoxanthomonas* and *Microbacterium* (Cébron et al. 2011). Sunflower (*Helianthus annuus* L.), when grown in the PAH contaminated soil, resulted in 93% degradation of PAH. The PAH-degrading efficient bacteria that was isolated from the sunflower planted soil includes *Xanthomonas, Sphingomonas* and *Oxalobacteria*. This proves that sunflower root exudates have the degrading capacity that in turn influences the microbial activity (Tejeda-Agredano et al. 2013). Some of the mesophilic and psychrophilic bacteria isolated from the rhizosphere of sweet flag (*Acorus calamus* L.) were capable of degrading atrazine in contaminated soil (Marecik et al. 2008). Similar to these observations, Fan et al. (2008) also reported the pyrene degrading capability of bacteria isolated from the roots of alfalfa close to the soil in pyrene contaminated soil. An enhanced microbial and peroxidase activity in the rhizosphere of legumes and grasses lead to the dissipation of phenanthrene and pyrene from the polluted soils (Lee et al. 2008).

Besides bacteria, fungi are also involved in the remediation of organic contaminants. Fungi forms a wide network of mycelium, have low specificity for catabolic enzymes and could use the organic chemicals for their growth. Therefore, fungi are well suited for bioremediation (Harms et al. 2011). Most commonly used fungi in the degradation of xenobiotic compounds include *Phanerochaete chrysosporium, Pleurotus ostreatus*, white rot fungi and *Trametes versicolor* (Moredo et al. 2003; Mangan et al. 2010). Species of *Aspergillus* and *Penicillium* are known to biodegrade wide range of xenobiotic compounds like pesticides, PAHs and synthetic dyes (Pinedo-Rilla et al. 2009; Scheibner et al. 1997). Fungi contribute to degradation of harmful organic compounds by production of extracellular enzymes. *Cladosporium cladosporioides* (Fresen.) G.A. de Vries has been reported to detoxify chlorpyrifos-contaminated soils (Chen et al. 2012).

Studies on Zucchini (*Cucurbita pepo* L.) cultivars showed that single AMF species were effective in degradation than the AMF species mixture. Nevertheless,

AMF association had a more positive effect on the removal of pollutants when compared to single species (White et al. 2006). In addition, microbial consortia contribute to higher levels of phytoremediation because it enhances competition and synergistic interactions between symbionts for pollutants degradation. The effect of microbial mixtures could be explained by a higher belowground allocation to sustain a higher proportion of AMF. In average, the mycorrhizal effect is highly variable depending on the contaminant, the mycorrhizal species, the plant type, the interaction with surrounding soil micro-organisms and soil conditions. The nutrient scavenging activity of the mycorrhizal fungal mycelium is responsible for low xenobiotic degradation potential as reported in some studies.

In places lacking AMF, the introduction of inoculums offers an interesting perspective on phytostabilization techniques. The process may be stimulated by appropriately selected fungal strains. For successful bioremediation, symbionts must be selected that can withstand the hostile environment of polluted sites. AMF enhance plant survival and growth by decreasing P deficiency (Joner and Leyval 2001) and water stress (Sanchez-Diaz and Honrubia 1994), improving membrane integrity (Graham et al. 1981) or by facilitating the production of an oxidative enzyme (Salzer et al. 1999). These phenomena are all responsible for the attenuation of stress due to pollution. It should be noted that the mechanisms involved in pollutant degradation may be indirect and mediated through stimulation of the associated rhizosphere microflora (Joner et al. 2001).

7 Mechanisms Involved in Rhizodegradation

7.1 Direct Enzymatic Degradation

Mycorrhizal fungi contribute to remediation by different mechanisms. The exact mechanisms behind the rhizodegradation of organic pollutants are yet obscure, but the possible explanations provided are relatively complex. The presumed mechanisms include the direct effect of the enzymes originating from roots (Gramss and Rudeschko 1998) and the activation of the metabolic precursors like the phenols exuded by roots which may induce the enzymatic activities in the metabolic pathways that may aggressively act against the pollutant (Curl and Truelove 1986). Nonetheless, in several instances degradation of the pollutants is moderated by cometabolism rather than direct metabolism. The mechanism in the involvement of AMF in the degradation of pollutants also does not involve fungal co-oxidation or catabolism because of the marginal saprophytic capability of the fungi involved. Other possible mechanistic explanations encompass the role of the mycorrhizal fungi on the activity of oxidative enzymes both in the roots and the rhizosphere. The contaminants in these cases are degraded by a mechanism involving non-specific free radicals involving lignin, laccases or class II peroxidases (Barr and Aust 1994).

Peroxidases can be found in the ECM fungal genera *Cortinariums*, *Lactarius* and *Russula* (Bodekar et al. 2009) and are not common in all mycorrhizal fungi (Kohler 2015). Members of the laccase gene family appear to participate in the degradation especially in the presence of redox mediators (Burke and Cairney 2002; Shah et al. 2015). The higher ligninolytic enzyme activity suggests that AMF have degrading potential as ERM and ECM do (Read et al. 2004). Nevertheless, the actual potential of the mycorrhizal fungi (when in association with the host plant) in the degradation of the pollutants and the impact of mycorrhization on the production or release of the exoenzymes by the host plant and its associated fungi are almost obscure. Courty et al. (2011) showed that the mycorrhization of the poplar (*Populus deltoids* Bartr. × *Populus trichocarpa* Torr. & A. Gray) roots by *Laccaria bicolor* (Maire) P.D.Orton significantly modified their capability to secrete enzymes that are involved in the breakdown of the organic matter or mobilization of the organic P. These authors further demonstrated that the ability of the ectomycorrhizal root tips to secrete enzymes was under the direct influence of the host genotype (Courty et al. 2011).

7.2 Indirect Mechanism

Indirectly, mycorrhiza can increase the ability of plants to withstand soil phytotoxicity by improving the nutrition, protecting the plants against pathogens and drought stress, enhance soil aggregation and ultimately increase the retention of xenobiotics due to their higher partition coefficient in mycelium than in root (Gao et al. 2010). Through altered root exudation, mycorrhiza affects the activity and microbial composition in the rhizosphere implying the microflora to be more effective in xenobiotic degradation. Moreover, mycorrhizal fungi help the plants to get rid off from build-up phytotoxic concentrations of pollutants by secreting some specific detoxifying compounds directly. Secretion of surfactants by bacteria increase the solubility of pollutants (Van Hamme et al. 2003), but the production of surfactant in mycorrhizal fungi has not investigated yet. Instead, AMF is known to produce ampiphilic peptides called glomalins which act as surfactants (Rillig and Mummey 2006).

The recalcitrant pollutants such as pesticide lindane that is co-metabolized are not efficiently mineralized but are transformed by microbes (Paul et al. 2005). The microbial transformation of organic compounds is driven by energy provided by root exudates allowing the spread of roots into deeper soil thereby accelerates the remediation process (Chaudhry et al. 2005). The dissipation of pollutant in the rhizosphere and mycorrhizosphere of ryegrass was due to biodegradation and biotransformation (Vivas et al. 2006).

8 Xenobiotic Metabolism

Metabolism of the xenobiotic is a form of biotransformation that detoxifies the compounds. The ability of mycorrhizal fungi to biotransform xenobiotics can be similar to the pollutant degradation by white rot fungi (Donnelly and Fletcher 1995). Mycorrhizal fungi detoxify the hyoxylate aromatic compounds through monooxygenase system such as cytochrome P-450 enzymes, cleaves aromatic rings via meta cleavage and oxidizes the aliphatic moiety to Krebs cycle intermediates, and further break down to CO_2 and H_2O (Higson 1992; Greń et al. 2008). Nitrogen starvation also induces lignin peroxidase production in mycorrhizal fungi (Harley 1989). However, there are no documented pathways for biotransformation of xenobiotics by the mycorrhizal fungi, but the products formed through meta cleavage indicate the presence of a typical route. Studies pertaining to mineralization of the organic compounds also suggest that the biotransformation pathways both in white rot and mycorrhizal fungi do involve a ring cleavage with the formation of CO_2 (Dietrich et al. 1995; Yadav et al. 1995).

9 Enzymes Involved in Rhizodegradation of Xenobiotics

Mycorrhizae not only influence the response of the host plant to pollutants but also affect the host plant growth. The mycorrhizal symbiosis can alleviate the toxicity of pollutants on plants through the stimulation of the host-oxidative stress response. ECM fungi can produce enzymes involved in various stages of xenobiotic decomposition in the soil (Barr and Aust 1994; Meharg and Cairney 2000). AMF colonization induces the production of extracellular enzymes which is of very low substrate specificity. This enables AMF to break down the aromatic compounds introduced into the environment. Accumulation of phenolics in the roots and rhizosphere soil could also induce degradation of complex organic compounds through the production of water-soluble molecules called quinones (Amrani et al. 2015). Nutrients derived from extraradical hyphae of AMF drive co-metabolic degradation within small soil pores making them unavailable to roots.

During the process of pollutant degradation, AMF stimulates the production and secretion of ascorbate peroxidase, superoxide dismutase and peroxidase in the mycorrhizal roots (Ibanez et al. 2011). Moreover, AMF also enhances the activities of the oxidoreductase both in the rhizosphere and in the roots (Ibanez et al. 2011). A study on the molecular cloning and limited characterization of the expression of a peroxidase gene in the roots of *Portulaca oleracea* L., suggested that peroxidase removes organic pollutants either through cross-linking them to the polysaccharides in the cell walls or binding them to proteins (Matsui et al. 2011). In addition, the fungal cytochrome P450 (CYP) monooxygenases are known to catabolize the aromatic compounds involving the incorporation of O_2 by CYP monooxygenase.

10 Molecular Techniques for Identifying Xenobiotic Organisms

Mycorrhizal fungi have a complicated genetic makeup as they can exist in heterokaryotic (genetically different nuclei) or in homokaryotic (identical nuclei) forms. A single fungal isolate can have very high diversity and one single root system might be colonized by different mycorrhizal fungi. Microsatellite-primed polymerase chain reaction, random amplification of polymorphic DNA and repeated

DNA probes are highly efficient approaches for the identification of distinct genotypes of AMF (Wyss and Bonfante 1993; Longato and Bonfante 1997).

The use of "omics" strategies deciphers the complex plant-mycorrhizal interactions in polluted environments. One of the methods used for the identification of the specific fungal assemblage in the root of *Pinus-Piloderma* symbiosis is through the cataloguing of the transcriptionally dominant fungal taxa by combining DNA/RNA extraction. Thus the availability of genomes of specific taxa in public databases allows the identification of patterns in gene content and transcript abundance (Liao et al. 2014). The evaluation of several DNA regions as barcodes specific for fungi by Fungal Barcoding Consortium contributed a lot to describe the association of plants–fungi communities of soils involved in the removal of hazardous chemicals in the environment (Schoch et al. 2012). Metagenomics can identify the functional potential and the taxonomic identity of all organisms in an environmental sample but provides no information on the activity of the members constituting the community. The isolation, screening and utilization of organisms with xenobiotic degradation potential as inoculums will be of immense value for bio-augmentation.

A genome-wide comparison of fungi for their bioremediation potential has become possible due to the availability of the complete sequences of fungal genomes in databases (Mougin et al. 2013). DNA sequencing allows the structural and functional investigations for assessing the role of catabolic processes involved in degradation of recalcitrant organic pollutant (Testa et al. 2012). The ability of the fungi for diverse metabolic adaptations due to diverse enzyme functions like for cytochrome P450 monooxygenase can be inferred by whole genome sequence analysis (Ichinose 2013).

11 Genetic Adaptations of Mycorrhizal Fungi to Xenobiotic Compounds

Mycorrhizal fungi possess a range of genetic mechanisms allowing the evolution of functional degradative pathways. Biodegradation never starts immediately after the exposure to xenobiotics. After a stipulated period of time, mineralization occurs. Various biochemical and molecular processes impart adaptive response to the growth of the microbial community, secretion of specific enzymes, acquisition and metabolization of substrates. One of the prime mechanisms of adaptation in xenobiotic-degraders to their substrates is the appearance of DNA rearrangements which results in the evolution of various pathways for the disintegration of xenobiotic substances in the native environment. The various processes of genetic adjustments include transposition, gene transfer, and genetic recombination and mutational drift. Involvement of these mechanisms overcomes the biochemical bottlenecks in natural pathways that prevent the degradation of novel substrates.

The process of adaptation includes choosing of mutants that have acquired either novel metabolic activities or altered enzymatic specificities that were absent at the onset of exposure to the introduced compounds (Spain and van Veld 1983; Barkay and Pritchard 1988). For understanding the adaptational process in nature, *in situ* genetic interactions among micro-organisms and the influence of environmental factors on selection dissemination of specific catabolic genes should be studied.

12 Mycorrhizal Fungi in Remediation of Xenobiotics Under Stress Conditions

Several abiotic factors like drought stress, climate change and pollution can affect the development of mycorrhizal relationships. The impact of AMF symbiosis on drought and pollutant degradation is well explained, but the exact mechanism is still unknown (Auge, 2001; Ruiz-Lozano, 2003; Khalvati 2005). Exposure of plants to drought or xenobiotics induces an oxidative stress which is responsible for many degenerative reactions caused by these stress factors. The AMF *Glomus hoi* S.M. Berch & Trappe moderate the adverse effects of restricted soil water and detoxify the barley (*Hordeum vulgare* L.) plant against pharmaceutical xenobiotics through increased production and activity of the antioxidant enzymes like catalase (Khalvati et al. 2010). The production of catalase and glutathione-S-transferase in roots enable plants to adapt to xenobiotic stress as these enzymes are capable of conjugating with xenobiotic molecules (Huber et al. 2009). Moreover, *Glomus* mycelia increase the water uptake and have water retention capacity in the rhizosphere under resource limiting conditions, thereby alleviating the stress.

Salinity is the major environmental stress that limits crop production in sustainable agriculture in most of the arid and semiarid regions (Chinnusamy et al. 2005; Rengasamy 2006; Munns and Tester 2008). Naturally, plants are associated with certain soil microorganisms that influence their growth and development and they are known as plant growth promoting microorganisms. Soil salinity reduces the uptake of major nutrients such as N, P, and potassium (K) by plants because of changes in nutrient metabolism and competition for binding sites (Evelin et al. 2009). Thus, AMF is the best remediator of salinity stress. AMF inoculated plants exhibit an increased chlorophyll content, higher uptake of N and Mg, but decreased Na-transport even under saline conditions (Borde et al. 2010; Talaat and Shawky 2014).

Several studies have shown that AMF symbiosis improves the plant tissue hydration, drought tolerance and physiology under water stress conditions (see Azcòn et al. 2013 and references therein). Water stresses affect plant growth and yield through three main mechanisms such as: decline the absorption of the photosynthetically active radiation, reduction in the efficiency of radiation use and lessened harvest index (Padmavathi 2017). Plants respond to abiotic stresses at morphological, cellular and metabolic levels with changes that allow them to either avoid the stress or to increase its tolerance against stress (Bray 1997). The intensity of water stress depends upon the occurrence and distribution of rainfall, evaporative demands and moisture retention capacity of the soils (Farooq et al. 2005).

Colonization of plant roots by AMF under heavy metal (HM) stress results in expression of specific genes responsible for the production of proteins (including metallothioneins) that increase the tolerance of plants to stress (Rivera-Becerril et al. 2005). There are many AMF and plant genes that are involved in the tolerance to HM stress, including metal transporter genes, which are expressed at different levels, and AM symbiosis can regulate the transcription of such genes (Lanfranco et al. 2002; Gonzalez-Guerrero et al. 2005; Hildebrandt et al. 2007). Nowadays following the physical and chemical methods for decontaminating a polluted environment are totally complicated and costly (Zamal et al. 2002). Microbial land remediation practices and the use of microorganisms as biocontrol agents in the place of chemical pesticides have gained considerable momentum for a sustainable agriculture. Asif and Bhabatosh (2013) demonstrated that the two plant species *Solanum melongena* L. and *Sorghum Sudanese* Staph., growing in the HM contaminated soil when inoculated with the AMF *Septoglomus deserticola* (Trappe, Bloss & J.A. Menge) G.A. Silva, Oehl & Sieverd. (=*Glomus deserticola*) resulted in good plant growth and alleviated the adverse effect of HM toxicity on plant growth.

13 Potential Use of AMF in Bioassays of Xenobiotics

Mycorrhizal fungi can be used as indicators of soil toxicity due to their effectiveness in remediation. The toxicity of pollutants can be determined by estimating spore germination, mycorrhizal infectivity and mycorrhizal colonization of roots (Weissenhorn and Leyval 1996; Jacquot et al. 2000). The germination of AMF spore aids in the indication of toxic compounds in the environment whereas, mycorrhizal colonization signifies the toxicity after symbiotic association between plant root and fungi has been established (Laheurte et al. 1990). AMF can even modify the chemical properties of root exudates and influence soil pH (Li et al. 1991), therefore influencing the soil microbial communities in the rhizosphere (Barea 1997) and improving soil structure (Rillig and Mummey 2006). Mycorrhizal fungi affect the uptake of metals in the by plants (Khan et al. 2000). The AMF contributes to the metal tolerance to plants in different ways (Joner et al. 2004). The mycorrhizal fungi eliminate the toxic metals by declining the metal translocation from root to shoot (Leyval et al. 1997). In a study carried out by Binet et al. (2000) the rye plants associated with mycorrhiza absorbed lower concentrations of anthracene and PAH in their root and shoots, while, non-mycorrhizal plants accumulated higher amounts of PAH which proves that AMF inoculated plants could thrive well in the contaminated soil.

Mycorrhizal fungi interact with some of the other favourable soil organisms in order to reach complete removal of the pollutants from the soils. The interaction between earthworm and mycorrhiza resulted in promoting remediation of HM contaminated soils. With the help of earthworm activities, mycorrhizal colonization of rye grassroots was quite rapid (Yu et al. 2011). This relationship resulted in significantly decreasing the cadmium content in the soil. Earthworm produces phytohormones, which may have also stimulated the mycorrhizal colonization of plant roots (Chibuike 2013). Gange (1993) showed that the number of infective propagules of mycorrhizal fungi was ten times higher in the presence of earthworm casts than in the neighbouring soils. The composite reaction of earthworm and mycorrhiza on remediation of HM in the soil is composite; the underlying mechanism involved in this relationship is not fully understood. Most of the microorganisms that are involved in the remediation of organic pollutants have the potential to biodegrade these pollutants; thus, other microorganisms used along with mycorrhiza, the remediation are rapid and more effective. The application of filamentous fungi (*Cunninghamella echinulata*) and bacteria (*Sphingomonas paucimobilis*) in combination with AMF have been reported in the degradation of polluted soil containing petroleum hydrocarbon (Alarcón et al. 2008).

14 AMF in Environmental Stabilization

The interaction of plant with beneficial rhizospheric microorganisms holds a great promise for low-cost tool for cleaning the environment (Karimi et al. 2011). Activities including usage of chemical pesticides, insecticides and mining activities that end up in the accumulation of HMs as soil contaminants pose threats to the soil environment (Atafar et al. 2010; Alloway 2013; Abd El-Ghany and Masmali 2016). Plants with mycorrhizal association play a key role in the elimination of soil pollutants by absorption and transformation of harmful pollutants in the soil (Mathur et al. 2007). The role of AMF in enhancing plant tolerance to pollutant depends on AMF species, plant genotype and type of pollutant (Sudova and Vosatka 2007). Therefore, suitable mycobiont for the plant could biodegrade the soil very effectively (Miransari 2011). Though, primary colonizers of polluted soils are presumed to be non-mycorrhizal (Shetty et al. 1994), enhancement of soil structure and aggregation chiefly rely on the presence of fungal symbionts (Rillig et al. 2015; Lehmann 2015).

The degraded lands are generally considered to have low levels of AMF propagules due to lack of plant roots for their proliferation (Brundrett and Abbott 2002), hence leading to soil disturbances that could pave a path for invasive plant species for their successful establishment (Mack et al. 2000). However, inoculation of mycorrhizal fungi could enhance or facilitate the restoration of degraded lands (Schnoor et al. 2011; Cardozo-Junior et al. 2016). The ruderal AMF species in the degraded lands are characterized by faster growth with short lifespan, sporulation and recolonization of the host plant roots (Chagnon and Bradley 2013). The fungal hyphae develop inside the soil matrix thus forming skeletal structures that grasp and holds the primary soil particles to form soil aggregations (Lehmann 2015). One of the chief features of AMF species is the production of glycoprotein, glomalin. Eventhough the structure of glomalin is still obscure, it is known to be composed of several monomeric structures that are bounded by hydrophobic interactions that adhere to soil particles and helps to stabilize the soil aggregate that is useful in improving soil fertility and aeration (Nichols 2003; Fokom et al. 2013). Glomalin extracted from the AMF species, *F. mosseae* from polluted soil proved to be potential in sequestering toxic elements (González-Chávez et al. 2004). The phosphate fertilizers are considered as an important source of soil contamination in the crop productivity (Nziguheba and Smolders 2008). Inoculation of native plants with *F. mosseae* improved the plant growth by increasing the P contents and reducing the concentrations of toxic compounds in the mining sites (Chen et al. 2007). These studies showed that the application of mycorrhizal fungi could be used as an efficient method for stabilizing the soil environment and also in ecological restoration.

Mycorrhizal interaction is important in remediating soil contamination because of the extraradical fungal mycelium radiating far from the colonized root into the surrounding soil. It was shown that mycorrhizal fungi along with rhizosphere bacterial strain degrade better the soil pollutants than the mycorrhizal fungus alone. Further, AMF also improves the bacterial motility in the soil and are therefore are called as bacterial highways. This increases the accessibility of pollutants to the degraders (Malachowska-Jutsz and Kalka 2010; Simon et al. 2015).

Although enormous studies have reported the beneficial aspects of AMF, there are only limited studies available on the degradation and detoxification of xenobiotics by the application of AMF. Apart from AMF, ECM fungi and ERM fungi can also degrade organic pollutants prevailing in the environment (Donnelly et al. 1993; Green et al. 1999). In some mycorrhizal species (alfalfa, ryegrass and corn), the presence of AMF reduced the pollutant concentration whereas ECM (pine) decreased the soil contaminant concentration in case of PAH, but not hydrocarbons. Joner et al. (2006) reported that ECM fungal inoculation reduced the remediation potential due to enhanced nutrient scavenging by the symbiotic fungi in the rhizosphere. The reason for fluctuation in degradation capacity may depend on nutrient and water availability. To the other extent, synthetic pesticides and herbicide residues along with organic matter had greatly affected the mycorrhizal colonization of the soil (Mariela et al. 2016).

15 Conclusion and Future Perspective

Xenobiotics pose one of the major threats to the environment and degrading such compounds is one of the challenging issues. Although most studies on the biodegradation of these harmful toxic compounds through biological processes involve soil microorganisms, the role of mycorrhizal fungi, especially AMF is lacking. Mycorrhizal fungi are able to detoxify pollutants mostly in combination or in interaction with other organisms; hence in future studies on the application of AMF in bioremediation should also involve other coexisting microorganisms. Of recent, transgenic plants are developed and tested for their efficacy to degrade xenobiotic compounds. However, these transgenic plants are not suitable for large scale field applications. Nevertheless, with the advent of new technologies, these issues could be resolved in future. Further, isolation and characterization of native and possibly stress-adapted mycorrhizal fungi could be a strategy for the successful restoration of soils contaminated with xenobiotics. Extensive experimental studies under field conditions are essential as studies conducted under laboratory or controlled conditions eliminate certain natural factors and are often conducted involving a single or few plant or microbial species. Additionally, identification of a suitable combination of plant and mycorrhizal fungi tolerant to a high concentration of pollutants like the xenobiotics should be the priority of future research in phytoremediation. One of the major bottlenecks in this line is the inadequate knowledge of the molecular mechanisms involved in the responses of mycorrhizal fungi to different types of stresses. In future, mycorrhizal fungi should not only be used as a tool to improve and restore vegetation in degraded ecosystems but also should be considered as a bioindicator of soil health and quality (Lenoir et al. 2016). The role of mycorrhizal fungi as a bioindicator of soil arises from the fact that these fungi act as a direct link connecting the root and the soil environment. Though modern techniques have helped to understand the effect of xenobiotics on mycorrhizal fungal diversity and activity to a certain extent, the total picture in this line is far from complete.

References

Abd El-Ghany, T. M., & Masmali, I. A. (2016). Fungal biodegradation of organophosphorus insecticides and their impact on soil microbial population. *Journal of Plant Pathology and Microbiology, 7*, 349. https://doi.org/10.4172/2157-7471.1000349.

Adesemoye, A. O., Torbert, H. A., & Kloepper, J. W. (2009). Plant growth-promoting rhizobacteria allow reduced application rates of chemical fertilizers. *Microbial Ecology, 58*, 921–929.

Agrawal, N., & Dixit, A. K. (2015). An environmental cleanup strategy-Microbial transformation of xenobiotic compounds. *International Journal of Current Microbiology and Applied Sciences, 4*, 429–461.

Alarcón, A., Davies, F. T., Autenrieth, R. L., & Zuberer, D. A. (2008). Arbuscular mycorrhiza and petroleum-degrading microorganisms enhance phytoremediation of petroleum-contaminated soil. *International Journal of Phytoremediation, 10*, 251–263.

Alexander, M. (1965). Biodegradation: Problems of molecular recalcitrance and microbial infallibility. *Advances in Applied Microbiology, 7*, 35–80.

Alloway, B. J. (2013). Sources of heavy metals and metalloids in soils. In B. Alloway (Ed.), *Heavy metals in soils. Environmental Pollution* (Vol. 22, pp. 11–50). Dordrecht: Springer.

Amrani, A. E., Dumas, A. S., Wick, L. Y., Yergeau, E., & Berthome, R. (2015). "Omics" insights into PAH degradation toward improved green remediation biotechnologies. *Environmental Science & Technology, 49*, 11281–11291.

Anderson TA, Coats JR (1995) An overview of microbial degradation in the rhizosphere and its implications for bioremediation. In: Skipper HD, Turco RF (eds) Bioremediation: Science and applications. SSSA Spec. Publ. 43. SSSA, ASA, and CSSA, Madison, WI, pp 135–143.

Anderson, T. A., Guthrie, E. A., & Walton, B. T. (1993). Bioremediation in the rhizosphere. *Environmental Science & Technology, 27*, 2630–2636.

Aranda, E., Scervino, J. M., Godoy, P., Reina, R., Ocampo, J. A., Wittich, R. M., & García-Romera, I. (2013). Role of arbuscular mycorrhizal fungus *Rhizophagus custos* in the dissipation of PAHs under root-organ culture conditions. *Environmental Pollution, 181*, 182–189.

Arora, P. K., Sasikala, C., & Ramana, C. V. (2012). Degradation of chlorinated nitroaromatic compounds. *Applied Microbiology and Biotechnology, 93*, 2265–2277.

Asif, M., & Bhabatosh, M. (2013). Effects of inoculation with stress adapted arbuscular mycorrhizal fungus *Glomus deserticola* on growth of *Solanum melogena* L. and *Sorghum sudanese* Staph., seedlings under salinity and heavy metal stress conditions. *Archives of Agronomy and Soil Science, 59*, 173–183.

Atafar, Z., Mesdaghinia, A. R., Nouri, J., Homaee, M., Yunesian, M., Ahmadimoghaddam, M., & Mahvi, A. H. (2010). Effect of fertilizer application on soil heavy metal concentration. *Environmental Monitoring and Assessment, 160*, 83–89.

Auge, R. M. (2001). Water relations drought and vesicular arbuscular mycorrhizal symbiosis. *Mycorrhiza, 11*, 3–42.

Azcón, R., Medina, A., Aroca, R., & Ruiz-Lozano, J. M. (2013). Abiotic stress remediation by the arbuscular mycorrhizal symbiosis and rhizosphere bacteria/yeast interactions. In F. J. de Bruijn (Ed.), *Molecular Microbial Ecology of the Rhizosphere* (Vol. 2, 1st ed., pp. 991–1002). New Jersey: John Wiley & Sons, Inc.

Baetz, U., & Martinoia, E. (2014). Root exudates: The hidden part of plant defense. *Trends in Plant Science, 19*, 90–98.

Bais, H. P., Weir, T. L., Perry, L. G., Gilroy, S., & Vivanco, J. M. (2006). The role of root exudates in rhizosphere interactions with plants and other organisms. *Annual Review of Plant Biology, 57*, 233–266.

Barea, J. M. (1997). Mycorrhiza/bacteria interactions on plant growth promotion. In A. Ogoshi, L. Kobayashi, Y. Homma, F. Kodama, N. Kondon, & S. Akino (Eds.), *Plant growth-promoting rhizobacteria, present status and future prospects* (pp. 150–158). Paris: OECD.

Barkay, T., & Pritchard, H. (1988). Adaptation of aquatic microbial communities to pollutant stress. *Microbiological Sciences, 5*, 165–169.

Barr, D. P., & Aust, S. D. (1994). Pollutant degradation by white rot fungi. *Reviews of Environmental Contamination and Toxicology, 138*, 49–72.

Bennet, T. W., Wnnch, K. G., & Faision, B. D. (2002). Use of fungi biodegradation. In C. J. Hurst (Ed.), *Manual of environmental microbiology* (pp. 960–971). Washington, DC: ASM Press.

Bhandari, G. (2018). Bioremediation of industrial waste using microbial metabolic diversity. In Pankaj & A. Sharma (Eds.), *Microbial biotechnology in environmental monitoring and cleanup* (pp. 286–304). Hershey: IGI Global: International Publisher of Information Science and Technology Research.

Binet, P., Portal, J. M., & Leyval, C. (2000). Fate of polycyclic aromatic hydrocarbons (PAH) in the rhizosphere and mycorrhizosphere of ryegrass. *Plant and Soil, 227*, 207–213.

Bodekar, I. T. M., Nygren, C. M. R., Taylor, A. F. S., Olson, A., & Lindahl, B. D. (2009). Class II peroxidase-encoding genes are present in a phylogenetically wide range of ectomycorrhizal fungi. *The ISME Journal, 3*, 1387–1395.

Borde, M., Dudhane, M., & Jite, P. K. (2010). AM fungi influences the photosynthetic activity, growth and antioxidant enzymes in *Allium sativum* L. under salinity condition. *Notulae Scientia Biologicae, 2*, 64–71.

Borie, F., Rubio, R., & Morales, A. (2008). Arbuscular mycorrhizal fungi and soil aggregation. *Journal of Soil Science and Plant Nutrition, 8*, 9–18.

Bray, D. E. (1997). Plant responses to water deficit. *Trends in Plant Science, 2*, 48–54.

Brundrett, M. C., & Abbott, L. K. (2002). Arbuscular mycorrhizas in plant communities. In K. Sivasithamparam, K. W. Dixon, & R. L. Barrett (Eds.), *Microorganisms in plant diversity conservation and biodiversity* (pp. 151–193). London: Kluwer Academica Publishers.

Bulucea, C. A., Rosen, M. A., Mastorakis, N. E., Bulucea, C. A., & Brindusa, C. (2012). Approaching resonant absorption of environmental xenobiotics harmonic oscillation by linear structure. *Sustainability, 4*, 561–573.

Burke, R. M., & Cairney, J. W. G. (2002). Laccases and other polyphenol oxidases in ecto- and ericoid mycorrhizal fungi. *Mycorrhiza, 12*, 105–116.

Campagnac, E., Sahraoui, L. A., Debiane, D., Fontaine, J., Laruclle, F., Garcon, G., Verdin, A., Durand, R., Shiralo, P., & Grandmougin, F. (2010). Arbuscular mycorrhiza partially protected chicory roots against oxidative stress induced by two fungicides, fenpropimorph and fenhexamid. *Mycorrhiza, 20*, 167–178.

Cardozo Junior, F. M., Carneiro, R. F. V., Rocha, S. M. B., Nunes, L. A. P. L., Santos, V. M., dos Feitoza, L. L., & Araújo, A. S. F. (2016). The impact of pasture systems on soil microbial biomass and community-level physiological profiles. *Land Degradation and Development, 29*, 284–291.

Cébron, A., Louvel, B., Faure, P., France-Lanord, C., Chen, Y., & Murrell, J. C. (2011). Root exudates modify bacterial diversity of phenanthrene degraders in PAH-polluted soil but not phenanthrene degradation rates. *Environmental Microbiology, 13*, 722–736.

Chagnon, P. L., & Bradley, R. L. (2013). Evidence that soil nutrient stoichiometry controls the competitive abilities of arbuscular mycorrhizal vs. root-borne non-mycorrhizal fungi. *Fungal Ecology, 6*, 557–660.

Channabasava, A., Lakshman, H. C., & Jorquera, M. A. (2015). Effect of fungicides on association of arbuscular mycorrhiza fungus Rhizophagus fasciculatus and growth of Proso millet (*Panicum miliaceum* L.). *Journal of Soil Science and Plant Nutrition, 15*, 35–45.

Chaudhry, Q., Blom-Zandstra, M., Gupta, S., & Joner, E. J. (2005). Utilising the synergy between plants and rhizosphere microorganisms to enhance breakdown of organic pollutants in the environment. *Environmental Science and Pollution Research, 12*, 34–48.

Chen, B. D., Zhu, Y. G., Duan, J., Xiao, X. Y., & Smith, S. E. (2007). Effects of the arbuscular mycorrhizal fungus *Glomus mosseae* on growth and metal uptake by four plant species in copper mine tailings. *Environmental Pollution, 147*, 374–380.

Chen, W., Consortium, F. B., & List, F. B. C. A. (2012). Nuclear ribosomal internal transcribed spacer (ITS) region as a universal DNA barcode marker for Fungi. *Proceedings of the National Academy of Sciences of the United States of America, 109*, 6241–6246.

Chibuike, G. (2013). Use of mycorrhiza in soil remediation: a review. *Scientific Research and Essays, 8*, 1679–1687.

Chinnusamy, V., Jagendorf, A., & Zhu, J. K. (2005). Understanding and improving salt tolerance in plants. *Crop Science, 45*, 437–448.

Corgié, S. C., Beguiristain, T., & Leyval, C. (2004). Spatial distribution of bacterial communities and phenanthrene degradation in the rhizosphere of *Lolium perenne* L. *Applied and Environmental Microbiology, 70*, 3552–3557.

Courty, P. E., Labbé, J., Kohler, A., Marçais, B., Bastien, C., Churin, J. L., & Garbaye, V. (2011). Effect of poplar genotypes on mycorrhizal infection and secreted enzyme activities in mycorrhizal and non-mycorrhizal roots. *Journal of Experimental Botany, 62*, 249–260.

Crowley, D. E., Brennerova, M. V., Irwin, C., Brenner, V., & Focht, D. D. (1996). Rhizosphere effects on biodegradation of 2 5-dichlorobenzoate by a bioluminescent strain of root colonizing *Pseudomonas fluorescens*. *FEMS Microbiology Ecology, 20*, 79–89.

Cunningham, S. D., Anderson, T. A., Schwab, A. P., & Hsu, F. C. (1996). Phytoremediation of soils contaminated with organic pollutants. *Advances in Agronomy, 56*, 55–114.

Curl, E. A., & Truelove, B. (1986). *The rhizosphere*. Berlin: Springer.

Dennis, P. G., Miller, A. J., & Hirsch, P. R. (2010). Are root exudates more important than other sources of rhizodeposits in structuring rhizosphere bacterial communities. *FEMS Microbiology Ecology, 72*, 313–327.

Díaz, E. (2004). Bacterial degradation of aromatic pollutants: a paradigm of metabolic versatility. *International Microbiology, 7*, 173–180.

Dietrich, D., Hickey, W. J., & Lamar, R. T. (1995). Degradation of 4,4′- dichlorobiphenyl, 3,3′, 4,4′-tetrachlorobiphenyl and 2,2′,4,4′,5,5′-hexachlorobiphenyl by the white-rot fungus *Phanerochaete chrysosporium*. *Applied and Environmental Microbiology, 61*, 3904–3909.

Donnelly, P. K., & Fletcher, J. S. (1995). PCB metabolism by ectomycorrhizal fungi. *Bulletin of Environmental Contamination and Toxicology, 54*, 507–513.

Donnelly, P. K., Entry, J. A., & Crawford, D. L. (1993). Degradation of atrazine and 2,4-dichlorophenoxyacetic acid by mycorrhizal fungi at three nitrogen concentrations *in vitro*. *Applied and Environmental Microbiology, 59*, 2642–2647.

Doty, S. L., James, C. A., Moore, A. L., Vajzovic, A., Singleton, G. L., Ma, C., Khan, Z., Xin, G., Kang, J. W., Park, J. Y., Meilan, R., Strauss, S. H., Wilkerson, J., Farin, F., & Strand, S. E. (2007). Enhanced phytoremediation of volatile environmental pollutants with transgenic trees. *Proceedings of the National Academy of Sciences of the United States of America, 104*, 16816–16821.

Druillea, M., Cabellob, M. N., Omacinia, M., & Golluscioa, R. A. (2013a). Glyphosate reduces spore viability and root colonization of arbuscular mycorrhizal fungi. *Applied Soil Ecology, 64*, 99–103.

Druillea, M., Omacinia, M., Golluscioa, R. A., & Cabellob, M. N. (2013b). Arbuscular mycorrhizal fungi are directly and indirectly affected by glyphosate application. *Applied Soil Ecology, 72*, 143–149.

Dubey, K. K., & Fulekar, M. H. (2011). Effect of pesticides on the seed germination of Cenchrus setigerus and *Pennisetum pedicellatum* as monocropping and co-cropping system: Implications for rhizospheric bioremediation. *Romanian Biotechnological Letters, 16*, 5908–5918.

Dubey, K. K., & Fulekar, M. H. (2013). Investigation of potential rhizospheric isolate for cypermethrin degradation. *Biotech, 3*, 33–43.

Duponnois, R., Kisa, M., & Plenchette, C. (2006). Phosphate solubilizing potential of the nematofungus *Arthrobotrys oligospora. Journal of Plant Nutrition and Soil Science, 169*, 280–282.

Edwards, S. J., & Kjellerup, B. V. (2013). Applications of biofilms in bioremediation and biotransformation of persistent organic pollutants, pharmaceuticals/personal care products, and heavy metals. *Applied Microbiology and Biotechnology, 97*, 9909–9921.

Evelin, H., Kapoor, R., & Giri, B. (2009). Arbuscular mycorrhizal fungi in alleviation of salt stress: a review. *Annals of Botany, 104*, 1263–1280.

Fan, S., Li, P., Gong, Z., Ren, W., & He, N. (2008). Promotion of pyrene degradation in rhizosphere of alfalfa (*Medicago sativa* L.). *Chemosphere, 71*, 1593–1598.

Fang, C., Radosevich, M., & Fuhrmann, J. J. (2001). Characterization of rhizosphere microbial community structure in five similar grass species using fame and biology analyses. *Soil Biology and Biochemistry, 33*, 679–682.

Farooq, M., Basra, S. M. A., Hafeez, K., & Ahmad, N. (2005). Thermal hardening a new seed vigor enhancement tool in rice. *Journal of Integrative Plant Biology, 47*, 187–193.

Farrar, J., Aawes, M., Tones, D., & Lindow, S. (2003). How roots control the flux of carbon to the rhizosphere. *Ecology, 84*, 827–837.

Flathman, P. E., & Lanza, G. R. (1998). Phytoremediation current views on an emerging green technology. *Journal of Soil Contamination, 7*, 415–432.

Fokom, R., Teugwa, M. C., Nana, W. L., Ngonkeu, M. E. L., Tchameni, S., Nwaga, D., Rillig, C. M., & Amvam, Z. P. H. (2013). Glomalin, carbon, nitrogen and soil aggregate stability as affected by land use changes in the humid forest zone in South Cameroon. *Applied Ecology and Environmental Research, 11*, 581–592.

Furukawa, K. (2018). Microbial degradation of polychlorinated biphenyls. In A. M. Chakrabarty (Ed.), *Biodegradation and detoxification of environmental pollutants* (pp. 33–58). New York: CRC Press.

Gange, A. (1993). Translocation of mycorrhizal fungi by earthworms during early succession. *Soil Biology and Biochemistry, 25*, 1021–1026.

Gao, Y., Cheng, Z., Ling, W., & Zhu, X. (2010). Arbuscular mycorrhizal fungal hyphae contribute to the uptake of polycyclic aromatic hydrocarbons by plant roots. *Bioresource Technology, 101*, 6895–6901.

Gianfreda, L. (2015). Enzymes of importance to rhizosphere processes. *Journal of Soil Science and Plant Nutrition, 15*, 283–306.

Gianfreda, L., & Rao, M. A. (2008). Interactions between xenobiotics and microbial and enzymatic soil activity. *Critical Reviews in Environmental Science and Technology, 38*, 269–310.

Gianinazzi, S., Gollotte, A., Marie-Noëlle, B., van Tuinen, D., & Redecker, W. D. (2010). Agroecology the key role of arbuscular mycorrhizas in ecosystem services. *Mycorrhiza, 20*, 519–530.

Giovannetti, M., Turrini, A., Strani, P., Sbrana, C., Avio, L., & Pietrangeli, B. (2006). Mycorrhizal fungi in ecotoxicological studies: Soil impact of fungicides, insecticides and herbicides. *Prevention Today, 2*, 47–61.

Gonzalez- Guerrero, M., Azcon- Aguilar, C., Mooney, M., Valderas, A., Macdiarmid, C. W., Eide, D. J., & Ferrol, N. (2005). Characterization of a *Glomus intraradices* gene encoding a putative Zn transporter of the cation diffusion facilitator family. *Fungal Genetics and Biology, 42*, 130–140.

Gonzalez, L. F., Sarria, V., & Sanchez, O. F. (2010). Degradation of chlorophenols by sequential biological-advanced oxidative process using *Trametes pubescens* and TiO2/U. *Bioresource Technology, 101*, 3493–3499.

González-Chávez, M. C., Carrillo-González, R., Wright, S. F., & Nichols, K. A. (2004). The role of glomalin, a protein produced by arbuscular mycorrhizal fungi, in sequestering potentially toxic elements. *Environmental Pollution, 130*, 317–323.

Graham, J. H., Leonard, R. T., & Menge, J. A. (1981). Membrane-mediated decrease in root exudation responsible for phosphorus inhibition of vesicular-arbuscular mycorrhiza formation. *Plant Physiology, 68*, 549–552.

Gramss, G., & Rudeschko, O. (1998). Activities of oxidoreductase enzymes in tissue extracts and sterile root exudates of three crop plants, and some properties of the peroxidase component. *The New Phytologist, 138*, 401–409.

Green, H., Larsen, J., Olsson, P. A., Jensen, D. F., & Jakobsen, I. (1999). Suppression of the biocontrol agent *Trichoderma harzianum* by mycelium of the arbuscular mycorrhizal fungus *Glomus intraradices* in root-free soil. *Applied and Environmental Microbiology, 65*, 1428–1434.

Greń, I., Guzik, U., Wojcieszyńska, D., & Łabużek, S. (2008). Molekularne podstawy rozkładu ksenobiotycznych związków aromatycznych. *Biotechnologia, 2*, 58–67.

Harley, J. L. (1989). The fourth benefactors' lecture the significance of mycorrhiza. *Mycological Research, 92*, 129–139.

Harms, H., Schlosser, D., & Wick, L. Y. (2011). Untapped potential: Exploiting fungi in bioremediation of hazardous chemicals. *Nature Reviews. Microbiology, 9*, 177–192.

He, Y., Xu, J., Tang, C., & Wu, Y. (2005). Facilitation of pentachlorophenol degradation in the rhizosphere of ryegrass (*Lolium perenne* L.). *Soil Biology and Biochemistry, 37*, 2017–2024.

Higson, F. K. (1992). Microbial degradation of nitroaromatic compounds. *Advances in Applied Microbiology, 37*, 1–19.

Hildebrandt, U., Regvar, M., & Bothe, H. (2007). Arbuscular mycorrhiza and heavy metal tolerance. *Phytochemistry, 68*, 139–146.

Hinsinger, P., Gobran, G. R., Gregory, P. J., & Wenzel, W. W. (2005). Rhizosphere geometry and heterogeneity arising from root-mediated physical and chemical processes. *The New Phytologist, 168*, 293–303.

Hinsinger, P., Bengough, A. G., Vetterlein, D., & Young, I. M. (2009). Rhizosphere: biophysics, biogeochemistry and ecological relevance. *Plant and Soil, 321*, 117–152.

Hsu, T. S., & Bartha, R. (1979). Accelerated mineralization of two organophosphate insecticides in the rhizosphere. *Applied and Environmental Microbiology, 37*, 36–41.

Huang, Y. L., Li, Q. B., Deng, X., Lu, Y. H., Liao, X. K., Hong, M. Y., & Wang, Y. (2005). Aerobic and anaerobic biodegradation of polyethylene glycols using sludge microbes. *Process Biochemistry, 40*, 207–211.

Huber, C., Bartha, B., Harpaintner, R., & Schröder, P. (2009). Metabolism of acetaminophen (paracetamol) in plants two independent pathways result in the formation of a glutathione and a glucose conjugate. *Environmental Science and Pollution Research, 16*, 206–213.

Husaini, A., Roslan, H. A., Hii, K. S. Y., & Ang, C. H. (2008). Biodegradation of aliphatic hydrocarbon by indigenous fungi isolated from used motor oil contaminated sites. *World Journal of Microbiology and Biotechnology, 24*, 2789–2797.

Ibanez, S. G., Medina, M. I., & Agostini, E. (2011). Phenol tolerance, changes of antioxidative enzymes and cellular damage in transgenic tobacco hairy roots colonized by arbuscular mycorrhizal fungi. *Chemosphere, 83*, 700–705.

Ichinose, H. (2013). Cytochrome P450 of wood-rotting basidiomycetes and biotechnological applications. *Biotechnology and Applied Biochemistry, 60*, 71–81.

Ipsilantis, I., Samourelis, C., & Karpouzas, D. G. (2012). The impact of biological pesticides on arbuscular mycorrhizal fungi. *Soil Biology and Biochemistry, 45*, 147–155.

Jacquot, E., van Tuinen, D., Gianinazzi, S., & Gianinazzi-Pearson, V. (2000). Monitoring species of arbuscular mycorrhizal fungi in planta and in soil by nested PCR: Application to the study of the impact of sewage sludge. *Plant and Soil, 226*, 179–188.

Jansa, J., Bukovská, P., & Gryndler, M. (2013). Mycorrhizal hyphae as ecological niche for highly specialized hypersymbionts – or just soil free-riders? *Frontiers in Plant Science, 4*, 134. https://doi.org/10.3389/fpls.2013.00134.

Johansson, J. F., Paul, L. R., & Finlay, R. D. (2004). Microbial interactions in the mycorrhizosphere and their significance for sustainable agriculture. *FEMS Microbiology Ecology, 48*, 1–13.

Joner, E., & Leyval, C. (2001). Influence of arbuscular mycorrhiza on clover and ryegrass grown together in a soil spiked with polycyclic aromatic hydrocarbons. *Mycorrhiza, 10*, 155–159.

Joner, E., & Leyval, C. (2009). Phytoremediation of organic pollutants using mycorrhizal plants: A new aspect of rhizosphere interactions. In E. Lichtfouse, M. Navarrete, P. Debaeke, S. Véronique, & C. Alberola (Eds.), *Sustainable agriculture* (pp. 885–894). Dordrecht: Springer.

Joner, E. J., Johansen, A., Loibner, A. P., de la Cruz, M. A., Szolar, O. H., Portal, J. M., & Leyval, C. (2001). Rhizosphere effects on microbial community structure and dissipation and toxicity of polycyclic aromatic hydrocarbons (PAHs) in spiked soil. *Environmental Science & Technology, 35*, 2773–2777.

Joner, E. J., Roos, P., Jansa, J., Frossard, E., Leyval, C., & Jakobsen, I. (2004). No significant contribution of arbuscular mycorrhizal fungi to transfer of radiocesium from soil to plants. *Applied and Environmental Microbiology, 70*, 6512–6517.

Joner, E. J., Leyval, C., & Colpaert, J. V. (2006). Ectomycorrhizas impede phytoremediation of polycyclic aromatic hydrocarbons (PAHs) both within and beyond the rhizosphere. *Environmental Pollution, 142*, 34–38.

Kaimi, E., Mukaidani, T., Miyoshi, S., & Tamaki, M. (2006). Ryegrass enhancement of biodegradation in diesel contaminated soil. *Environmental and Experimental Botany, 55*, 110–119.

Karimi, A., Khodaverdiloo, H., Sepehri, M., & Sadaghiani, M. R. (2011). Arbuscular mycorrhizal fungi and heavy metal contaminated soils. *African Journal of Microbiology Research, 5*, 1571–1576.

Karpouzas, D. G., & Singh, B. K. (2006). Microbial degradation of organophosphorus xenobiotics metabolic pathways and molecular basis. *Physics of Bacterial Morphogenesis, 51*, 119–185.

Karpouzas, D. G., Papadopoulou, E., Ipsilantis, I., Friedel, I., Petric, I., Udikovic-Kolic, N., Djuric, S., Kandeler, E., Menkissoglu-Spiroudi, U., & Martin-Laurent, F. (2014). Effects of nicosulfuron on the abundance and diversity of arbuscular mycorrhizal fungi used as indicators of pesticide soil microbial toxicity. *Ecological Indicators, 39*, 44–53.

Khalvati, M. A. (2005). Quantification of water uptake of hyphae contributing to barley subjected to drought conditions. Doctoral Dissertation, Technical University of Munich. Academic, pp 89.

Khalvati, M., Barth, B., Dupigny, A., & Schröder, P. (2010). Arbuscular mycorrhizal association is beneficial for growth and detoxification of xenobiotics of barley under drought stress. *Journal of Soils and Sediments, 10*, 54–64.

Khan, A. G., Kuek, C., Chaudhry, T. M., Khoo, C. S., & Hayes, W. J. (2000). Plants, mycorrhizae and phytochelators in heavy metal contaminated land remediation. *Chemosphere, 41*, 197–207.

Kohler, A. (2015). Convergent losses of decay mechanisms and rapid turnover of symbiosis genes in mycorrhizal mutualists. *Nature Genetics, 47*, 410–415.

Kuiper, I., Lagendijk, E. L., Bloemberg, G. V., & Lugtenberg, B. J. (2004). Rhizoremediation: A beneficial plant–microbe interaction. *Molecular Plant-Microbe Interactions, 17*, 6–15.

Laheurte, F., Leyval, C., & Berthelin, J. (1990). Root exudates of maize, pine and beech seedlings influenced by mycorrhizal and bacterial inoculation. *Symbiosis, 9*, 111–116.

Lanfranco, L., Bolchi, A., Ros, E. C., Ottonello, S., & Bonfante, P. (2002). Differential expression of a metallothionein gene during the presymbiotic versus the symbiotic phase of an arbuscular mycorrhizal fungus. *Plant Physiology, 130*, 58–67.

Leake, J. R., Johnson, D., Donnelly, D. P., Muckle, G. E., Boddy, L., & Read, D. J. (2004). Networks of power and influence: the role of mycorrhizal mycelium in controlling plant communities and agroecosystem functioning. *Canadian Journal of Botany, 82*, 1016–1045.

Lee, S.-H., Lee, W.-S., Lee, C.-H., & Kim, J.-G. (2008). Degradation of phenanthrene and pyrene in rhizosphere of grasses and legumes. *Journal of Hazardous Materials, 153*, 892–898.

Lehmann, A. (2015). Plant root and mycorrhizal fungal traits for understanding soil aggregation. *The New Phytologist, 205*, 1385–1388.
Leja, K., & Lewandowicz, G. (2010). Polymer biodegradation and biodegradable polymer sea review. *Polish Journal of Environmental Studies, 19*, 255–266.
Lenoir, I., Fontaine, J., & Lounès-Hadj Sahraoui, A. (2016). Arbuscular mycorrhizal fungal responses to abiotic stresses: A review. *Phytochemistry, 123*, 4–15.
Leyval, C., Turnau, K., & Haselwandter. (1997). Effect of heavy metal pollution on mycorrhizal colonization and function: physiological, ecological and applied aspects. *Mycorrhiza, 7*, 139–153.
Li, X.-L., George, E., & Marschner, H. (1991). Phosphorus depletion and pH decrease at the root-soil and hyphae-soil interfaces of VA mycorrhizal white clover fertilized with ammonium. *The New Phytologist, 119*, 397–404.
Liao, H. L., Chen, Y., Bruns, T. D., Peay, K. G., Taylor, J. W., Branco, S., Talbot, J. M., & Vilgalys, R. (2014). Metatranscriptomic analysis of ectomycorrhizal roots reveal genes associated with Piloderma-Pinus symbiosis: Improved methodologies for assessing gene expression in situ. *Environmental Microbiology, 16*, 3730–3742.
Longato, S., & Bonfante, P. (1997). Molecular identification of mycorrhizal fungi by direct amplification of microsatellite regions. *Mycological Research, 101*, 425–432.
Mack, R. N., Simberloff, D., Mark Lonsdale, W., Evans, H., Clout, M., & Bazzaz, F. A. (2000). Biotic invasions: Causes, epidemiology, global consequences and control. *Ecological Applications, 10*, 689–710.
Malachowska-Jutsz, A., & Kalka, J. (2010). Influence of mycorrhizal fungi on remediation of soil contaminated by petroleum hydrocarbons. *Bulletin of Environment, 19*, 3217–3223.
Mangan, S. A., Schnitzer, S. A., Herre, E. A., Mack, K. M. L., & Valencia, M. C. (2010). Negative plant-soil feedback predicts tree-species relative abundance in a tropical forest. *Nature, 466*, 752–755.
Marco-Urrea, E., & Reddy, C. A. (2012). Degradation of chloro organic pollutants by white rot fungi. In S. N. Singh (Ed.), *Microbial degradation of xenobiotics* (pp. 31–66). Heidelberg: Springer.
Marecik, R., Kroliczak, P., Czaczyk, K., Bialas, W., Olejnik, A., & Cyplik, P. (2008). Atrazine degradation by aerobic microorganisms isolated from the rhizosphere of sweet flag (*Acorus calamus* L.). *Biodegradation, 19*, 293–301.
Mariela, F. P., Pável, M. E. I., Manuel, S. R. L., Jesús, F. G. M., & Reyes, L. O. (2016). Dehydrogenase and mycorrhizal colonization: tools for monitoring agrosystem soil quality. *Applied Soil Ecology, 100*, 144–153.
Marschner, H. (1995). *Mineral nutrition of higher plants* (2nd ed.). London: Academic.
Mathur, N., Bhatnagar, P., Mohan, K., Bakre, P., Nagar, P., & Bijarnia, M. (2007). Mutagenicity evaluation of industrial sludge from common effluent treatment plant. *Chemosphere, 67*, 1229–1235.
Matsui, T., Nomura, Y., Takano, M., Imai, S., Nakayama, H., Miyasaka, H., Okuhata, H., Tanaka, S., Matsuura, H., Harada, K., Bamba, T., Hirata, K., & Kato, K. (2011). Molecular cloning and partial characterization of a peroxidase gene expressed in the roots of *Portulaca oleracea* cv., one potentially useful in the remediation of phenolic pollutants. *Bioscience, Biotechnology, and Biochemistry, 75*, 882–890.
Maurya, P. K., & Malik, D. S. (2016). Distribution of heavy metals in water, sediments and fish tissue (*Heteropneustis fossilis*) in Kali River of western UP India. *International Journal of Fisheries and Aquatic Studies, 4*, 208–215.
Meharg, A. A., & Cairney, J. W. G. (2000). Ectomycorrhizas- extending the capabilities of rhizosphere remediation? *Soil Biology and Biochemistry, 32*, 1475–1484.
Mendes, R., Garbeva, P., & Raaijmakers, J. M. (2013). The rhizosphere microbiome: Significance of plant beneficial, plant pathogenic, and human pathogenic microorganisms. *FEMS Microbiology Reviews, 37*, 634–663.
Miransari, M. (2011). Hyperaccumulators, arbuscular mycorrhizal fungi and stress of heavy metals. *Biotechnology Advances, 29*, 645–653.

Moredo, N., Lorenzo, M., Domínguez, A., Moldes, D., Cameselle, C., & Sanroman, A. (2003). Enhanced ligninolytic enzyme production and degrading capability of *Phanerochaete chrysosporium* and *Trametes versicolor*. *World Journal of Microbiology and Biotechnology, 19*, 665–669.

Mougin, C., Cheviron, N., Pinheiro, M., Lebrun, J. D., & Boukcim, H. (2013). New insights into the use of filamentous fungi and their degradative enzymes as tools for assessing the ecotoxicity of contaminated soils during bioremediation processes. In E. Goltapeh, Y. Danesh, & A. Varma (Eds.), *Fungi as Bioremediators. Soil Biology* (Vol. 32, pp. 419–432). Berlin/Heidelberg: Springer.

Munns, R., & Tester, M. (2008). Mechanisms of salinity tolerance. *Annual Review of Plant Biology, 59*, 651–681.

Nichols, K. A. (2003). Characterization of glomalin, a glycoprotein produced by arbuscular mycorrhizal fungi. Ph.D., Dissertation, University of Maryland, USA, http://hdl.handle.net/1903/86

Nziguheba, G., & Smolders, E. (2008). Inputs of trace elements in agricultural soils via phosphate fertilizers in European countries. *Science of the Total Environment, 390*, 53–57.

Padmavathi, T. (2017). The role of arbuscular mycorrhizal fungi in salt and drought stresses. In D. J. Bagyaraj & Jamaluddin (Eds.), *Microbes for plant stress management* (pp. 183–204). New Delhi: New India Publishing Agency.

Paul, D., Pandey, G., Pandey, J., & Jain, R. K. (2005). Accessing microbial diversity for bioremediation and environmental restoration. *Trends in Biotechnology, 23*, 135–142.

Peng, G., Xiaofen, W., Wanbin, Z., Hongyan, Y., Xu, C., & Zongjun, C. (2008). Degradation of corn stalk by the composite microbial system of MC1. *Journal of Environmental Sciences, 20*, 109–111.

Philippot, L., Raaijmakers, J. M., Lemanceau, P., & van der Putten, W. H. (2013). Going back to the roots: The microbial ecology of the rhizosphere. *Nature Reviews. Microbiology, 11*, 789–799.

Pinedo-Rilla, C., Aleu, J., & Collado, I. G. (2009). Pollutants biodegradation by fungi. *Current Organic Chemistry, 13*, 1194–1214.

Raaijmakers, J. M., Paulitz, T. C., Steinberg, C., Alabouvette, C., & Mo€enne-Loccoz, Y. (2009). The rhizosphere: A playground and battlefield for soil borne pathogens and beneficial microorganisms. *Plant and Soil, 321*, 341–361.

Rambelli, A. (1973). The rhizosphere of mycorrhizae. In G. C. Marks & T. T. Kozlowski (Eds.), *Ectomycorrhizae: Their ecology and physiology* (pp. 299–343). New York: Academic.

Read, D. J., Leake, J. R., & Perez-Moreno, J. (2004). Mycorrhizal fungi as drivers of ecosystem process in heathland and boreal forest biomes. *Canadian Journal of Botany, 82*, 1243–1263.

Reiger, P. G., Meier, H.-M., Gerle, M., Vogt, U., Groth, T., & Knackmuss, H.-J. (2002). Xenobiotics in the environment: Present and future strategies to obviate the problem of biological persistence. *Journal of Biotechnology, 94*, 101–123.

Rengasamy, P. (2006). World salinization with emphasis on Australia. *Journal of Experimental Botany, 57*, 1017–1023.

Rillig, M. C., & Mummey, D. L. (2006). Mycorrhizas and soil structure. *The New Phytologist, 171*, 41–53.

Rillig, M. C., Wright, S. F., & Eviner, V. T. (2002). The role of arbuscular mycorrhizal fungi and glomalin in soil aggregation: Comparing effects of five plant species. *Plant and Soil, 238*, 325–333.

Rillig, M. C., Aguilar-Trigueros, C. A., Bergmann, J., Verbruggen, E., Veresoglou, S. D., & Lehmann, A. (2015). Plant root and mycorrhizal fungal traits for understanding soil aggregation. *The New Phytologist, 205*, 1385–1388.

Rivera-Becerril, F., van Tuinen, D., Martin-Laurent, F., Metwally, A., Dietz, K. J., Gianinazzi, S., & Gianinazzi-Pearson, V. (2005). Molecular changes in *Pisum sativum* L. roots during arbuscular mycorrhiza buffering of cadmium stress. *Mycorrhiza, 16*, 51–60.

Rohrbacher, F., & St-Arnaud, M. (2016). Root exudation: The ecological driver of hydrocarbon rhizoremediation. *Agronomy, 6*, 19. https://doi.org/10.3390/agronomy6010019.

Roth-Bejerano, N., Navarro-Ródenas, A., & Gutiérrez, A. (2014). Types of mycorrhizal association. In V. Kagan-Zur, N. Roth-Bejerano, Y. Sitrit, & A. Morte (Eds.), *Desert Truffles. Soil Biology* (Vol. 38, pp. 69–80). Berlin/Heidelberg: Springer.

Ruiz-Lozano, J. M. (2003). Arbuscular mycorrhizal symbiosis and alleviation of osmotic stress. New perspectives for molecular studies. *Mycorrhiza, 13*, 309–317.

Sainz, M. J., Gonzalez, P. B., & Vilarino, A. (2006). Effects of hexachlorocyclohexane on rhizosphere fungal propagules and root colonization by arbuscular mycorrhizal fungi in *Plantago lanceolata*. *European Journal of Soil Science, 57*, 83–90.

Salzer, P., Corbere, H., & Boller, T. (1999). Hydrogen peroxide accumulation in *Medicago truncatula* roots colonized by the arbuscular mycorrhiza-forming fungus *Glomus intraradices*. *Planta, 208*, 319–325.

Sanchez-Diaz, M., & Honrubia, M. (1994). Water relations and alleviation of drought stress in mycorrhizal plants. In S. Gianinazzi & H. Schepp (Eds.), *Impact of arbuscular mycorrhizas on sustainable agriculture and natural ecosystems* (pp. 167–178). Basel: Birkhuser.

Savci, S. (2012). Investigation of effect of chemical fertilizers on environment. *APCBEE Procedia, 1*, 287–292.

Scheibner, K., Hofrichter, M., & Fritsche, W. (1997). Mineralization of 2-amino-4,6-dinitrotoluene by manganese peroxidase of the white-rot fungus *Nematoloma frowardii*. *Biotechnology Letters, 19*, 835–839.

Scheublin, T. R., Sanders, I. R., Keel, C., & van der Meer, J. R. (2010). Characterisation of microbial communities colonising the hyphal surfaces of arbuscular mycorrhizal fungi. *The ISME Journal, 4*, 752–763.

Schnoor, T. K., Lekberg, Y., Rosendahl, S., & Olsson, P. A. (2011). Mechanical soil disturbance as a determinant of arbuscular mycorrhizal fungal communities in semi-natural grassland. *Mycorrhiza, 21*, 211–220.

Schoch, C. L., Seifert, K. A., Huhndorf, S., Robert, V., Spouge, J. L., Levesque, C. A., Chen, W., Bolchacova, E., Voigt, K., & Crous, P. W. (2012). Nuclear ribosomal internal transcribed spacer (ITS) region as a universal DNA barcode marker for fungi. *Proceedings of the National Academy of Sciences of the United States of America, 109*, 6241–6246.

Schreiner, R. P., & Bethlenfalvay, G. J. (1997). Plant and soil response to single and mixed species of arbuscular mycorrhizal fungi under fungicide stress. *Applied Soil Ecology, 7*, 93–102.

Schweiger, P. F., & Jakobsen, I. (1998). Dose-response relationships between four pesticides and phosphorus uptake by hyphae of arbuscular mycorrhizas. *Soil Biology and Biochemistry, 30*, 1415–1422.

Serra, A.-A., Nuttens, A., Larvor, V., Renault, D., Couée, I., Sulmon, C., & Gouesbet, G. (2013). Low environmentally relevant levels of bioactive xenobiotics and associated degradation products cause cryptic perturbations of metabolism and molecular stress responses in *Arabidopsis thaliana*. *Journal of Experimental Botany, 64*, 2753–2766.

Shah, F., Nicolas, C., Bentzer, J., Ellstrom, M., Smits, M., Rineau, F., Canbäck, B., Floudas, D., Carleer, R., Lackner, G., Braesel, J., Hoffmeister, D., Henrissat, B., Ahrén, D., Johansson, T., Hibbett, D. S., Martin, F., Persson, P., & Tunlid, A. (2015). Ectomycorrhizal fungi decompose soil organic matter using oxidative mechanisms adapted from saptrotrophic acncestors. *The New Phytologist, 209*, 1705–1719.

Shann, J. R., & Boyle, J. J. (1994). Influence of plant species on in situ rhizosphere degradation. In T. A. Anderson & J. R. Coats (Eds.), *Bioremediation through rhizosphere technology* (pp. 70–81). Washington, DC: American Chemical Society.

Shaw, L. J., & Burns, R. G. (2007). Influence of the rhizosphere on the biodegradation of organic xenobiotics—a case study with 2,4-dichlorophenoxyacetic acid. In H. J. Heipieper (Ed.), *Bioremediation of soils contaminated with aromatic compounds: Effects of rhizosphere, bioavailability, gene regulation and stress adaptation* (pp. 5–30). Berlin: Springer.

Shetty, K. G., Hetrick, B. A. D., Figge, D. A. H., & Schwab, A. P. (1994). Effects of mycorrhizae and other soil microbes on revegetation of heavy metal contaminated mine spoil. *Environmental Pollution, 86*, 181–188.

Sicliliano, S. D., & Germide, J. T. (1999). Enhanced phytoremediation of chlorobenzoates in rhizosphere soil. *Soil Biology and Biochemistry, 31*, 299–305.
Silambarasan, S., & Abraham, J. (2013). Mycoremediation of endosulfan and its metabolites in aqueous medium and soil by *Botryosphaeria laricina* JAS6 and *Aspergillus tamarii* JAS9. *PLoS One, 8*, 77–170.
Simon, A., Bindshedler, S., Job, D., Wick, L. Y., Filippidou, S., Kooli, W. M., & Junier, P. (2015). Exploiting the fungal highway: Development of a novel tool for the in situ isolation of bacteria migrating along fungal mycelium. *FEMS Microbiology Ecology, 91*, 1–13.
Singh, R. (2017). Biodegradation of xenobiotics- a way for environmental detoxification. *IJDR, 7*, 14082–14087.
Singh, D. K. (2008). Biodegradation and bioremediation of pesticide in soil: Concept, method and recent developments. *Indian Journal of Microbiology, 48*, 35–40.
Singh, B., & Walker, A. (2006). Microbial degradation of organophosphorus compounds. *FEMS Microbiology Reviews, 30*, 428–471.
Smith, S. E., & Read, D. J. (2008). *Mycorrhizal symbiosis*. New York: Academic.
Smith, S. E., & Smith, F. A. (2011). Roles of arbuscular mycorrhizas in plant nutrition and growth: new paradigms from cellular to ecosytems scales. *Annual Review of Plant Biology, 62*, 227–250.
Sonon, L. S., & Schwab, A. P. (2004). Transport and persistence of nitrate atrazine and alachlor in large intact soil columns under two levels of moisture content. *Soil Science, 8*, 541–553.
Spain, J. C., & van Veld, P. A. (1983). Adaptation of natural microbial communities to degradation of xenobiotic compounds: Effects of concentration, exposure time, inoculum, and chemical structure. *Applied and Environmental Microbiology, 45*, 428–435.
Spokes, J. R., MacDonald, R. M., & Hayman, D. M. (1981). Effects of plant protection chemicals on vesicular-arbuscular mycorrhizas. *Pesticide Science, 12*, 346–350.
Sudova, R., & Vosatka, M. (2007). Difference in the effects of three arbuscular mycorrhizal fungal strains on P and Pb accumulation by maize plants. *Plant and Soil, 296*, 77–83.
Talaat, N. B., & Shawky. (2014). Protective effect of arbuscular mycorrhizal fungi on wheat (*Triticum aestivum* L.) plants exposed to salinity. *Environment and Experimental Botany, 98*, 20–31.
Tejeda-Agredano, M. C., Gallego, S., Vila, J., Grifoll, M., Ortega-Calvo, J. J., & Cantos, M. (2013). Influence of the sunflower rhizosphere on the biodegradation of PAHs in soil. *Soil Biology and Biochemistry, 57*, 830–840.
Teng, Y., Luo, Y., Sun, X., Tu, C., Xu, L., Liu, W., Li, Z., & Christie, P. (2010). Influence of arbuscular mycorrhiza and *Rhizobium* on phytoremediation by alfalfa of an agricultural soil contaminated with weathered PCBs: A field study. *International Journal of Phytoremediation, 12*, 516–533.
Testa, A., Di Matteo, A., Rao, M. A., Monti, M. M., Pedata, P. A., & Van Der Lee, T. A. J. (2012). A genomic approach for identification of fungal genes involved in pentachlorophenol degradation. *Advance Research Science Areas, 9*, 1386–1389.
Toljander, J. F., Lindahl, B. D., Paul, L. R., Elfstrand, M., & Finlay, R. D. (2007). Influence of arbuscular mycorrhizal mycelial exudates on soil bacterial growth and community structure. *FEMS Microbiology Ecology, 61*, 295–304.
Turnau, K., Mleczko, P., Blaudez, D., Chalot, M., & Botton, B. (2002). Heavy metal binding properties of *Pinus sylvestris* mycorrhizas from industrial wastes. *Acta Societatis Botanicorum Poloniae, 71*, 253–261.
Van Elsas, J. D., Costa, R., Jansson, J., Sjöling, S., Bailey, M., & Nalin, R. (2008). The metagenomics of disease-suppressive soils– experiences from the metacontrol project. *Trends in Biotechnology, 26*, 591–601.
Van Hamme, J. D., Singh, A., & Ward, O. P. (2003). Recent advances in petroleum microbiology. *Microbiology and Molecular Biology Reviews, 67*, 503–549.
Varsha, Y. M., Naga Deepthi, C. H., & Chenna, S. (2011). An emphasis on xenobiotic degradation in environmental cleanup. *Journal of Bioremediation and Biodegradation, 11*, 001. https://doi.org/10.4172/2155-6199.S11-001.

Visioli, F. (2015). Xenobiotics and human health: A new view of their pharma-nutritional role. *Pharma Nutrition, 3*, 60–64.
Vivas, A., Barea, J. M., Biro, B., & Azcon, R. (2006). Effectiveness of autochthonous bacterium and mycorrhizal fungus on *Trifolium* growth, symbiotic development and soil enzymatic activities in Zn contaminated soil. *Journal of Applied Microbiology, 100*, 587–598.
Volante, A., Lingua, G., Cesaro, P., Cresta, A., Puppo, M., Ariati, L., & Berta, G. (2005). Influence of three species of arbuscular mycorrhizal fungi on the persistence of aromatic hydrocarbons in contaminated substrates. *Mycorrhiza, 16*, 43–50.
Wang, F. Y., Shia, Z. Y., Tongb, R. J., & Xua, X. F. (2011a). Dynamics of phoxim residues in green onion and soil as influenced by arbuscular mycorrhizal fungi. *Journal of Hazardous Materials, 185*, 112–116.
Wang, F. Y., Tong, R. J., Shi, Z. Y., Xu, X. F., & He, X. H. (2011b). Inoculations with arbuscular mycorrhizal fungi increase vegetable yields and decrease phoxim concentrations in carrot and green onion and their soils. *PLoS One, 6*, e16949. https://doi.org/10.1371/journal.pone.0016949.
Wang, Y., Ye, X., Ding, G., & Xu, F. (2013). Overexpression of phyA and appA genes improves soil organic phosphorus utilisation and seed phytase activity in *Brassica napus*. *PLoS One, 8*(4), e60801. https://doi.org/10.1371/journal.pone.0060801.
Weissenhorn, I., & Leyval, C. (1996). Spore germination of arbuscular mycorrhizal fungi in soils differing in heavy metal content and other parameters. *European Journal of Soil Biology, 32*, 165–172.
White, J. C., Ross, D. W., Gent, M. P., Eitzer, B. D., & Mattina, M. I. (2006). Effect of mycorrhizal fungi on the phytoextraction of weathered p,p-DDE by *Cucurbita pepo*. *Journal of Hazardous Materials, 137*, 1750–1757.
Wyss, P., & Bonfante, P. (1993). Amplification of genomic DNA of arbuscular mycorrhizal (AM) fungi by PCR using short arbitrary primers. *Mycological Research, 97*, 1351–1357.
Yadav, J. S., Quensen, J. F., III, Tiedje, J. M., & Reddy, C. A. (1995). Degradation of polychlorinated biphenyl mixtures (Aroclors 1242, 1254 and 1260) by the white-rot fungus Phanerochaete chrysosporium as evidenced by congener specific analysis. *Applied and Environmental Microbiology, 61*, 2560–2565.
Yu, X. Z., Wu, S. C., Wu, F. Y., & Wong, M. H. (2011). Earthworm- mycorrhiza interaction on Cd uptake and growth of ryegrass. *Soil Biology and Biochemistry, 37*, 195–201.
Zamal, A., Ayub, N., Usman, M., & Khan, A. G. (2002). Arbuscular mycorrhizal fungi enhance zinc and nickel uptake from contaminated soil by soybean and lentil. *International Journal of Phytoremediation, 4*, 205–221.
Zelenev, V. V., van Bruggen, A. H. C., & Semenov, A. M. (2005). Modeling wave-like dynamics of oligotrophic and copiotrophic bacteria along wheat roots in response to nutrient input from a growing root tip. *Ecological Modelling, 188*, 404–417.
Zhang, Y., Liu, J., Zhou, Y., Gong, T., Wang, J., & Ge, Y. (2013). Enhanced phytoremediation of mixed heavy metal mercury organic pollutants trichloroethylene with transgenic alfalfa co-expressing glutathione S-transferase and human P4502 E1. *Journal of Hazardous Materials, 15*, 1100–1107.
Zhang, L., Fan, J. Q., Ding, X. D., He, X. H., Zhang, F. S., & Feng, G. (2014). Hyphosphere interactions between an arbuscular mycorrhizal fungus and a phosphate solubilizing bacterium promote phytate mineralization in soil. *Soil Biology and Biochemistry, 74*, 177–183.
Zocco, D., Van Aarle, I. M., Oger, E., Lanfranco, L., & Declerck, S. (2011). Fenpropimorph and fenhexamid impact phosphorus translocation by arbuscular mycorrhizal fungi. *Mycorrhiza, 21*, 363–374.

Soil Genesis, Survey and Classification

8

Narendra Kumar, Saurabh Gangola, Pankaj Bhatt,
Neha Jeena, and Rowndel Khwairakpam

1 Soil Genesis

1.1 Introduction of Soil

Unlike minerals, plants and animals, soils are not exactly definable. Soils might be described as border-like phenomenon of the earth's surface. They belong to the pedosphere, in which lithosphere, atmosphere, hydrosphere and biosphere overlap and interact. Definition of the soil varies, according to a Pedologist, soil is a natural body, forms at the surface by the biogeochemical and physical processes which is capable of supporting life, and can be mapped at correct scale. According to engineers it is an unconsolidated material. According to geologists it is the natural medium on land surface for the growth of plants. According to the soil microbiologists it is group which is regulated by soil microorganism. According to the FAO, "Soil is a natural body consisting soil horizons and medium for the plant growth. In other words we can say soil is the result of weathering and erosion of rock into smaller particles. (Brevik and Hartemink 2010; Hartemink 2016).

N. Kumar (✉)
Department of Crop Physiology, Uttaranchal University, Dehradun, Uttarakhand, India

S. Gangola
Department of Microbiology, G.B Pant University of Agriculture and Technology, Pantnagar, Uttarakhand, India

P. Bhatt
Department of Microbiology, Dolphin (P.G) Institute of Biomedical and Natural Sciences, Dehradun, Uttarakhand, India

N. Jeena
Department of Biotechnology, Bhimtal Kumaun University, Nainital, Uttarakhand, India

R. Khwairakpam
Department of Crop Physiology, Assam Agriculture University, Jorhat, Assam, India

© Springer Nature Singapore Pte Ltd. 2019

A. Varma, D. K. Choudhary (eds.), *Mycorrhizosphere and Pedogenesis*,
https://doi.org/10.1007/978-981-13-6480-8_8

1.2 Soil Genesis

Soil genesis also termed as Pedogenesis (Greek pedo- meaning 'soil, earth,' and genesis, meaning 'origin, birth') is the process of soil formation which is regulated by the effects of place and environment, it initially starts from rocks or 'parent material'. It might be deep into the soil, or far away if water and glacier are come in contact with the soil. In genesis of soil climatic factors influences the physical and chemical weathering of initial materials. It is a combination of structural development, differentiation into horizons and its translocation (Wakatsuki and Rasyidin 1992).

1.3 Factors of Soil Genesis

There are different factors of soil genesis (identified by Dokuchaev, V V, 1800s) which are as follows:

(a) Soil formation
(b) Climatic (including water)
(c) Topography
(d) Biological
(e) Time

Jenny 1941 gave a mathematical expression of these factors which are designated as cl, o, r, p and t these factors provide huge information about genesis of soil.

1.3.1 Soil formation

These are starting materials by which the chemical and physical properties of soil are influenced. In organic soil it may be plant debris or mineral matter. It is material transported through running water (alluvium), gravity (colluviums), wind (Aeolian) and glacial deposits) (Brewer 1954; Bockheim et al. 2014). It may be in the form of bedrock, glacial deposits, old soil surface, peat, mine waste etc. Parent materials have important mineral, chemical as well as physical properties.

1.3.2 Climatic (Including Water)

Rate and type of soil formation controlled by the climate. Wind, temperature, water and solar radiation affect directly the soil formation. They determine presence of water and its temperature in the environment. These climatic factors influences

(a) Break down of parent material
(b) Breakdown of soil particles
(c) Deposition of clays
(d) Leaching of soluble cations
(e) Pedoturbation by freeze-thaw of soil water
(f) Microbial activity

There are different types of climates such as humid (more precipitation occur), arid (usually a desert), Oceanic (variation in the rainfall), Mediterranean (precipitation is moderate), continental (cold winters and warm summers) as well as tropical (Cold humid along with warm summers condition) and subtropical climates which is a warm along with hot humid. These different dominant climates influence the soil genesis (Stephens 1965; Jones et al. 2009).

Water in different forms supply to the soil and required for biochemical activity, transport of materials and the most important weathering. Atmospheric humidity (i.e. 90–100%) is directly affected the Surface soil which is a micro soil climate but it is more moist when the humidity percentage is 60%. Besides these climatic factors precipitation, evaporation and transpiration are important variable of the climate. The equality between these three (precipitation, evaporation and transpiration) determine the climatic effect on soil moisture. Solar radiation as heat flux is also a climatic variable in soil genesis which drives photosynthesis, involve in carbon fixation and finally fix into the soil (Karmakar et al. 2016).

1.3.3 Topographic

Topography decides the configuration of soil surface areas. It designate the difference between land surface area and the slope and influences the drainage characteristics i.e. water flowing onto and off of the soil (Table 8.1).

Formation of topography is a combination of different features which includes stream incision, dunes, moraines, mass wasting etc. Topographic soil has impaired drainage conditions and distinct horizons but rain water and other form of water easily percolate in the soil. Percolation of water limit water for plant growth and accountable for erosion as well as formation of soil. In this particular soil generally stony, shallow and poorly developed profiles with fewer horizons are found and because of erosion, soil formation adversely affected. The moisture content is less in topographic position of soil and high in semi arid and sub humid areas. It might be due to water received by these areas as runoff and this condition determine the rate of organic matter decay and plant growth. These conditions finally form dark-colored soils as well as rich in organic matter (Matejkov et al. 2008).

Table 8.1 Different topographic distributions (FAO 1990)

S. No	Land Surface	With the slope of
1.	Leveled surface	0–2%
2.	Gently undulating	2–5%
3.	Undulating topography	5–10%
4.	Rolling topography	10–15%
5.	Hilly topography	15–3 0%
6.	Steeply dissect	>30% with an elevation of <300 m
7.	Mountainous	>30% with an elevation of >300 m

1.3.4 Biological

Soil ecosystem constituent's microorganism, plants as well as animal in the micro environment of the soil. Organic matter degradation, nitrogen fixation, mineralization and formation of humic substances are the process which is basically done by soil microorganism. Animals are involved in the dig and mixing the soil mass which will directly affect the parent material. In soil genesis men are actively participated and manipulated natural vegetation as well as agricultural practices etc. Due to manipulation of vegetation infiltration rate of water in soil is decreased thereby increasing runoff rate as well as soil erosion.

Vegetation directly affects the property of soils by number of ways which are as follows

- Soil surface covered by protective cover which reduces movement of soil material by the process of erosion.
- For improving the soil structure growing roots are important and it will form a network of miniature channels and pores for plant growth.
- Legumes with the nodulation system or nod gene involve in the nitrogen fixation and improve soil fertility.
- In wet environments accumulation of plant matter on the surface of soil, leads to formation of organic acids.
- Transpiration is important process in plants through which water removed from the soil thus prevent the water logging and Stalinization condition of the soil (Bidwell and Hole 1965; Breemen 2004).

1.3.5 Time

Different factors of soil genesis are correlated with time. It is a period from stage to top soil of the stage of fully developed horizons. In many years there is formation of an inch of soil. Soil genesis is directly and indirectly the combination of movement of particles, formation of soil as well as the process of leaching which ultimately depends on the time. Longer the time periods for all these factors more differentiated and distinctive the soil will become. With the passes of time, formation of soil along with leaching mechanism continues leads to divergence of chemical and mineralogical composition of the soil. With longer time duration the soil finally reaches in the stability constituted more iron, silica along with high content of aluminum (Jenny 1994).

1.4 Soil Forming Process

There are four process of soil genesis

1. **Additions:** Decomposed vegetation and organisms or new mineral materials are deposited in the soil through wind or water.

2. **Losses**: The soil chemical and physical property has been altered as the organic matter, clay, silt etc. are harvested from the soil. It might be due to the wind or the excess of water in the surface of the soil.
3. **Transformations:** It is a type of chemical weathering in which organic matter transformed into humus.
4. **Translocations**: In this there is translocation of organic or mineral compounds in between horizons. Over time, there are more noticeable changes in texture of soil, structure as well as color of the soil (Velbel 1988; Phillips 2004).

2 Classification of Soil

The main purpose of soil classification is to understand and elaborate the connection within the soils, among the soil as well as most important factors which decides the soil physical and chemical character. Soil classification initiated in the mid of 1800 based on geologic concept and parent material and then emphasis on climate and vegetation. ICAR categorized soil in different group of soil according to the nature, chemical, physical, fertility, similar properties and different properties of the soil. Classification decides the uses and management practices of the soils.

Nationally and internationally soil classification was a matter of controversy because of the lack of agreement for a common classification system of the soil. Although there is a existence of two soil classification systems which are widely used (*viz.* The USDA Soil Taxonomy and the FAO/UNESCO legend). Soil classification begins with soil profiles. Different series of horizon forms a soil. These horizons have specific properties such as appearance, thickness which are the action of the different processes of the soil formation. Horizons (O, A, B, C horizons) are the building block of the soil profile (i.e. vertical section of the soil).

2.1 Modern Soil Classification

After mid of nineteenth century activity related to soil survey confirmed that Indian agriculture system affected by the global market expansion and economic reconstruction. It gives the idea about soil conservation and alternative land uses. Identification of the new soil commenced by many scientists for that there is a need of classification system. National classification of soil was developed by the many countries which is important for the diagnostic horizons and features of the particular soil. Major soil types contain alkaline soil, Peat Soil, Permafrost, Serpentine soil, Volcanic Soil etc.

2.2 Soil Classification System in India

Indian soil classified on the basis of the particle sizes. It ranges from the very large (300 mm) to very small particles (<0.002 mm). The particle size ranges from <0.001

Table 8.2 Soil classification according to different groups and size

S. No.	Soil Types	Size
1.	Boulder	>300 mm
2.	Cobble	80–300 mm
3.	Gravel	4.75–80 mm
4.	Sand	0.075–4.75 mm
5.	Silt	0.002–0.075 mm
6.	Clay	<0.002 mm

to less than 0.002 mm was considered in the clay particles. According to **Indian Standard Soil Classification System (ISSCS),** soils are classified into groups on the basis and groups further divided into coarse, medium and fine sub-groups. The particles size are of different types i.e. boulder, cobble, gravel, sand, silt or clay (Table 8.2).

2.3 Microbiological Basis of Soil Classification

Soil contains the rich amount of macro and micro-nutrients, these nutrients are recycled by microorganisms in environment through biochemical cycles. We can divide the soil microbiologically in three categories.

2.3.1 Soil Free from Microbes

This is the rare situation where microbes not present. Such types of soil can be obtained by complete sterilization of natural soil which contains indigenous microbial population.

2.3.2 Soil with Free Living Microorganism

Free living microorganism means they can present anywhere in the soil system.

2.3.3 Soil with Rhizospheric Region of Plant

Plants secrete root exudates which attract bacteria towards the rhizospheric region. Due to plant microbe interaction increased microbial activity found in this region. These interactions also helpful for aggregation of the soil particles.

3 Soil Survey

3.1 Introduction to Soil Survey

In reference to soil science glossary, Soil survey is interpreted as the evaluation, description of soil, classification of soil, and plotting of soils within a specified area. It is classified on the basis of (i) magnitude and types of field investigation. (ii) Augmentation and application of standards technique to describe particular soils of a specific area are undertaken by the program of the National Cooperative Soil

Survey. It is a science in which a descriptive, systematic examination, elaborative classification, and plotting of the soils in a specific area." Brady and Weil (1996)

A soil survey is the systematic description of soils in the specified area, clustering them into clearly-defined mapping units such as soil series; phases etc. in attempting to establish their best suited use demarcate boundaries and show their location on the map. The objective of soil survey can be listed below:

1. Classification of soils into clearly defined mapping units i.e. soil series, phases etc.
2. Locating their distributions and demarcations in the field on the map.
3. Find out their best possible use.
4. Predicting their performance under different management package and practices i.e. yield of different crops under different management practices.

3.2 Types of Soil Surveys

Soil surveys are of three types:

3.2.1 Detail Soil Survey

In this, type of survey soils are studied in detail. Soil possessing contain common physical properties and also morphological properties such as texture, structure, colour, pH, carbonate, natural vegetation, slope, erosion, depth of soil, natural vegetation etc., are clustered into units which can be readily recognized in the specified area. The soil units are commonly defined based on the above-mentioned characteristics of the surface soil, soil depth, slope and erosion and the soil profile. The soil units are the soil series. This type of survey extensively uses Cadastral (village) maps, which are the base maps (Scale 1: 10,000 to 1: 5,000). Soil boundaries are demarcated on a larger scale (1:10,000 to 1:5,000) base map called cadastral map. Mapping units is a soil series.

3.2.2 Reconnaissance Soil Survey

Here observations are taken at an extended period of interval and recorded on the Survey of India toposheets of scale i.e. 1: 2, 50,000 to 1: 50,000. Detailed study of soil is not necessary for this type of survey.

3.2.3 Detailed Reconnaissance

A detailed survey of 15% of the area where research projects are to be established follows reconnaissance soil survey

3.3 Land Capability Classification

Land capability classification is an interpretative clustering of soil plotting units mainly based on distinctive soil characteristics, apparent land features and environmental factors that predominantly restrains the use of land for agriculture, pasture, or other uses on a sustained basis (IARI 1971). It is a systematic of soil grouping based on their potential for production of commonly cultivated crops in a land without causing any adverse effect for a long time. A capability class is assigned with a soil map unit throughout Class I to VIII, and for agricultural purposes few classes such as Class II and Class VII are hazardous. In brief it is the potential of the land surface to perform a land-use without causing any unfavorable consequences to the soil although the nutrient may be dislodged (Fenton 2006).

Different organizations have classified land capability classification (Sitorus 2012). Some of these land capability classification that can be mentioned are:

1. U.S. soil conservation service
2. Cornell system of economic land classification
3. Britain L.D. Stamp's land classification
4. U.S.S.R; Japan; China Land classification.
5. Iraq Land-use capability classification
6. land classification of Northern Island system
7. Reclamation of U. S. Bureau
8. Department of Agriculture U.S.
9. National resource planning Board of U.S.

Land capability classifications by Soil and Land Use Survey of India and ICAR organization Institutes.

The basis on which land capability is classified environmental and geomorphic attributes, landscape ecology, variety of potential crops, productivity, and simplicity of management and possibility of degradation. These methods which are listed above are cited, widely accepted and have relevant to some extent in modern Land-use practices.

3.3.1 The Three Major Categories of Soil Grouping of Land Capability Classification

LCC further subdivided based on the main hazardous operation on the land.

(i) Capability unit,
(ii) Capability sub-classes,
(iii) Capability classes.

3.3.1.1 Capability Unit

It is a collection of soil-mapping units which has identical characteristic hazards of a particular land. It is the lowest and most detailed group of classification and it is used for clustering records, which responds identically to same management, adaptation to same kind and condition of crops or vegetation, having nearly comparable yield potential and similar management practices and.

3.3.1.2 Capability Subclass

There are different subclasses and subclass-s a part of soil that are prevalent in shallow soil predominantly, such type of soil has root zone is shallow, hard pans stones, insufficient capacity of moisture-holding, fertility is poor i.e. tough to reclaim, unfavorable soil texture and structure, salinity and toxicity etc. Subclass-c includes soils for which the limitations and major exploits affecting their utilization is the climate (the temperature or moisture deficit). There is a priority in the use denotation in the order of e, w, s, c, in cases where soils have two kinds of limitations. For example, soils with susceptibility as well as erosion and excessive water hazard, e has more precedence than w but sometimes, both the limitations are necessary. According to the priority e showed first followed by w and then s, etc. Use of multiple symbols in general should be avoided as much as possible as it deviates from the objectives of classification and preserve the simplicity. With alight modification the land capability sub-classes have been used in some parts of the earth by incorporating local hazards.

3.3.1.3 Capability Class

It is the most extended unit in which class I to VIII are used to designate irrigated land capability and non-irrigated land capability with progressing degree of soil damage and limitation. There are no subdivisions within Class. It is helpful for the map user and provide the correct informatics data of the soil which is directly or indirectly play important role in agricultural and forest utilization. It provides the information only regarding agricultural limitations in soil. According to Soil Conservation Service, Department of Agriculture, U.S. land classes are divided into eight classes based on the topographic situations.

3.4 Land Class I

Soils with some limitation that restrict their usability are considered in this class. Generally, soils of this class very rich with some hazards. They can be used for multiple purposes or can be used to safely cultivate a wide range of crops, pastures, and forestry with conventional farming methods. This kind of soils has properties such as good depth, highly productive, furrow, and has very low erosion i.e. flat (0–3°). It is subjected to the puddle erosion and fertility but not with overflow damage. The soil of this class has been utilized for cultivation practices i.e. for maintaining fertility and structure of the soil. Class I land has been confined to alluvial areas and well drained.

3.5 Land Class II

With the moderate limitations this class of soil reduced plants choice with less conservation practices. These soils are subjected to hazards and les risk of damage. It is rich soils with simple cultivation by applying simple recommended management like irrigation-control devices and conservation soil. There are numerous difference between Class I and Class II soils but only the difference is undulating (0–7°) slopes which are subjected to moderate erosion, also are of moderately deep to slightly shallow that are subjected to flooding and well drained to moderately drained in need of drainage.

3.6 Land Class III

Class III soils have severe demerits that restrict the planting materials choice or practices related to the conservation as well. It is subjected to more intensive restrictions used in arable lands due to higher risks or damages that are moderately good soils. It is highly limited than the Class II towards a greater hazard. This class of soil is good for the crop and the management practices with good rotations. This class of soils has a elevation of 0–11° and are subject towards the severe erosion. They are intrinsically of low fertility. In this kind of soils crop choice, planting time and the time of tillage operations are limited due to the above-mentioned limitations. Cover crops with adequate plant cover are the choice of plants in this class of soil, which also helps in protection of soil from erosion and to the preservation of soil structure (Sanchez et al. 2003).

3.7 Land Class IV

The soils of this class comprises with those which causes very extreme limitation for crop cultivation although the soil may be considered fairly good. Growing crops in these soils may be possible only if handled with great care under right conditions, as the crops located in low elevations and prone to extreme erosion of soil. This type of soil should remain permanently covered by suitable pastoral vegetation but may be cultivated with a grain crop every five to six years due to its inherently poor fertility. They are also shallow soil or moderately deep on moderate slopes.

3.8 Land Class V

Soils of this class kept in lifelong vegetative covered, therefore, limiting their use to pastures or forests. In these soils, the permanent limitations are none existent and slight hazards. Because of the some limitations cultivation is not possible in any kind of crop. The land has no slope and is almost leveled and subjected to erosion by water or wind if good management practice is followed. Interference by human and mulch animals must be operated for the undistribution of plant cover.

3.9 Land Class VI

In this soils are recommended for use in grazing and forestry. They risk moderate hazard when utilized. These soils are not suitable for the cultivation and subjected to some moderate hazard. When the slope is perpendicular animal grazing should be completely restricted in order to not spoil the plant cover. Under proper management, this class is capable of forage producing. In this class land has some elevation and is subjected to erosion of wind which is greater than Class IV soils.

3.10 Land Class VII

The characteristics as abrupt, rough, drought-prone soils make this class of VII put through extremely lifelong hazards when utilized for pasture. In these cases it is used for pasture or forest it must be handled carefully with limited interferences. Where there is sufficient precipitation or available water, land area must be utilized for forestry while limited for animal grazing with strict management practices.

3.11 Land Class VIII

It is not suitable for forestry or pasture. These are usually reserved for recreational, watershed projects or wildlife reserves.

4 Conclusion

This is a brief introductory section where we have summarized soil classification, formation of soil usually considered as pedology. Formation of soil comprises of concepts, studies, factors, theories and the methodologies play important role in changes and soil development. Classification of soil elaborate with the scientific or technical soil categorization divided in clusters, and the process of acquiring the knowledge and understanding through thought of this categorization. It is established fact that for land-use planning, study and classification related to the capability of land are necessary. It certainly, provides and glimpse of the land type productivity as well as a guideline for prediction of the design and use of land. The definitions and identification of a plane surface, profile of soil, individual soil, establishment and categorization are very important for the use, identification of soil and soil mapping.

References

Bidwell, O. W., & Hole, F. D. (1965). Man as a factor of soil formation. *Soil Science, 99*, 65–72.

Bockheim, J. G., Gennadiyev, A. N., Hartemink, A. E., & Brevik, E. C. (2014). Soil-forming factors and soil taxonomy. *Geoderma, 227*, 231–237.

Brady, N. C., & Weil, R. R. (1996). *The nature and properties of soils* (13th ed., p. 960). Upper Saddle River: Prentice-Hall.

Breemen, N. V. (2004). The formation of soils. *Developments in Soil Science, 29*, 21–40.

Brevik, E. C., & Hartemink, A. E. (2010). Early soil knowledge and the birth and development of soil science. *Catena, 83*, 23–33.

Brewer, R. (1954). Soil parent material. *The Australian Journal of Science, 16*, 134–138.

FAO. (1990). *Guidelines for soil profile description. Third edition (revised). Soil resources, management and conservation service, land and water development division.* Rome: FAO.

Fenton, T. E. (2006). Land capability classification. *Encyclopedia of Soil Science, 2*, 962–964.

Hartemink, A. E. (2016). The definition of soil since the early 1800s. *Advances in Agronomy, 137*, 73–126.

IARI. (1971). *Soil survey manual* (p. 121). New Delhi: IARI Publ.

Jenny, H. (1994). *Factors of soil formation*. New York: McGraw-Hill Book Company Inc.

Jones, A., Stolbovoy, V., Ruscol, E., Gentile, A. R., Gardi, C., Marechal, B., & Montanarella, L. (2009). Climate change in Europe. 2. Impact on soil. Review. *Agronomy for Sustainable Development, 29*, 423–443.

Karmakar, R., Das, I., Dutta, D., & Rakshit, A. (2016). Potential effects of climate change on soil properties. Review. *Science International, 4*(2), 51–73.

Matejkova, S., Fifernova, M., Lipavsky, J., & Kumhala, F. (2008). Topography impact on nutrition content in soil and yield. *Plant, Soil and Environment, 54*(6), 255–261.

Phillips, J. D. (2004). Geogenesis, pedogenesis, and multiple causality in the formation of texture-contrast soils. *Catena, 58*, 275–229.

Sanchez, P. A., Palm, C. A., & Buol, S. W. (2003). Fertility capability soil classification: A tool to help assess soil quality in the tropics. *Geoderma, 114*(3–4), 157–185.

Sitorus, S. R. (2012). Land capability classification for land evaluation: A review. *JSumberLah, 4*(02).

Stephens, C. G. (1965). Climate as a factor of soil formation through the quaternary. *Soil Science, 99*(1), 9–14.

Velbel, M. A. (1988). Weathering and soil-forming processes. In *Forest hydrology and ecology at Coweeta* (pp. 93–102). New York: Springer.

Wakatsuki, T., & Rasyidin, A. (1992). Rates of weathering and soil formation. *Geoderma, 52*(3–4), 51–263.

The Beneficial Influence of Microbial Interactions on Plant Diseases and Plant Growth Promoting Effect

9

Ömür Baysal and Ragıp Soner Silme

1 Introduction

In insight of data collected by molecular biology and evidence from certain fossil plants, Rhynia and Asteroxylon are similar to present day arbuscular mycorrhizal fungi. The data on microflora shows the soil dynamics are changeable and display very complicated feature. Due to its role as a medium for the growth and activities of plants and soil microorganisms, many interactions take place in assessing the properties of soil. Most of the terrestrial plants have a mutual and beneficial symbiosis effect both on the soil fungi and arbuscular mycorrhiza (AM). It's a fact that AM fungi has been in a continuous interaction between various type of soil microorganisms including nonbacterial soil microorganisms, plant growth promoting rhizobacteria besides mycorrhiza helper bacteria and deleterious bacteria. Some examples of these interactions between the AM fungi and soil bacteria includes the binding of soil bacteria to the fungal spore, the injection of molecules by bacteria into the fungal spore, the production of volatiles by bacteria and the degradation of fungal cellular wall. Understanding of these interactions can have important implications in agriculture. The expression of genes in AM fungi and hence their performance and ecosystem compatibility can be affected such mechanisms. Hence, consideration of such synergistic behaviour is very important. In this review, some of the most significant last findings about the interactions between AM fungi and soil bacteria with some new concepts are presented.

Ö. Baysal (✉)
Faculty of Science, Department of Molecular Biology and Genetics, Muğla Sıtkı Koçman University, Muğla, Turkey
e-mail: omurbaysal@mu.edu.tr

R. S. Silme
Center of Biotechnology and Genetic Engineering Research and Practice, Istanbul University, Istanbul, Turkey
e-mail: rsoner@istanbul.edu.tr

© Springer Nature Singapore Pte Ltd. 2019

A. Varma, D. K. Choudhary (eds.), *Mycorrhizosphere and Pedogenesis*,
https://doi.org/10.1007/978-981-13-6480-8_9

2 Effect of Mycorrhizal Fungi on Soil Microbial Structure

AM fungi colonisation affects bacterial communities on root micro-flora both in direct and indirect ways. Host plant directly assimilates carbon derived compounds through help of mycorrhizosphere provision via fungal hyphae, which changes in pH in soil, competition for nutrients, and induces secretion of other inhibitory and stimulant compounds affecting hormone metabolism of plant. Indirect interactions have effect on host plant growth, root exudation, and soil structure. These effects have been reported by Ames et al. (1984) using a simple dilution plate with single bacterial isolate and mycorrhizosphere of *Glomus mosseae*. The results showed even AM colonisation never exceeded 5.5%. Similar studies have been carried out on sweet corn (*Zea mays*) and subterranean clover (*Trifolium subterraneum*) plants (Meyer and Linderman 1986a). Secilia and Bagyaraj (1987) showed increased amino acid exudation by P-deficient mycorrhizal plants by stimulated the growth of amino acid-requiring bacteria after colonisation of *Glomus fasciculatum*, *Gigaspora margarita* and *Sclerocystis dussi* in plants.

Organic compounds produced by AM play a role in converting of soil particles into aggregates form (Tisdall and Oades 1979) that provides micro niche for microbial colonisation and growth. Forster and Nicolson (1981) analysed the microbial composition of soil aggregates and identified a range of bacteria, actinomycetes and algae. In another study qualitative and quantitative effects of AM on microbial communities in the mycorrhizosphere and the stability of soil aggregates suggested a higher number of P-solubilising bacteria in water-stable soil-aggregate (Andrade et al. 1998). These findings were confirmed by Schreiner et al. (1997) that they found increases in water-stable soil-aggregate in mycorrhizal soybean (*Glycine max*) that may influence bacterial composition. Many studies have demonstrated the effects of AM fungi on bacterial communities but the underlying mechanisms are unclear.

There are still few or no studies on the compounds released from AM mycelium. Although some studies showed chancing on bacterial community structure associated with the mycorrhizosphere. Until now no direct effect of compounds produced by AM fungi that stimulate or inhibit bacteria has been reported. Some studies by PLFA (phospholipid fatty acid) analysis and BIOLOG™ test methods implied the effect of mycorrhizal colonisation on bacterial populations dependent on plants (Söderberg et al. 2002) but it remained unclear due to complexity of bacterial community structure.

3 Interactions of AM Fungi with Fungal Pathogens and Nitrogen Transforming Bacteria

AM fungi interact with other root-associated pathogenic fungi. The postulated mechanism of interaction was mentioned before in the previous part. On these inhibitory effects, Filion et al. (1999), have tested crude extract from the growth medium of the AM fungus *G. intraradices* on the sporulation of two pathogenic fungi and followed the growth of two bacterial species in vitro. In these studies, while plant root pathogen *Fusarium oxysporum* was reducing, conidial germination of the mycoparasitic fungus *Trichoderma harzianum* and the growth of *Pseudomonas chlororaphis* were stimulated. *Clavibacter michiganensis* did not show any recession. There was no significant influence of pH on growth or germination. Findings showed the secretion of unspecified substances by the AM fungus is main factor into the growth medium that affects growth of the tested microorganisms. In a similar study Citernesi et al. (1996) tested 17 year old *G. mosseae* to bacteria by pot experiments that antagonistic effect mycorrhizosphere against soil-borne pathogens *Fusarium* and *Phytophthora* was detected in vitro.

Many researchers claimed the ability of AM colonisation on plants changes according to increased nutritional status in the host plant. In field experiments, Newsham et al. (1995) has demonstrated AM-inoculated seedlings did not change P concentrations on *Vulpia ciliate* but seedling shoot and root growth were protected from negative effects of *Fusarium oxysporum*. In another study AM inoculated plants had naturally less infection of *F. oxysporum* on *Embellisia chlamydospora* (Newsham et al. 1995). Potential biocontrol role of AM fungi on soil-borne diseases has been observed by AM inoculated tomato (*Lycopersicon esculentum*) roots to *Fusarium* populations in the soil (Caron 1989).

On the other hand, AM fungi enhances nodulation and N fixation by legumes. There is a positive synergistic correlation between mycorrhizal colonisation and nodule symbioses on pathogen suppression and uptake of mineral nutrition and plant growth (Amora-Lazcano et al. 1998). AM symbiosis is beneficial for the functioning of the nitrogenase enzyme that increases P uptake, leading to increased N fixation and induces root and mycorrhizal development (Puppi et al. 1994). Amora-Lazcano et al. (1998) have acknowledged these positive effects on N-transforming microorganisms to two different *Glomus* species. However, contrary to these findings autotrophic nitrifying bacteria (ammonifying and denitrifying bacterial populations) has shown significantly reduction on sweet corn after AM colonisation. These studies prove complexity of interactions between N-transforming microorganisms and AM fungi in soil. The most important point drawing attention is to distinguish between direct effects on the pathogens and indirect effects resulted from nutritional status of the mycorrhizal plants.

4 Interactions of Mycorrhizosphere Bacteria with AM Fungi

On mode of action of mycorrhizosphere bacteria affecting AM fungi mainly 5 concepts have been suggested. Most of the findings have been obtained in ectomycorrhizal fungi (Garbaye 1994), but (1) reception of the root status and nutritional balance and anionic cationic changes in soil, (2) recognition phase ongoing between fungi and roots, (3) the colonisation of AM fungi, (4) characteristic property of soil chemistry and structural changes in soil rhizosphere; and (5) germination capacity depending on inheritance property of propagules. In a previous study, much higher fungi germination ratio has been determined on agar with growth of bacteria (Mosse 1959). A quick germination of *Glomus versiforme* in non-sterile soils was observed compared to not in autoclaved soil (Daniels and Trappe 1980). Germination is enhanced by kaolin or activated charcoal addition, which were associated with active degradation role of soil bacteria, or immobilisation by substances with a high ion exchange capacity. Similarly, non-surface sterilised spores of *G. versiforme* have germinated much vigorously than surface-sterilised spores that show isolation of bacteria from these spores, including *Pseudomonas* and *Corynebacterium*, enhances spore germination (Mayo et al. 1986). Carpenter-Boggs et al. (1995) have shown higher AM spore germination depending on amounts of volatile compounds produced by the *Actinomycetes* and *Streptomyces*. However, there is also inhibitory effect of fumigation or sterilisation of soils on spore germination at presence of some soil bacteria (Tommerup 1985; Ross 1980; Wilson et al. 1988).

Nitrogen fixing bacteria affects AM fungi colonisation. Even existence of genes for N fixation has been proved in endosymbiotic *Burkholderia* sp. (Minerdi et al. 2001), there is no clear data to show influencing of bacterial growth on mycorrhizal growth. *Rhizobium* spp. has synergistic effect on AM fungi and host plant. AM colonisation increases nodulation and N fixation due to the mycorrhiza using P that is required element for the enzymes involved in the N fixation process (Puppi et al. 1994).

Its known that AM fungi increase the absorptive area (Bolan 1991; Smith and Read 1997). The fine and thinner structure of the fungal hyphae can provide better access to soil pores and explore larger soil volumes, which results in more efficient mining ability for phosphate sources (Rosewarne et al. 1999; Drew et al. 2003; Smith and Read 2008; Schnepf et al. 2011).

The mutual interaction increases the growth of the host plant (Meyer and Linderman 1986b). Even many studies have shown potential effects on AM fungi and mycorrhizosphere bacteria under laboratory conditions the effects should be observed in the field experiments. Cultivation and isolation of whole living soil microorganisms is not possible due to presence of a large number of non-culturable microorganisms in the soil microflora (Amann et al. 1995), to identify and isolate them are also necessary in view of understanding for the function of these organisms in soil ecosystems (Baysal and Silme 2018). Even cultivation and identification of bacterium-like organisms (BLOs) not grown on cell-free media is quite difficult, new studies have shown their existence in AM fungi that AM cytoplasm shelters these microorganisms (MacDonald et al. 1982). Morphological and molecular studies have shown bacterial endosymbionts in cytoplasm of the AM fungus *Gigaspora margarita* (Bianciotto et al. 1996a). Analysis of the small-subunit rRNA gene sequence of the BLOs in *G. margarita* spores is the evident of very close relation of endosymbionts with genus *Burkholderia*. (Bianciotto et al. 1996a) The interaction of the endosymbionts with AM remains to be explored, however, the results indicate these bacteria contain a component of the fungal cytoplasm that should be considered for extent of microbial diversity in ecosystems as much as other noncultivated microorganism. Metagenomics approaches should be extended in more detail for this purpose to understand of microbial balance and ecological diversity

(Bianciotto et al. 1996a, b, 2000, 2001; Baysal and Silme 2018). Many of the mycorrhizosphere microbial interactions were not fully understood and elucidated. A microorganism, which plays active role in plant growth, plant growth promoting rhizobacteria (PGPR) such as rhizobia and pseudomonads attach to spores and hyphae of the AM fungus *Gigaspora margarita*, and this attachment changes according to bacterial strains and inorganic surfaces (Bianciotto et al. 1996b). These data suggest electrostatic attraction is the reason for a surface variety governed by general physiochemical parameters and bacterial secretion plays role in diversification on mechanism in phase while attaching to the fungal structures. Microorganisms green fluorescent protein (gfp)-labelled *Pseudomonas fluorescens* WCS365 and *P. chlororaphis* PCL1391 applied to tomato seedlings showed higher biocontrol activity on pathogenic fungus and formed protective biofilms (Bianciotto et al. 2001; Chin-A-Woeng et al. 2000). Filippi et al. (1998) have shown bacterial cells embedded in spore walls and micro-niches formation by TEM observations. On molecular studies using chromosomal insertion of Tn5, luxAB gene in a phosphate starvation-inducible locus and a control were followed using *Glomus intraradices* on *Pseudomonas fluorescens* DF57 in terms of luxAB gene expression, the results showed P starvation response is not induced and there is no metabolic activity of the bacterium at presence of AM in hyphosphere where the microbes interact within each other. Intercellular interactions associated bacteria have also been reported by rapid exchange of energy and nutrients between plant and roots with mycorrhizal fungi (Wamberg et al. 2003). These findings show AM provides a dynamic microflora containing bacterial colonies, which may in turn have positive effects on nutrient uptake and control of pathogens on host plant.

5 AM Fungi and Plant Hormones

Soil fungi in plant rhizosphere affect the level of symbiotic living between AM fungi and the host plant, which changes according to the properties of soil fungi. That is to say, fungi can induce systemic resistance through hormone pathways on host plant. Plant resistance can be stimulated by AM fungi (Harman et al. 2004; Kloepper et al. 2004; Waller et al. 2005; Van Wees et al. 2008).

Plant hormones have directly influence on induction of systemic resistance (Van Wees et al. 2008) that AM fungi alter level of plant hormones indirectly which changes jasmonic acid (JA) and abscisic acid (ABA) level as well (Ludwig-Muller 2000; Hause et al. 2007; Grunwald et al. 2009). These hormones are required for realization of AM symbiosis (Isayenkov et al. 2005). In addition, the newly classified plant hormones can also affect mycorrhization in plants, for instance; strigolactones which is also influenced by mitochondrial activities (Akiyama et al. 2005; Akiyama and Hayashi 2006; Besserer et al. 2009). Nutrient deficiency increases the level of strigolactones, plants under stress conditions, resulting in the reduction of shoot and AM growth. Detailed studies on interactions between auxin, cytokinins and strigolactones are necessary regarding root growth (Shimizu-Sato et al. 2009) and mycorrhizal establishment. Interestingly, AM fungi regulate stress on plant and

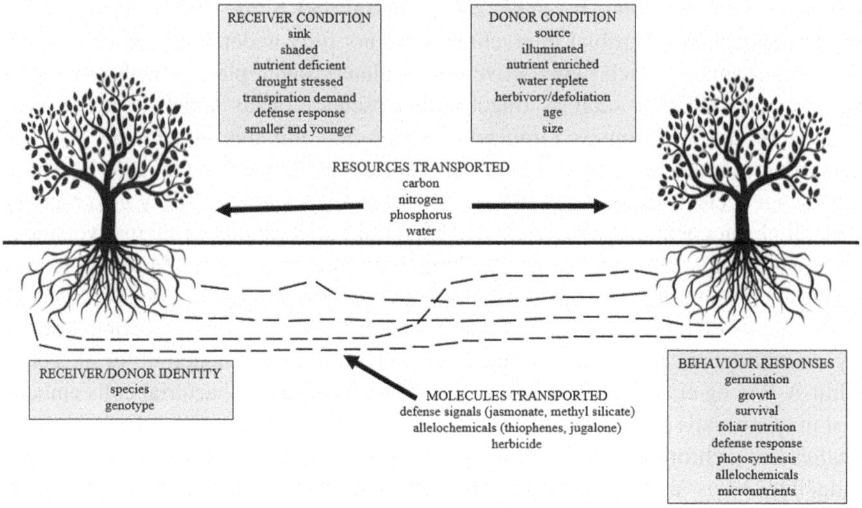

Fig. 9.1 Interactions for resources and signals through an MN, as well as some of the stimuli that elicit transfer of these molecules in donor and receiver plants

alter the activity of ABA alleviating negative effects on plant growth (Miransari and Smith 2008).

In a recent study, mycorrhizal network (MN) conducting links to neighbouring plants, which are concerned with gene regulation and defence response, defence signals/and or allelochemicals depending on biochemical pathways in cell, has affected adaptive behaviour of plants. These connections provide interplant communication via transfer of nutrients depending on mycorrhizal fungal colonization (Gorzelak et al. 2015) (Fig. 9.1). They have also found the behavioural changes in ectomycorrhizal plants are environmental cues, for the characteristics property of mycorrhizal network. This integration ascribed as 'tree talk' in the complex adaptive nature of forestry ecosystem.

6 Use of Mycorrhizal Fungi in Agricultural Practice

Microbial interactions and microbial dynamism provide synergistically viability while AM fungi and soil microbes are mutually in challenge with microflora that it provides agriculturally higher yield sustainable production. The most important implications are the alleviation of different soil stresses including salinity, drought, acidity, compaction and heavy metals by convenient microbial inoculants. Berg (2009); Joner and Leyval (2009) have suggested application of arbuscular mycorrhizal fungi to alleviate the negative effects of heavy metals residues in soil which influences plant growth.

In another study AM fungi has been tested to diminish negative effect of heavy metals accumulated in the vacuoles of their vesicles which turn in production of the insoluble glycoprotein, glomalin and this application increased plant growth (Khan 2005). Barea et al. (2005) has shown AM fungi as a factor altering the microbial combination that enhance plant resistance to plant pathogens in soil with bacterial activities and their production. Soil microbes, AM fungi and bacteria change soil structure by the production of bacterial metabolites including polysaccharides, formation of AM hyphae and glycoprotein, glomalin (Rillig and Mummey 2006) that they bind soil particles and form aggregates that have positive effect on improvement of soil structure (Andrade et al. 1998; Barea et al. 2005). These applications resulted in remarkable change in soil structures, and enhance fertility in desert areas that several tolerant legume species by co-inoculation with AM fungi and rhizobium enhanced N fixation as well as improved soil structure under drought conditions (Jeffries and Barea 2001). Efficient bacterial strains, particularly ones interacting with AM fungi, displayed very positive results (Hartmann et al. 2009; Franzini et al. 2010). Bacterial strain *Paenibacillus brasilensis*, which has suppressing effects on the activity of plant pathogens stimulate AM species (Von der Weid et al. 2005). These have been demonstrated using gfp by tagging of bacteria that production of green fluorescent has been suggested as a good tool to investigate soil microbes including pathogens interacting with AM fungi. Furthermore, advanced molecular techniques (DNA sequencing, PCR, staple isotope probing) were also used (Johnson et al. 2001; Griffiths et al. 2004).

Although different studies have been carried out on the interactive activities between AM fungi and bacteria, new researches are necessary to clearly elucidate the processes involved in the interactions between AM fungi and soil bacteria for assessing optimum bio-inoculant combination in purpose of sustainable agricultural production strategies (Artursson et al. 2006). Successfully singly and mixing application of AMF, PGPR, or PGPF which have commercial potential bio-inoculants, to control of tomato, celery, bell pepper and citrus diseases have been reported in Florida (Nemec et al. 1996; Datnoff et al. 1995). The effective control using co-inoculant form of commercial formulations of AMF *G. intraradices* and PGPF with *T. harzianum* was provided against to *Fusarium* crown and root rot of tomato in Florida. Different Pseudomonad strains and AMF synergistically improved grain quality on wheat at the rain-fed in India (Roesti et al. 2006). Moreover, fertilizer program and plant's growth stage have been associated with rhizobacterial community as well. Dual inoculation of AMF and PGPR inoculants have given positive result in controlling of *Verticillium* wilt of strawberry (Tahmatsidou et al. 2006). Chemical and biological fertilizers (*Azotobacter* and *Aspergillus*) and native AMF combination have been suggested as effective method which leads to 50% control of *Fusarium* wilt and Sigatoka disease in the frame of IPNM (Integrated Pest and Nutrient Management) strategies as well (Phirke et al. 2008).

7 Biocontrol Efficiency of Mycorrhizal Fungi

PGPR enhances plant growth through indirect way and suppress plant pathogenic microorganisms (Zehnder et al. 2001; Compant et al. 2005). PGPR like *Pseudomonas*, *Streptomyces* (Schrey and Tarkka 2008), and *Bacillus* strains (Kloepper et al. 2004; Baysal et al. 2008; Baysal et al. 2013), and non-streptomycete actinomycetes (El-Tarabily and Sivasithamparam 2006) have high potential as biocontrol agents. Even many studies (Azcón-Aguilar and Barea 1996; Xavier and Boyetchko 2002; Hyakumachi and Kubota 2004a; Harman et al. 2004; Vinale et al. 2008) have reported positive effect of AMF and PGPR strains, several mechanisms induced by signalling, cell signal trafficking cascades and proteome data of commercial PGPR strains are still not clearly understood (Baysal et al. 2013). Antagonistic properties of PGPR and PGPF are hyperparasitism, antibiosis and competition in view of nutrition and localization. Nevertheless, induction of systemic resistance on plants after exogenously application PGPR and PGPF strains has also been observed (Wei et al. 1991; Shivanna et al. 1996; Pieterse et al. 2003; Van Loon 2007). The suppressive effect of beneficial soil microorganisms on plant pathogens and their interactions with AMF have been acknowledged in many studies (Kaye et al. 1984; Krishna and Bagyaraj 1983; Caron et al. 1986; Garcia-Garrido and Ocampo 1988; Rosendahl and Rosendahl 1990; Trotta et al. 1996; Bødker et al. 2002; Karagiannidis et al. 2002; De la Peña et al. 2006; Garmendia et al. 2006; Alejo-Iturvide et al. 2008). Interestingly AMF colonization give rise to positive effect on nutritional status and uptake of organic and inorganic compounds of plant and higher physiological activities (Dehne 1982). Correspondingly, a combination of *G. mosseae* and *T. harzianum*, *P. oxalicum* or *B. subtilis* has reduced Geranium root rot caused by *Fusarium solani* and *Macrophomina phaseolina* in artificial- or naturally-infested soils (Haggag and Abd-El Latif 2001). A similar experiment has been carried out on cucumber seedlings co inoculated with *F. equiseti* and *G. mosseae* that treatment was suppressed damping-off symptom development (Saldajeno and Hyakumachi 2011).

All these findings showed that pre-inoculated AMF strains have protective capability and they have synergistically positive effects on competition, altered root exudation, anatomical and morphological changes in the root system and plant defence of host plant (Azcón-Aguilar and Barea 1996; (Chandanie et al. 2006; Pozo and Azcón-Aguilar 2007; Saldajeno and Hyakumachi 2011). PGPF-mediated ISR has been confirmed following lignin deposition at the point of attempted penetration by the pathogen *C. orbiculare* on cucumber hypocotyls (Shivanna et al. 1996) and also superoxide generation (Koike et al. 2001). Accumulation of salicylic acid and increased activities of chitinase, ß-1,3-glucanases and peroxidase in cucumber plants induced by PGPF. Biochemical and molecular analyses have revealed multiple defence mechanisms including expression of pathogenesis-related genes and signalling pathways involved in PGPF-mediated ISR in *Arabidopsis* plants (Hyakumachi and Kubota 2004b). Furthermore, some antibiotics secreted by the PGPR strains fluorescent *Pseudomonas* and *G. fasciculatum* have been associated with the diminishing the negative effects of *F. oxysporum* and *R. solani* on the plants (Basu and Santhaguru 2009).

The AMF-plant nematode interaction were reported due to biocontrol efficiency. The group of Siddiqui and Akhtar (2008a, b, 2009) in India has many studies on the AMF-PGPR or AMF-PGPF combination against plant nematodes. These studies show that various PGPR or saprotrophic fungi when applied together with AMF with or without organic amendments or fertilizer in the presence or absence of a pathogen (*M. phaseolina*) could decrease galling and nematode multiplication in tomato or chickpea plants. There results indicate that AMF can be used as effective biocontrol agents against plant parasitic nematodes, and the combinations with PGPR or PGPF can enhance its efficiency in plant parasitic nematode biocontrol. Conversely, the interactions between the AMF *G. coronatum* and the non-pathogenic *F. oxysporum* in the control of *Meloidogyne incognita* on tomato is also studied by Diedhiou et al. (2003). Their studies indicate that pre-inoculation of tomato plants with either *G. coronatum* or *F. oxysporum* induced plant growth and reduced *M. incognita* infestation but combined application of the AMF and *F. oxysporum* only induced mycorrhization of tomato roots but did not affect overall nematode control or plant growth.

8 Outlook

Up to now some prominent studies have announced beneficial effects of AMF-PGPR, AMF-PGPF, or AMF-PGPR-PGPF combinations on plant growth promotion or disease control. This situation maintains its validity under stress conditions causing of salinity, drought or metal toxicity (Saleem et al. 2007). ACC deaminase-producing bacteria and AMF increase tolerance to salt stress (Gamalero et al. 2010). *G. mosseae* and *Brevibacillus* renders plant to tolerant cadmium (Cd)- contamination in soil (Vivas et al. 2003a, b). Additionally, growth of *Eucalyptus globulus* became possible by inoculation of some AMF and saprotrophic fungi in sewage sludge and heavy metals contaminated soils (Arriagada et al. 2009). In recent years an AMF and rhizobacteria (PGPR or PGPF) combinations are commonly used to protect the biodiversity and ecosystem balance particularly, at forest regeneration, where endangered plant species conserved, phytoremediation of contaminated soils, and detoxification of biological wastes (Zubek et al. 2009). Beneficial effect of AMF *G. deserticola* has also reported due to its alleviating effect on phytotoxicity of organic olive mill residue and studies showed it converts toxic biological waste to a non-hazardous bio-fertilizer form to use for growth of tomato plants (Aranda et al. 2009).

9 Conclusion

In the last decades, alternative control measurements for effective management of plant pathogens causing major diseases have attracted attention by many researchers cause of globally public concern with environmental and health issues resulted from chemical pesticide and fertilizer.

Therefore, biological control of plant pathogens has regarded as a correct and accurate management of disease without giving damage on ecosystem and biodiversity (Baysal and Silme 2018). Even many studies have reported the suppressive effect of singly and mix combination of bio-inoculants, beneficial soil microorganisms like the AMF, PGPR and PGPF and their role in reducing plant diseases are not well understood or on a very limited level.

We should admit that nowadays with the advance techniques, we are also not able to totally understand soil micro-flora dynamism and their ongoing interactions within each other due to its huge complexity. Further studies to be carried out with the help of metagenome and proteome and metabolic data will provide insight of all cases remained unclear points that these findings will show us correct way in large scale or commercial application of AMF, PGPR or PGPF as bio-inoculants to reduce the use of toxic chemical fertilizers and pesticides.

References

Akiyama, K., & Hayashi, H. (2006). Strigolactones: chemical signals for fungal symbiosis and parasitic weeds in plant roots. *Annals of Botany, 97*, 925–931.

Akiyama, K., Matsuzaki, K., & Hayashi, H. (2005). Plant sesquiterpenes induce hyphal branching in arbuscular mycorrhizal fungi. *Nature, 435*, 824–827.

Alejo-Iturvide, F., Márquez-Lucio, M. A., Morales-Ramírez, I., Vázquez-Garcidueñas, M. S., & Olalde-Portugal, V. (2008). Mycorrhizal protection of chili plants challenged with *Phytophthora capsici*. *European Journal of Plant Pathology, 120*, 13–20.

Amann, R. I., Ludwig, W., & Schleifer, K. H. (1995). Phylogenetic identification and in-situ detection of individual microbial cells without cultivation. *Microbiological Reviews, 59*, 143–169.

Ames, R. N., Reid, C. P. P., & Ingham, E. R. (1984). Rhizosphere bacterial population responses to root colonization by a vesicular-arbuscular mycorrhizal fungus. *The New Phytologist, 96*, 555–563.

Amora-Lazcano, E., Vazquez, M. M., & Azcon, R. (1998). Response of nitrogen-transforming microorganisms to arbuscular mycorrhizal fungi. *Biology and Fertility of Soils, 27*, 65–70.

Andrade, G., Mihara, K. L., Linderman, R. G., & Bethlenfalvay, G. J. (1998). Soil aggregation status and rhizobacteria in the mycorrhizosphere. *Plant Soil, 202*, 89–96.

Aranda, E., Sampedro, I., Díaz, R., García-Sánchez, M., Arriagada, C. A., Ocampo, J. A., & García-Romera, I. (2009). The effects of the arbuscular mycorrhizal fungus *Glomus deserticola* on growth of tomato plants grown in the presence of olive mill residues modified by treatment with saprophytic fungi. *Symbiosis, 47*, 133–140.

Arriagada, C., Sampedro, I., Garcia-Romera, I., & Ocampo, J. (2009). Improvement of growth of *Eucalyptus globulus* and soil biological parameters by amendment with sewage sludge and inoculation with arbuscular mycorrhizal and saprobe fungi. *Science of the Total Environment, 407*, 4799–4806.

Artursson, V., Finlay, R. D., & Jansson, J. K. (2006). Interactions between arbuscular mycorrhizal fungi and bacteria and their potential for stimulating plant growth. *Environmental Microbiology, 8*, 1–10.

Azcón-Aguilar, C., & Barea, J. M. (1996). Arbuscular mycorrhizas and biological control of soilborne plant pathogens – an overview of the mechanisms involved. *Mycorrhiza, 6*, 457–464.

Barea, J., Pozo, M., Azcon, R., & Aguilar, C. (2005). Microbial co-operation in the rhizosphere. *Journal of Experimental Botany, 56*, 1761–1778.

Basu, M. J., & Santhaguru, K. (2009). Impact of *Glomus fasciculatum* and fluorescent pseudomonads on growth performance of *Vigna radiata* (L.) Wilczek challenged with phytopathogens. *Journal of Plant Protection Research, 49*, 190–194.

Baysal Ö, Silme RS (2018) The ecological role of biodiversity for crop protection. In: Dunea D (eds) Plant Competition in Cropping Systems. Intech, (in press).

Baysal, Ö., Calışkan, M., & Yeşilova, Ö. (2008). An inhibitory effect of a new *Bacillus subtilis* strain (EU07) against *Fusarium oxysporum* f. sp. *radicis lycopersici*. *Physiological and Molecular Plant Pathology, 73*, 25–32.

Baysal, Ö., Lai, D., Xu, H. H., Siragusa, M., Calışkan, M., Carimi, F., Teixeira da Silva, J. A., & Tor, M. (2013). A proteomic approach provides new insights into the control of soil-borne plant pathogens by *Bacillus* species. *PLoS One, 8*, e53182.

Berg, G. (2009). Plant–microbe interactions promoting plant growth and health: perspectives for controlled use of microorganisms in agriculture. *Applied Microbiology and Biotechnology, 84*, 11–18.

Besserer, A., Bécard, G., Roux, C., & Séjalon-Delmas, N. (2009). Role of mitochondria in the response of arbuscular mycorrhizal fungi to strigolactones. *Plant Signaling & Behavior, 4*, 75–77.

Bianciotto, V., Bandi, C., Minerdi, D., Sironi, M., Tichy, H. V., & Bonfante, P. (1996a). An obligately endosymbiotic mycorrhizal fungus itself harbors obligately intracellular bacteria. *Applied and Environmental Microbiology, 62*, 3005–3010.

Bianciotto, V., Minerdi, D., Perotto, S., & Bonfante, P. (1996b). Cellular interactions between arbuscular mycorrhizal fungi and rhizosphere bacteria. *Protoplasma, 193*, 123–131.

Bianciotto, V., Lumini, E., Lanfranco, L., Minerdi, D., Bonfante, P., & Perotto, S. (2000). Detection and identification of bacterial endosymbionts in arbuscular mycorrhizal fungi belonging to the family *gigasporaceae*. *Applied and Environmental Microbiology, 66*, 4503–4509.

Bianciotto, V., Andreotti, S., Balestrini, R., Bonfante, P., & Perotto, S. (2001). Extracellular polysaccharides are involved in the attachment of *Azospirillum brasilense* and *Rhizobium leguminosarum* to arbuscular mycorrhizal structures. *European Journal of Histochemistry, 45*, 39–49.

Bødker, L., Kjøller, R., Kristensen, K., & Rosendahl, S. (2002). Interactions between indigenous arbuscular mycorrhizal fungi and *Aphanomyces euteiches* in field-grown pea. *Mycorrhiza, 12*, 7–12.

Bolan, N. S. (1991). A critical-review on the role of mycorrhizal fungi in the uptake of phosphorus by plants. *Plant and Soil, 134*, 189–207.

Caron, M. (1989). Potential use of mycorrhizae in control of soilborne diseases. *Canadian Journal of Plant Pathology, 11*, 177–179.

Caron, M., Fortin, J. A., & Richard, C. (1986). Effect of inoculation sequence on the interaction between Glomus intraradices and *Fusarium oxysporum* f. sp. *radices lycopersici* in tomatoes. *Canadian Journal of Plant Pathology, 8*, 12–16.

Carpenter-Boggs, L., Loynachan, T. E., & Stahl, P. D. (1995). Spore germination of *Gigaspora margarita* stimulated by volatiles of soil-isolated actinomycetes. *Soil Biology and Biochemistry, 27*, 1445–1451.

Chandanie, W. A., Kubota, M., & Hyakumachi, M. (2006). Interactions between plant growth promoting fungi and arbuscular mycorrhizal fungus *Glomus mosseae* and induction of systemic resistance to anthracnose disease in cucumber. *Plant and Soil, 286*, 209–217.

Chin-A-Woeng, T. F. C., Bloemberg, G. V., Mulders, I. H. M., Dekkers, L. C., & Lugtenberg, B. J. J. (2000). Root colonization by phenazine-1-carboxamide-producing bacterium *Pseudomonas*

chlororaphis PCL1391 is essential for biocontrol of tomato foot and root rot. *Molecular Plant-Microbe Interactions, 13*, 1340–1345.
Citernesi, A. S., Fortuna, P., Filippi, C., Bagnoli, G., & Giovannetti, M. (1996). The occurrence of antagonistic bacteria in *Glomus mosseae* pot cultures. *Agronomie, 16*, 671–677.
Compant, S., Duffy, B., Nowak, J., Clément, C., & Barka, E. A. (2005). Use of plant growth-promoting bacteria for biocontrol of plant diseases: principles, mechanisms of action, and future prospects. *Applied and Environmental Microbiology, 71*, 4951–4959.
Daniels, B. A., & Trappe, J. M. (1980). Factors affecting spore germination of the vesicular-arbuscular mycorrhizal fungus, *Glomus epigaeus*. *Mycologia, 72*, 457–471.
Datnoff, L. E., Nemec, S., & Pernezny, K. (1995). Biological control of Fusarium crown and root rot of tomato in Florida using *Trichoderma harzianum* and *Glomus intraradices*. *Biological Control, 5*, 427–431.
De la Peña, E., Rodríguez-Echeverría, S., van der Putten, W. H., Freitas, H., & Moens, M. (2006). Mechanism of control of root-feeding nematodes by mycorrhizal fungi in the dune grass *Ammophila arenaria*. *The New Phytologist, 169*, 829–840.
Dehne, H. W. (1982). Interaction between vesicular-arbuscular mycorrhizal fungi and plant pathogens. *Pytopathology, 72*, 1115–1119.
Diedhiou, P. M., Hallmann, J., Oerke, E. C., & Dehne, H. W. (2003). Effects of arbuscular mycorrhizal fungi and a non-pathogenic *Fusarium oxysporum* on *Meloidogyne incognita* infestation of tomato. *Mycorrhiza, 13*, 199–204.
Drew, E. A., Murray, R. S., Smith, S. E., & Jakobsen, I. (2003). Beyond the rhizosphere: growth and function of arbuscular mycorrhizal external hyphae in sands of varying pore sizes. *Plant Soil, 251*, 105–114.
El-Tarabily, K. A., & Sivasithamparam, K. (2006). Non-streptomycete actinomycetes as biocontrol agents of soil-borne fungal plant pathogens and as plant growth promoters. *Soil Biology and Biochemistry, 38*, 1505–1520.
Filion, M., St-Arnaud, M., & Fortin, J. A. (1999). Direct interaction between the arbuscular mycorrhizal fungus *Glomus intraradices* and different rhizosphere microorganisms. *The New Phytologist, 141*, 525–533.
Filippi, C., Bagnoli, G., Citernesi, A. S., & Giovannetti, M. (1998). Ultrastructural spatial distribution of bacteria associated with sporocarps of *Glomus mosseae*. *Symbiosis, 24*, 1–12.
Forster, S. M., & Nicolson, T. H. (1981). Aggregation of sand from maritime embryo sand dune by microorganisms and higher plants. *Soil Biology and Biochemistry, 13*, 199–203.
Franzini, V., Azcon, R., Mendes, F., & Aroca, R. (2010). Interactions between *Glomus* species and *Rhizobium* strains affect the nutritional physiology of drought-stressed legume hosts. *Journal of Plant Physiology, 167*, 614–619.
Gamalero, E., Berta, G., Massa, N., Glick, B. R., & Lingua, G. (2010). Interactions between *Pseudomonas putida* UW4 and *Gigaspora rosea* BEG9 and their consequences for the growth of cucumber under salt-stress conditions. *Journal of Applied Microbiology, 108*, 236–245.
Garbaye, J. (1994). Helper bacteria – a new dimension to the mycorrhizal symbiosis. *The New Phytologist, 128*, 197–210.
Garcia-Garrido, J. M., & Ocampo, J. A. (1988). Interaction between *Glomus mosseae* and *Erwinia carotovora* and its effect on the growth of tomato plants. *The New Phytologist, 110*, 551–555.
Garmendia, I., Aguirreolea, J., & Goicoechea, N. (2006). Defence-related enzymes in pepper roots during interactions with arbuscular mycorrhizal fungi and/or *Verticillium dahliae*. *BioControl, 51*, 293–310.
Gorzelak, M., Asay, A., Pickles, B., & Simard, S. (2015). Inter-plant communication through mycorrhizal networks mediates complex adaptive behaviour in plant communities. *AoB plants, 7*. https://doi.org/10.1093/aobpla/plv050.
Griffiths, R. I., Manefield, M., Ostle, N., McNamara, N., O'Donnell, A. G., Bailey, M. J., & Whiteley, A. (2004). 13CO2 pulse labelling of plant in tandem with stable isotope probing: methodological considerations for examining microbial function in the rhizosphere. *Journal of Microbiological Methods, 58*, 119–129.

Grunwald, U., Guo, W., Fischer, K., Isayenkov, S., Ludwig-Müller, J., Hause, B., Yan, X., Küster, H., & Franken, P. (2009). Overlapping expression patterns and differential transcript levels of phosphate transporter genes in arbuscular mycorrhizal, Pi-fertilised and phytohormone-treated *Medicago truncatula* roots. *Planta, 229*, 1023–1034.

Haggag, W. M., & Abd-El Latif, F. M. (2001). Interaction between vesicular arbuscular mycorrhizae and antagonistic biocontrol microorganisms on controlling root rot disease incidence of *geranium* plants. *OnLine Journal of Biological Sciences, 1*, 1147–1153.

Harman, G. E., Howell, C. R., Viterbo, A., Chet, I., & Lorito, M. (2004). *Trichoderma* species—opportunistic, avirulent plant symbionts. *Nature Reviews Microbiology, 2*, 43–56.

Hartmann, A., Schmid, M., van Tuinen, D., & Berg, G. (2009). Plant-driven selection of microbes. *Plant and Soil, 321*, 235–257.

Hause, B., Mrosk, C., Isayenkov, S., & Strack, D. (2007). Jasmonates in arbuscular mycorrhizal interactions. *Phytochemistry, 68*, 101–110.

Hyakumachi, M., & Kubota, M. (2004a). Fungi as plant growth promoter and disease suppressor. In D. K. Arora (Ed.), *Fungal biotechnology in agricultural, food, and environmental applications* (Vol. 21, pp. 101–110). New York: Marcel Dekker Inc.

Hyakumachi M, Kubota M (2004b) *Biological control of plant diseases by plant growth promoting fungi*. Proceedings of International Seminar Biological Control Soil Borne Plant Diseases (pp. 87–123). Japan-Argentina Joint Study.

Isayenkov, S., Mrosk, C., Stenzel, I., Strack, D., & Hause, B. (2005). Suppression of allene oxide cyclase in hairy roots of *Medicago truncatula* reduces jasmonate levels and the degree of mycorrhization with *Glomus intraradices*. *Plant Physiology, 139*, 1401–1410.

Jeffries, P., & Barea, J. M. (2001). Arbuscular mycorrhiza: a key component of sustainable plant–soil ecosystems. In B. Hock (Ed.), *The Mycota: fungal associations* (Vol. 9, pp. 95–113). Berlin: Springer.

Johnson, D., Leake, J. R., & Read, D. J. (2001). Novel in-growth core system enables functional studies of grassland mycorrhizal mycelial networks. *The New Phytologist, 152*, 555–562.

Joner, E., & Leyval, C. (2009). Phytoremediation of organic pollutants using mycorrhizal plants: a new aspect of rhizosphere interactions. In E. Lichtfouse, M. Navarrete, P. Debaeke, S. Véronique, & C. Alberola (Eds.), *Sustainable agriculture* (pp. 885–894). Dordrecht: Springer.

Karagiannidis, N., Bletsos, F., & Stavropoulos, N. (2002). Effect of *Verticillium* wilt (*Verticillium dahliae* Kleb.) and mycorrhiza (*Glomus mosseae*) on root colonization, growth and nutrient uptake in tomato and eggplant seedlings. *Scientia Horticulturae, 94*, 145–156.

Kaye, J. W., Pfleger, F. L., & Stewart, E. L. (1984). Interaction of *Glomus fasciculatum* and *Pythium ultimum* on greenhouse-grown poinsettia. *Canadian Journal of Botany, 62*, 1575–1579.

Khan, A. (2005). Role of soil microbes in the rhizospheres of plants growing on trace metal contaminated soils in phytoremediation. *Journal of Trace Elements in Medicine and Biology, 18*, 355–364.

Kloepper, J. W., Ryu, C.-M., & Zhang, S. (2004). SA: Induced systemic resistance and promotion of plant growth by *Bacillus* spp. *Phytopathology, 94*, 1259–1266.

Koike, N., Hyakumachi, M., Kageyama, K., Tsuyumu, S., & Doke, N. (2001). Induction of systemic resistance in cucumber against several diseases by plant growth promoting fungi: lignification and superoxide generation. *European Journal of Plant Pathology, 107*, 523–533.

Krishna, K. R., & Bagyaraj, D. J. (1983). Interaction between *Glomus fasciculatum* and *Sclerotium rolfsii* in peanut. *Canadian Journal of Botany, 61*, 2349–2351.

Ludwig-Muller, J. (2000). Indole-3-butyric acid in plant growth and development. *Plant Growth Regulation, 32*, 219–230.

MacDonald, R. M., Chandler, M. R., & Mosse, B. (1982). The occurrence of bacterium-like organelles in vesicular-arbuscular mycorrhizal fungi. *The New Phytologist, 90*, 659–663.

Mayo, K., Davis, R. E., & Motta, J. (1986). Stimulation of germination of spores of *Glomus versiforme* by spore-associated bacteria. *Mycologia, 78*, 426–431.

Meyer, J. R., & Linderman, R. G. (1986a). Response of subterranean clover to dual Inoculation with vesicular arbuscular mycorrhizal fungi and a plant growth-promoting bacterium, *Pseudomonas putida*. *Soil Biology and Biochemistry, 18*, 185–190.

Meyer, J. R., & Linderman, R. G. (1986b). Selective influence on populations of rhizosphere or rhizoplane bacteria and actinomycetes by mycorrhizas formed by *Glomus fasciculatum*. *Soil Biology and Biochemistry, 18*, 191–196.

Minerdi, D., Fani, R., Gallo, R., Boarino, A., & Bonfante, P. (2001). Nitrogen fixation genes in an endosymbiotic *Burkholderia* strain. *Applied and Environmental Microbiology, 67*, 725–732.

Miransari, M., & Smith, D. L. (2008). Using signal molecule genistein to alleviate the stress of suboptimal root zone temperature on soybean–*Bradyrhizobium* symbiosis under different soil textures. *Journal of Plant Interactions, 3*, 287–295.

Mosse, B. (1959). The regular germination of resting spores and some observations on the growth requirements of an *Endogone* sp. causing vesicular-arbuscular mycorrhiza. *Transactions of the British Mycological Society, 42*, 273–286.

Nemec, S., Datnoff, L. E., & Strandberg, J. (1996). Efficacy of biocontrol agents in planting mixes to colonize plant roots and control root diseases of vegetables and citrus. *Crop Protection, 15*, 735–742.

Newsham, K. K., Fitter, A. H., & Watkinson, A. R. (1995). Arbuscular mycorrhiza protect an annual grass from root pathogenic fungi in the field. *Journal of Ecology, 83*, 991–1000.

Phirke, N. V., Kothari, R. M., & Chincholkar, S. B. (2008). Rhizobacteria in mycorrhizosphere improved plant health and yield of banana by offering proper nourishment and protection against diseases. *Applied Biochemistry and Biotechnology, 151*, 441–451.

Pieterse, C. M. J., van Pelt, J. A., Verhagen, B. W. M., Ton, J., van Wees, S. C. M., Léon-Kloosterziel, K. M., & van Loon, L. C. (2003). Induced systemic resistance by plant growth promoting rhizobacteria. *Symbiosis, 35*, 39–54.

Pozo, M. J., & Azcón-Aguilar, C. (2007). Unravelling mycorrhiza-induced resistance. *Current Opinion in Plant Biology, 10*, 393–398.

Puppi, G., Azcón, R., & Höflich, G. (1994). Management of positive interactions of arbuscular mycorrhizal fungi with essential groups of soil microorganisms. In S. Gianinazzi & H. Schüepp (Eds.), *Impact of arbuscular mycorrhizas on sustainable agriculture and natural ecosystems* (pp. 201–215). Basel: Birkhäuser Verlag.

Rillig, M. C., & Mummey, D. L. (2006). Mycorrhizas and soil structure. *The New Phytologist, 171*, 41–53.

Roesti, D., Gaur, R., Johri, B. N., Imfeld, G., Sharma, S., Kawaljeet, K., & Aragno, M. (2006). Plant growth stage, fertilizer management and bio-inoculation of arbuscular mycorrhizal fungi and plant growth promoting rhizobacteria affect the rhizobacterial community structure in rain-fed wheat fields. *Soil Biology and Biochemistry, 38*, 1111–1120.

Rosendahl, C. N., & Rosendahl, S. (1990). The role of vesicular-arbuscular mycorrhiza in controlling damping-off and growth reduction in cucumber caused by *Pythium ultimum*. *Symbiosis, 9*, 363–366.

Rosewarne, G. M., Barker, S. J., Smith, S. E., Smith, A. F., Schachtman, D. P. A., & Schachtman, D. P. (1999). A *Lycopersicon esculentum* phosphate transporter (LePT1) involved in phosphorous uptake from a vesicular-arbuscular mycorrhizal fungus. *The New Phytologist, 144*, 507–516.

Ross, J. P. (1980). Effect of nontreated soil on sporulation of vesicular-arbuscular mycorrhizal fungi associated with soybean. *Phytopathology, 70*, 1200–1205.

Saldajeno, M. G. B., & Hyakumachi, M. (2011). The plant growthpromoting fungus *Fusarium equiseti* and the arbuscular mycorrhizal fungus *Glomus mosseae* stimulate plant growth and reduce severity of anthracnose and damping-off diseases in cucumber (*Cucumis sativus*) seedlings. *The Annals of Applied Biology, 159*, 28–40.

Saleem, M., Arshad, M., Hussain, S., & Bhatti, A. S. (2007). Perspective of plant growth promoting rhizobacteria (PGPR) containing ACC deaminase in stress agriculture. *Journal of Industrial Microbiology & Biotechnology, 34*, 635–648.

Schnepf, A., Leitner, D., Klepsch, S., Pellerin, S., & Mollier, A. (2011). Modelling phosphorus dynamics in the soil-plant system. In E. K. Bünemann, A. Obserson, & E. Frossard (Eds.), *Phosphorus in action: biological processes in soil phosphorus cycling* (pp. 113–133). Heidelberg: Springer.

Schreiner, R. P., Mihara, K. L., McDaniel, H., & Bethlenfalvay, G. J. (1997). Mycorrhizal fungi influence plant and soil functions and interactions. *Plant and Soil, 188*, 199–209.
Schrey, S. D., & Tarkka, M. T. (2008). Friends and foes: streptomycetes as modulators of plant disease and symbiosis. *Antonie Van Leeuwenhoek, 94*, 11–19.
Secilia, J., & Bagyaraj, D. J. (1987). Bacteria and actinomycetes associated with pot cultures of vesicular arbuscular mycorrhizas. *Canadian Journal of Microbiology, 33*, 1069–1073.
Shimizu-Sato, S., Tanaka, M., & Mori, H. (2009). Auxin–cytokinin interactions in the control of shoot branching. *Plant Molecular Biology, 69*, 429–435.
Shivanna, M. B., Meera, M. S., Kageyama, K., & Hyakumachi, M. (1996). Plant growth promoting fungi induce systemic disease resistance in cucumber. In T. Wenhua, R. J. Cook, & A. Rovira (Eds.), *Advances in biological control of plant diseases* (pp. 175–184). Beijing: China Agricultural University Press.
Siddiqui, Z. A., & Akhtar, M. S. (2008a). Synergistic effects of antagonistic fungi and a plant growth promoting rhizobacterium, an arbuscular mycorrhizal fungus, or composted cow manure on populations of *Meloidogyne incognita* and growth of tomato. *Biocontrol Science and Technology, 18*, 279–290.
Siddiqui, Z. A., & Akhtar, M. S. (2008b). Effects of fertilizers, AM fungus and plant growth promoting rhizobacterium on the growth of tomato and on the reproduction of root-knot nematode *Meloidogyne incognita*. *Journal of Plant Interactions, 3*, 263–271.
Siddiqui, Z. A., & Akhtar, M. S. (2009). Effects of antagonistic fungi, plant growth promoting rhizobacteria, and arbuscular mycorrhizal fungi alone and in combination on the reproduction of *Meloidogyne incognita* and growth of tomato. *Journal of General Plant Pathology, 75*, 144–153.
Smith, S. E., & Read, D. J. (1997). *Mycorrhizal symbiosis* (2nd ed.). London: Academic.
Smith, S. E., & Read, D. J. (2008). *Mycorrhizal symbiosis* (3rd ed.). New York: Academic.
Söderberg, K. H., Olsson, P. A., & Baath, E. (2002). Structure and activity of the bacterial community in the rhizosphere of different plant species and the effect of arbuscular mycorrhizal colonisation. *FEMS Microbiology Ecology, 40*, 223–231.
Tahmatsidou, V., O'Sullivan, J., Cassells, A. C., Voyiatzis, D., & Paroussi, G. (2006). Comparison of AMF and PGPR inoculants for the suppression of *Verticillium* wilt of strawberry (*Fragaria x ananassa* cv. *Selva*). *Applied Soil Ecology, 32*, 316–324.
Tisdall, J. M., & Oades, J. M. (1979). Stabilization of soil aggregates by the root systems of rye grass. *Australian Journal of Soil Research, 17*, 429–441.
Tommerup, I. C. (1985). Inhibition of spore germination of vesicular-abuscular mycorrhizal fungi in soil. *Transactions of the British Mycological Society, 85*, 267–278.
Trotta, A., Varese, G. C., Gnavi, E., Fusconi, A., Sampò, S., & Berta, G. (1996). Interactions between the soilborne root pathogen *Phytophthora nicotianae* var. *parasitica* and the arbuscular mycorrhizal fungus *Glomus mosseae* in tomato plants. *Plant and Soil, 185*, 199–209.
Van Loon, L. C. (2007). Plant responses to plant growth promoting rhizobacteria. *European Journal of Plant Pathology, 119*, 243–254.
Van Wees, S., Van der Ent, S., & Pieterse, C. (2008). Plant immune responses triggered by beneficial microbes. *Current Opinion in Plant Biology, 11*, 443–448.
Vinale, F., Sivasithamparam, K., Ghisalberti, E. L., Marra, R., Woo, S. L., & Lorito, M. (2008). *Trichoderma*–plant–pathogen interactions. *Soil Biology and Biochemistry, 40*, 1–10.
Vivas, A., Vörös, I., Biró, B., Barea, J. M., Ruiz-Lozano, J. M., & Azcón, R. (2003a). Beneficial effects of indigenous Cd-tolerant and Cd-sensitive *Glomus mosseae* associated with a Cd-adapted strain of *Brevibacillus* sp. in improving plant tolerance to Cd contamination. *Applied Soil Ecology, 24*, 177–186.
Vivas, A., Vörös, I., Biró, B., Campos, E., Barea, J. M., & Azcón, R. (2003b). Symbiotic efficiency of autochthonous arbuscular mycorrhizal fungus (*G. mosseae*) and *Brevibacillus* sp. isolated from cadmium polluted soil under increasing cadmium levels. *Environment Pollution, 126*, 179–189.

Von der Weid, I., Artursson, V., Seldin, L., & Jansson, J. K. (2005). Antifungal and root surface colonization properties of GFP-tagged *Paenibacillus bransilensis* PB177. *World Journal of Microbiology and Biotechnology, 21*, 1591–1597.

Waller, F., Achatz, B., Baltruschat, H., Fodor, J., Becker, K., Fischer, M., Heier, T., Hückelhoven, R., Neumann, C., von Wettstein, D., Franken, P., & Kogel, K. H. (2005). The endophytic fungus *Piriformospora indica* reprograms barley to salt-stress tolerance, disease resistance, and higher yield. *Proceedings of the National Academy of Sciences of the United States of America, 102*, 13386–13391.

Wamberg, C., Christensen, S., Jakobsen, I., Muller, A. K., & Sorensen, S. J. (2003). The mycorrhizal fungus (*Glomus intraradices*) affects microbial activity in the rhizosphere of pea plants (*Pisum sativum*). *Soil Biology and Biochemistry, 35*, 1349–1357.

Wei, G., Kloepper, J. W., & Tuzun, S. (1991). Induction of systemic resistance of cucumber to *Colletotrichum orbiculare* by select strains of plant growth-promoting rhizobacteria. *Phytopathology, 81*, 1508–1512.

Wilson, G. W. T., Hetrick, B. A. D., & Kitt, D. G. (1988). Suppression of mycorrhizal growth-response of big bluestem by nonsterile soil. *Mycologia, 80*, 338–343.

Xavier, L. J. C., & Boyetchko, S. M. (2002). Mycorrhizae as biocontrol agents. In K. G. Mukerji, C. Manoharachary, & B. P. Chamola (Eds.), *Techniques in mycorrhizal studies* (pp. 493–536). Dordrecht: Kluwer Academic Publishers.

Zehnder, G. W., Murphy, J. F., Sikora, E. J., & Kloepper, J. W. (2001). Application to rhizobacteria for induced resistance. *European Journal of Plant Pathology, 107*, 39–50.

Zubek, S., Turnau, K., Tsimilli-Michael, M., & Strasser, R. J. (2009). Response of endangered plant species to inoculation with arbuscular mycorrhizal fungi and soil bacteria. *Mycorrhiza, 19*, 113–123.

Role of Bacteria in Pedogenesis

10

Palika Sharma and Gaurav Bhakri

1 Introduction

Soil can be defined as the solid material on the surface of Earth resulting from the interaction of weathering and biological activity on the parent material or underlying hard rock. Pedogenesis (Greek term where pedo means 'soil' and genesis, means 'origin) is the process of soil formation which is under regulation of effects of place, environment and history. Pedogenesis is a branch of pedology that is study of soil in its natural environment. Soil synthesis is a result of biogeochemical processes taking place within the soil. Such activities result in development of layers called soil horizons, differentiated by colour, texture, structure and chemistry (Buol et al. 1973). Pedogenesis basically involves physical, chemical and microbial weathering of rocks out of which physical and chemical processes are well studied. To our knowledge, microbial weathering is functionally and taxonomically least investigated. Till date, only lichens (symbiotic associations between fungi and photosynthetic algae or cyanobacteria) have been regarded as the first mineral weathering pioneer organisms (Banfield et al. 1999). Lichens colonized the same mineral spot on the surface of a rock or monument for decades and carried out its weathering. Interestingly, in lichens, complex microbial communities have been identified, such as bacteria belonging to *Anabaena, Bradyrhizobium, Burkholderia* and *Collimonas* (Gorbushina 2007; Seneviratne and Indrasena 2006). An indirect effect of *Bradyrhizobium elkanii* on the mineral-weathering ability of lichen has been reported (Seneviratne and Indrasena 2006): rather than directly weathering the mineral substrate the bacterium, fixes and supplies nitrogen to the fungus, enhancing

P. Sharma (✉)
Department of Microbiology, Punjab Agricultural University, Ludhiana, Punjab, India

G. Bhakri
Department of Biotechnology, National Bureau of Animal Genetic Resources, Karnal, Haryana, India

© Springer Nature Singapore Pte Ltd. 2019

A. Varma, D. K. Choudhary (eds.), *Mycorrhizosphere and Pedogenesis*,
https://doi.org/10.1007/978-981-13-6480-8_10

fungal organic acid production. Howbeit, the relative impact of these types of bacteria on the weathering process is poorly understood.

1.1 Factors Involved in Soil-Formation

The soil-formation process include two active factors, climate and organisms, which carry out catalysis in pedogenesis and three passive factors (parent material, topography, and time) that respond to the forces exerted by active factors (Fig. 10.1).

1.2 Passive Soil Forming Factors

The passive soil forming factors are those which represent the source of soil forming mass and conditions affecting it. These provide a base on which the active soil forming factors work or act for the development of soil.

Parent material is that consolidated matter from which the soil is formed.

Climatic factors, temperature and rainfall directly affect soil. Weathering of rocks occurs quickly in warm and moist climates where temperature affects the rate of chemical as well as biological activity (Huddleston 1984). As far as organisms are concerned they exist mainly in two significant groups:

1. Macroorganisms
2. Microorganisms

2 Macroorganisms

Macroorganisms include living or dead large plants and animals. Dead and decaying plant matter build up organic matter in the soil. Live macroorganisms are the source of nearly all organic matter. Plants in the form of grasses, woody vegetation and trees contribute largely towards organic matter production. The positive effect of organic matter in the soil cannot be overemphasized (Thompson and Troeh 1973).

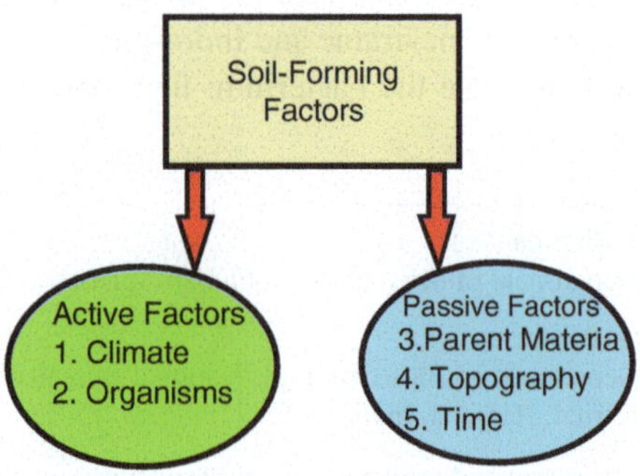

Fig. 10.1 Soil-Forming Factors

3 Microorganisms

Microorganisms or microbes are tiny entities that are not visible with naked eyes but only seen through microscope. Microbes play an important role in pedogenesis process where existing soil, rocks, water, air and organisms interact with each other (Bin et al. 2010). There exists a wide knowledge gap in the biological weathering of rocks, especially through microbes which is sparsely studied. Therefore in this chapter we have emphasized on the role of microbes mainly bacteria in pedogenesis. Bacteria are tiny, single-celled organisms – generally 4/100,000 of an inch wide (1 μm) and somewhat longer in length. What bacteria lack in size, they make up in numbers. In a teaspoon of productive soil bacteria ranges between 100 million and 1 billion in number which equals to mass of two cows per acre (Ingham 2009).

Bacteria, fungi, protozoa, nematodes, and algae are the primary decomposers of organic matter (Khatoon et al. 2017). They convert raw plant and animal residues into a complex, dark brown or black substance called humus, improve soil tilth and release soil nitrogen as an essential nutrient for plants Mosier et al. (2004). Microbes and the humus they produce, makes topsoil rich and fertile by acting as a glue to hold soil particles together in form of aggregates thus minimise soil erosion Coleman. Well-aggregated soil ideally provides the rightful combination of air and water to plant roots.

3.1 Mechanisms Involved in Pedogenesis By Bacteria

The soil is where living organisms, or the biosphere, interact with rocks and minerals (geosphere), water (hydrosphere), atmosphere and dead organic matter (detritosphere).

> Without microbes, soil would be a virtually inert (lifeless) body. With them, soil is truly a living, dynamic system

The first assumption about bacteria playing major role in soil fertility and decomposition was typified by the book of Löhnis and Fred (1923). The nitrogen fixing *Rhizobium*, *Mesorhizobium* and *Bradyrhizobium* along with methanogens (*Methylobacter* and *Methylophilus)* have been implicated in weathering of minerals. Chitinase producing *Collimonas* may degrade live hyphae (de Boer et al. 2004) alone but along with *Burkholderia* they weather minerals aswell (Uroz et al. 2007). *Nitrospira* and *Nitrobacter* are the ammonia oxidisers, *Thiobacillus*- the iron oxidiser and *Methylomonas* is the methanotroph (Aislabie and Deslippe 2013).

3.2 Bacterial Weathering of Minerals:

Bare rock surfaces being in contact with air, could be a special habitat to microbes, even though they are featured by enormous fluctuating harsh environments including solar radiation, drought, nutrient deficiency and temperature, still various types of microorganisms exist in the cracks as well as on the rock surfaces (Gorbushina, 2007). Earlier some kinds of autotrophic photosynthetic nitrogen-fixers such as algae, cyanobacteria and lichens were described as rock microorganisms but later, some heterotrophic bacteria and fungi were also observed on exposed rocks surface. Although the rock surface is unsuitable for microbial survival, autotrophic microorganisms cope up with adverse conditions via photosynthesis and N fixation, whereas heterotrophic bacteria interact symbiotically with autotrophs (e.g. lichen type fungi) or intercept smaller soil particles to thrive nutrients which were occasionally brought in by air and rainwater (Viles and Gorbushina, 2003). These rocky microorganisms are of collaborative or symbiotic type, and differ from soil microbes that usually exist in competitive or predatory type of relationship thus are pioneers of species responsible for weathering of rocks (Burford et al. 2006; Gorbushina et al. 2003). Some bacteria, fungi, lichens and algae were identified or isolated from the surface of Triassic limestone and dolomite in Guizhou, Southwest China (Fig. 10.2) (Lian et al. 2008). Different microbes mainly work on retainment of water and trace nutrients for sustaining life activities and reproduction (Gorbushina, 2007). Under nutrient deficiency these microrganisms bore into the rock resulting in to small cave or tunnel, strengthening the colonization on the rock surface to form biofilm or biological crust under suitable conditions (such as favourable temperature and humidity) (Lian et al. 2008). Prokaryotic microorganisms usually have spores and exopolysaccharides to protect cells against desiccating conditions on the rock surface.

Recent research shows that in order to acquire energy, bacteria may be directly responsible for driving cascade of reactions that reduce rocks to soil and free biologically important minerals (Shelobolina et al. 2012). It has been reported that number of bacterial strains belonging to diverse genera (*Pseudomonas, Streptomyces, Staphylococcus, Frateuria, Rhanella, Sphingomonas, Aminobacter, Burkholderia, Enterobacter, Agrobacterium, Achromobacter, Collimonas, Acinetobacter, Azotobacter, Citrobacter, Shewanella, Serratia, Bacillus, Mycobacterium, Arthrobacter and Rhizobium*) are capable of mineral-weathering (Uroz et al. 2009). Such bacteria are able to effect mineral stability alone or in combination with other microorganisms by making complex microbial communities that colonize mineral surfaces. Albeit most of the functional studies reported have emphasized on

Fig. 10.2 Microbes including lichens on the surface of Triassic carbonate rocks. (Bin et al. 2010)

bacterial isolates from soil. Stones act as primary ecosystems due to their exclusive mineral composition where only a few adapted microbes possessing mineral-weathering abilities, can grow and survive (Calvaruso et al. 2007; Uroz et al. 2007).

Bacterial weathering of minerals in soil has remained the target of most functional studies concerned with bacterial abilities for mineral weathering. These rock surfaces are complex environments and diversified in composition that are usually colonized by specific bacteria (Fig. 10.3) that vary from those inhabiting the surrounding soil (Uroz et al. 2009). Furthermore, the surface and the core of soil mineral particles seems to be inhabited by non similar bacterial communities: in limestone, the endolithic bacterial community seemed to be composed mainly of Gram-positive bacteria and acidobacteria, whereas the epilithic population was composed of approximately 50% proteobacteria (McNamara et al. 2006). Mineral particles contain inorganic nutrients (aluminium, silica and calcium) which are used by these bacteria thus mineral composition is another key factor influencing bacterial communities. Bacteria have been reported by Gleeson et al. (2006) and Carson et al. (2007) for colonization of different primary minerals such as granite, limestone, apatite, plagioclase, quartz, however, the fingerprints of bacterial communities colonizing granite varied with mineral inclusion (muscovite, plagioclase, Kfeldspar and quartz). All these observations showing correlation between mineral composition and bacterial communities prompts us to propose a new concept, the mineralosphere.

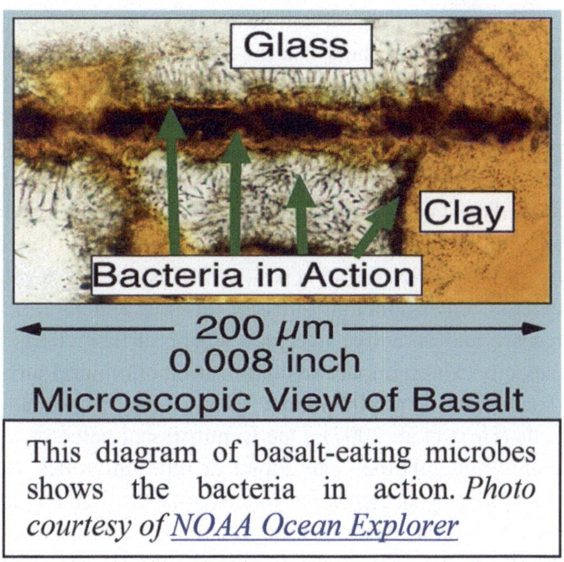

Fig. 10.3 Microscopic view of rock eating microbe in volcanic rock

3.3 Rhizosphere and Mineralosphere

It is well established that, in the rhizosphere, only 1–2% of bacteria promote plant growth and act as biofertilizers (Antoun and Kloepper 2001). The rhizosphere effect is well known from the beginning of the twentieth century (Hiltner 1904). The proliferation of soil microorganisms in the vicinity of plant roots get influenced via root exudation which are preffered food source for microbes (Walker et al. 2003). The rhizosphere is a hot-spot of plant microbe interactions leading to efficient geochemical cycling of nutrients. Rhizospheric region of the soil is rich in primary and secondary metabolites that orchestrate almost every type of rhizospheric interaction where plant roots communicate with their below-ground microbial residents. A biochemical signal is conducted between rhizobacteria and plant roots resulting in dynamic interactions flourishing either symbiosis or pathogenicity (Vessey and Buss 2002). Rhizosphere contains innumerable secondary metabolites where particularly flavonoids and auxins are documented to be the most important signalling elements in plant-microbe interactions. Numerous rhizobacterial species which are known to facilitate plant growth by exerting beneficial effects are generally referred to as plant growth promoting rhizobacteria (PGPR) (Vera et al. 2013). These are most widely studied beneficial, saprophytic, heterogenous group of rhizospheric bacteria which aids in the plant growth through direct and indirect mechanisms via nitrogen fixation, solubilization of zinc and phosphorus, lowering of ethylene concentration through ACC-deaminase activity under abiotic and biotic stressed conditions, production of plant hormones such as indole acetic acid (IAA), gibberellins and cytokinins, production of siderophores and competitive exclusion of pathogens or elimination of substances toxic to plants (Prasad et al. 2017).

The surroundings of soil minerals, where microorganisms are selected for their ability to preferentially use the inorganic nutrients released by soil minerals, such mineral-influenced habitats are called as 'mineralosphere'. In accordance to Uroz et al. (2007) the 'mineralosphere' signifies the specific interface and habitat comprising the rock surfaces and the surrounding soil, which are physically, chemically and biologically influenced by minerals. Physically, the mineralosphere is characterized by several zones, including pores and cracks which modify water circulation and can be considered as microbial sanctuaries. Bacteria get accumulated in such mineralospheric zones via passive diffusion, and develop relative protection against external environmental stresses (abiotic and biotic). Chemically, it is a nutrient reserve and an active interface where surface charges and the minerals exchange capacity exhibit impact on colonization of mineral surfaces. Indeed, positive charges (such as in the phyllo silicate inter layers) can attract negatively charged bacterial cells (Uroz et al. 2007). Due to nutritional value or toxicity of nutrients contained within minerals they can attract or repel microbes. Released nutrients have direct availability to bacteria, but in case of unavailable nutrients microbes can carry out precipitatation (oxides) through solubilization to make them available (Uroz et al. 2009). Biologically, this habitat is enriched in low-carbon and mineral-rich environments adapted microorganisms which potentially contribute towards mineral weathering. In mineralosphere, the mineral-weathering capability of bacteria may be regulated by their nutritional requirements, nutrient availability, and/or the mineral type. Undoubtedly, this habitat is under great influence of environmental factors, including soil parameters such as pH and water availability, or the organic and inorganic nutrient inputs. Generally, the carbonate rocks are enriched in Ca, Mg and lack of Si, Al, and Fe, but soil inorganic substances are mainly Si, Al, Ca, Mg, Fe, etc. Rocks cannot be weathered easily to supply a large number of soil minerals. The microorganisms in such areas erode the rocks by forming micro-colonies, biofilms, and biological crust on the rock surface or in micro-cracks through the chemical degradation (organic acids secreted by the microbial metabolism to promote calcium carbonate dissolution and weathering), the biological effect (the mineral particles are broken down via microbial growth along with interspersed fungal hyphae to erode rock surface more easily) and the enhancement of erosion via bacterial metabolites or enzymes (microorganisms secrete enzymes such as carbonic anhydrase enzymes etc.) to speed up the weathering of calcium carbonate (Dou and Lian 2009; Chen et al. 2008). Lepleux et al. (2012) performed a BIOLOG analysis to highlight potent mineral-weathering bacterial isolates in contrast to those of the surrounding bulk soil or the mycorrhizo-sphere, where mineral-associated (mineralosphere) bacteria exhibited oligotrophic behaviour by metabolizing only few substrates and that too with a very low intensity. Interestingly, the most intensively and unique substrate utilized by the mineralospheric bacteria appeared to be glucose. On the other hand, bacteria from the bulk soil prefer to metabolize amino and carboxylic acids with high intensity, with comparatively poor glucose metabolism. It has been reported that the most efficient mineral-weathering bacteria produce high concentrations of oxalate (Frey et al. 2010). These observations demonstrate that mineral-associated bacterial isolates are physiologically active, metabolize organic substrates, and produce metabolites, suggesting that they may participate in mineral weathering and nutrient cycling and finally in pedogensis.

Rhizosphere and mineralosphere both are important zones of biotic interactions and both provide physical support to life (Fig. 10.4). Due to some common traits such as nutrient bioavailability and uptake (Nitrogen fixation, P, S, K and Zn solubilisation by rhizospheric bacteria and inorganic mineral nutrients Ca, Mg, K, Na, Al, Fe, Mn by mineralospheric bacteria through production of organic acids and metal chelators along with siderophores production for iron acquisition), biofilm production, antagonism and decomposition of organic compounds and mineral precipitation. All these factors suggest that mineralopshere is the inorganic twin to rhizosphere (Uroz et al. 2009).

4 Minerals Regulating Bacterial Gene Expression

From above studies it is apparent that bacterial communities colonize minerals, but this question also prompts in our mind that is there any effect of these minerals on the physiology of the bacteria under the influence of mineralosphere. To answer this question many attempts have been made using chemoheterotrophic bacteria able to respire in metals contained within minerals. Expression of different genes and

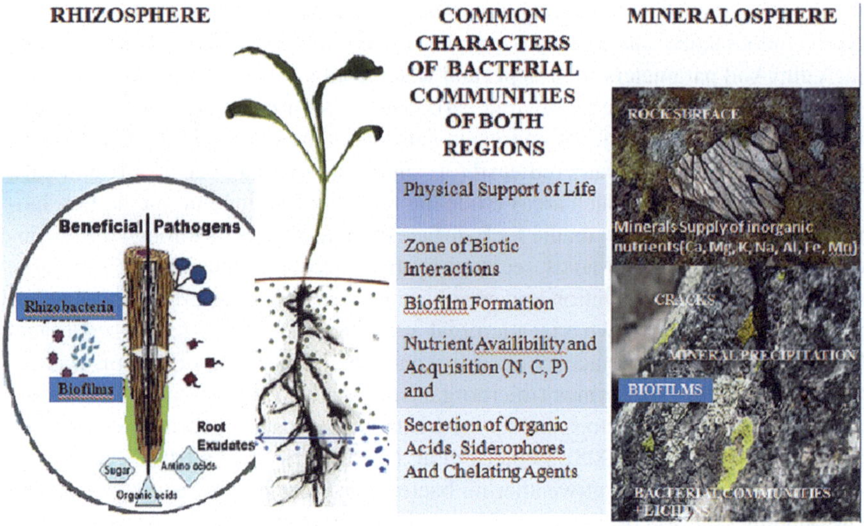

Fig. 10.4 The Rhizosphere and Mineralosphere

protein production was observed when the bacterial cells formed biofilms on minerals (Vera et al. 2013). To address the above question Olsson-Francis et al. (2010) used a microarray approach to decrypt the molecular mechanisms by using *Cupriavidus metallidurans* CH34 to weather basalt in a minimal medium lacking iron. Their microarray analyses revealed that only in the absence of basalt siderophores were produced and other functional genes were up- or down regulated. It was noticed that when basalt was present in the minimal medium, multiple genes involved in transport and motility were upregulated whereas genes encoding TonB-dependent outer membrane transporter and ostensive cytochromes were downregulated. In similar context, Almario et al. (2013) reported that during interaction of *Gaeumannomyces graminis–Pseudomonas*, production of 2,4-diacetylphloroglucinol was induced significantly higher in the presence of iron-rich vermiculite than in the presence of illite when they were trying to decipher the effect of iron availability on the 2,4-diacetylphloroglucinol production by *Pseudomonas* CHA0. Hence, these above mentioned results suggest that, physico- chemical properties of minerals, influence gene expression of bacteria residing in mineralosphere.

5 Bacteria and Nutrient Cycling

Soils constitute highly complex ecosystems where different biogeochemical cycles interact with each other such as carbon, nitrogen, phosphorus and sulphur. And, microbes add to soil's complexity by mediating most of these biogeochemical interactions. Soil is the site for organic matter decomposition and nutrient mobilisation via oxidation and reduction reactions of nutrient elements, symbiotic N-fixation and photoautotrophic activity. These activities of soil are carried out by bacteria, archaea and fungi which together drive nutrient cycling and weathering of minerals. In soils, microbes play a pivotal role in nutrient cycling through processes like decomposition and mineralisation. Soil microbes carry out decomposition by degrading non-living organic matter to get energy for growth. Mineralisation takes place when organic components get completely degraded into inorganic products such as carbon dioxide, ammonia, and water.

5.1 Bacteria and Soil Nitrogen

Among the autotrophs are nitrite oxidisers in the genera *Nitrospira* and *Nitrobacter,* and phototrophs in *Rhodospirillum* and *Rhodobacter*. Members of Burkholderia are also reported to fix nitrogen and promote plant growth (Aislabie and Deslippe 2013).

5.2 Bacteria and Soil Carbon

When we stand on soil, we are standing on an important reservoir of the carbon cycle from where large amount of carbon is added to the atmosphere. Microbes play major roles in the cycling of carbon- the key constituent of all living organisms. In terrestrial ecosystems, CO_2 gets fixed in to organic matter by primary producers (plants, algae and cyanobacteria). Within soil, autotrophic microbes can also fix carbon dioxide. Heterotrophic bacteria and fungi degrade complex organic molecules that higher organisms cannot do hence they are the ultimate recyclers of non-living organic matter. Numerous Actino and Proteobacteria, degrade soluble organic acids, amino acids, and sugars (Eilers et al. 2010). Some bacteria, such as *Bacteroidetes* target recalcitrant carbon compounds (cellulose, lignin and chitin) and they perform well in N rich environments in order to support the production of extracellular and transport enzymes (Treseder et al. 2011). In contrast, bacteria adapted to low levels of N are more likely to metabolise nitrogenous organic compounds such as amino acids. It has been reported that overall mineralisation of soil's carbon is positively correlated with abundance of β-Proteobacteria and Bacteroidetes but exhibit negative correlation with Acidobacteria (Fierer et al. 2007). Fermentative microbes can anaerobically degrade organic compounds in to organic acids resulting in generation of gases such as H_2 and CO_2. Further under strict anaerobic environment the hydrogen may be utilized by methanogens to reduce CO_2 in to methane CH_4 gas. Some methanogens can metabolise methanol, acetate or methylamine to methane and carbon dioxide.

6 Bacteria and Soil Bioremediation

Absorption, detoxification and recycling of applied wastes (e.g. effluent disposal), agrochemicals and oil spills takes place in soil attributed to its microbial activities which makes soil healthy and reduces potential harm to humans and ecosystem. Microbial processes like mineralisation and immobilisation are responsible for these services. This detoxification of soil through microbial intervention is known as soil bioremediation. Detoxifying microbes may be limited by the availability of soil nutrients (e.g. N or P), which in turn depends on soil microbial activities. The heterotrophic bacteria *Sphingomonas* are known to degrade a range of toxic compounds (pentachlorophenol and polyaromatic hydrocarbons) (Aislabie and Deslippe 2013). More oftenly implicated bacterial genera reported for oil degradation belong to spp. *Pseudomonas*, *Sphingomonas* and *Mycobacterium*. Amongst all *Pseudomonas* have been studied well for utilizing alkanes, monoaromatics, naphthalene, and phenanthrene as a sole carbon source under aerobic conditions via degradation through enzymes (Kuran et al. 2014). The mechanisms being employed by these oil degrading bacteria have been applied in situ. For example, enhancing oil degradation in soil typically involves addition of nutrients (N and P) and sometimes oxygen and water. Due to ubiquitous nature of hydrocarbon-degrading bacteria there is no need to add them to oil-contaminated sites in soil besides they increase in number when oil is spilled. Howbeit, elevated hydrocarbon concentration leads to depletion of available N and P due to their assimilation during biodegradation; consequently, activity of the hydrocarbon degraders may become limited by these nutrients (Lang et al. 2016).

Bacteria and fungi also degrade pesticides. Example of the bacteria degrading pesticide *is Arthrobacter nicotinovorans* HIM that utilized atrazine as a sole source of C and N and also degraded the related triazine compounds simazine, terbuylazine, propazine, and cyanazine (Aislabie et al. 2005). Biodegraded pesticides in soil are usually ineffective to control pests. Pesticides like DDT are not readily degradable so persist in soil and when aerobic conditions are available DDT is converted to DDE, which has been regarded as a dead-end metabolite. *Terrabacter* sp. Strain DDE-1, metabolised DDE when grown on biphenyl (Aislabie et al. 1999). Leaching of pollutants like excess of nutrients, heavy metals and organic compounds into soils is another environmental issue which can contaminate ground water and aquatic ecosystems which are life threatening events for humans. Soils absorb and retain solutes and pollutants, avoiding their release into water. Microbial products contribute to soil's hydrophobicity and wettability, both that impacts on the soils ability to filter contaminants (Aislabie et al. 2005).

7 Future Prospective

Based on recent studies presented above a new concept the 'mineralosphere' has came in to the picture which forms the basis of pedogenesis through action of microbes mainly bacteria. Exact mechanisms and their survival tendency in mineralosphere is yet to be investigated thoroughly so it is perquisite to implement recent molecular techniques to study diversity analysis and to unravel 99% undiscovered microbiota from the environment. These tiny creatures ensure the permanent existence of nutrients in soil. Due to their role in pedogenesis and improvement of soil fertility these minute entities have become major subject of investigation in recent past. Further research is desired for comparison of mineral-weathering potentials of bacterial isolates from the mineralosphere to those of rhizospheric bacteria to characterize this unexplored niche.

References

Aislabie, J., & Deslippe, J. R. (2013). Soil microbes and their contribution to soil services. In J. R. Dymond (Ed.), *Ecosystem services in New Zealand – conditions and trends*. Lincoln: Manaaki Whenua Press.

Aislabie, J., Davison, A. D., Boul, H. L., Franzmann, P. D., Jardine, D. R., & Karuso, P. (1999). Isolation of *Terrabacter* sp. strain DDE-1 which metabolises DDE when induced with biphenyl. *Applied and Environmental Microbiology, 65*, 5607–5611.

Aislabie, J., Bej, A. K., Ryburn, J., Lloyd, N., & Wilkins, A. (2005). Characterization of *Arthrobacter nicotinovorans* HIM, an atrazine-degrading bacterium, from agricultural soil New Zealand. *FEMS Microbiology Ecology, 52*, 279–286.

Almario, J., et al. (2013). Effect of clay mineralogy on iron bioavailability and rhizosphere transcription of 2,4-diacetylphloroglucinol biosynthetic genes in biocontrol *Pseudomonas protegens*. *Molecular Plant-Microbe Interactions, 26*, 566–574.

Antoun, H., & Kloepper, J. W. (2001). Plant growth promoting rhizobacteria. In S. Brenner & J. H. Miller (Eds.), *Encyclopedia of genetics* (pp. 1477–1480). New York: Academic.

Banfield, J. F., et al. (1999). Biological impact on mineral dissolution: application of the lichen model to understanding mineral weathering in the rhizosphere. *Proceedings of the National Academy of Sciences of the United States of America, 96*, 3404–3411.

Bin, L., Ye, C., & Yuan, T. (2010). Microbes on carbonate rocks and pedogenesis in karst regions. *Journal of Earth Science, 21*, 293–296.

Buol, S. W., Hole, F. D., & McCracken, R. J. (1973). *Soil genesis and classification* (1st ed.). Ames: Iowa State University Press. ISBN 978-0-8138-1460-5.

Burford, E. P., Hillier, S., & Gadd, G. M. (2006). Biomineralization of fungal hyphae with calcite ($CaCO_3$) and calcium oxalate mono- and dihydrate in carboniferous limestone microcosms. *Geomicrobiology Journal, 23*(8), 599–611.

Calvaruso, C., et al. (2007). Impact of ectomycorrhizosphere on the functional diversity of soil bacterial and fungal communities from a forest stand in relation to nutrient mobilization processes. *Microbiol Ecology, 54*, 567–577.

Carson, J. K., et al. (2007). Altering the mineral composition of soil causes a shift in microbial community structure. *FEMS Microbiology Ecology, 61*, 414–423.

Chen, S., Lian, B., & Liu, C. Q. (2008). Effect of *Bacillus mucilaginosus* on weathering of phosphorite and a preliminary analysis of bacterial proteins. *Chinese Journal of Geochemistry, 27*(2), 209–216.

De Boer, W., Leveau, J. H. J., Kowalchuk, G. A., Klein Gunnewiek, P. J. A., Abeln, E. C. A., Figgge, M. J., et al. (2004). *Collimonas fungivorans* gen. nov., sp. nov., a chitinolytic soil bacterium with the ability to grow on living fungal hyphae. *International Journal of Systematic and Evolutionary Microbiology, 54*, 857–864.

Dou, C. W., & Lian, B. (2009). Microbial weathering of calcite by rock fungi. *Acta Mineralogical Sinica, 29*(3), 387–391.

Eilers, K. G., Lauber, C. L., Knight, R., & Fierer, N. (2010). Shifts in bacterial community structure associated with inputs of low molecular weight carbon compounds to soil. *Soil Biology and Biochemistry, 42*, 896–903.
Fierer, N., Bradford, M. A., & Jackson, R. B. (2007). Towards an ecological classification of soil bacteria. *Ecol, 88*, 1354–1364.
Frey, B., et al. (2010). Weathering-associated bacteria from the Damma glacier for efield: Physiologica lcapabilities and impact on granite dissolution. *Applied and Environmental Microbiology, 76*, 4788–4796.
Gleeson, D. B., et al. (2006). Characterization of bacterial community structure on a weathered pegmatitic granite. *Microbial Ecology, 51*, 526–534.
Gorbushina, A. A. (2007). Life on the rocks. *Environmental Microbiology, 9*, 1613–1631.
Gorbushina, A. A., Whitehead, K., Dornieden, T., et al. (2003). Black fungal colonies as units of survival: Hyphal mycosporines synthesized by rock dwelling microcolonial fungi. *Canadian Journal of Botany, 81*(2), 131–138.
Hiltner, L. (1904). Über neuere erfahrungen und probleme auf dem gebiete der bodenbakteriologie unter besonderer berücksichtigung der gründüngung und brache. *Arb Landwirt Dtsch Ges, 98*, 59–78.
Huddleston, J. H., & Kling, G. F. (1984). *Manual for judging oregon soils*. Corvallis: Oregon State University Extension Service.
Ingham, E. R. (2009). *Soil biology primer, Chapter 4: Soil fungus* (pp. 22–23). Ankeny: Soil & Water Conservation Society. soils.usda.gov/sqi/concepts/soil_biology.
Khatoon, H., Solanki, P., Narayan, M., Tewari, L., & Rai, J. P. N. (2017). Role of microbes in organic carbon decomposition and maintenance of soil ecosystem. *International Journal of Chemical Studies, 5*(6), 1648–1656.
Kuráň, P., Trögl, J., Nováková, J. et al. (2014). Biodegradation of spilled diesel fuel in agricultural soil: Effect of humates, zeolite, and bioaugmentation. *Scientific W J,* Article ID 642427, 8 pages. https://doi.org/10.1155/2014/642427.
Lang, F. S., Tarayre, C., Destain, J., Delvigne, F., Druart, P., Ongena, M., & Thonart, P. (2016). The effect of nutrients on the degradation of hydrocarbons in mangrove ecosystems by microorganisms semboung. *International Journal of Environmental Research, 10*(4), 583–592.
Lepleux, C., et al. (2012). Correlation of the abundance of beta- proteobacteria on mineral surfaces with mineral weathering in forest soils. *Applied and Environmental Microbiology, 78*, 7114–7119.
Lian, B., Chen, Y., Zhu, L. J., et al. (2008). Progress in the study of the weathering of carbonate rock by microbes. *Earth Science Frontiers, 15*(6), 90–99.
Löhnis, F., & Fred, E. B. (1923). *Textbook of agricultural bacteriology*. New York: McGraw–Hill.
McNamara, C. J., et al. (2006). Epilithic and endolithic bacterial communities in limestone from a Maya archaeological site. *Microbial Ecology, 51*, 51–64.
Mosier, A. R., Syers, J. K., & Freney, J. R. (Eds.). (2004). *Agriculture and the Nitrogen Cycle: Assessing the Impacts of Fertilizer Use on Food Production and the Environment* (SCOPE Series 65). Washington, DC: Island Press.
Olsso-Francis, K., et al. (2010). Microarray analysis of a microbe– mineral interaction. *Geobiology, 8*, 446–456.
Prasad, J. K., Gupta, S. K., & Raghuwanshi, R. (2017). Screening multifunctional plant growth promoting rhizobacteria strains for enhancing seed germination in wheat (Triticum aestivum L.). *International Journal of Agricultural Research, 12*, 64–72.
Seneviratne, G., & Indrasena, I. K. (2006). Nitrogen fixation in lichens is important for improved rock weathering. *Journal of Biosciences, 31*, 639–643.
Shelobolina, E. S., Xu, H., Konishi, H., Kukkadapu, R. K., Wu, T., Blothe, M., & Roden, E. E. (2012). Microbial Lithotrophic Oxidation of Structural Fe(II) in Biotite. *Applied and Environmental Microbiology, 78*(16), 5746–5752.
Thompson, L. M., & Troeh, F. R. (1973). *Soils and soil fertility* (3rd ed.). New York: McGraw-Hill.
Treseder, K. K., Kivlin, S. N., & Hawkes, C. V. (2011). Evolutionary trade-offs among decomposers determine responses to nitrogen enrichment. *Ecology Letters, 14*, 933–938.

Uroz, S., et al. (2007). Effect of the mycorrhizosphere on the genotypic and metabolic diversity of the bacterial communities involved in mineral weathering in a forest soil. *Applied and Environmental Microbiology, 73*, 3019–3027.

Uroz, S., Calvaruso, C., Turpault, M. P., & Frey-Klett, P. (2009). Mineral weathering by bacteria: ecology, actors and mechanisms. *Trends in Microbiology, 17*(8), 378–387.

Vera, M., et al. (2013). Shotgun proteomics study of early biofilm formation process of *Acidithiobacillus ferrooxidans* ATCC 23270 on pyrite. *Proteomics, 13*, 1133–1144.

Vessey, J. K., & Buss, T. J. (2002). Bacillus cereus UW85 inoculation effects on growth, nodulation and N accumulation in grain legumes. Controlled-environment studies. *Canadian Journal of Plant Science, 82*, 282–290.

Viles, H. A., & Gorbushina, A. A. (2003). Soiling and Microbial Colonisation on Urban Roadside Limestone: A Three Year Study in Oxford England. *Building and Environment, 38*(9–10), 1217–1224.

Walker, T. S., Bais, H. P., Grotewold, E., & Vivanco, J. M. (2003). Root exudation and rhizosphere biology. *Plant Physiology, 132*, 44–51.

Recent Trends to Study the Functional Analysis of Mycorrhizosphere

11

Pankaj Bhatt, Divya Joshi, Nitin Kumar, and Narendra Kumar

1 Introduction

Soil contain rich diversity of microbial populations (Bacteria, fungi, Protozoa, Viruses), that affect the plant growth and soil fertility. *Arbuscular Mycorrhiza* fungi (AMF) make mutualistic interaction with approximately 80% terrestrial plants. These group of fungi is important, for natural ecosystem and sustainable agricultural production. *Arbuscular Mycorrhizae* (AM) play critical role in plant nutrition and it is the key component of microbial population. This group of fungi increases the surface area of plant root system that help in absorption of nutrient from soil. The hyphae of these mutualistic group of fungi help to make interaction with other group of microorganisms, and also make an important pathway for the translocation of energy rich plant assimilates to the soil (Gianinazzi and Schuepp 1994). The term rhizosphere given by Hiltner in, 1904, it is the region surrounds the plant roots. Rhizosphere is characterized by increased microbial activity because plant secrete root exudates. Mycorrhizosphere is the term used commonly for plant fungus mutual relationship. Mycorrhizosphere is the region influenced by both the root and the mycorrhizal fungus, similarly term hyphosphere, includes the region surrounding individual fungal hyphae. Hyphosphere can be created in plant adhered soil or in plant free soil. Mycorrhiza and fungal hyphae present in everywhere in soil, on the basis of biomass it is highest in number in earth (Rambelli 1973; Johansson

P. Bhatt (✉)
Department of Microbiology, Dolphin (P.G) Institute of Biomedical and Natural Sciences, Dehradun, Uttarakhand, India

D. Joshi
Department of Microbiology, G.B.P.U.A&T, Pantnagar, Uttarakhand, India

N. Kumar
Department of Plant Physiology, G.B.P.U.A&T, Pantnagar, Uttarakhand, India

N. Kumar
Department of Crop Physiology, Uttaranchal University, Dehradun, Uttarakhand, India

© Springer Nature Singapore Pte Ltd. 2019

A. Varma, D. K. Choudhary (eds.), *Mycorrhizosphere and Pedogenesis*,
https://doi.org/10.1007/978-981-13-6480-8_11

et al. 2004). Plants play an important role, to the maintenance of microbial ecological niches due to root-based systems. Plant secretes chemical compounds from their roots due to which specific microbial populations exist in this condition (Berg and Smalla 2009). Plant roots secrete variety of chemical compounds that are diverse in nature. These compounds are organic in nature like organic acids, amino acids, phytohormones etc, that attract microbial population towards it. These microbes have the ability to colonize the root surface of the plants (Bais et al. 2006; Pothier et al. 2007; Badri et al. 2009; Shukla et al. 2011; Drogue et al. 2013).

Carbon is important element for microbial cell to make its cellular constituent. Root exudates act as source of carbon for microbes surrounding the roots, due to which rich microbial diversity occurs, i.e. upto 10^{10} bacteria per gram in soil (Gans et al. 2005; Roesch et al. 2007) and consist of large part of each taxa (Kyselková et al. 2009; Gomes et al. 2010). The microbial community associated with the plant root is known as rhizomicrobiome (Chaparro et al. 2013). Due to competition for nutrition in the rhizospheric region automatic difference created in rhizosphere as compare to normal soil microorganisms. The nearby area of root contains high microbial diversity as compare to the free living organism (Raynaud et al. 2008; Bulgarelli et al. 2013; Chaparro et al. 2013). Chemical composition of root exudates changes throughout the root system, as per different development stages of plant genotypes, so this is a reason rhizomicrobiome differs for plant to plant (Berg and Smalla 2009; Aira et al. 2010; Bouffaud et al. 2012; Bulgarelli et al. 2013; Chaparro et al. 2013). In mycorrhizosphere some beneficial fungal species can promote the growth of plant by various direct and indirect methods (Couillerot et al. 2009; Richardson et al. 2009). During symbiotic relationship each plant and microbes get benefitted due to exchange of chemicals (Odum and Barrett 2005; Bulgarelli et al. 2013) these mechanisms can be categorized in two important categories (Drogue et al. 2012). The first one is symbiotic interactions in which obligate relationship occurs between host and plant. They form special structures during symbiosis called nodules and arbuscules due to the presence of bacteria and fungi respectively (Parniske 2008; Masson-Boivin et al. 2009). The second type interaction is associative symbiosis in which no obligatory relationship is necessary (Barea et al. 2005; Drogue et al. 2012). If bacteria are present in the root surface and stimulate plant growth referred to as plant growth promotory rhizobacteria (PGPRs) (Barea et al. 2005). The bacteria colonize on the root system heterogeneously. PGPRs interact with large group of host plant species and creates lot of taxonomic diversity in nature, when they attached with roots of proteobacteria and firmicutes (Lugtenberg and Kamilova 2009; Drogue et al. 2012). PGPR can enhance plant nutrition via siderophore production, nitrogen fixation, phosphate solubilization (Richardson et al. 2009; Pankaj et al. 2014). They have ability to improve the roots by producing the phytohormones, enzymatic activities support the initial establishment of rhizobial or mycorrhizal symbioses. It is reported that PGPRs help in protection from plant pathogens due to rapid development of induced systemic resistance (Fig. 11.1) (Couillerot et al. 2009; Lugtenberg and Kamilova 2009).

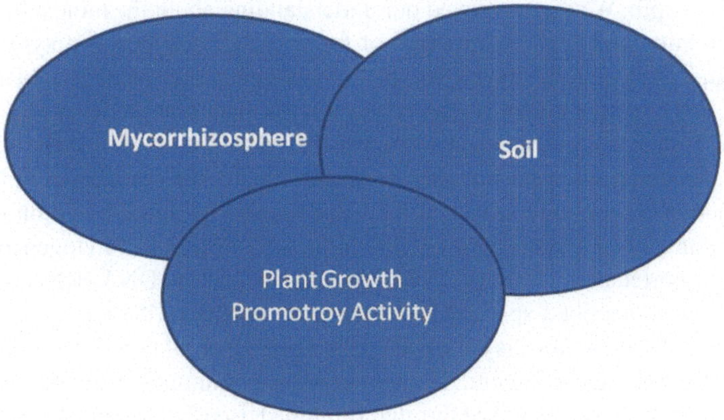

Fig. 11.1 Interaction of different component in soil

2 Arbuscular Mycorrhizal Fungi (AMF)

It is well established that Arbuscular mycorrhizal fungi (AMF) and PGPR bacteria makes important component of plant root biology. These microbes have ability to influence many physiological processes, mainly affect plant under stress condition. Besides this AMF facilitates the nutrient uptake by alternative pathway, that is important during low nutrient conditions. In alkaline soil phosphate mobilization is low so the phosphate zones depleted after its use that create phosphate limiting condition. Previously it is reported that AM fungi root colonization help in nitrate uptake in tomato plant as compare to uninoculated plant. This mechanism is mediated by higher expression of NRT2.3, that is critically responsible for translocation of nitrogen in other species that is far away. The detailed mechanism behind this translocation was further confirmed by an increased expression of four different AMF-related nitrate transporter genes in mycorrhizal Medicago truncatula roots. Conclusively we can say that AMF, PGPR can improve the availability of nutrients for plants by various mechanisms. These mechanisms include acidification of soil, organic acid biosynthesis, chelation and exchange reactions. (Lugtenberg and Kamilova 2009). The microbial interaction with crop plant is beneficial. Different microbial strains have the ability to improve the plants by mobilization of various nutritional elements (Bhatt and Barh 2018; Pankaj et al. 2016).

3 Genomic Analysis of Mycorrhizosphere

Due to development of next generation highthroughput DNA sequencing methods, the detailed study of mycorrhizosphere is possible in short time. Advancement of recent sequencing tools our understanding towards Mycorrhizosphere can be explored in depth. We can improved our understanding about the molecular phylogeny, functioning and genome structure of fungi. Single nucleotide polymorphisms (SNPs) and phylogenetically important marker genes can be extracted from the genomes and can be used for phylogenetic tree formation at the level of isolates to kingdom. Traditional genomics analysis was based on the single culture dependent approach in which we were not able to handled many data at a time. Polymerase chain reaction were commonly used for identification of fungi, based on their ITS (Internal transcribed spacer) region with unique set of primers. Phylogenetic analysis earlier reported in the form of, DNA:DNA hybridization, DNA:rRNA hybridization, Internal transcribed spacer(ITS), Intergenic spacer (IGS) sequence (Krishna et al. 2006). Metagenomics is a term used for analysis of all types of genomics data at a time from the environment. The low coverage genomics data and analysis of DNA bar-coding can be analysed through Illumina Hiseq 2×150 paired end sequencing technology (Tedersoo et al. 2016). Besides this proteomics and transcriptomics study is also useful for the complete mycorrhizosphere study. Many of the previous reports concluded the study such data in details using insilico and wet labs (Pankaj et al. 2016; Bhatt and Barh 2018).

4 Proteomic Analysis of Mycorrhizosphere

Proteins are expressed in mycorrhizopshere can be studied by various tools. These tools include from traditional SDS-PAGE (sodium dodecyl sulfate-polyacrylamide gel electrophoresis) to recent sequencing tools. Bioinformatics based analysis added the advantage in these technologies towards clear cut analysis. Proteome is the study of all proteins in mycorrhizosphere in the particular time and environment. All expressed proteins can be analysed for similarity using NCBI database. Structure and functions of these proteins can be determined by homology modeling and molecular docking studies. These *insilico* tools are less expensive and giving the depth of knowledge as compared to lab experiment (Fig. 11.2).

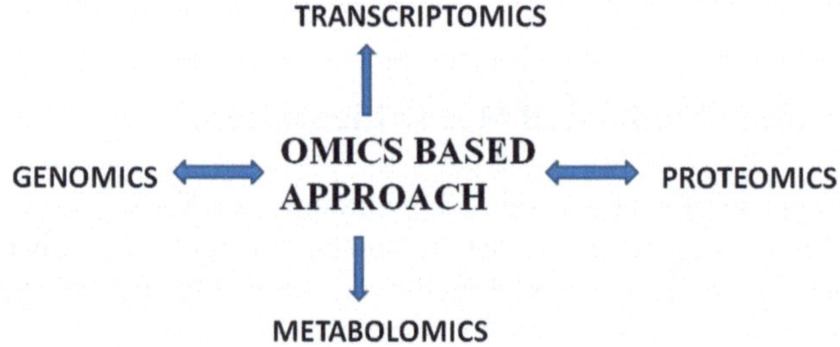

Fig. 11.2 Omics approaches for mycorrhizosphere

5 Transcriptomic Analysis of Mycorrhizosphere

Till date many of the fungal clades known for establishment of mycorrhizospheric symbiosis like phylum Glomeromycota, Ascomycota, Basidiomycota. The mycorrhizal symbiosis is both ancient and prevalent in nature (Tedersoo et al. 2010). They help in metal detoxification and stress tolerance in host plant species. (Kuo et al. 2014; González-Guerrero et al. 2016), in this mutualism plant exchange sugar which is synthesized by photosynthesis. Fungus acquired this sugar from plant due to symbiotic relationship (Symanczik et al. 2017). But now we need to focus on the molecular level of symbiosis, i.e. identification of the genes, mRNA and proteins responsible for this symbiosis. Ultimately these biomolecules have the information in coded form. Many of the researchers already conducted research on the fungal phylum Basiodiomycota, Ascomycota and Glomeromycota which represent majority of mycorrhizal flora. It is concluded that the symbiosis occurs in Basidiomycota and Ascomycota evolved independently (Tedersoo et al. 2010). Different types of diverse mycorrhizae provides the new methods and application of symbiotic relationship.

RNA molecules are essential components of all living cells. They also serve as the basis of modern day transcriptomic analysis. Studies conducted for identification and to check abundance of each RNA molecule in a given cell under a specific set of condition is the primary objective of RNA based research. Transcriptomics studies mainly focused on registration and quantification of the RNA content of a plant and microbial cell in specific condition. When we target the whole RNA from a cell means that our goal is specific to transcriptomics. Through which we can understand the real expression pattern (Hrdlickova et al. 2016). Almost majority of land plants including crop plant root system makes the symbiosis with arbuscular mycorrhizal fungus, that help in nutrient uptake from soil (Vangelisti et al. 2018). Several studies concluded the establishment and reprogramming of arbuscular mycorrhizal symbioses that decipher the molecular level changes occurs in different parts of plant species, like *Medicago truncatula, Lotus japonicus, Pisum sativum*, tomato, potato, rice and soybean (Schaarschmidt et al. 2013; Gallou et al. 2012; Hogekamp and Küster 2013; Grunwald et al. 2004; Handa et al. 2015; Fiorilli et al. 2009, 2015; Vangelisti et al. 2018). AM fungi releases the signaling molecules lipochitooligosaccharides and chitooligosaccharides which help in activation of symbiotic pathways in plant root (Genre et al. 2013). This pathway is necessary for establishment for initial symbiosis and further activate the downstream genes coding for the receptor kinases and ion channels. (Charpentier et al. 2016), nucleoporins (Groth et al. 2010), calcium and calcium dependent kinases (Sugimura and Saito 2017), and CYCLOPS/IPD3 (Singh et al. 2014). Many of the genes have the ability to encode transcription factor, kinases and proteins, that are highly upregulated during AM symbiosis. These studies are confirmed by transcriptomic microarray and RNA-seq analyses of legumes and non-legumes (Hogekamp and Küster 2013; Handa et al. 2015; Dutta and Podile 2010; Hartman et al. 2009). Microarray study on *Solanum lycopersicum* L. and the model legume *M. truncatula* showed comparative analysis that only few orthologous genes were highly expressed in AM roots (Fiorilli et al. 2009; Sugimura and Saito 2017). Large variation in gene expression was detected in presence of fungal symbiont and plant roots. Trancriptomics study confirmed revealed changes in plant and fungal cells. These sequences of mRNA confirmed the fungal -plant symbiosis establishment (Shu et al. 2016). More study on tomato mycorrhizal roots has identified plant genes as functional markers for AM symbiosis and concluded that genes express differentially in cells (Zouari et al. 2014). Except roots mycorrhizae have ability to make interaction with shoots, leaves and activate genes responsible for various metabolic process like defence, transport and hormonal metabolism (Fiorilli et al. 2009; Lopez-Raez et al. 2010; Doornbos et al. 2012). The mycorrhizal symbiosis is a complex biological process. Differential expression of genes observed in several steps of host plant and mycorrhizae interaction. Handa et al. 2015 performed *de novo* transcriptome assembly using RNA-seq data from the model legume *L. japonicus* and identified thousands of genes that were differentially expressed in AM roots as compared to non-mycorrhizal (NM) roots. Arbuscular mycorrhizae symbiosis also facilitated by bacterial communities associated with them. These bacteria associated with AMF spores and mycelium (Desirò et al. 2014; Agnolucci et al. 2015; Bhatt and Nailwal 2018; Cjaza et al. 2012; Compant et al. 2010). Such bacteria have all beneficial PGPR properties like nitrogen fixation, siderophore production, antibiotic production, phosphate solubilization and phytohormone like indole acetic acid production (Rouphael et al. 2015; Battini et al. 2016).

6 Conclusion

Mycorrhizosphere is important concepts of study. If we imagine the whole world with microscope there is a sheet of hyphae throughout the earth. So we can understand the importance of this valuable topic. Due to development of high throughput sequencing methods we have huge omics information in the form of bioinformatics. So it's a time to analyse the whole omics things through bioinformatic and latest tools. This will increase our understanding for study of mycorrhizosphere and mechanisms in depth. By compiling the all mechanisms we can developed more potential role of mycorrhizosphere towards biofertilizer production.

References

Agnolucci, M., Battini, F., Cristani, C., & Giovannetti, M. (2015). Diverse bacterial communities are recruited on spores of different arbuscular mycorrhizal fungal isolates. *Biology and Ferility of Soils, 51*, 379–389. https://doi.org/10.1007/s00374-014-0989-5.

Aira, M., Gómez-Brandón, M., Lazcano, C., Bååth, E., & Domïnguez, J. (2010). Plant genotype strongly modifies the structure and growth of maize rhizosphere microbial communities. *Soil Biology and Biochemistry, 42*, 2276–2281. https://doi.org/10.1016/j.soilbio.2010.08.029.

Badri, D. V., Weir, T. L., van der Lelie, D., & Vivanco, J. M. (2009). Rhizosphere chemical dialogues: Plant-microbe interactions. *Current Opinion in Biotechnology, 20*, 642–650. https://doi.org/10.1016/j.copbio.2009.09.014.

Bais, H. P., Weir, T. L., Perry, L. G., Gilroy, S., & Vivanco, J. M. (2006). The role of root exudates in rhizosphere interactions with plants and other organisms. *Annual Review of Plant Biology, 57*, 233–266. https://doi.org/10.1146/annurev.arplant.57.032905.105159.

Barea, J. M., Pozo, M. J., Azcón, R., & Azcón-Aguilar, C. (2005). Microbial co-operation in the rhizosphere. *Journal of Experimental Botany, 56*, 1761–1778. https://doi.org/10.1093/jxb/eri197.

Battini, F., et al. (2016). Dual inoculation with AMF and associated bacteria improves nutraceutical value of sweet basil grown under commercial conditions. *Agrochimica, 60*, 81–99.

Berg, G., & Smalla, K. (2009). Plant species and soil type cooperatively shape the structure and function of microbial communities in the rhizosphere. *FEMS Microbiology Ecology, 68*, 1–13. https://doi.org/10.1111/j.1574-6941.2009.00654.x.

Bhatt, P., & Barh, A. (2018). Bioinformatic tools to study the soil microorganisms: An in silico approach for sustainable agriculture. In D. K. Choudhary et al. (Eds.), *In silico approach for sustainable agriculture* (pp. 169–182). https://doi.org/10.1007/978-981-13-0347-0_10.

Bhatt, P., & Nailwal, T. K. (2018). Crop improvement through microbial technology: A step toward sustainable agriculture. In R. Prasad, S. S. Gill, & N. Tuteja (Eds.), *Crop improvement through microbial biotechnology* (pp. 245–253). https://doi.org/10.1016/B978-0-444-63987-5.00011-6.

Bouffaud, M. L., Kyselkova, M., Gouesnard, B., Grundmann, G., Muller, D., & Mënne-Loccoz, Y. (2012). Is diversification history of maize influencing selection of soil bacteria by roots? *Molecular Ecology, 21*, 195–206. https://doi.org/10.1111/j.1365-294X.2011.05359.x.

Bulgarelli, D., Schlaeppi, K., Spaepen, S., Ver Loren van Themaat, E., & Schulze-Lefert, P. (2013). Structure and functions of the bacterial microbiota of plants. *Annual Review of Plant Biology, 64*, 807–838. https://doi.org/10.1146/annurev-arplant-050312-120106.

Chaparro, J. M., Badri, D. V., Bakker, M. G., Sugiyama, A., Manter, D. K., & Vivanco, J. M. (2013). Root exudation of phytochemicals in Arabidopsis follows specific patterns that are developmentally programmed and correlate with soil microbial functions. *PLoS ONE, 8*, e55731. https://doi.org/10.1371/journal.pone.0055731.

Charpentier, M., Sun, J., Martins, T. V., Radhakrishnan, G. V., Findlay, K., Soumpourou, E., Thouin, J., Very, A. A., Sanders, D., Morris, R. J., & Oldroyd, G. E. D. (2016). Nuclear-localized cyclic nucleotide–gated channels mediate symbiotic calcium oscillations. *Science, 352*(6289), 1102–1105. https://doi.org/10.1126/science.aae0109.

Cjaza, L. F., Hogekamp, C., Lamm, P., Maillet, F., Martinez, E. A., Samain, E., Denarie, J., Kuster, H., & Hohnjec, N. (2012). Transcriptional responses towards diffusible signals from symbiotic microbes reveal MtNFP and MtDMI3-dependent reprogramming of host gene expression by AM fungal LCOs. *Plant Physiology Preview, 159*, 1–35. https://doi.org/10.1104/pp.112.195990.

Compant, S., Clément, C., & Sessitsch, A. (2010). Plant growth-promoting bacteria in the rhizo- and endosphere of plants: Their role, colonization, mechanisms involved and prospects for utilization. *Soil Biology and Biochemistry, 42*, 669–678. https://doi.org/10.1016/j.soilbio.2009.11.024.

Couillerot, O., Prigent-Combaret, C., Caballero-Mellado, J., & Mënne-Loccoz, Y. (2009). Pseudomonas fluorescens and closely-related fluorescent pseudomonads as biocontrol agents of soil-borne phytopathogens. *Letters in Applied Microbiology, 48*, 505–512. https://doi.org/10.1111/j.1472-765X.2009.02566.x.

Desirò, A., Salvioli, A., Ngonkeu, E. L., Mondo, S. J., Epis, S., Facio, A., Kaech, A., Pawlowska, T. A., & Bonfante, P. (2014). Detection of a novel intracellular microbiome hosted in arbuscular mycorrhizal fungi. *The ISME Journal, 8*, 257–270. https://doi.org/10.1038/ismej.2013.151.

Doornbos, R. F., van Loon, L. C., & Bakker, P. A. H. M. (2012). Impact of root exudates and plant defense signaling on bacterial communities in the rhizosphere. A review. *Agronomy for Sustainable Development, 32*, 227–243. https://doi.org/10.1007/s13593-011-0028-y.

Drogue, B., Dore, H., Borland, S., Wisniewski-Dyé, F., & Prigent-Combaret, C. (2012). Which specificity in cooperation between phytostimulating rhizobacteria and plants? *Research in Microbiology, 163*, 500–510. https://doi.org/10.1016/j.resmic.2012.08.006.

Drogue, B., Combes-Meynet, E., Mënne-Loccoz, Y., Wisniewski-Dyé, F., & Prigent-Combaret, C. (2013). Control of the cooperation between plant growth-promoting rhizobacteria and crops by rhizosphere signals. In F. J. de Bruijn (Ed.), *Vol. 1 and 2 Molecular microbial ecology of the rhizosphere* (pp. 281–294). Hoboken: Wiley. https://doi.org/10.1002/9781118297674.ch27.

Dutta, S., & Podile, A. R. (2010). Plant growth promoting rhizobacteria (PGPR): the bugs to debug the root zone. *Critical Reviews in Microbiology, 36*, 232–244. https://doi.org/10.3109/10408411003766806.

Fiorilli, V., Catoni, M., Miozzi, L., Novero, M., Accotto, G. P., & Lanfranco, L. (2009). Global and cell-type gene expression profiles in tomato plants colonized by an arbuscular mycorrhizal fungus. *New Phytologist Journal., 184*, 975–987. https://doi.org/10.1111/j.1469-8137.2009.03031.x.

Fiorilli, V., et al. (2015). Host and non-host roots in rice: Cellular and molecular approaches reveal differential responses to arbuscular mycorrhizal fungi. *Frontiers in Plant Science, 6*, 636. https://doi.org/10.3389/fpls.2015.00636.

Gallou, A., Declerck, S., & Cranenbrouck, S. (2012). Transcriptional regulation of defence genes and involvement of the WRKY transcription factor in arbuscular mycorrhizal potato root colonization. *Functional & Integrative Genomics, 12*, 183–198. https://doi.org/10.1007/s10142-011-0241-4.

Gans, J., Wolinsky, M., & Dunbar, J. (2005). Computational improvements reveal great bacterial diversity and high metal toxicity in soil. *Science, 309*, 1387–1390. https://doi.org/10.1126/science.1112665.

Genre, A., Chabaud, M., Balzergue, C., Puech-Pagès, V., Novero, M., Rey, T., et al. (2013). Short-chain chitin oligomers from arbuscular mycorrhizal fungi trigger nuclear Ca2+ spiking in *Medicago truncatula* roots and their production is enhanced by strigolactone. *New Phytologist Journal., 198*, 190–202. https://doi.org/10.1111/nph.12146.

Gianinazzi, S., & Schuepp, H. (1994). *Impact of arbuscular mycorrhizas on sustainable agriculture and natural ecosystems* (p. 226). Basel: Birkhauser Verlag.

Gomes, N. C. M., Cleary, D. F. R., Pinto, F. N., Egas, C., Almeida, A., Cunha, A., et al. (2010). Taking root: Enduring effect of rhizosphere bacterial colonization in mangroves. *PLoS ONE, 5*, e14065. https://doi.org/10.1371/journal.pone.0014065.

González-Guerrero, M., Escudero, V., Saéz, Á., & Tejada-Jiménez, M. (2016). Transition metal transport in plants and associated endosymbionts: Arbuscular mycorrhizal fungi and rhizobia. *Frontiers in Plant Science, 7*, 1088. https://doi.org/10.3389/fpls.2016.01088.

Grunwald, U., et al. (2004). Identification of mycorrhiza-regulated genes with arbuscule development-related expression profile. *Plant Molecular Biology, 55*, 553–566. https://doi.org/10.1007/s11103-004-1303-y.

Handa, Y., Nishide, H., Takeda, N., Suzuki, Y., Kawaguchi, M., & Saito, K. (2015). RNA-seq transcriptional profiling of an arbuscular mycorrhiza provides insights into regulated and coordinated gene expression in *Lotus japonicus* and *Rhizophagus irregularis*. *Plant and Cell Physiology, 56*, 1490–1511. https://doi.org/10.1093/pcp/pcv071.

Hartman, A., Schmid, M., Tuinen, D., & Berg, G. (2009). Plant-driven selection of microbes. *Plant and Soil, 321*, 235–257. https://doi.org/10.1007/s11104-008-9814-y.

Hiltner, L. (1904). Uber neuere Erfahrungen und Probleme auf dem Gebiet der Bodenbakteriologie unter besonderer Ber uksichtigung der Ground ungung und Brache [On recent insights and problems in the area of soil bacteriology under special consideration of the use of green manure and fallowing]. *Arb Dtsch Landwirt Ges, 98*, 59–78.

Hogekamp, C., & Küster, H. (2013). A roadmap of cell-type specific gene expression during sequential stages of the arbuscular mycorrhiza symbiosis. *BMC Genomics, 14*, 306. https://doi.org/10.1186/1471-2164-14-306.

Hrdlickova, R., Toloue, M., & Tian, B. (2016). RNA-Seq methods for transcriptome analysis. *Wiley Interdisciplinary Reviews: RNA, 8*(1), 1364. https://doi.org/10.1002/wrna.1364.

Johansson, J. F., Paul, L. R., & Finlay, R. D. (2004). Microbial interactions in the mycorrhizosphere and their significance for sustainable agriculture. *FEMS Microbiology Ecology, 48*, 1–13.

Krishna, P., Reddy, M. S., & Satyanarayana, T. (2006). Molecular Techniques for understanding the microbial community structure in mycorrhizosphere. In K. G. Mukerji, C. Manoharachary, & J. Singh (Eds.), *Microbial activity in the rhizosphere* (Soil Biology, Vol. 7). Berlin/Heidelberg: Springer. https://doi.org/10.1007/3-540-29420-1_10.

Kuo, A., Kohler, A., Martin, F. M., & Grigoriev, I. V. (2014). Expanding genomics of mycorrhizal symbiosis. *Frontiers in Microbiology, 5*, 1–7. https://doi.org/10.3389/fmicb.2014.00582.

Kyselková, M., Kopecký, J., Frapolli, M., Défago, G., Ságová-Marečková, M., Grundmann, G. L., et al. (2009). Comparison of rhizobacterial community composition in soil suppressive or conducive to tobacco black root rot disease. *The ISME Journal, 3*, 1127–1138. https://doi.org/10.1038/ismej.2009.61.

Lopez-Raez, J. A., Verhage, A., Fernandez, I., Garcia, J. M., zcon-Aguilar, C., Flors, V., & Pozo, M. J. (2010). Hormonal and transcriptional profiles highlight common and differential host responses to arbuscular mycorrhizal fungi and the regulation of the oxylipin pathway. *Journal of Experimental Botany, 61*, 2589–2601. https://doi.org/10.1093/jxb/erq089.

Lugtenberg, B. J., & Kamilova, F. (2009). Plant-growth-promoting rhizobacteria. *Annual Review of Microbiology, 63*, 541–556. https://doi.org/10.1146/annurev.micro.62.081307.162918.

Masson-Boivin, C., Giraud, E., Perret, X., & Batut, J. (2009). Establishing nitrogen-fixing symbiosis with legumes: how many rhizobium recipes? *Trends in Microbiology, 17*, 458–466. https://doi.org/10.1016/j.tim.2009.07.004.

Odum, E. P., & Barrett, G. W. (2005). *Fundamentals of ecology* (5th ed.). Belmont: Thomson Brooks/Cole.

Pankaj, Nailwal, T. K., Singh, L., & Panwar, A. (2014). Isolation and Characterization of rhizobial isolates from the rhizospheric soil of an endangered plant *Meizotropis pellita*. *Asian Journal of Microbiology Biotechnology Environmental Sciences, 16*(2), 301–306.

Pankaj, Negi, G., Gangola, S., Khati, P., Kumar, G., Srivastava, A., & Sharma, A. (2016). Differential expression and characterization of cypermethrin-degrading potential proteins in Bacillus thuringiensis strain, SG4. *3 Biotech, 6*(2), 225.

Parniske, M. (2008). Arbuscular mycorrhiza: The mother of plant root endosymbioses. *Nature Reviews. Microbiology, 6*, 763–775. https://doi.org/10.1038/nrmicro1987.

Pothier, J. F., Wisniewski-Dyé, F., Weiss-Gayet, M., Mënne-Loccoz, Y., & Prigent-Combaret, C. (2007). Promoter-trap identification of wheat seed extract-induced genes in the plant-growth-promoting rhizobacterium Azospirillum brasilense Sp245. *Microbiology, 153*, 3608–3622. https://doi.org/10.1099/mic.0.2007/009381-0.

Rambelli, A. (1973). The rhizosphere of mycorrhizae. In G. L. Marks & T. T. Koslowski (Eds.), *Ectomycorrhizae* (pp. 299–343). New York: Academic Press.

Raynaud, X., Jaillard, B., & Leadley, P. W. (2008). Plants may alter competition by modifying nutrient bioavailability in rhizosphere: A modeling approach. *The American Naturalist, 171*, 44–58. https://doi.org/10.1086/523951.

Richardson, A. E., Baréa, J. M., McNeill, A. M., & Prigent-Combaret, C. (2009). Acquisition of phosphorus and nitrogen in the rhizosphere and plant growth promotion by microorganisms. *Plant and Soil, 321*, 305–339. https://doi.org/10.1007/s11104-009-9895-2.

Roesch, L. F. W., Camargo, F. A. O., Bento, F. M., & Triplett, E. W. (2007). Biodiversity of diazotrophic bacteria within the soil, root and stem of field-grown maize. *Plant and Soil, 302*, 91–104. https://doi.org/10.1007/s11104-007-9458-3.

Rouphael, Y., Franken, P., Schneider, C., Schwarz, D., Giovannetti, M., & Agnolucci, M. (2015). Arbuscular mycorrhizal fungi act as biostimulants in horticultural crops. *Scientia Horticulturae, 196*, 91–108.

Schaarschmidt, S., Gresshoff, P. M., & Hause, B. (2013). Analyzing the soybean transcriptome during autoregulation of mycorrhization identifies the transcription factors GmNF-YA1a/b as positive regulators of arbuscular mycorrhization. *Genome Biology, 14*, R62. https://doi.org/10.1186/gb-2013-14-6-r62.

Shu, B., Li, W., Liu, L., Wei, Y., & Shi, S. (2016). Transcriptomes of arbuscular mycorrhizal fungi and litchi host interaction after tree girdling. *Frontiers in Microbiology, 7*, 408. https://doi.org/10.3389/fmicb.2016.00408.

Shukla, K. P., Sharma, S., Singh, N. K., Singh, V., Tiwari, K., & Singh, S. (2011). Nature and role of root exudates: Efficacy in bioremediation. *African Journal of Biotechnology, 10*, 9717–9724.

Singh, S., Katzer, K., Lambert, J., Cerri, M., & Parniske, M. (2014). CYCLOPS, a DNA-binding transcriptional activator, orchestrates symbiotic root nodule development. *Cell Host & Microbe, 15*, 139–152. https://doi.org/10.1016/j.chom.2014.01.011.

Sugimura, Y., & Saito, K. (2017). Transcriptional profiling of arbuscular mycorrhizal roots exposed to high levels of phosphate reveals the repression of cell cycle-related genes and secreted protein genes in *Rhizophagus irregularis*. *Mycorrhiza, 27*, 139–146. https://doi.org/10.1007/s00572-016-0735-y.

Symanczik, S., Michelle, M., Thonar, C., Schlaeppi, K., Heijden, M. V., Kahmen, A., Boller, T., & Mader, P. (2017). Application of mycorrhiza and soil from a permaculture system improved phosphorus acquisition in Naranjilla. *Frontiers in Plant Science, 8*, 1263. https://doi.org/10.3389/fpls.2017.01263.

Tedersoo, L., May, T. W., & Smith, M. E. (2010). Ectomycorrhizal lifestyle in fungi: Global diversity, distribution, and evolution of phylogenetic lineages. *Mycorrhiza, 20*, 217–263. https://doi.org/10.1007/s00572-009-0274-x.

Tedersoo, L., Liiv, I., Kivistik, P. A., Anslan, S., Koljalg, U., & Bahram, M. (2016). Genomics and metagenomics technologies to recover ribosomal DNA and single-copy genes from old fruit-body and ectomycorrhiza specimens. *MycoKeys, 13*, 1–20. https://doi.org/10.3897/mycokeys.13.8140.

Vangelisti, A., Natali, L., Bernardi, R., Sbrana, C., Turrini, A., Hassani-Pak, K., Hughes, D., Cavallini, A., Giovanetti, A., & Giordani, T. (2018). Transcriptome changes induced by arbuscular mycorrhizal fungi in sunflower (*Helianthus annuus* L.) roots. *Scientific Reports, 8*(4), 1–14. https://doi.org/10.1038/s41598-017-18445-0.

Zouari, I., Salvioli, A., Chialva, M., Novero, M., Miozzi, L., Tenore, G. C., Bagnaresi, P., & Bonfante, P. (2014). From root to fruit: RNA-Seq analysis shows that arbuscular mycorrhizal symbiosis may affect tomato fruit metabolism. *BMC Genomics, 15*(1), 221. https://doi.org/10.1186/1471-2164-15-221.

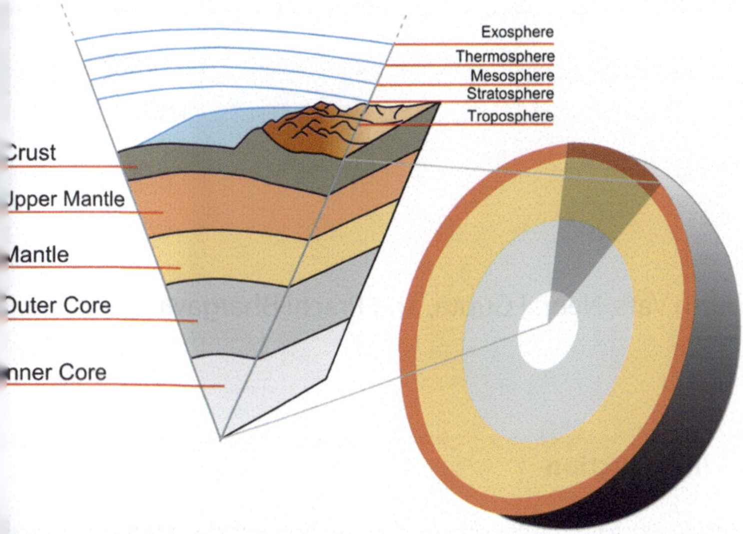

.1 Pedosphere and the earth's Lithosphere. (Source: https://commons.wikimedia.org/wiki/ rth-crust-cutaway-english.svg)

Pedospheric Soil

ormation is a very complex process because soil cannot be considered as a cal entity only. Since time immemorial natural soil has supported the life on (Liu et al. 2010). With rise in population the earth is under immense pressure et the need of food and resources, therefore to increase the production new lture techniques are being used. With the beginning of the twentieth century, cal fertilizers have found their acceptability for agricultural practices all over orld (Fowler 2013, Kaur et al. 2010). The last two decades have witnessed the mption of fertilizers with a multiple fold rise (Vats and Mishra 2016). The ve effects of chemical fertilizers are not just confined to soil but the left over cal fertilizers, are being added to air, water or land and which finally end up soil (Vats and Kumar 2015). Soil has its specific biological, physical and cal characteristics and varies in their structure, origin, chemical nature ic/inorganic components), color, texture, water holding capacity at different aphical locations (Kaur et al. 2010; Gupta et al. 2018).

Pedosphere and Its Organization

rganization of pedosphere is very complex and hierarchical, based on thermo- ic systems, harmoniously embedded and finds exchange of substance and among themselves as well as with the outer environment. These systems are tructures, macro structures, mesostructures and microstructures respectively. iversity of soil composition, its functionality and spatial regularities are based morpho-lithologic and bioclimatic factors. And both these factors have direct lirect influence from microbiota of soil. Pedosphere (soil) quantitatively

Vulnerability of Soil Micro Biota T Natural and Anthropogenic Induc Changes and Loss of Pedospheric Functionality

Siddharth Vats, Neeraj Gupta, and Prachi Bharga

1 Introduction

Survival and existence of the very human life depends up lenges and simultaneously providing protection to natural r judicious exploitation in a situation where, rising pollut threatening the sustainable development (Khaleel et al. 198 of environment, lithosphere provides an anchorage to life. tant, valuable and limited natural resource whose, top mos sphere (Goudie 2018). Soil is an important component microflora and micro fauna in maintaining the fertility of th culture, and ecological environment is very well understood tion. Natural process takes place in the soil (Zhongiun et al. the pedosphere is factor of microflora and micro fauna pres

1.1 Pedosphere, Soil and Present Scenario

In its simplest form pedosphere is defined as the outermost skin of layer is made up of soil and this layer is home to biotic compon micro fauna, microflora, insects etc. and abiotic components like et al. 2014). This is the layer, where, soil formation process is car among atmosphere (air in the soil), biosphere (microflora, microfa sphere (water trapped in the soil, water over and above, in and out (bed rock and regolith). This is the layer where all the layers of lithosphere, biosphere, hydrosphere and atmosphere (Van Passen e

S. Vats · N. Gupta · P. Bhargava (✉)
Institute of Biosciences and Technology, Shri Ramswaroop Memorial
Barabanki, Uttar Pradesh, India
e-mail: prachi.bio@srmu.ac.in

© Springer Nature Singapore Pte Ltd. 2019
A. Varma, D. K. Choudhary (eds.), *Mycorrhizosphere and Pedogenesi*
https://doi.org/10.1007/978-981-13-6480-8_12

contains far large carbon (C) than the amount present in land vegetation and atmosphere. And the reason is the higher availability of soil microbiota which includes bacterial, protists, fungi, archaea and viruses in the pedosphere. This large pool of microbiota process inputs from living and dead plants and animals (Zhongjun et al. 2017) (Table 12.1).

Table 12.1 Twelve order of soil taxonomy (United State Department of Agriculture)

S.No.	Soil types	Characteristics
1	Alfisols	Soil which have a kandic, argillic and natric horizon with 35% or above base saturation and typical ochric epipedon and sometime umbric epidpedon.
2	Andisols	Major component of the soil are short range order minerals in the semi weathered soils to highly weathered soils, rich in glass component of volcanic origin. Andic soils forms the major components between mineral layer and organic layer.
3	Aridisols	This is dry soil type with an aridic moisture, ochric/anthropic epipedon100 cm of its upper boundary a calcic, natic, cambic, gypsic,. Petro calcic, petrogypsic or salic/ duripan/ argillic horizon. Its salic horizon is saturated with water upto 100 cm for few months. Its aridic regime has no moisture in the normal years for the growth of life.
4	Entisols	Soil with no pedogenic horizons, most with either ochric epipedon or anthropic epipedon, sandy to shallo epipedon.
5	Gelisols	Soils surface upto 100 cm is rich in permafrost or gelic materials uptill 100 cm followed by permafrost uptill 200 cm. The active layer is rich in gelic material are rich in organic material with frost churning or ice segregation. The upper part of the surface have permafrost.
6	Histosols	Soil type which is dominantly organic in nature and are called as moors, peats or bogs. It does not have any permafrost.
7	Inceptisols	Majorly find in humid and subhumid regions, with loss of metals like aluminium and iron. Less amount of weathered minerals. Main types of horizons are natric kandic, spodic, oxic and argillic
8	Mollisols	Surface is dark colored; base is rich in minerals. This type of soil has mollic epipedon. Horizon are majorly calcic while petrocalci and duripan horizon are also found.
9	Oxisols	Mainly soil of tropics and subtropics. Slopes are found, mixture of kaolin, quartz, organic matter and free oxides. It is also called featureless soil. Properties are very gradual and arbitrary boundaries.
10	Spodosols	Contains organic matter of amorphous nature mixed with aluminum. Iron may or may not present. Gray color appears due to quartz. Silicate content is very low. On the basis of particle size, it can be classified as coarse silty, loamy, coarse loamy, loamy skeletal, sandy skeletal.
11	Ultisols	Soil contains clay (translocated silicate) with kandic or argillic horizon. Base saturation is 35%.
12	Vertisols	Made up of expanding clays, on loss of water it shrink while swells up on wet

2 Functions of Pedosphere

This is the most active subaqueous layer of the earth. Functions of this layer are interconnected with all the others spheres of the earth. The pedospheric functions associated with the biosphere is to provide a suitable environment and habitat for terrestrial species and reason for soil fertility, providing a life line to the biodiversity and forest and the lignocellulose material released from the forest (Vats 2017) (Fig. 12.2, Table 12.2).

Soil beneath the pedosphere, is Ferrasols are highly weathered soil having high concentrations of ferrous, kaolinite and oxides of aluminum. These are the soil found deep inside up to 20 m. The soil contains clay minerals and have highly leached characteristics (Hou et al. 2015).

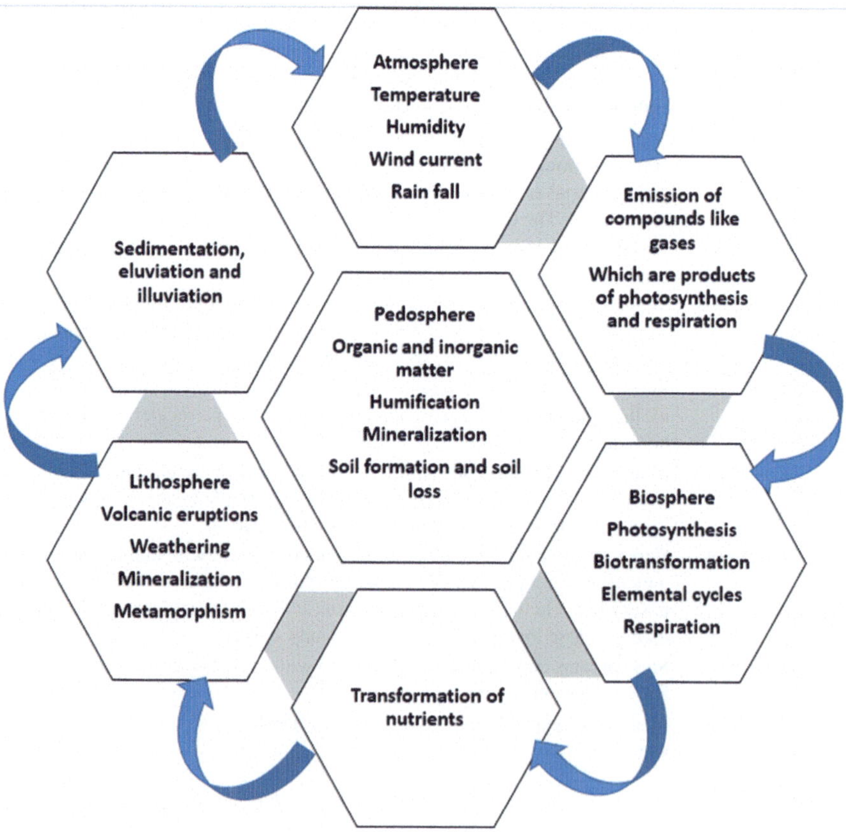

Fig. 12.2 Functions of pedosphere

Table 12.2 Ecological functions of Pedosphere

S.No.	Characteristics of pedosphere	Ecological function
1	Physical characteristics	Providing a habitat to micro-fauna and micro-flora. This ecological room varies based on structures, properties and forms of pedosphere
2	Chemical characteristics	Presence of minerals and moisture in the pedosphere effects the chemicals.
3	Biological characteristics	Habitat to all forms of life (plants, animals, microflora and microfauna). Preservation of seeds, embryos to providing living habitat.
4	Physical-chemical characteristics	Helps in formation of various niches and habitat of varying properties and forms. Different forms of pedosphere have different absorptive power for nutrients and water for plants and soil animals. Recycling by destruction of minerals and biophilic and organic elements and their resynthesis.
5	Bio-chemical	Preservation of seed and embryos of animals by providing accumulation space for moisture and gasses. Resynthesis of organic molecules.

2.1 Natural Process and Their Effects on the Pedosphere

The term pedosphere was introduced by the A.A. Yarilov in year 1905. Pedosphere forms 1–2 m vertically of the earth's skin. This almost insignificant layer of the earth surface holds the maximum density of life on earth with maximum diversity. There are numbers of chemical, biological and physical changes continuously taking place in the soil all the time at micro and macro level. Chemical processes have strong effects on the composition of soil and its resources. The water present in the ground in all different forms, weathering of the rocks, formation of crusts, and layers of air in the atmosphere are all influenced by the chemical processes taking place in the pedosphere (Brady et al. 2008). One of the most important biological function of the pedosphere is to maintain biological productivity and fertility by providing nutrition to plants. Pedosphere act as an important interface between air and water besides serving as the limiting factor which prevents the dilution and loss of elements. Functional values of pedosphere to the biosphere can be summarized into the table mentioned below (Table 12.3).

3 Anthropogenic Activity

Pedosphere is not immune to human activities. Every human activity like mining, deforestation, land use, agriculture has some direct or indirect effect on the physical, chemical and biological properties of pedosphere (Vats et al. 2011). Degradation of top most layer of the soil like its acidification, salinization, erosion, compaction, and fertility loss.

Table 12.3 Pedosphere and its characteristics

S. No.	Physical Role of pedosphere	Role of pedosphere for living organisms	Chemical and physico-chemical role of pedosphere	General role of pedosphere
1	Holds up the moisture and water	Enhances the fertility of the soil	Reservoir of biochemical energy	Helps in understanding the ecosystem.
2	Act as a life space	Support ecosystem	Home to biophillic elements	Help in regulation of structural components of ecosystem.
3	Protective towards ecological		Sorption and absorption of elements.	

3.1 Anthropogenic Factors Negatively Affecting Soil Functionality

Pedosphere is vital for our existence as it preserves biodiversity, helps in regulation and maintenance of the water, carbon and all types of biogeochemical cycles, acts as store house of natural resources, serves as the ground for housing settlement for humans and terrestrial animals and their transportation. Different microorganisms play various pivotal roles in the smooth functioning of all biogeochemical cycles that are responsible for the environment of soils and oceans (Gupta et al. 2018).

A study conducted by UN has observed that more than 20% of usable land has lost its natural capacity and fertility due to unsustainable actions of humans like deforestation, overgrazing, miss managed agricultural practices with over use of chemical based fertilizers and pesticides, positive pollution, heavy use of machinery, industrialization and loss of flora and fauna (Bhargava et al. 2017b).

3.1.1 Agriculture and Its Impact on the Soil Microflora and Micro Fauna

Agricultural activities help mankind to feed huge population and its domestic animals. The use of only few crops and our dependence on some staple crops (like grains, maize and other crops consumed by human beings) has drastically affected the agricultural biodiversity. Lack of rotational farming (monocropping) cause negative effect on the biodiversity and the soil micro flora and fauna of any area. Agroecosystem is very complex involving various factors and players. The reduction in biodiversity negatively impacts on the soil microflora and fauna which in turn reduces the functionality of the pedosphere (McDaniel et al. 2014). It further reduces the N and C content of the pedosphere eventually degrading the natural power of that soil to sustain crops. This gives rise to a vicious circle which require the use of synthetic fertilizers to increase the fertility of that soil. For sustainable development this complex ecosystem must be used synergistically. Use of crop rotation system can enhance the pedospheric functionality of that soil. McDaniel et al. 2014 reported that crop rotation of cereals and legumes have positive impacts on C and N content of the soil. C content increased by 3.6% and N content by 5.3% by addition of strategy of crop rotation. The increase in C and N content was 8.5% and 12.8% respectively when a crop was grown with monoculture grown with cover crops. Total microbial C biomass and N biomass content also got increased by 20.7% and 26.1% respectively. Due to this benefits, strategy can be used for agro-ecosystem sustainability (McDaniel et al. 2014).

3.1.2 Effect of Anthropogenic Activity on Soil Organic Carbon (SOC)

Microbiota in soil transforms the dead and living wastes into SOC and also get back this SOC into the feed of terrestrial needs. Microbes present in the pedosphere act as engine to recycle minerals and elements and influence each cycle type, this process eventually affects the global climate (Schimel and Schaeffer 2012). Miro-flora and micro-fauna present in the soil are responsible for fixation of atmospheric CO_2, conversion and decomposition of organic carbon of plants and animal origin. Their characteristics influence the quality and quantity of the of SOC and its physical and chemical properties (Schmidt et al. 2011). New agriculture techniques affect the soil quality. Intensified agriculture practices and management have brought qualitative and quantitative negative changes in the SOC. SOC is an important factor which governs and defines the soil fertility. But due to the use of pesticides and chemical based fertilizers, microbiota of the soil has seen a reduction in number. Which has negatively affected the SOC and consequently affected the fertility of the soil. Microbiota present in the soil produces 2300 gega tons of C. This safeguard the sustainability of the life on earth (Stockmann et al. 2013).

3.2 Microflora and Microfauna of the Soil

All the animals belonging to a specific geological period, habitat or particular region constitute the fauna of that niche. The smallest fauna reported is of size 0.1 mm or less are visible only by microscope comes under the category of microfauna. Among various micro fauna, protozoa and nematodes are the most important for the functionality of the soil (Kaur et al. 2010). Habitat for nematodes is generally sandy soils and their motions depend upon the water film formed around the sand particles. On the other hand, protozoas have variability in their shape and size. Protozoans feed on bacteria and have well adapted themselves to feed on nutrients available in soil particles and slide over think water film on the roots and soil particles. Bacteria, viruses and fungi are the three important forms of microflora in soils. And out of these three Bacteria the single cells, tiny microbes with billions to trillions in numbers and thousands in species variation just only in 1 g of soil sample. Bacteria performs transformations and weathering of minerals rich rocks, decomposition of biomass and in major player in geochemical and nutrient cycles. Similarly, Fungi in the form of filaments, globules and spores, and are also very common in soils, taking nutrients from living and dead matter. Out of various fungi, mycorrhizal fungi have symbiotic relationship with plants and help in stabilizing soil and its aggregates with decomposed organic matter (Gupta et al. 2018; Auge et al. 2004). Plants secure nutrients from these fungi as they are dependent upon these relations. Viruses, the smallest and simplest multiplying entities, are all parasitic in nature with flora and fauna present in pedosphere. Moisture, organic components and soil aggregates with mycorrhizal fungi are the factors that controls the growth of viruses in the pedosphere.

3.3 Soil Fauna Population and Their Role in Fertility of Pedosphere

Fertility of soil is the outcome of synergistic activities of microflora and microfauna present in the soil. It is believed that estimation of metabolic activities of soil animal population and their role in processing activities and other functions can be extrapolated from the biomass and density (Petersen and Luxton 1982). Biomass produced by different fauna population at different sites provide information for animal communities residing in those sites. The presence of communities and their ability to produce biomass depends upon moisture content of the soil, pH, biome and organic content present in the top soil. Panphytophages feed on both dead plant and microbial communities. Substrate utilization, food selectivity, amelioration among intra and inter species, and niche and habitat exploitation, microbial competition, respiration, reproduction and assimilation has been affected by anthropogenic activities.

Many strategies are used to study the diversity of the microbes present in different soil types. Techniques like Phospholiped Fatty Acid analyses (PLFA), 16S rRNA, and Biolog, Denaturating Gradient Gel Electrophoresis (DGGE) etc. are used to asses' microbial community in the soil. A study was carried out by Dong et al. (2008) found that it's the anthropogenic factors not the age of the soil that diversify the species. Tea gardens where agriculture is being carried out by humans involving various human activities and interferences have high gram positive to gram negative bacteria compared to the land which is not cultivated or a forest. Also in the tea gardens the ratio of fungal to bacteria was higher in the tea gardens.

3.4 Effects of Heavy Metals Released by Human Activity

The concentration of heavy metals like cadmium is increasing in the soil all over the world alarmingly (Bhargava et al. 2017a). This rise in heavy metals have negatively affected the soil fertility because of toxic effect on the soil microorganisms. Eco toxicity caused by heavy metals like Zn, Cr, Pb, Cd, Cu and Hg etc. due to industrial waste release in the form of solid, liquid and gaseous wastes and have negative effects on the pedosphere because of their accumulation (Han et al. 2002). Bioavailability of Cd on soil micro-fauna and microflora are associated with the soil type, time, season, speciation, ageing, environmental multifactor, presence of other heavy metals, source of other heavy metals, and other microbes (Vig et al. 2003). It has been reported that the rise of heavy metals like Pb, Cd, and Hg in the pedosphere are comparatively higher than lithosphere. Cd, Hg, and Pb are 5.2, 6.1 and 9.6 times higher in concentration in pedosphere. Heavy metals accumulation has increased the per capita heavy metal burden to 17.3, 0.10, 38.6, 0.18, 74.2, 5.9 and 58.2 kg for Cr, Hg, Pb, Cd, Cu, Ni and Zn respectively. Toxicity and bioavailability of these health affecting heavy metals get increased with the acidification (Han et al. 2002). Enzymes are the biocatalyst responsible for carrying out all different types of metabolic functions in the living organisms. Presence of pollutant like heavy metals in the soil have negative effect on the enzyme activity. Enzyme synthesis decreases drastically in the presence of heavy metals in the soil. Gadd and Griffiths (1977) found that the decrease in the synthesis of alfaglucosidase and amylase 50% and 75% respectively on the addition of 2000 Pb μg/g of soil enriched with maltose and glucose.

3.5 Effect of pH on the Soil

Red soils are suitable for the production of pulses, potatoes, citrus fruits, jowar, tobacco etc. and for the growth of the crops pH of the soil is an important factor. pH of the soil also effects the microbes growing in that soil. One indicator for the analysis the impact of pH on the soil microbial population is the microbial biomass. Microbial biomass is quantified on the carbon, nitrogen and phosphorous content. Li and Chen (2004) studied the impact of pH on the soil microbial biomass under cultivation for citrus fruits for different time period. Content of microbial C and P are significantly affected by the variation of pH of the soil. There are critical values for acidic and alkaline extreme for pH beyond which microbes are not able to survive. At alkaline scale it is 8.0 while on acidic side it is 3.0.

3.6 Effects of Pesticides on Soil Microflora and Pedospheric Functionality

The fertility of the soil is influenced by the structure of the soil and the microflora and micro-fauna residing in it. But the use of pesticides has negatively affected the diversity of the soil microorganisms (Bhargava et al. 2017a, b). Pesticides protects crops from pests but have been found responsible for altering the biological population of the soil, especially the micro-organisms. The knowledge of the effects of the microbes on pesticides and pesticides on microbes is still at its infant stage. This is because only 1% of the soil microbes are cultural able. Molecular methods have their own limitations while, applications of techniques like denaturing gradient gel electrophoresis (DGGE) is not possible by many of the labs. While, some of the pesticides have depressive effects on the growth of microbes while some stimulates the growth of certain types of microbes. Pesticides are classified into herbicides, insecticides, fungicides etc. and have different mode of actions based on different chemical structures (Karpachevskii 2011). A detailed study was carried out by Lo in 2011 on the effects of pesticides on the soil sample from different soil types (Table 12.4).

Table 12.4 Effects of various pesticides on the biological activity of the pedosphere. (Ref Lo.C.C et al. 2010)

S. No.	Type of pesticides	Name	Crops	Soil types	Effects on biological activity of the pedosphere
1.	Herbicides	Butachlor (NButoxymethyl-2-chloro-2′,6′-diethylacetanilide))	Rice	Agricultural	Upto 20 μg/g) reduced the population of *Azospirillum* and anerobic nitrogen fixers.
2.	Herbicides	Carbaryl (1-naphthyl methylcarbamate)	Rice	Agricultural	Up to 10 (μg/g) has no effect on the nitrogenase activity.
3.	Insecticides	Carbofuran (I)	Rice	Agricultural	Upto 2 (μg/g) reduced the population of *Azospirillum* and anerobic nitrogen fixers.
					Upto (4 μg/g) it stimulated the *Azospirillum* and anerobic nitrogen fixers.
4.	Insecticides	Diflubenzuron (I)	Maize	Agricultural	(100~500 μg/g) stimulated the growth of *Azotobacter vinelandii*
5.	Insecticides	Methylpyrimifos (I)	Maize	Agricultural	(100~300 μg/g) decreased the population of aerobic microbes. No effects on fungal populations and denitrifying bacteria
6.	Insecticides	Chlorpyrifos (I)	–	Agricultural	(10~300 μg/g) decreased aerobic dinitrogen fixing bacteria.
					But no effects on fungal populations and denitrifying bacteria
7.	Herbicides	Metsulfuron methyl (H	–	Agricultural	Reduced the growth of fluorescent psendomonads

(continued)

Table 12.4 (continued)

S. No.	Type of pesticides	Name	Crops	Soil types	Effects on biological activity of the pedosphere
8.	Herbicides	Chlorsulfuron (H)	–	Agricultural	Reduced the growth of fluorescent psendomonads
9.	Herbicides	Thifensulfuron methyl (H)	–	Agricultural	Reduced the growth of fluorescent psendomonads
10.	Herbicides	Diuron (H),	Orchard	Agricultural	DGGE pattern shows difference in community pattern prior and post treatment
11.	Herbicides	Linuron (H),	Orchard	Agricultural	DGGE pattern that microbial community structures of the treated and nontreated soils were significantly different.
12.	Herbicides	Chlorotoluron (H)	Orchard	Agricultural	DGGE pattern that microbial community structures of the treated and nontreated soils were significantly different.
13.	Herbicides	Fenpropimorph	–	Research land	For first 10 days growth of fungus inhibited, at 17 days bacterial population almost nil, 56 days large bacterial growth.
14.	Herbicides	Propanil (H),	Sugar beet	Agricultural	No effects on bacterial count, less persistence level
15.	Herbicides	Prometryne (H)	Sugar beet	Agricultural	No effect on bacterial count, more persistent level

(continued)

Table 12.4 (continued)

S. No.	Type of pesticides	Name	Crops	Soil types	Effects on biological activity of the pedosphere
16.	Fungicides	Iprodione (F)	Strawberry	Agricultural	At 30 °C more growth less at 15 °C
					At 30 °C and with 50 μg/g iprodione treatment, the amounts of soil bacterial communities increased quickly and remain high for 23 days.
17.	Herbicides	Glyphosate (H)	Ponderosa pine plantations	Agricultural	At high concentration increased the growth of bacteria.
18.	Insecticides	Methamidophos (I)	–	Agrochemical free soil	Biomass decreased (41–83%) but CFU increased (86–88%).
19.	Insecticides	Methylparathion (I)	–	20 years contaminated soil	Increase in the number of two members of γ -porteobacteria, which were closely related to the *Pseudomonas stutzeri* (similarity 99%) and *Pseudomonas putida* (similarity 99%).
20.	Fungicides	Azoxystrobin (F)	–	Conventional farm	No change in bacterial species, flagellate protozoan *Paraflabellula hoguae* was affected by azoxystrobin
21.	Fungicides	Chlorothalonil (F)	–	And organic management	No change in bacterial species but ciliate protozoan *Arcuospathidium* sp. or *Bresslaua vorax*, was affected by chlorothalonil.,

(continued)

Table 12.4 (continued)

S. No.	Type of pesticides	Name	Crops	Soil types	Effects on biological activity of the pedosphere
22.	Fungicides	Tebuconazole (F)	–	Farm	No change in bacterial species ascomycete fungus *Cladosporium tenuissimum* (was affected by tebuconazole.
23.	Herbicides	Isoproturon (H)	–	Agricultural	Degradation in sub-soil between 40–50 and 70–80 cm depths and proliferation of *Sphingomonas* spp.
24.	Insecticides	Methamidophos (I)		Black soil	High concentrations of methamidophos (250 mg/kg) could significantly stimulate fungal populations.
25.	Herbicides	Butachlor (H)	Paddy	Agricultural	Decrease in diversity of microbes.
26.	Insecticides	Fenamiphos (I)		Agricultural	Not effects on dehydrogenase or urease. Reduces the actiitity of nitrification taking place in the soil.

4 Conclusion

Pedosphere is the layer of the earth used for the agriculture and is of utmost importance for a country like India where agriculture is the main occupation and has its major share in the country's GDP. Green revolution in India was started in the 1960s by introducing higher yielding crops/strains of crops, advanced machinery, technology, fertilizers and chemical pesticides and Punjab has been the forefront runner state and is known as India's granary. Green revolution in Punjab was implemented as "take it or leave it" strategy. The whole implementation was with unprecedented energy. Though the package had many drawbacks but due to the support price farmers took it and due to the high yields brought prosperity with rise in income, farmers, began to cultivate each and every available piece of land with rice and wheat and replacing traditional practices and crops. But the dark side of the story is loss of natural fertility and capacity of the soil and with that the compulsion of use of fertilizers and pesticides have put them in deep debt. Rise in cancer patients, loss of flora and fauna. Indian subcontinent has laterite, black, alluvial, red, deserts and mountain soil types as the major soil types. Rate of desertification and loss of fertility due to over use of pesticides, chemical based fertilizers, over exploitation of water in the arid and semi arid regions has been alarmingly fast. Changing climate conditions which are outcome of collective effects of various pollutions has degraded almost a third of the total agricultural land and per year a stagnant figure of 12 million hectares of land gets converted into barren land. We can change the fate of the fertility and functionality of the soil by changing the methods of agriculture moving from chemical based to organic based. This will lead to a better sustainable and healthy ecosystem to live in.

Acknowledgements PB thanks DST-SERB: SB/YS/LS-213/2013 for the financial support.

References

Augé, R. M. (2004). Arbuscular mycorrhizae and soil/plant water relations. *Canadian Journal of Soil Science, 84*(4), 373–381.

Bhargava, P., Gupta, N., Vats, S., & Goel, R. (2017a). Health issues and heavy metals. *Austin Journal of Environmental Toxicology, 3*(1), 3018.

Bhargava, P., Singh, A. K., & Goel, R. (2017b). Microbes: Bioresource in agriculture and environmental sustainability. In *Plant-microbe interactions in agro-ecological perspectives* (pp. 361–376). Singapore: Springer.

Brady, N. C., & Weil, R. R. (2008). *The nature and properties of soils* (Vol. 360). Upper Saddle River: Pearson Prentice Hall.

Dong, X. U. E., Huai-Ying, Y. A. O., De-Yong, G. E., & Huang, C. Y. (2008). Soil microbial community structure in diverse land use systems: A comparative study using biolog, DGGE, and PLFA analyses. *Pedosphere, 18*(5), 653–663.

Fowler, B. A. (Ed.). (2013). *Biological and environmental effects of arsenic* (Vol. 6). Amsterdam: Elsevier.

Gadd, G. M., & Griffiths, A. J. (1977). Microorganisms and heavy metal toxicity. *Microbial Ecology, 4*(4), 303–317.

Goudie, A. S. (2018). *Human impact on the natural environment.* Hoboken: Wiley.

Gupta, N., Vats, S., & Bhargava, P. (2018). Sustainable agriculture: Role of metagenomics and metabolomics in exploring the soil microbiota. In *Silico approach for sustainable agriculture* (pp. 183–199). Singapore: Springer.

Han, F. X., Banin, A., Su, Y., Monts, D. L., Plodinec, J. M., Kingery, W. L., & Triplett, G. E. (2002). Industrial age anthropogenic inputs of heavy metals into the pedosphere. *Naturwissenschaften, 89*(11), 497–504.

Hou, E. Q., Xiang, H. M., Li, J. L., Li, J., & Wen, D. Z. (2015). Soil acidification and heavy metals in urban parks as affected by reconstruction intensity in a humid subtropical environment. *Pedosphere, 25*(1), 82–92.

Karpachevskii, L. O. (2011). A book on the pedosphere of the earth. *Eurasian Soil Science, 44*(7), 832–833.

Kaur, A., Vats, S., Rekhi, S., Bhardwaj, A., Goel, J., Tanwar, R. S., & Gaur, K. K. (2010). Physicochemical analysis of the industrial effluents and their impact on the soil microflora. *Procedia Environmental Sciences, 2*, 595–599.

Khaleel, R., Reddy, K. R., & Overcash, M. R. (1981). Changes in soil physical properties due to organic waste applications: A review 1. *Journal of Environmental Quality, 10*(2), 133–141.

Li, X., & Chen, Z. (2004). Soil microbial biomass C and N along a climatic transect in the Mongolian steppe. *Biology and Fertility of Soils, 39*(5), 344–351.
Liu, E., Yan, C., Mei, X., He, W., Bing, S. H., Ding, L., et al. (2010). Long-term effect of chemical fertilizer, straw, and manure on soil chemical and biological properties in northwest China. *Geoderma, 158*(3–4), 173–180.
McDaniel, M. D., Tiemann, L. K., & Grandy, A. S. (2014). Does agricultural crop diversity enhance soil microbial biomass and organic matter dynamics? A meta-analysis. *Ecological Applications, 24*(3), 560–570.
Petersen, H., & Luxton, M. (1982). A comparative analysis of soil fauna populations and their role in decomposition processes. *Oikos, 39*, 288–388.
Renella, G., Ogunseitan, O., Giagnoni, L., & Arenella, M. (2014). Environmental proteomics: A long march in the pedosphere. *Soil Biology and Biochemistry, 69*, 34–37.
Schimel, J., & Schaeffer, S. (2012). Microbial control over carbon cycling in soil. *Frontiers in Microbiology, 3*, 348.
Schmidt, M. W. I., Torn, M. S., Abiven, S., Dittmar, T., Guggenberger, G., Janssens, I. A., Kleber, M., K¨ogel-Knabner, I., Lehmann, J., Manning, D. A. C., Nannipieri, P., Rasse, D. P., Weiner, S., & Trumbore, S. E. (2011). Persistence of soil organic matter as an ecosystem property. *Nature, 478*, 49–56.
Stockmann, U., Adams, M. A., Crawford, J. W., Field, D. J., Henakaarchchi, N., Jenkins, M., Minasny, B., McBratney, A. B., de Remy de Courcelles, V., Singh, K., Wheeler, I., Abbott, L., Angers, D. A., Baldock, J., Bird, M., Brookes, P. C., Chenu, C., Jastrow, J. D., Lal, R., Lehmann, J., O'Donnell, A. G., Parton, W. J., Whitehead, D., & Zimmermann, M. (2013). The knowns, known unknowns and unknowns of sequestration of soil organic carbon. *Agriculture, Ecosystems & Environment, 164*, 80–99.
Van Paassen, L. A., Daza, C. M., Staal, M., Sorokin, D. Y., van der Zon, W., & van Loosdrecht, M. C. (2010). Potential soil reinforcement by biological denitrification. *Ecological Engineering, 36*(2), 168–175.
Vats, S. (2017). Methods for extractions of value-added nutraceuticals from lignocellulosic wastes and their health application. In *Ingredients extraction by physicochemical methods in food* (pp. 1–64). Washington, DC: American Chemical Society.
Vats, S., & Kumar, R. (2015). Amylolytic-extremoenzymes: Saviour of environments. *European Journal of Biomedical and Pharmaceutical Sciences, 2*(5), 694–702.
Vats, R., & Mishra, A. (2016). Soil agro-ecological management by vermicompost a potential organic nutrient source for the state of Uttar Pradesh. *European Journal of Pharmaceutical and Medical Research, 3*(9), 604–609.
Vats, S., Kumar, R., & Miglani, A. K. (2011). Isolation, characterization and identification of high salinity tolerant, heavy metal contaminant and antibiotics resistant amylolytic-thermophilic pseudomonas Sp. *International Journal of Pharmaceutical Sciences Review and Research, 10*(2), 125–129.
Vig, K., Megharaj, M., Sethunathan, N., & Naidu, R. (2003). Bioavailability and toxicity of cadmium to microorganisms and their activities in soil: A review. *Advances in Environmental Research, 8*(1), 121–135.
Zhongjun, J. I. A., Kuzyakov, Y., Myrold, D., & Tiedje, J. (2017). Soil organic carbon in a changing world. *Pedosphere, 27*(5), 789–791.

Metagenomics as a Tool to Explore Mycorrhizal Fungal Communities

13

Prachi Bhargava, Siddharth Vats, and Neeraj Gupta

1 What Is Mycorrhiza?

A mycorrhiza is a symbiotic connection or relationship between a fungus and the roots of a vascular host plant. Mycorrhizae play important roles in soil biosphere. When the symbiotic fungi grows inside the plant's roots, it is called as endomycorrhizae and when it grows on the surfaces of the roots, it is termed as ectomycorrhizae.

2 Types of Mycorrhiza

On the basis of mode of nutrition fungi are divided into three classes viz. saprophytic fungi, pathogenic fungi and mycorrhizal fungi. Mycorrhizal fungi are classified into two types ectomycorrhizae and endomycorrhizae; whether fungus forms a mantle or sheath over the surface of fine lateral roots of the host trees or enter into the root cells which are generally restricted to the cortical region rarely crossing the endodermis. In all different classes of mycorrhiza, endomycorrhizal fungi are the major types of reported fungal community. Figure 13.1 summarizes the broad classification of mycorrhiza.

P. Bhargava (✉) · S. Vats · N. Gupta
Institute of Biosciences and Technology, Shri Ramswaroop Memorial University, Barabanki, Uttar Pradesh, India
e-mail: prachi.bio@srmu.ac.in

© Springer Nature Singapore Pte Ltd. 2019

A. Varma, D. K. Choudhary (eds.), *Mycorrhizosphere and Pedogenesis*,
https://doi.org/10.1007/978-981-13-6480-8_13

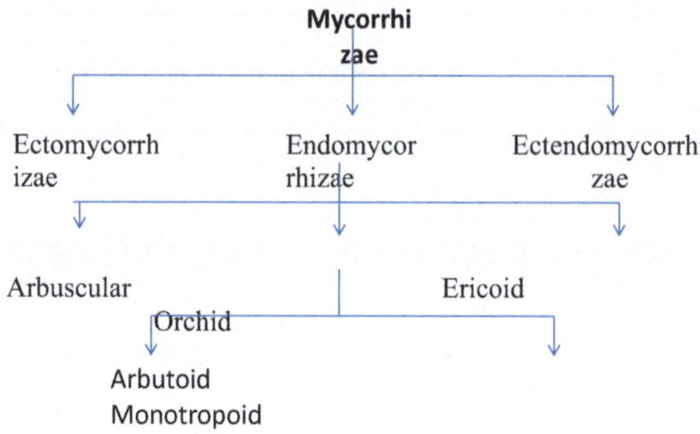

Fig. 13.1 Classification of mycorrhiza

3 Arbuscular

Arbuscular mycorrhiza is the most common type of symbiotic relationship between a fungus and the roots of a vascular plant. Fungi improve the supply of water and nutrients, such as phosphate and nitrogen, to the host plant (Parniske 2008). In Arbuscular mycorrhiza fungi are members of the glomeromycota.

4 Ericoid

Ericoid mycorrhizas are mutual relationships between fungi and plant (ericaceous species) in which fungal coils are formed in the epidermal cells of the fine hair roots of plant. (Read 1996).

5 Orchid

Orchid mycorrhizae are found in Orchidaceae family in which there are symbiotic relationships between the roots of plants e family and a variety of fungi. All orchids are mycoheterotrophic at some point in their life cycle.

6 Role of Mycorrhiza in Environmental Sustainability

Mycorrhiza performs many beneficial roles for the plants and in return get place and nutrients from its host. It protect the plants against root pathogens and toxic stresses and therefore produce more strong and healthy plants. It helps in increasing the crop yields and its quality by increasing the plant's tolerance to soil salinity and droughts. It helps in maintaining the soil quality and nutrient cycling thereby enhancing flowering and fruiting. It also helps in the optimal ustilization of fertilizers especially phosphorous. Studies have revealed that it binds to the soil and thus helps in reducing soil erosion. The major benefits of mycorrhiza to the environment has been summarized in Fig. 13.2.

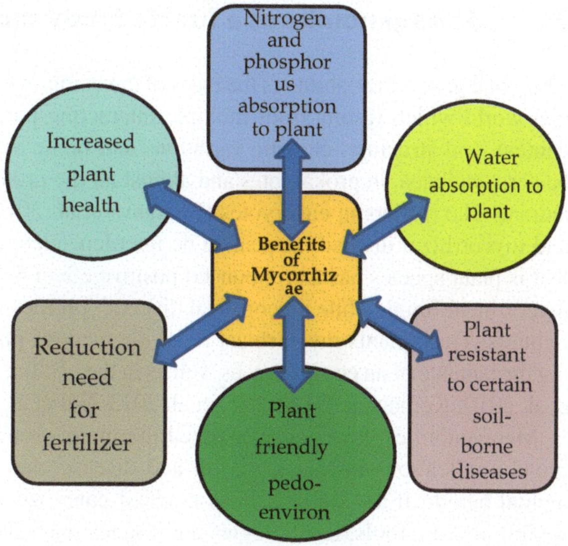

Fig. 13.2 Benefits of mycorrhizae

It has been proved by various studies that the diversity and the richness of different taxa of AMF have a positive effect on the productivity of the plant community residing in a particular niche (Wang and Qiu 2006). Thus environmentalists working on environmental sustainability have a major goal to explore the AMF composition and its contribution towards healthy ecosystems. It also helps to understand the factors which affect the interrelationship between plant communities, their respective AMF counterparts and the functioning of these symbiotic ecosystems. There are various external factors which indirectly affect the hereditary material of mycorrhiza like manures, pesticides, and heavy metals. The overall genome of all the plants including AMF is governed by the balance of all the internal and external factors and failure in the control of any of these factors result in severe complications not only in plant health but also in the future environment and wellbeing of all organisms (Bhargava et al. 2017a, b).

Mycorrhizal Arbuscular Fungi are ubiquitous in their geographical distribution and their multidimensional niche is determined by the edaphic factors. Studies on mycorrhizal diversity have relied on morphologic and other phenotypic characteristics, and these have been the main criteria for their classification and observation. Different taxa tend to show different morphological traits which again vary as a component of time and phase of life cycle. This instability and overlapping of phenotypic characters between different taxa does not allow scientists to rely on a single method for identification of fungi. However molecular taxonomy has proved itself as a blessing but now a days the use hybrid approaches is very common to study and explore new traits in AMF communities.

7 Metagenomics as a Tool to Study the AMF Community

Microbial genetic variability is the basis of the complexity and sustainability of microbial world which results from multiple interacting parameters including pH, water content, soil structure, climatic variations and biotic activity. Microbes comprise of bacteria, archaea, in prokaryotes and almost all the protozoans, algae, fungi, and few animals like rotifers in eukaryotes (Bhargava et al. 2017a, b). Plant species richness and mycorrhizal diversity appear to be interdependent as the richness of first factor that is plant species has a pronounced positive effect on the richness and diversity of mycorrhizal fungi, while mycorrhizal diversity promotes an increase in plant species richness. Thus, plants-mycorrhizal fungi interactions play a crucial role in maintaining the balance of an ecosystem, as well as in the cycling of C and N sources (Hollister et al. 2010; Kernaghan 2005; Peay et al. 2013; Wu et al. 2007; Zak et al. 2003).

Metagenomics has revolutionalized the major thrust areas of research as it has broad applications in human health and disease, animal production and environmental health. It has opened those magical gates where at the other end lies huge wealth of data, tools, technologies and usages that allow us to explore majority of organisms that are uncultivable (an estimated 99% of microbial life). Numerous research groups are developing various strategies, methodologies, tools and applications to benefit from this new field, as larger data sets from environments including the human body, the oceans and soils are being generated. The field of metagenomics continues to boom with more and more datasets from various metagenomes. A concerted effort is needed to collate all this information in a centralized place so that all the people working in this field may have easy access to this vast pool of data.

Metagenomics has been used to explore newer species of different microorganisms, although studying AMF with this tool has few challenges. Inspite of the fact that they can spend part of their life cycle as free-living organisms, mycorrhizal fungi always associate with the roots of higher plants, indeed over 90% of plant species, including forest trees, wild grasses and many crops. Both partners benefit from the relationship: mycorrhizal fungi improve the nutrient status of their host plants, influencing mineral nutrition, water absorption, growth and disease resistance, whereas in exchange, the host plant is necessary for fungal growth and reproduction. As we know that more than 98% of the microbes are unexplored till yet, studies reveal that most of the obligate symbiotic fungi are also difficult to maintain in culture, therefore there is a growing need for alternative approaches like metagenomics to obtain tissue and DNA and subsequent genomic assemblies from such species. Figure 13.3 beautifully depicts the different metagenomic strategy to explore any metagenome.

8 Metagenomic Sample Collection and Its Handling

Mycorrhizas have garnered good amount of attention owing to a combination of cell biology and genetics and the availability of genome sequences from both mycorrhizal plants and fungi. Tools, like phylogenomics and metabolomics, explore

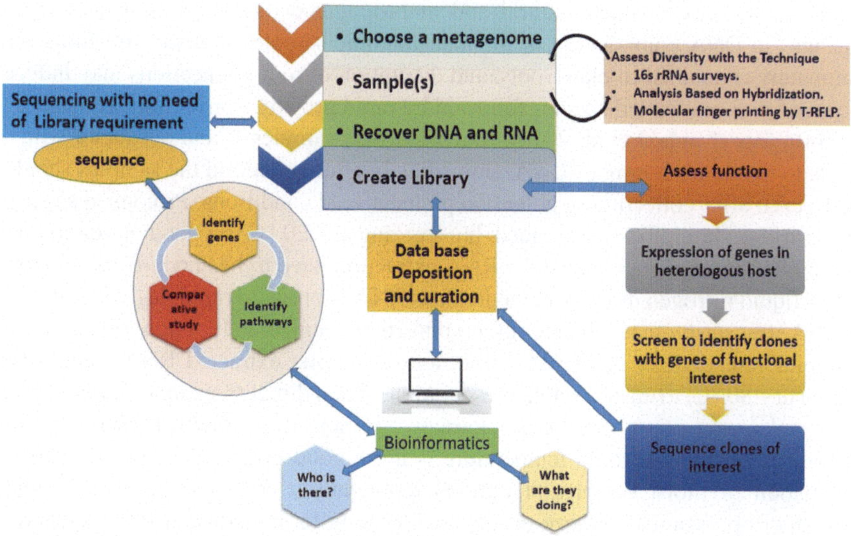

Fig. 13.3 Flowchart showing the different approaches of metagenomics

symbiont communication, development, and diversity and reveal the contribution of each partner in the association. Direct cultivation or indirect molecular approaches can be used to explore and exploit the microbial diversity present in soil. Cultivation and isolation of microorganisms is the traditional method but, as the percentage of cultivable microbes is only 0.1% to 1.0% using standard cultivation methods the diversity of soil microbial communities has been mainly unexplored and only a tiny portion of the gene pool has been characterized using cultivation and isolation.

The first step to study any metagenome is to decide the location and type of sample. Sampling is filled with challenges as the conclusions and inferences which can be drawn from a metagenomic analysis depend on the type, size, number, and timing of sampling. The major hurdle lies to ensure that the sample must be a representative of the habitat which is to be studied as soil communities' change on a micrometer scale, owing to the physical and chemical heterogeneity of the mineral and biological materials that make up the soil. The process of collection of field samples to cover the targeted variation and unbiased contribution represents a major challenge, and n number of strategies concerning the number and spatial distribution of samples have been incorporated in a number of experiments. (Prosser 2010; Lennon 2011). Collection of mycorrhizal sample require certain precautionary considerations, owing to the indeterminate growth of mycelia and the amount of contrasting morphologies and trophic strategies that coexist and interact in the communities. It is important to employ a minimum distance between samples that exceeds the largest expected size of fungal mycelia as individual mycelia vary in size (Douhan et al. 2011) and to avoid spatial autocorrelation as a result of repeated sampling of single individuals.

There are cases where natural and anthropogenic disturbances trigger rapid changes in DNA composition of the metagenomic sample, as many soil fungi are intimately connected to plant roots, and disruption of root connections may induce death of root-associated species followed by rapid growth of mycelium-consuming opportunists (Lindahl et al. 2010). Sieving of soil samples is another factor which leads to biased screening of DNA. Immediately after collection the samples should be freezed after collection or at least kept under cold conditions. Prolonged storage in the fridge should best be avoided, but freezing at −20 °C is a good option to preserve DNA. Samples collected for RNA extraction have to be shock-frozen on dry ice or liquid nitrogen directly in the field, as RNA is prone to rapid degradation, and mRNA transcriptomes change in composition immediately upon disturbance. Samples intended for RNA extraction, as well as the extracted RNA, should be stored at −80 °C, to ensure stable preservation. When direct freezing is not possible, chemical preservation may be an alternative (Grant et al. 2006). Preservation of samples by drying at room temperature is not a good option, because it involves incubation of moist samples at optimal temperatures for sporulation and rapid growth of opportunists. Freeze-drying enables long-term storage at room temperature, and may also aid later sample homogenization.

Most of the protocols for metagenomic DNA and RNA extraction are based on small amounts (mg to g) of sample material. The key to success for both targeted and shotgun studies is the quality of environmental DNA (Reigstad et al. 2011). Sample size calculation and design, as well as standardized methods for the isolation of high quality DNA have been proposed and validated, including the use of commercial kits as tailor made solutions for different types of samples. In spite of this, there is still room for significant development in this area, as evidenced by the low representativeness of fungi in the main metagenomic databases compared to bacterial sequences. The output of fungal metagenomic studies is dependent on the methodological strategy used, but also on the computational tools chosen for sequence analysis. Not surprisingly, bioinformatics has quickly turned into one of the main challenges and a bottleneck in metagenomic research.

The common techniques used are bead beating and crushing in liquid nitrogen. Subsampling and homogenization have to be adapted to each specific substrate and study, but a basic rule is that, when the size of the subsample decreases in relation to the entire sample, careful homogenization becomes more critical.

9 Soil DNA Extraction: Indirect and Direct Methods

As you sow so shall you reap. This proverb fits perfectly on the connection between the quality and completeness of data obtained from metagenomic analysis of any community and the procedures used for the extraction of DNA from a sample. The major challenge is to extract such DNA which represents all the members of a metagenome. Metagenomic DNA to study AMF from a microbial population can be extracted by direct and indirect extraction methods. The ideal methodology should include unbiased lysis of source of DNA and its extraction. It should be able to overcome the problems of lysis of the more recalcitrant cells in a community while not being too harsh to cause degradation of DNA from other community members. Another issue that needs to be addressed while extracting DNA is the strategy which discriminates between DNA from viable and dead cells in a given sample—a distinction that may be important in drawing conclusions about the overall metabolic capabilities of a microbial community.

Indirect Method In this method the source of DNA is the spore which is a dormant phase of the life cycle of fungus. Spore isolation from any soil requires taxonomic expertise and most studies exploring AMF diversity have to rely on morphological identification (Gai et al. 2006). The method involves trap cultures which are soil samples being used as inocula to propagate the local AMF species in pot cultures (Mathimaran et al. 2005). However this method has been criticized as it reflects only one stage of the microorganism's life cycle and the spore density assessed in the field samples may or may not necessarily reflect the true participation figure of the AMF population actually colonizing the roots. Moreover it is almost impossible to distinguish between spores of current season and previous seasons and it may be a possibility that a certain species sporulate occasionally. Various studies reveal that the spores of Glomus species dominate in most of the soil types as the abundance and distribution of spores also depends on the type of farming.

Direct Method The increasing use of molecular tools to study fungi and explore fungal diversity in a phylogenetic context has revolutionized the fungal ecology and phylogenetics. Metagenomic soil DNA and root DNA can be extracted using manual or various commercial kits available. They can be further purified using optional addition of RNase-A to remove unwanted RNA. The effect of contaminants on the recovery of DNA or RNA of interest presents another technical problem. Even minor contaminations of a sample reduce the effectiveness of the DNA or RNA for sequencing thereby increasing the cost of the experiment. It also reduces the chances of recovering those members of the community which are less in number. PCR amplification with primers that hybridize to highly conserved regions in genes in AMF community is followed by cloning and sequencing yields an initial description of any AMF community. Significant insights into species richness, structure, composition, and membership of microbial communities have been gained with the help of this methodology.

The use of the r RNA gene and its variable regions in prokaryotes as taxonomic markers for the classification had been established long back but their use for eukaryotes was still debatable. Efforts have been made since long to establish similar universal molecular markers for fungal taxa. Fungal molecular taxonomic studies were in full bloom in the early 1990s which relied heavily on the analysis of the nuclear ribosomal gene cluster, which comprises the 18S or small subunit (SSU), the 5.8S subunit, and the 28S or large subunit (LSU) genes. Small Subunit Ribosomal RNA (SSU-rRNA) and Internal Transcribed Spacer (ITS) are used as a barcode DNA fragment to identify a particular species or group of organisms. Several primers that target such regions are AMF specific. Inclusion of ITS and large sub unit rDNA (LSU-rDNA) helps in strong phylogenetic analysis and high species level resolution.

The SSU and LSU have proved to be very efficient in the differentiation of high taxonomic levels. The former two markers have a shortcoming for not being good for intraspecific resolution. The ITS1 and ITS2 regions are more suitable markers for fungal phylogenetic studies due to their high degree of interspecific variability, conserved primer sites, and multicopy nature in the genome. The utilization of the ITS regions as universal DNA barcode markers for fungi have been validated by the study based on testing the potential of four markers (ITS, LSU, SSU, and *rpb1*), where ITS was observed to have a superior species resolution for a broad range of taxonomic groups along with their use in intra-specific differentiation.

10 Strategies to Study AMF Community

The main aim of metagenomics is to identify and explore all the members of a community irrespective of their number and cultivability in the metagenome. The various strategies used in metagenomics try to overcome this challenge. There are two important methods which are popularly used depending on the purpose of studies-targeted metagenomics and shotgun metagenomics.

In targeted metagenomics the isolated environmental DNA is used as the template and is amplified using primers against known microorganisms. Finally the PCR product is identified using one or more molecular markers. This technique helps to know the presence and composition of a particular community in the soil by amplifying the target region of the DNA extracted from mycorrhizal roots. The amplicons are then cloned into suitable vectors and hosts to isolate the individual fragments and identify them by sequencing. A large number of AM fungal taxa have been identified using the clone library analysis technique which were not identified using the spore techniques by direct strategies. Sequence-based metagenomics captures a massive amount of information on the microbial community under study and has revolutionized the quantitative metagenomics. It has become the main strategy to explore and study the AMF diversity in soil as well as roots coupled with RFLP, reduction in sequencing costs and shorter analysis time.

Another approach is the random shotgun sequencing of the metagenome, also called shotgun metagenomics which allows the evaluation of the whole metagenome, and thus the assessment of the whole community structure and gene content. In function-based metagenomics, billions of random DNA fragments of a library are translated into proteins by bacteria that grow in the laboratory. Clones producing “foreign” proteins after gene expression are then screened for various capabilities, such as vitamin production or antibiotic resistance. This helps and enables the researchers to access and explore the vast genetic diversity in any community without having any knowledge about the respective gene sequence, the structure of the desired protein, or the microbe of origin. This technique has facilitated the discovery of many new antibiotics and enzymes. Table 13.1 summarizes the different AMF communities by various molecular methods.

Table 13.1 Summary of few AMF taxa which have been studied by molecular methods

S. no.	Ecosystem type	AM fungus Texa/ species	Methods	References
1	**Field samples, Brazil**	62 tested	Direct analysis of field samples and trap cultures	Leal et al. (2017)
2	Subtropical hilly area of southern China	–	Cloning and sequencing	Caihuan Wang et al. (2015)
3	**Ulleungdo and Dokdo volcanic islands**	–	Pyrosequencing	Nam et al. (2015)
4	Sissle valley (Frick, Switzerland	53 tested	–	Säle et al. (2015)
5	Pampa Ondulada region (Argentina)	29 morphological species of AM fungi tested	Pyrosequencing	Colombo et al. (2014)
6	Wheat fields of the Canadian prairie	33 tested	Pyrosequencing	Dai M et al. (2012)
7	Sardinian soil (Italy)	117 tested	Pyrosequencing	Lumini et al. (2010)
8	British grassland	–	Pyrosequencing	Dumbrell et al. (2011)
9	Wood land United Kingdom	37 tested	Cloning and sequencing	Balestrini et al. (2010)
10	Semiarid Mediterranean areas of Southeastern Spain	9 tested	Cloning and sequencing	Alguacil et al. (2009)
11	Mediterranean semiarid soil of Spain	21 tested	Cloning and sequencing	Alguacil et al. (2009)
12	Serpentine soil of United State of America	19 tested	Cloning and sequencing	Schechter and Bruns (2008)
13	Liverworts worldwide	10 tested	Cloning and sequencing	Ligrone et al. (2007)
14	Volcanic desert in Japan	11 tested	Cloning and sequencing	Wu et al. (2007)
15	**Contaminated site of Northern Italy**	12 tested	Cloning and sequencing	Vallino et al. (2006)
16	**Wetland habitat Germany**	–	Cloning and sequencing	Wirsel (2004)
17	Tropical forest from Panama	18	Cloning and sequencing	Husband et al. (2002)
18	Arable field (UK)	8	Cloning and sequence	Daniell et al. (2001)
19	Semisynthetic wood land (UK)	6/8 [P]	Cloning and sequence	Helgason et al. (1999)
20	Wood land UK	6/10 [N]	Cloning and sequence	Helgason et al. (1998)

Many powerful computational tools like (FastGroup and DOTUR are used to assess species richness in a sample and the similarity between two communities in membership (SONS) or structure (AMOVA, LIBSHUFF, UNIFRAC, and TreeClimber). Most of the communities have uneven abundance of various species (that is, some species are abundant). This presents a sampling issue: how many samples need to be taken to find members of the sparser groups? However, recent estimates based on 16S rRNA sequencing and statistical modeling of soil communities indicate that with decreasing sequencing costs, it is possible to conduct a fairly complete census of soil communities even though these are the most species-rich and uneven in structure of communities studied so far.

11 Metabarcoding and Pyrosequescing Studies

DNA metabarcoding infers to the isolation of eDNA (environmental DNA from soil, air or water) and iDNA (invertebrate DNA) its amplifying, sequencing and analysing target genomic regions. Metabarcoding is a high-throughput screening method of biodiversity assessment and AMF screening. It comprises of two technologies: DNA based identification and high-throughput DNA sequencing. DNA based identification uses universal PCR primers to mass-amplify DNA. Figure 13.4 depicts the working strategy of metabarcoding and metagenomic approach for high-throughput screening of Arbuscular mycorrhizal fungi. DNA-based species identification changed the long-established approach to the study of biodiversity science (Cristescu 2014). Arbuscular mycorrhizal fungi (AMF) are currently studied by

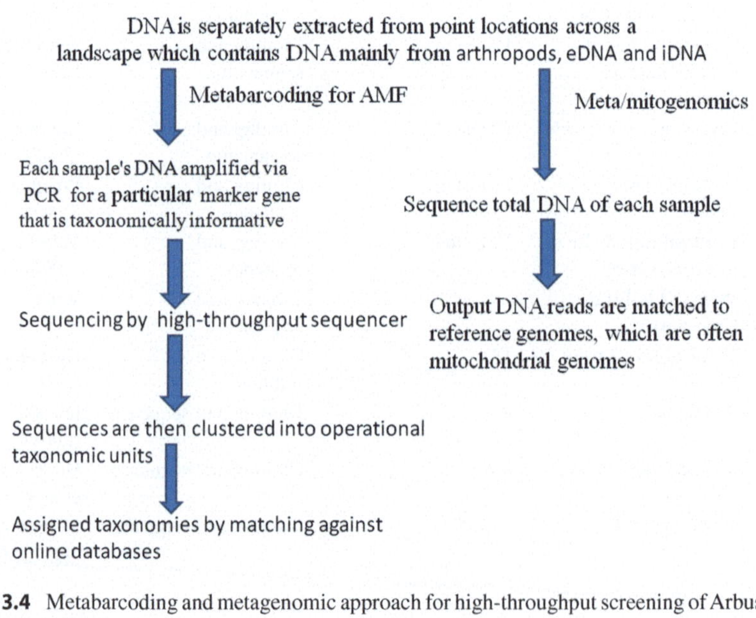

Fig. 13.4 Metabarcoding and metagenomic approach for high-throughput screening of Arbuscular mycorrhizal fungi

using Next Generation Sequencing (NGS) approaches. Buée et al. (2009) has discovered the diverse taxa of fungus in forest soils using 454 pyrosequencing. Parallel 454 sequencing has also been used to study hyperdiverse fungal communities in temperate (Jumpponen and Jones 2009). Pyrosequencing, was used to study the extensive fungal communities in soils of three islands in the Yellow Sea of Korea, between Korea and China (Lim et al. 2010). Opik et al. (2009) suggested that partner specificity in arbuscular mycorrhiza symbiosis may occur at the point of ecological groups, rather than at the species level, of both plant and fungal partners. Assessment of AMF diversity in boreonemoral forest has also been done by using 454 sequencing (Öpik et al. 2009). Pyrosequencing based approach was used to study the biodiversity of AMF communities present in five Sardinian soils (Italy) subjected to different land-use (tilled vineyard, covered vineyard, pasture, managed meadow and cork-oak formation) (Lumini et al. 2010).

12 Conclusion

Arbuscular mycorrhiza (AM) is the predominant and ancestral type of mycorrhiza in land plants. Its occurrence in a vast majority of land plants and early-diverging lineages of liverworts suggests that the origin of AM probably coincided with the origin of land plants. Molecular techniques have been very useful to study and explore the structure, function, abundance of AMF and its relationship with its environment. Metagenomics has paved the path for the identification and study of uncultivable microrganisms. Metagenomics along with metabarcoding and bioinformatics will surely help to explore the mycorrhizal fungal communities more accurately and quickly.

Acknowledgements PB thanks DST-SERB: SB/YS/LS-213/2013 for the financial support.

References

Alguacil, M. M., Roldán, A., & Torres, M. P. (2009). Complexity of semiarid gypsophilous shrub communities mediates the AMF biodiversity at the plant species level. *Microbial Ecology, 57*(4), 718–727.

Balestrini, R., Magurno, F., Walker, C., Lumini, E., & Bianciotto, V. (2010). Cohorts of arbuscular mycorrhizal fungi (AMF) in Vitis vinifera, a typical Mediterranean fruit crop. *Environmental Microbiology Reports, 2*(4), 594–604.

Bhargava, P., Gupta, N., Vats, S., & Goel, R. (2017a). Health issues and heavy metals. *Austin Journal of Environmental Toxicology, 3*(1), 3018.

Bhargava, P., Singh, A. K., & Goel, R. (2017b). Microbes: Bioresource in agriculture and environmental sustainability. In *Plant-microbe interactions in agro-ecological perspectives* (pp. 361–376). Singapore: Springer.

Buée, M., Reich, M., Murat, C., Morin, E., Nilsson, R. H., Uroz, S., & Martin, F. (2009). 454 pyrosequencing analyses of forest soils reveal an unexpectedly high fungal diversity. *New Phytologist, 184*(2), 449–456.

Colombo, R. P., Fernandez Bidondo, L., Silvani, V. A., Carbonetto, M. B., Rascovan, N., Bompadre, M. J., & Godeas, A. M. (2014). Diversity of arbuscular mycorrhizal fungi in soil from the Pampa Ondulada, Argentina, assessed by pyrosequencing and morphological techniques. *Canadian Journal of Microbiology, 60*(12), 819–827.

Cristescu, M. E. (2014). From barcoding single individuals to metabarcoding biological communities: Towards an integrative approach to the study of global biodiversity. *Trends in Ecology & Evolution, 29*(10), 566–571.

Dai, M., Hamel, C., St. Arnaud, M., He, Y., Grant, C., Lupwayi, N., & Zhou, Z. (2012). Arbuscular mycorrhizal fungi assemblages in Chernozem great groups revealed by massively parallel pyrosequencing. *Canadian Journal of Microbiology, 58*(1), 81–92.

Daniell, T. J., Husband, R., Fitter, A. H., & Young, J. P. W. (2001). Molecular diversity of arbuscularmycorrhizal fungi colonising arable crops. *FEMS Microbiology Ecology, 36*(2–3), 203–209.

Douhan, G. W., Vincenot, L., Gryta, H., & Selosse, M. A. (2011). Population genetics of ectomycorrhizal fungi: From current knowledge to emerging directions. *Fungal Biology, 115*(7), 569–597.

Dumbrell, A. J., Ashton, P. D., Aziz, N., Feng, G., Nelson, M., Dytham, C., & Helgason, T. (2011). Distinct seasonal assemblages of arbuscular mycorrhizal fungi revealed by massively parallel pyrosequencing. *New Phytologist, 190*(3), 794–804.

Gai, J. P., Christie, P., Feng, G., & Li, X. L. (2006). Twenty years of research on community composition and species distribution of arbuscular mycorrhizal fungi in China: A review. *Mycorrhiza, 16*(4), 229–239.

Grant, S., Grant, W. D., Cowan, D. A., Jones, B. E., Ma, Y., Ventosa, A., & Heaphy, S. (2006). Identification of eukaryotic open reading frames in metagenomic cDNA libraries made from environmental samples. *Applied and Environmental Microbiology, 72*(1), 135–143.

Helgason, T., Daniell, T. J., Husband, R., Fitter, A. H., & Young, J. P. W. (1998). Ploughing up the wood-wide web? *Nature, 394*(6692), 431.

Helgason, T., Fitter, A. H., & Young, J. P. W. (1999). Molecular diversity of arbuscularmycorrhizal fungi colonisingHyacinthoides non-scripta (bluebell) in a seminatural woodland. *Molecular Ecology, 8*(4), 659–666.

Hollister, E. B., Schadt, C. W., Palumbo, A. V., Ansley, R. J., & Boutton, T. W. (2010). Structural and functional diversity of soil bacterial and fungal communities following woody plant encroachment in the southern Great Plains. *Soil Biology and Biochemistry, 42*(10), 1816–1824.

Husband, R., Herre, E. A., Turner, S. L., Gallery, R., & Young, J. P. (2002). Molecular diversity of arbuscular mycorrhizal fungi and patterns of host association over time and space in a tropical forest. *Molecular Ecology, 11*(12), 2669–2678.

Jumpponen, A., & Jones, K. L. (2009). Massively parallel 454 sequencing indicates hyperdiverse fungal communities in temperate Quercus macrocarpa phyllosphere. *New Phytologist, 184*(2), 438–448.

Kernaghan, G. (2005). Mycorrhizal diversity: Cause and effect? *Pedobiologia, 49*(6), 511–520.

Leal, P. L., Carvalho, T. S., Siqueira, J. O., & Moreira, F. (2017). Assessment of the occurrence and richness of arbuscular mycorrhizal fungal spores by direct analysis of field samples and trap culture-a comparative study. Anais da Academia Brasileira de Ciências, (AHEAD), 0–0.

Lennon, J. T. (2011). Replication, lies and lesser-known truths regarding experimental design in environmental microbiology. *Environmental Microbiology, 13*(6), 1383–1386.

Ligrone, R., Carafa, A., Lumini, E., Bianciotto, V., Bonfante, P., & Duckett, J. G. (2007). Glomeromycotean associations in liverworts: A molecular, cellular, and taxonomic analysis. *American Journal of Botany, 94*(11), 1756–1777.

Lim, Y. W., Kim, B. K., Kim, C., Jung, H. S., Kim, B. S., Lee, J. H., & Chun, J. (2010). Assessment of soil fungal communities using pyrosequencing. *The Journal of Microbiology, 48*(3), 284–289.

Lindahl, B. D., De Boer, W., & Finlay, R. D. (2010). Disruption of root carbon transport into forest humus stimulates fungal opportunists at the expense of mycorrhizal fungi. *The ISME Journal, 4*(7), 872–881.

Lumini, E., Orgiazzi, A., Borriello, R., Bonfante, P., & Bianciotto, V. (2010). Disclosing arbuscular mycorrhizal fungal biodiversity in soil through a land-use gradient using a pyrosequencing approach. *Environmental Microbiology, 12*(8), 2165–2179.

Mathimaran, N., Ruh, R., Vullioud, P., Frossard, E., & Jansa, J. (2005). Glomus intraradices dominates arbuscular mycorrhizal communities in a heavy textured agricultural soil. *Mycorrhiza, 16*(1), 61–66.

Nam, Y. J., Kim, H., Lee, J. H., Yoon, H., & Kim, J. G. (2015). Metagenomic analysis of soil fungal communities on Ulleungdo and Dokdo Islands. *The Journal of General and Applied Microbiology, 61*(3), 67–74.

Öpik, M., Metsis, M., Daniell, T. J., Zobel, M., & Moora, M. (2009). Large-scale parallel 454 sequencing reveals host ecological group specificity of arbuscular mycorrhizal fungi in a boreonemoral forest. *New Phytologist, 184*(2), 424–437.

Parniske, M. (2008). Arbuscular mycorrhiza: The mother of plant root endosymbioses. *Nature Reviews Microbiology, 6*(10), 763–775.

Peay, K. G., Baraloto, C., & Fine, P. V. (2013). Strong coupling of plant and fungal community structure across western Amazonian rainforests. *The ISME Journal, 7*(9), 1852–1861.

Prosser, J. I. (2010). Replicate or lie. *Environmental Microbiology, 12*(7), 1806–1810.

Read, D. J. (1996). The structure and function of the ericoid mycorrhizal root. *Annals of Botany, 77*(4), 365–374.

Reigstad, L. J., Bartossek, R., & Schleper, C. (2011). Preparation of high-molecular weight DNA and metagenomic libraries from soils and hot springs. In *Methods in enzymology* (Vol. 496, pp. 319–344). San Diego: Academic.

Säle, V., Aguilera, P., Laczko, E., Mäder, P., Berner, A., Zihlmann, U., & Oehl, F. (2015). Impact of conservation tillage and organic farming on the diversity of arbuscular mycorrhizal fungi. *Soil Biology and Biochemistry, 84*, 38–52.

Schechter, S. P., & Bruns, T. D. (2008). Serpentine and non-serpentine ecotypes of Collinsia sparsiflora associate with distinct arbuscular mycorrhizal fungal assemblages. *Molecular Ecology, 17*(13), 3198–3210.

Vallino, M., Massa, N., Lumini, E., Bianciotto, V., Berta, G., & Bonfante, P. (2006). Assessment of arbuscular mycorrhizal fungal diversity in roots of Solidago gigantea growing in a polluted soil in Northern Italy. *Environmental Microbiology, 8*(6), 971–983.

Wang, B., & Qiu, Y. L. (2006). Phylogenetic distribution and evolution of mycorrhizas in land plants. *Mycorrhiza, 16*(5), 299–363.

Wang, C., Gu, Z., Cui, H., Zhu, H., Fu, S., & Yao, Q. (2015). Differences in arbuscular mycorrhizal fungal community composition in soils of three land use types in subtropical hilly area of southern China. *PLoS One, 10*(6), e0130983.

Wirsel, S. G. (2004). Homogenous stands of a wetland grass harbour diverse consortia of arbuscular mycorrhizal fungi. *FEMS Microbiology Ecology, 48*(2), 129–138.

Wu, B., Hogetsu, T., Isobe, K., & Ishii, R. (2007). Community structure of arbuscular mycorrhizal fungi in a primary successional volcanic desert on the southeast slope of Mount Fuji. *Mycorrhiza, 17*, 495–506.

Zak, D. R., Holmes, W. E., White, D. C., Peacock, A. D., & Tilman, D. (2003). Plant diversity, soil microbial communities, and ecosystem function: Are there any links? *Ecology, 84*(8), 2042–2050.

Biocontrol of Soil Phytopathogens by Arbuscular Mycorrhiza – A Review

14

Pranay Jain and Ram Kumar Pundir

1 Introduction

The use of benign microbes as control mechanism so called 'Biocontrol' to kill phytopathogens has been extensively studied wherein biocontrol implies to likely enemies of pests or pathogens to eradicate or control their population. It involves the prologue of foreign species that exists as expected in the environment. It has been considered as environmentally safe and the easier option accessible to protect plants against detrimental flora and fauna (Azcon-Aguilar and Barea 1992). Arbuscular mycorrhizal fungi (AMF) are organisms that have been used as biocontrol agents of plants. Mycorrhizae are ubiquitous soil-borne fungi and serve as prospective tools for sustainable agriculture. Mycorrhizae are generally associated with most terrestrial vascular plant species worldwide (Srnith and Read 2008: Brundrett 2009), being beneficial in improving plant growth and development (Jeffries et al. 2003). They belong to the Glomeromycota phylum (Schübler et al. 2001) and originated approximately 450my ago (Schübler and Walker 2011). They improve the growth of plant-root system and control plant pathogens (Gianinazzi and Schuepp 1994). Arbuscular mycorrhizal (AM) fungi influence plant augmentation and improvement. Their interactions with rhizosphere microorganisms influence the overall development of plants (Azcon-Aguilar and Barea 1992; Fitter and Sanders 1992). A harmful involvement between the host plant and the indigenous mycrorrhizal fungi leads to solemn fatalities in crop yields, which indicate the connotation of AMF in agriculture (Caron 1989; Ravnskov and Jakobsen 1995; St-Arnaud et al. 1995; Frankenberger and Arshad 1995).

P. Jain (✉)
Faculty in Department of Biotechnology, University Institute of Engineering and Technology (UIET), Kurukshetra University, Kurukshetra, Haryana, India

R. K. Pundir
Faculty in Department of Biotechnology Engineering, Ambala College of Engineering and Applied Research (ACE), Ambala, Haryana, India

© Springer Nature Singapore Pte Ltd. 2019

A. Varma, D. K. Choudhary (eds.), *Mycorrhizosphere and Pedogenesis*,
https://doi.org/10.1007/978-981-13-6480-8_14

Mycorrhizal fungi are the most influential group of soil microflora with reference to sustainability of ecosystem, once they establish mutualistic relationship with plants (Jeffries and Barea 2012). The rhizosphere is characterised by improved microbial activity owed to the root exudates (Grayston et al. 1997). Mycorrhizosphere include the fungal component of the symbiosis while plant roots in normal and semi-normal ecosystems are found to have mycorrhizal relations (Rambelli 1973).

The mycorrhizosphere is the area surrounding mycorrhizal fungus where the nutrients on the rampage as of the fungus raise the microbial actions (Linderman 1988).The mycorrhizosphere effect indicates the provoked changes in the plant biochemistry as a result of mycorrhizal-root immigration which causes a shift in the rhizosphere microflora that favours the absence or presence of pathogens (Paulitz and Linderman 1989). The mycorrhizosphere effect causes changes in root exudate composition mainly because of root membrane permeability. The root colonization with AM fungi has been shown to suppress harmful effects of fungi, stramenopiles, nematodes and bacteria ((Graham et al. 1981).

2 The Arbuscular Mycorrhizal Fungi

Mycorrhizal associations diverge broadly in structures and functionalities, but the AM are the most common interactions (Harrier 2001). These fungi are nonculturable and are obligate biotrophs, in view of the fact that these fungi can not inclusive their life cycle devoid of congregate a host. The study of these fungi and their biology and biotechnological applications has been hampered because of nonculturability (Schübler and Walker 2011; Barea et al. 2013). Six genera of AM fungi have been recognized based on phenetic characteristics of sexual spores and also based on various biochemical studies and molecular methods (Peterson et al. 2004). Various biochemical, molecular and immunological characteristics criteria employed for identification of AM fungi (Mukerji et al. 2002). AM fungi include genera such as *Glomus, Gigaspora, Sclerocystis, Acaulospora, Entrophospora* and *Scutellospora* (Garbaye 1994).

3 The AM Symbiosis

Srnith and Read (2008) reported that AM symbiosis is the most numerous type of mycorrhizal relationship wherein worldwide approximately 250 k species of plant, including angiosperms, petersengymnosperms and pteridophytes, tend to form such association. Herein, AM symbiosis initiates with the fungal infiltration in the root cortical cell walls followed by configuration of arbuscules -like structures (haustoria or coils) that interface with the host cytoplasm. These fungal structures help to augment exterior area for swap over of metabolites flanked by the plant and the fungus. Several mycorrhizal fungi are recognized to construct vesicles for storage. It has been revealed that in natural ecosystems plants colonised with mycorrhizal fungi may incur 10–20% of the photosynthetically fixed carbon for their fungal symbionts (Johnson et al. 2002a, b).

The mycorrhizal fungi interact directly with the soil by producing extraradical hyphae that extend deep into the soil (Rhodes and Gerdemann 1975). Extra-radical hyphae raise the potential for nutrient and water uptake (Augé 2001). Hyphae of AM fungi form soil aggregates which play an important role in soil stabilisation (Tisdall and Oades 1979). The extraradical hyphae are responsible for acquisition of phosphorus and other mineral nutrients by plants (Read and Perez-Moreno 2003). These hyphae also improve mobilisation of organically bound nitrogen from plant litter (Hodge et al. 2001). Mycorrhizal fungi also alleviate negative effects of plant pathogens and toxic metals (Khan et al. 2000). The extraradical hyphae interact with other soil organisms either directly by physically and/or metabolically interacting with other organisms in the mycorrhizosphere or indirectly by changing host plant physiology. Extra-radicular hyphae are surrounded by complex microbial communities that interact with the plant-mycorrhiza and sustain this relationship (Frey-Klett and Garbaye 2005). Thus, the establishment of the arbuscular mycorrhizal symbiosis affects the structure and diversity of microorganisms not only in the rhizosphere but also in other soil microhabitats.

4 Establishment of Arbuscular Mycorrhiza Fungi

Most vascular plants have exhibited mycorrhizal associations in both natural and agro-ecosystems (van der Heijden et al. 2015; Brundrett 2009; Jeffries and Barea 2012; Bonfante and Desirò 2015; López-Ráez et al. 2011a, b; Maillet et al. 2011). Gutjahr and Parniske 2013; Bonfante and Desirò 2015). Upon root colonization, the extraradical mycelium (ERM) is formed which is frequently considered as "branching absorbing structures", (Bago et al. 1998). It is able to absorb and transport nutrients up to 25 cm distance (Jansa et al. 2003; Smith and Smith 2012).

5 Biocontrol of Phytopathogenic Fungi by AM Fungi

Phytopathogenic fungi contribute substantially to the overall loss in crop yield followed by plant pathogenic bacteria and viruses. The control of phytopathogens has always been practiced by agrochemical application, which are applied at various sites of plants. However, the constant use of such chemicals results in negative effects on the environment that affects water bodies, soil, plants, animals and human health (Bodker et al. 2002). Phytopathogenic microorganisms also develop resistance against these agrochemicals with the passage of time which makes it more difficult to control. Therefore, biological control as part of integrate pathogen management has been regarded as the most sustainable and a viable alternative to the indiscriminate use of agrochemicals.

The convenience of AM Fungi as biocontrol for controlling various phytopathigenic fungi has been widely accepted (Cordier et al. 1996; Bodker et al. 2002;

Harrier and Watson 2004; Azcon-Aguilar et al. 2002; Jaizme-Vega et al. 1998; Li et al. 1997; Pozo et al. 1999; Kulkarni et al. 1997), Prashanthi et al. (1997; Sharma et al. 1997). Feldmann and Boyle (1998) suggested that the crop loss due to phytopathogenic fungi could be reduced by an aggressively root colonizing AM Fungi. They observed an inverse relationship between *G. etunicatum* root colonization of begonia species and susceptibility to the powdery mildew fungus *Erysiphe cichoracearum.* Filion et al. (1999) found that extraradical mycelium of *G. intraradices* reduced the growth of *F. oxysporum* f. sp. *chrysanthemi*. They suggested that the chemical equilibrium of the mycorrhizosphere resulted in control of pathogen. In another study, Slezack et al. (2000) challenged pea with *Aphanomyces euteiches* and found that a fully established AMF symbiosis essential for protection against the pathogen. *Phytophthora* spp., have been commonly used as model fungi for demonstrating AMF-mediated plant disease control (Trotta et al. 1996). Caron and co-workers (1985) in their studies used the AMF species *G. intraradices* and pathogen *F. oxysporum* f. sp. *lycopersici* on tomato, and revealed that the combination of growth medium used, the application of Phosphorus and pretreatment with AM fungi could reduce disease severity. Newsham et al. (1995) reported that on pre-inoculating the annual grass *Vulpia ciliata* var. *ambigua* with *Glomus* sp. and re-introducing the grass into a natural grass population, there was a reduction in indigenous *F. oxysporum.* Torres-Barragan et al. (1996) in their study found that onion pretreated with *Glomus* sp. delayed the onset of onion white rot caused by *Sclerotium cepivorum* by two weeks in the field.

Hwang (1988) carried out a detailed study on interactions of mycorrhizal fungi and two wilt pathogens of alfalfa (*Medicago sativa*), *Verticillium albo-atrum* and *Fusarium oxysporum* f. sp. *medicaginis*, under controlled conditions over a 6-month period.

6 Biocontrol of Phytopathogenic Bacteria by AM Fungi

The AM fungi have been found to interact with diazotrophic bacteria, biological control agents, and other rhizosphere inhabitants (Nemec 1994) that often result insignificant alterations to plant growth and development. Filion et al. (1999) and Shalaby and Hanna (1998) suggested that interactions between mycorrhizal fungi and bacteria may have negative or beneficial effects or have neutral effect at all on the plant pathogens. Sharma et al. (1995) found that on inoculation of mulberry with *Glomus fasciculatum* or *G. mosseae* in combination with phosphorus the incidence of bacterial blight caused by *P. syringae* pv. *mori* was found to significantly reduce. In a study by Shalaby and Hanna (1998), it was found that *Glomus mosseae* prevented the infection of soybean plants by *P. syringae* by suppressing pathogen population in soybean. Li et al. (1997) also found in their study that *G. macrocarpum* alleviated the infection caused by *P. lacrymans*in eggplant and cucumber. Waschkies et al. (1994) reported that on AMF inoculation of grapevines, the fluorescent pseudomonads on the rhizoplane were reduced which in turn reduced the incidence of grapevine replant disease. Similarly, root colonization by AMF caused a reduction in the colonization of apple seedling rootlets by actinomycetes causing replant disease (Otto and Winkler 1995).

7 Biocontrol of Phytopathogenic Viruses by AM Fungi

Mycorrhizae-mediated biocontrol of plant pathogenic viruses has been least studied. Earlier, Nemec and Myhre (1984) demonstrated that mycorrhizal plants increase the rate of multiplication of viruses, increased leaf lesions are found on mycorrhizal plants than on nonmycorrhizal plants and the number of AMF spores in the rhizosphere are reduced considerably. (Shaul et al. 1999). Schonbeck and Spengler (1979) reported that mycorrhizal tobacco plants (*Nicotiana glutinosa* L.) exhibited higher levels of tobacco mosaic virus colonization following the inoculation of mycorrhizal as compared to nonmycorrhizal tobacco. Contrary, Ferraz and Brown (2002) reported that mung bean yellow mosaic bigeminivirus reduced the AM colonization and yield of mycorrhizal plants, while Takahashi et al. (1994) reported lack of response to viral infection by a mycorrhizal host. Jabaji-Hare and Stobbs (1984) used electron microscopy to observe interaction of AMF with plant viruses.

8 Biocontrol of Plant-Parasitic Nematodes by AM Fungi

Many species of plant-parasitic nematodes could be potential pests on agricultural crops (Ferraz and Brown (2002). They are frequently found in the soil, but *Ditylenchus* spp. could act as aboveground pests and classified based on their feeding patterns (Perry and Moens 2011). The AMF has been deployed as biocontrol agents for nematodes (Jones et al. 2013; Gheysen and Mitchum 2011; Wesemael et al. 2011; Hao et al. 2012; Nicol et al. 2011; Alban et al. 2013; Salvioli and Bonfante 2013; Salvioli and Bonfante 2013).

9 Mechanisms of Mycorrhizae-Mediated Biocontrol

9.1 Higher Nutrient Uptake

The AMF has been suggested to improve phosphorus nutrition, enhance nitrogen uptake, or improve disease resistance in their host plants (Baum et al. 2015; Smith and Smith 2011; Gianinazzi et al. 2010; Singh et al. 2011; Fritz et al. 2006; Smith and Smith 2011). Nitrogen fixing bacteria or Phosphate solubilising bacteria have been found to synergistically interact with AM fungi and benefit plant development and growth (Puppi et al. 1994). Hodge et al. (2001) demonstrated the improved decay of plant litter in soil and N capture from the litter (^{15}N–^{13}C labelled *Lolium perenne* leaves) in the presence of the AM symbiont *Glomus*. Minerdi et al. (2001)

reported the presence of genes for Nitrogen fixation in endosymbiotic *Burkholderia* in AM which makes it apparent that there may be a potential for enhanced nitrogen supply to mycorrhizal plants all the way through fixation of atmospheric Nitrogen.

9.2 Altered Root Morphology

The AMF symbiotic plants often show increased root growth and branching (Gutjahr and Paszkowski 2013). The root morphology responses resulting from AMF colonization depend on plant characteristics, with tap roots profit more from AM fungi than fibrous roots in terms of gained biomass and nutrient acquisition (Yang et al. 2014). Increased root branching observed in mycorrhizal plants have implications for pathogen infection as well (Vos et al. 2014). The mycorrhizal fungi increase host tolerance of pathogen attack by compensating for the loss of root biomass and functions caused by soilborne pathogenic fungi and nematodes which could be an indirect contribution to the biological control through the conservation of root system function through mycorrhizal arbuscules formation (Linderman 1994; Stoffelen et al. 2000; Norman et al. 1996; Elsen et al. 2003)

9.3 Competition for Nutrients and Space

The basis for interface between AMF and soil microorganisms is largely the physical opposition between mycorrhizal fungi and rhizosphere microorganisms to occupy more space in the roots (Bansal and Mukerji 1996). The pathogen suppression in mycorrhizal plants is mainly due to the competition for nutrients such as carbon by mycorrhiza fungi and rhizosphere soil microorganisms with the same physiological requirements (Jung et al. 2012; Vos et al. 2014). Hammer et al. (2011) stated that there is 4–20% carbon transfer of the total assimilated carbon from the host plant to the AMF. Cordier et al. (1998) reported that *Phytophthora* could not penetrate in arbuscule containing tomato plant. Lerat et al. (2003) reported that different AMF species mediate different levels of biocontrol as there is a difference in carbon sink strength between different AMF species. Vos (2012) reported that the AM fungus *Rhizophagus irregularis* was not having a stronger biocontrol effect on plant parasitic nematodes *Rhadopholus similis* and *Pratylenchus coffeae* in banana nor on *Meloidogyne incognita* in tomato despite its higher carbon sink strength compared to *Funneliformis mosseae*.

9.4 Systemic Induced Resistance

From the biocontrol point of view AMF has been used to develop systemic induced resistance (SIR) in plants (Trotta et al. 1996; Cordier et al. 1998). The SIR is defined as the unrelenting induction of resistance or tolerance to infection in plants by inoculating with a pathogen, exposing to an environmental influence or treating with a chemical, with or without antimicrobial activity (Handelsman and Stabb 1996). Jones and Dangl (2006) and Zamioudis and Pieterse (2012) demonstrated that the disease resistance by AMF is mainly due to action of MAMPs. Bodker et al. (1998) reported SIR factor by *G. intraradices* in pea plant for resistance to *A. euteiches*. The AMF-mediated SIR protected potatoes against post-harvest suppression of potato dry rot, wherein dry rot in *G. intraradix*-inoculated potato was reduced by up to 90% compared to uninoculated control (Brendan et al. 1996).

9.5 Altered Rhizosphere Interactions

The AMF symbiosis leads to an changed root exudation composition and distribution in host plants rhizosphere (Jones et al. 2004; Hage-Ahmed et al. 2013; Harrier and Watson 2004; McArthur and Knowles 1992; Steinkellner et al. 2007; López-Ráez et al. 2011a, b. The root exudation may or may not be AMF specific (Kobra et al. 2009; Lioussanne et al. 2008). It helps in autoregulation of symbiosis interaction between plant and AMF (Schaarschmidt et al. 2013; Vierheilig et al. 2008; Pozo and Azcón-Aguilar 2007). Lioussanne et al. (2008) observed that the depending on the maturity level of the AM fungi colonization the attraction of *Phytophthora nicotianae* zoospores toward *R. irregularis* colonized root exudates changed to repellency. The bacterial colonization in rhizosphere induced by AMF reported in recent period (Nuccio et al. 2013; Philippot et al. 2013; Zamioudis and Pieterse 2012; Sood 2003; Druzhinina et al. 2011; Sikora et al. 2008).

9.6 Phytoalexins and Phytoanticipins

Under response to pathogen attack plants produce phytoalexins which are natural products and exhibited antagonistic activity against microflora and –fauna and plant *per se*. They are lipophilic in natures that have the ability to cross the plasma membrane and act inside the cell (Braga et al. 1991). Based on earlier researches it has been demonstrated that phytoalexins are produced in response to microbial infection (Paxton 1981; Wyss et al. 1991) whereas phytoanticipins considered as the storage products in plant cells that produced in anticipation of or prior to pathogen attack (VanEtten et al. 1995). Upon mycorrhiza fungal colonization of roots there is an increase in the level of lignin, syringic, ferulic or coumaric acids and phenolics namely, isoflavonoids or flavonoids (Morandi 1996).

As a result of pathogen invasion (*F. oxysporum*), Dehne and Schonbeck (1979) explored the phytoalexins synthesis in mycorrhizal tomato plants where the plants were inoculated with *G. mosseae*. Upon treatment it has been reported that plants showed greater resistance to the *F. oxysporum* which lead to enhanced phenylalanine and beta-glucosidase activity along with total phenol content in their roots compared to control plants (Dehne and Schonbeck (1979). Sundaresan et al. (1993) reported *in vitro* inhibition of *F. oxysporum* by a purified ethanol root fraction of mycorrhizal cowpea. Caron et al. 1986 recommended that phytoalexins neutralize the antagonistic effects of pathogens in mycorrhizal plants as compared to control

9.7 Hydrolases

The AMF mediated biocontrol has explored the subsistence of defense-related genes in mycorrhizal plants (Lambais and Mehdy 1995). Pozo et al. (2002) reported that entry of mycorrhizal fungi into tomato roots induced fabrication of hydrolytic enzymes such as chitinase, chitosanase, b-glucanase, and superoxide dismutase to host defense mechanism against *Pseudomonas parasitica.* Graham and Graham (1991) reported constructive relationship between the level of glucanase activity in host tissues and resistance to phytopathogens.

9.8 Antibiosis

Earlier it has been reported that under non influential impact of pH the AMF namely, *G. intraradices* produced unidentified antimicrobial substance that helps in control of conidial germination of *F. oxysporum* f. sp. *Chrysanthemi* (Filion et al. 1999). Likewise, Budi et al. (1999) recovered a bacterium viz., *Paenibacillus* sp. strain from the rhizosphere of *Sorghum bicolor* plants inoculated with *G. mosseae* that showed noteworthy inhibitory activity against *Phytophthora parasitica.*

10 Challenges and Future Perspectives in AM Fungi Mediated Biocontrol

The worth of AMF for controlling of phytopathogens usually measured to be high, but there are restrictions in use of AMF as biological agents for control of phytodiseases under field conditions. Budi et al. (1999) reported that there are few important consideration to deploy AMF in the field that include firstly, the production of large quantities of AMF quorum and secondly, occurrence of negative interactions between the introduced AMF and the indigenous AMF and microbial community. A host which is greatly mycotrophic or host cultivar may be considered as more appropriate for AMF propagation and imitation than one that is not highly mycotrophic (Bever et al. 1996; Xavier 1999). High soil Phosphorous levels also affect AM fungal colonization in host plants (Ratnayake et al. 1978; Bever et al. 1996).Bever et al. (1996) demonstrated that abundantly diverse AMF community ensures efficient biocontrol of phytopathogens. The diversity of AMF in soils has affected by the preference of host genotype and rotation, levels of fertilizer deployment

(McGonigle and Miller 1996), tillage (McGonigle and Miller 1993), pesticide submission (Schreiner and Bethlenfalvay 1997), and the effect of associated quorum of microflora (Xavier and Germida 2003). Further, Johnson et al. (1992) emphasized that continuous cropping selectively improves the proliferation of AMF that lead to alterations in mycorrhizal biodiversity in the rhizosphere. Likewise, Xavier (1999) observed that use of sole meticulous AMF host out of an indigenous AMF residents resulting in the selective fortification of certain AMF species above others.

The approaches involve AM fungi have been deployed as biocontrol of phytopathogens. Sikora (1997) has been anticipated a holistic approach "natural system administration that derived biologically" for humanizing plant roots that adopts specific cropping patterns that uphold plant protection mechanisms such as tolerance and/or resistance to phytopathogens. This has been considered as practicable substitute to integrated pest administration and inundative approaches to the nonrhizospheric soil for biological control purposes, and underlines the implication of mycorrhizae in root growth and development. In addition, Bagyaraj (1984) recommended that assortment of AMF species for a preferred activity must be based on their capability for continued existence, forceful colonization of host roots and efficacy. It has been shown that AMF species originally recovered from test host roots are benign for numerous plant species (Vinayak and Bagyaraj 1990). It has been observed that inoculating plants with AM fungi induce resistance in plants. Cordier et al. (1998) pointed out that "priming" plants against phytopathogens by AMF inocula helps in protection of plants by employing systemic induced resistance. Herein, the inoculum wants to be functional to plantlets fashioned all the way through tissue culture technique. Boyetchko (1996) reported that an appliance of the bioagent prior to transplanting eliminates the requirement for composite formulations and relevance techniques then exhibits greater biocontrol commotion, reduces costs whereby reflects environment-friendly approach.

11 Conclusions

The AMF not only act as biocontrol of phytopathogens caused by detrimental flora and fauna, it also enhances crop efficiency using offered assets, avoiding battle development to chemicals and maintaining effluence conforming to sustenance of agroecosystem. It is speculated that in the near future, task of AM fungi must become one of the practicable and ecosystem friendly solutions to supervise plant diseases and reducing pathogen occurrence and quorum.

Acknowledgement The authors are grateful to the Director, University Institute of Engineering and Technology, Kurukshetra University, Kurukshetra (Haryana, India) and Management of Ambala College of Engineering and Applied Research, Ambala (Haryana, India), for providing necessary infrastructure to carry out literature search.

References

Alban, R., Guerrero, R., & Toro, M. (2013). Interactions between a root knot nematode (*Meloidogyne exigua*) and arbuscular mycorrhizae in coffee plant development (*Coffea arabica*). *American Journal of Plant Sciences, 4*, 19–23.

Augé, R. M. (2001). Water relations, drought and vesicular-arbuscular mycorrhizal symbiosis. *Mycorrhiza, 11*, 3–42.

Azcon-Aguilar, C., & Barea, J. M. (1992). Interactions between mycorrhizal fungi and other rhizosphere microorganisms. In M. F. Allen (Ed.), *Mycorrhizal functioning: An integrative plant-fungal process* (pp. 163–198). New York: Chapman and Hall.

Azcon-Aguilar, C., Jaizme-Vega, M. C., & Calvet, C. (2002). The contribution of arbuscular mycorrhizal fungi for bioremediation. In S. Gianinazzi, H. Schuepp, J. M. Barea, & K. Haselwandter (Eds.), *Mycorrhizal technology in agriculture. From genes to bioproducts* (pp. 187–197). Berlin: Birkhauser Verlag.

Bago, B., Azcón-Aguilar, C., Goulet, A., & Piché, Y. (1998). Branched adsorbing structure (BAS): a feature of the extraradical mycelium of symbiotic arbuscular mycorrhizal fungi. *The New Phytologist, 139*, 375–388.

Bagyaraj, D. J. (1984). Biological interactions with VA mycorrhizal fungi. In C. L. Powell & D. J. Bagyaraj (Eds.), *VA Mycorrhiza* (pp. 131–153). Florida: CRC Press.

Bansal, M., & Mukerji, K. G. (1996). Root exudates and its rhizosphere biology. In K. G. Mukerji, V. P. Singh, & S. Dwivedi (Eds.), *Concepts in applied microbiology and biotechnology* (pp. 79–119). New Delhi: Aditya Books Pvt Ltd.

Barea, J. M., Pozo, M. J., Azcón, R., & Azcón-Aguilar, C. (2013). Microbial interactions in the rhizosphere. In F. de Bruijn (Ed.), *Molecular microbial ecology of the rhizosphere* (pp. 29–44). Hoboken: Wiley-Blackwell.

Baum, C., El-Tohamy, W., & Gruda, N. (2015). Increasing the productivity and product quality of vegetable crops using arbuscular mycorrhizal fungi: A review. *Scientia Horticulturae (Amsterdam), 187*, 131–141.

Bever, J. D., Morton, J. B., Antonovics, J., & Schultz, P. A. (1996). Host dependent sporulation and species diversity of arbuscular mycorrhizal fungi in a mown grassland. *Journal of Ecology, 84*, 71–82.

Bodker, L., Kjoller, R., & Rosendahl, S. (1998). Effect of phosphate and the arbuscular mycorrhizal fungus *Glomus intraradices* on disease severity of root rot of peas (*Pisum sativum*) caused by *Aphanomyces euteiches*. *Mycorrhiza, 8*, 169–174.

Bodker, L., Kjoller, R., Kristensen, K., & Rosendahl, S. (2002). Interactions between indigenous arbuscular mycorrhizal fungi and Aphanomyces euteiches in field-grown pea. *Mycorrhiza, 12*, 7–12.

Bonfante, P., & Desirò, A. (2015). Arbuscular mycorrhizas: The lives of beneficial fungi and their plant host. In B. Lugtenberg (Ed.), *Principles of plant-microbe interactions* (pp. 235–245). Cham: Springer.

Boyetchko, S. M. (1996). Impact of soil microorganisms on weed biology and ecology. *Phytoprotection, 77*, 41–56.

Braga, M. R., et al. (1991). Phytoalexins induction in Rubiacea. *Journal of Chemical Ecology, 17*, 1079–1090.

Brendan, N. A., Hammerschmidt, R., & Safir, G. R. (1996). Postharvest suppression of potato dry rot (*Fusarium sambucinum*) in prenuclear minitubers by arbuscular mycorrhizal fungal inoculum. *American Potato Journal, 73*, 509–515.

Brundrett, M. C. (2009). Mycorrhizal associations and other means of nutrition of vascular plants: Understanding the global diversity of host plants by resolving conflicting information and developing reliable means of diagnosis. *Plant and Soil, 320*, 37–77.

Budi, S. W., van Tuinen, D., Martinotti, G., & Gianinazzi, S. (1999). Isolation from *the Sorghum bicolor* mycorrhizosphere of a bacterium compatible with arbuscular mycorrhiza development and antagonistic towards soilborne fungal pathogens. *Applied and Environmental Microbiology, 65*, 5148–5150.

Caron, M. (1989). Potential use of mycorrhizae in control of soil-borne diseases. *Canadian Journal of Plant Pathology, 11*, 177–179.

Caron, M., Fortin, J. A., & Richard, C. (1985). Influence of substrate on the interaction of *Glomus intraradices* and *Fusarium oxysporum* f. sp. *radicis-lycopersici* on tomatoes. *Plant and Soil, 87*, 233–239.

Caron, M., Fortin, J. A., & Richard, C. (1986). Effect of phosphorus concentration and *Glomus intraradices* on *Fusarium* crown and root rot of tomatoes. *Phytopathology, 76*, 942–946.

Cordier, C., Gianinazzi, S., & Gianinazzi-Pearson, V. (1996). Colonisation patterns of root tissues by *Phytophthora nicotianae* var. parasitica related to reduced disease in mycorrhizal tomato. *Plant and Soil, 185*, 223–232.

Cordier, C., Pozo, M. J., Barea, J. M., Gianiniazzi, S., & Gianinazzi-Pearson, V. (1998). Cell defense responses associated with localized and systemic resistance to Phytophthora parasitica induced in tomato by an arbuscular mycorrhizal fungus. *Molecular Plant Microbe Interactions, 11*, 1017–1028.

Dehne, H. W., & Schonbeck, F. (1979). The influence of endotrophic mycorrhiza on plant diseases. II. Phenol metabolism and lignification Fusarium oxysporum. Untersuchungen zum Einfluss der endotrophen Mycorrhiza auf Pflanzenkrankheiten. II. Phenolstoffwechsel und Lignifizierung. *Phytopathologische Zeitschrift, 95*, 210–216.

Druzhinina, I. S., Seidl-Seiboth, V., Herrera-Estrella, A., Horwitz, B. A., Kenerley, C. M., Monte, E., et al. (2011). *Trichoderma*: The genomics of opportunistic success. *Nature Reviews Microbiology, 9*, 749–759.

Elsen, A., Baimey, H., Swennen, R., & DeWaele, D. (2003). Relative mycorrhizal dependency and mycorrhiza nematode interaction in banana cultivars (*Mus* spp.) differing in nematode susceptibility. *Plant and Soil, 256*, 303–313.

Feldmann, F., & Boyle, C. (1998). Concurrent development of arbuscular mycorrhizal colonization and powdery mildew infection on three Begonia hiemalis cultivars. *Zeitschrift für Pflanzenkrankheiten und Pflanzenschutz, 105*, 121–129.

Ferraz, L., & Brown, D. (2002). *An introduction to nematodes—plant nematology*. Sofia: Pensoft.

Filion, M., St. Arnaud, M., & Fortin, J. A. (1999). Direct interaction between the arbuscular mycorrhizal fungus *Glomus intraradices* and different rhizosphere microorganisms. *The New Phytologist, 141*, 525–533.

Fitter, A. H., & Sanders, I. R. (1992). Interactions with the soil fauna. In M. F. Allen (Ed.), *Mycorrhizal functioning: An integrative plant- fungal process* (pp. 333–354). New York: Chapman and Hall.

Frankenberger, W. T., & Arshad, M. (1995). Microbial biosynthesis of auxins. In W. T. Frankenberger & M. Arshad (Eds.), *Phytohormones in soil* (pp. 35–71). New York: Marcel Dekker.

Frey-Klett, P., & Garbaye, J. (2005). Mycorrhiza helper bacteria: a promising model for the genomic analysis of fungal–bacterial interactions. *The New Phytologist, 168*, 4–8.

Fritz, M., Jakobsen, I., Lyngkjaer, M. F., Thordal-Christensen, H., & Pons- Kühnemann, J. (2006). Arbuscular mycorrhiza reduces susceptibility of tomatoto *Alternaria solani*. *Mycorrhiza, 16*, 413–419.

Garbaye, J. (1994). Tansley review No. 76 Helper bacteria: A new dimension to the mycorrhizal symbiosis. *The New Phytologist, 128*, 197–210.

Gheysen, G., & Mitchum, M. G. (2011). How nematodes manipulate plant development pathways for infection. *Current Opinion in Plant Biology, 14*, 415–421.

Gianinazzi, S., & Schuepp, H. (1994). *Impact of arbuscular mycorrhizas on sustainable agriculture and natural ecosystems* (p. 226). Basel: Birkhauser Verlag.

Gianinazzi, S., Gollotte, A., Binet, M.-N., van Tuinen, D., Redecker, D., & Wipf, D. (2010). Agroecology: The key role of arbuscular mycorrhizas in ecosystem services. *Mycorrhiza, 20*, 519–530.

Graham, T. L., & Graham, M. Y. (1991). Cellular coordination of molecular responses in plant defense. *Molecular Plant-Microbe Interactions, 4*, 415–422.
Graham, J. H., Leonard, R. T., & Menge, J. A. (1981). Membrane mediated decrease in root exudation responsible for inhibition of vesicular-arbuscular mycorrhiza formation. *Plant Physiology, 68*, 548–552.
Grayston, S. J., Vaughan, D., & Jones, D. (1997). Rhizosphere carbon flow in trees, in comparison with annual plants: The importance of root exudation and its impact on microbial activity and nutrient availability. *Applied Soil Ecology, 5*, 29–56.
Gutjahr, C., & Parniske, M. (2013). Cell and developmental biology of arbuscular mycorrhiza symbiosis. *Annual Review of Cell and Developmental Biology, 29*, 593–617.
Gutjahr, C., & Paszkowski, U. (2013). Multiple control levels of root system remodeling in arbuscular mycorrhizal symbiosis. *Frontiers in Plant Science, 4*, 204.
Hage-Ahmed, K., Moyses, A., Voglgruber, A., Hadacek, F., & Steinkellner, S. (2013). Alterations in root exudation of intercropped tomato mediated by the arbuscular mycorrhizal fungus *Glomus mosseae* and the soil borne pathogen *Fusarium oxysporum* f. sp. *lycopersici*. *Journal of Phytopathology, 161*, 763–773.
Hammer, E. C., Pallon, J., Wallander, H., & Olsson, P. A. (2011). Tit for tat? A mycorrhizal fungus accumulates phosphorus under low plant carbon availability. *FEMS Microbiology Ecology, 76*, 236–244.
Handelsman, J., & Stabb, E. V. (1996). Biocontrol of soilborne plant pathogens. *Plant Cell, 8*, 1855–1869.
Hao, Z., Fayolle, L., Van Tuinen, D., Chatagnier, O., Li, X., & Gianinazzi, S. (2012). Local and systemic mycorrhiza-induced protection against the ectoparasitic nematode *Xiphinema index* involves priming of defence gene responses in grapevine. *Journal of Experimental Botany, 63*, 3657–3672.
Harrier, L. A. (2001). The arbuscular mycorrhizal symbiosis: A molecular review of the fungal dimension. *The Journal of Experimental Medicine, 52*, 469–478.
Harrier, L. A., & Watson, C. A. (2004). The potential role of arbuscular mycorrhizal (AM) fungi in the bioprotection of plants against soil-borne pathogens in organic and/or other sustainable farming systems (Special issue: Current research at the Scottish Agricultural College). *Pest Management Science, 60*, 149–157.
Hodge, A., Campbell, C. D., & Fitter, A. H. (2001). An arbuscular mycorrhizal fungus accelerates decomposition and acquires nitrogen directly from organic matter. *Nature, 413*, 297–299.
Hwang, S. F. (1988). Effect of VA mycorrhizae and metalaxyl on growth of alfalfa seedlings in soils from fields with "alfalfa sickness" in Alberta. *Plant Disease, 72*, 448–452.
Jabaji-Hare, S. H., & Stobbs, L. W. (1984). Electron microscopic examination of tomato roots coinfected with *Glomus* sp. and tobacco mosaic virus. *Phytopathology, 74*, 277–279.
Jaizme-Vega, M. C., Sosa-Hernandez, B., Hernandez, J. M., & Galan- Sauco, V. (1998). Interaction of arbuscular mycorrhizal fungi and the soil pathogen *Fusarium oxysporum* f. sp. *cubense* on the first stages of micropropagated Grande Naine banana. *Acta Horticulturae, 490*, 285–295.
Jansa, J., Mozafar, A., & Frossard, E. (2003). Long distance transport of P and Zn through the hyphae of an arbuscular mycorrhizal fungus in symbiosis with maize. *Agronomie, 23*, 481–488.
Jeffries, P., & Barea, J. M. (2012). Arbuscular mycorrhiza-a key component of sustainable plant-soil ecosystems. In B. Hock (Ed.), *The mycota* (pp. 51–75). Berlin/Heidelberg: Springer.
Jeffries, P., Gianinazzi, S., Perotto, S., Turnau, K., & Barea, J. M. (2003). The contribution of arbuscular mycorrhizal fungí in sustainable maintenance of plant health and soil fertility. *Biology and Fertility of Soils, 37*, 1–16.
Johnson, N. C., Copeland, P. J., Crookston, R. K., & Pfleger, F. L. (1992). Mycorrhizae: Possible explanation for yield decline with continuous corn and soybean. *Agronomy Journal, 84*, 387–390.
Johnson, D., Leake, J. R., Ostle, N., Ineson, P., & Read, D. J. (2002a). In situ (CO2)–C-13 pulse-labelling of upland grassland demonstrates a rapid pathway of carbon flux from arbuscular mycorrhizal mycelia to the soil. *New Phytologist, 153*, 327–334.

Johnson, D., Leake, J. R., & Read, D. J. (2002b). Transfer of recent photosynthate into mycorrhizal mycelium of an upland grassland: Short-term respiratory losses and accumulation of ^{14}C. *Soil Biology and Biochemistry, 34*, 1521–1524.

Jones, J. D. G., & Dangl, J. L. (2006). The plant immune system. *Nature, 444*, 323–329.

Jones, D. L., Hodge, A., & Kuzyakov, Y. (2004). Plant and mycorrhizal regulation of rhizo deposition. *The New Phytologist, 163*, 459–480.

Jones, J. T., Haegeman, A., Danchin, E. G. J., Gaur, H. S., Helder, J., MGK, J., et al. (2013). Top10plant-parasiticnematodesinmolecularplantpathology. *Molecular Plant Pathology, 14*, 946–961.

Jung, S. C., Martinez-Medina, A., Lopez-Raez, J. A., & Pozo, M. J. (2012). Mycorrhiza-induced resistance and priming of plant defenses. *Journal of Chemical Ecology, 38*, 651–664.

Khan, A. G., Kuek, C., Chaudhry, T. M., Khoo, C. S., & Hayes, W. J. (2000). Role of plants, mycorrhizae and phytochelators in heavy metal contaminated land remediation. *Chemosphere, 41*, 197–207.

Kobra, N., Jalil, K., & Youbert, G. (2009). Effects of three Glomus species as biocontrol agents against *Verticillium*-induced wilt in cotton. *Journal of Plant Protection Research, 49*, 185–189.

Kulkarni, S. A., Kulkarni, S., Sreenivas, M. N., & Kulkarni, S. (1997). Interaction between vesicular-arbuscular (VA) mycorrhizae and *Sclerotium rolfsii Sacc.* in groundnut. *Karnataka Journal of Agricultural Science, 10*, 919–921.

Lambais, M. R., & Mehdy, M. C. (1995). Differential expression of defense-related genes in arbuscular mycorrhiza. *Canadian Journal of Botany, 73*, S533–S540.

Lerat, S., Lapointe, L., Piché, Y., & Vierheilig, H. (2003). Strains colonizing barley roots. *Canadian Journal of Botany, 81*, 886–889.

Li, S. L., Zhao, S. J., Zhao, L. Z., Li, S. L., Zhao, S. J., & Zhao, L. Z. (1997). Effects of VA mycorrhizae on the growth of eggplant and cucumber and control of diseases. *Acta Phytophylacica Sinica, 24*, 117–120.

Linderman, R. G. (1988). Mycorrhizal interactions with the rhizosphere microflora-The Mycorrhizosphere effect. *Proceedings of the American Phytopathology Society, 78*(3), 366–371.

Linderman, R. G. (1994). Role of VAM fungi in biocontrol. In F. L. Pfleger & R. G. Linderman (Eds.), *Mycorrhizae and plant health* (pp. 1–27). St. Paul: The American Phytopathological Society.

Lioussanne, L., Jolicoeur, M., & St-Arnaud, M. (2008). Mycorrhizal colonization with *Glomus intraradices* and development stage of transformed tomato roots significantly modify the chemotactic response of zoospores of the pathogen *Phytophthora nicotianae*. *Soil Biology and Biochemistry, 40*, 2217–2224.

López-Ráez, J. A., Charnikhova, T., Fernández, I., Bouwmeester, H., & Pozo, M. J. (2011a). Arbuscular mycorrhizal symbiosis decreases strigolactone production in tomato. *Journal of Plant Physiology, 168*, 294–297.

López-Ráez, J. A., Pozo, M. J., & García-Garrido, J. M. (2011b). Strigolactones: A cry for help in the rhizosphere. *Botany, 89*, 513–522.

Maillet, F., Poinsot, V., André, O., Puech-Pagés, V., Haouy, A., Gueunier, M., Cromer, L., Giraudet, D., FormeyD, N. A., Martinez, E. A., Driguez, H., Bécard, G., & Dénarié, J. (2011). Fungal lipochitooligosaccharide symbiotic signals in arbuscular mycorrhiza. *Nature, 469*, 58–64.

McArthur, D. A., & Knowles, N. R. (1992). Resistance responses of potato to vesicular-arbuscular mycorrhizal fungi under varying abiotic phosphorus levels. *Plant Physiology, 100*, 341–351.

McGonigle, T. P., & Miller, M. H. (1993). Mycorrhizal development and phosphorus absorption in maize under conventional and reduced tillage. *Soil Science Society of America Journal, 57*, 1002–1006.

McGonigle, T. P., & Miller, M. H. (1996). Mycorrhizae, phosphorus absorption, and yield of maize in response to tillage. *Soil Science Society of America Journal, 60*, 1856–1861.

Minerdi, D., Fani, R., Gallo, R., Boarino, A., & Bonfante, P. (2001). Nitrogen fixation genes in an endosymbiotic *Burkholderia* strain. *Applied and Environmental Microbiology, 67*, 725–732.

Morandi, D. (1996). Occurrence of phytoalexins and phenolic compounds in endomycorrhizal interactions, and their potential role in biological control. *Plant and Soil, 185*, 241–251.

Mukerji, K. G., Manoharachary, C., & Chamola, B. P. (2002). *Techniques in mycorrhizal studies* (1st ed., pp. 285–296). London/Dordrecht: Kluwer Academic Publishers.

Nemec, S. (1994). Soil microflora associated with pot cultures of *Glomus intraradix*-infected *Citrus reticulata. Agriculture, Ecosystems and Environment, 1*, 299–306.

Nemec, S., & Myhre, D. (1984). Virus–Glomus etunicatum interactions in citrus rootstocks [Sour orange, *Citrus macrophylla*, Duncan grapefruit, potential of mycorrhizal citrus rootstock seedlings to protect against growth suppression by viruses]. *Journal of Plant Disease, 68*, 311–314.

Newsham, K. K., Fitter, A. H., & Watkinson, A. R. (1995). Arbuscular mycorrhiza protect an annual grass from root pathogenic fungi in the field. *Journal of Ecology, 83*, 991–1000.

Nicol, J. M., Turner, S. J., Coyne, D. L., den Nijs, L., Hockland, S., & Tahna Maafi, Z. (2011). Current nematode threats to world agriculture. In J. Jones, G. Gheysen, & C. Fenoll (Eds.), *Genomics and molecular genetics of plant-Nematode interactions* (pp. 347–367). Heidelberg: Springer.

Norman, J., Atkinson, D., & Hooker, J. (1996). Arbuscular mycorrhizal fungal- induced alteration to root Architecture in strawberry and induced resistance to the root pathogen *Phytophthora fragariae. Plant and Soil, 185*, 191–198.

Nuccio, E. E., Hodge, A., Pett-Ridge, J., Herman, D. J., Weber, P. K., & Firestone, M. K. (2013). An arbuscular mycorrhizal fungus significantly modifies the soil bacterial community and nitrogen cycling during litter decomposition. *Environmental Microbiology, 15*, 1870–1881.

Otto, G., & Winkler, H. (1995). Colonization of rootlets of some species of Rosaceae by actinomycetes, endotrophic mycorrhiza, and endophytic nematodes in a soil conducive to specific cherry replant disease. *Zeitschrift fur Pflanzenkrankheiten und Pflanzenschutz, 102*, 63–68.

Paulitz, T. C., & Linderman, R. G. (1989). Interactions between fluorescent pseudomonads and VA mycorrhizal fungi. *The New Phytologist, 113*, 37–45.

Paxton, J. D. (1981). Phytoalexins- A working redefinition. *Journal of Phytopathology, 101*(2), 106–109.

Perry, R. N., & Moens, M. (2011). Introduction to plant-parasitic nematodes; modes of parasitism. In T. Jones, L. Gheysen, & C. Fenoll (Eds.), *Genomics and molecular genetics of plant–nematode interactions* (pp. 3–20). Heidelberg: Springer.

Peterson, R. L., Massicotte, H. B., & Melville, L. H. (2004). *Mycorrhizas: Anatomy and cell biology*. Ottawa: NRC Research Press.

Philippot, L., Raaijmakers, J. M., Lemanceau, P., & van der Putten, W. H. (2013). Going back to the roots: The microbial ecology of the rhizosphere. *Nature Reviews. Microbiology, 11*, 789–799.

Pozo, M. J., & Azcón-Aguilar, C. (2007). Unraveling mycorrhiza-induced resistance. *Current Opinion in Plant Biology, 10*, 393–398.

Pozo, M. J., Azcon-Aguilar, C., Dumas-Gaudot, E., & Barea, J. M. (1999). β-1,3-glucanase activities in tomato roots inoculated with arbuscular mycorrhizal fungi and/or *Phytophthora parasitica* and their possible involvement in bioprotection. *Plant Science, 141*, 149–157.

Pozo, M. J., Cordier, C., Dumas-Gaudot, E., Gianinazzi, S., Barea, J. M., & Azcon-Aguilar, C. (2002). Localized versus systemic effect of arbuscular mycorrhizal fungi on defence responses to *Phytophthora* infection in tomato plants. *Journal of Experimental Botany, 53*, 525–534.

Prashanthi, S. K., Kulkarni, S., Sreenivasa, M. N., & Kulkarni, S. (1997). Integrated management of root rot disease of safflower caused by Rhizoctonia bataticola. *Environment and Ecology, 15*, 800–802.

Puppi, G., Azcón, R., & Hoflich, G. (1994). Management of positive interactions of arbuscular mycorrhizal fungi with essential groups of soil microorganisms. In S. Gianinazzi & H. Schouepp (Eds.), *Impact of arbuscular mycorrhizas on sustainable agriculture and natural ecosystems* (pp. 201–215). Basel: Birkhäuser.

Rambelli, A. (1973). The rhizosphere of mycorrhizae. In G. L. Marks & T. T. Koslowski (Eds.), *Ectomycorrhizae* (pp. 299–343). New York: Academic.

Ratnayake, M., Leonard, R. T., & Menge, J. A. (1978). Root exudation in relation to supply of phosphorus and its possible relevance to mycorrhiza formation. *The New Phytologist, 81*, 543–552.

Ravnskov, S., & Jakobsen, I. (1995). Functional compatibility in arbuscular mycorrhizas measured as hyphal P transport to the plant. *The New Phytologist, 129*, 611–618.

Read, D. J., & Perez-Moreno, J. (2003). Mycorrhizas and nutrient cycling in ecosystems – A journey towards relevance? *The New Phytologist, 157*, 475–492.

Rhodes, L. H., & Gerdemann, J. W. (1975). Phosphate uptake zones of mycorrhizal and non-mycorrhizal onions. *The New Phytologist, 75*, 555–561.

Salvioli, A., & Bonfante, P. (2013). Systems biology and omics tools: A cooperation for next-generation mycorrhizal studies. *Plant Science, 203–204*, 107–114.

Schaarschmidt, S., Gresshoff, P. M., & Hause, B. (2013). Analyzing the soybeantranscriptomeduringautoregulationofmycorrhizationidentifies the transcription factorsGmNF-YA1a/bas positive regulators of arbuscular mycorrhization. *Genome Biology, 14*, R62.

Schonbeck, F., & Spengler, G. (1979). The detection of TMV in mycorrhizal cells of the tomato plant by means of immunofluorescence. *Phytopathologische Zeitschrift, 94*, 84–86.

Schreiner, R. P., & Bethlenfalvay, G. J. (1997). Mycorrhizae, biocides, and biocontrol: 3. Effects of three different fungicides on developmental stages of three AM fungi. *Biology and Fertility of Soils, 24*, 18–26.

Schübler, A., & Walker, C. (2011). Evolution of the 'plant-symbiotic' fungal phylum, Glomeromycota. In S. Póggeler & J. Wostemeyer (Eds.), *Evolution of fungi and fungal like organisms* (pp. 163–185). Berlin/Heidelberg: Springer.

Schübler, A., Gehrig, H., Schwarzott, D., & Walker, E. (2001). Analysis of partial *Glomales* SSU rRNA gene sequences: Implications for primer design and phylogeny. *Mycological Research, 105*, 5–15.

Shalaby, A. M., & Hanna, M. M. (1998). Preliminary studies on interactions between VA mycorrhizal fungus *Glomus mosseae, Bradyrhizobium japonicum* and *Pseudomonas syringae* in soybean plants. *Acta Microbiologica Polonica, 47*, 385–391.

Sharma, D. D., Govindaiah, S., Katiyar, R. S., Das, P. K., Janardhan, L., Bajpai, A. K., Choudhry, P. C., & Janardhan, L. (1995). Effect of VA-mycorrhizal fungi on the incidence of major mulberry diseases. *Indian Journal of Sericulture, 34*, 34–37.

Sharma, S., Dohroo, N. P., & Sharma, S. (1997). Management of ginger yellows through organic amendment, fungicide seed treatment and biological methods. *Indian Cocoa Arecanut Spice Journal, 21*, 29–30.

Shaul, O., Galili, S., Volpin, H., Ginzberg, I., Elad, Y., Chet, I., & Kapulnik, Y. (1999). Mycorrhiza-induced changes in disease severity and PR protein expression in tobacco leaves. *Molecular Plant-Microbe Interactions, 12*, 1000–1007.

Sikora, R. A. (1997). Biological system management in the rhizosphere an inside-out/outside-in perspective. *Mededelingen – Faculteit Landbouwkundige en Toegepaste Biologische Wetenschappen, Universiteit Gent, 62*, 105–112.

Sikora, R. A., Pocasangre, L., FeldeZum, A., Niere, B., Vu, T. T., & Dababat, A. A. (2008). Mutualistic endophytic fungi and in planta suppressiveness to plant-parasitic nematodes. *Biological Control, 46*, 15–23.

Singh, L. P., Gill, S. S., & Tuteja, N. (2011). Unraveling the role of fungal symbionts in plant abiotic stress tolerance. *Plant Signaling & Behavior, 6*, 175–191.

Slezack, S., Dumas-Gaudot, E., Paynot, M., & Gianinazzi, S. (2000). Is a fully established arbuscular mycorrhizal symbiosis required for bioprotection of Pisum sativum roots against *Aphanomyces euteiches. Molecular Plant-Microbe Interactions, 13*, 238–241.

Smith, F. A., & Smith, S. E. (2011). What is the significance of the arbuscular mycorrhizal colonization of many economically important crop plants? *Plant and Soil, 348*, 63–79.

Smith, S. E., & Smith, F. A. (2012). Fresh perspectives on the roles of arbuscular mycorrhizal fungi in plant nutrition and growth. *Mycologia, 104*, 1–13.

Sood, G. S. (2003). Chemotactic response of plant-growth-promoting bacteria towards roots of vesicular-arbuscular mycorrhizal tomato plants. *FEMS Microbiology Ecology, 45*, 219–227.

Srnith, S. E., & Read, D. I. (2008). *Mycorrhizal symbiosis* (3rd ed.). New York: Elsevier/Academic.

St-Arnaud, M., Hamel, C., Vimard, B., Caron, M., & Fortin, J. A. (1995). Altered growth of *Fusarium oxysporum* f. sp. *chrysanthemi*in an *in vitro* dual culture system with the vesicular ar366buscular mycorrhizal fungus *Glomus intraradices* growing on *Daucus carota* transformed roots. *Mycorrhiza, 5*, 431–438.

Steinkellner, S., Lendzemo, V., Langer, I., Schweiger, P., Khaosaad, T., Toussaint, J.-P., et al. (2007). Flavonoids and strigolactones in root exudates as signals in symbiotic and pathogenic plant-fungus interactions. *Molecules, 12*, 1290–1306.

Stoffelen, R., Verlinden, R., Xuyen, N. T., DeWaele, D., & Swennen, R. (2000). Host plant response of Eumusa and Australimusa bananas(*Musa* spp.) to migratory endoparasitic and root-knot nematodes. *Nematology, 2*, 907–916.

Sundaresan, P., Raja, N. U., & Gunasekaran, P. (1993). Induction and accumulation of phytoalexins in cowpea roots infected with a mycorrhizal fungus *Glomus fasciculatum* and their resistance to *Fusarium* wilt disease. *Journal of Biosciences, 18*, 291–301.

Takahashi, T., Katano, H., & Yoshikawa, N. (1994). Evidence for vesicular-arbuscular mycorrhizal infection in viroid-infected hop root tissues. *Zeitschrift fur Pflanzenkrankheiten und Pflanzenschutz, 101*, 267–271.

Tisdall, J. M., & Oades, J. M. (1979). Stabilization of soil aggregates by the root systems of rye grass. *Australian Journal of Soil Research, 17*, 429–441.

Torres-Barragan, A., Zavaleta-Mejia, E., Gonzalez-Chavez, C., & Ferrera-Cerrato, R. (1996). The use of arbuscular mycorrhizae to control onion white rot (*Sclerotium cepivorum* Berk.) underfield conditions. *Mycorrhiza, 6*, 253–258.

Trotta, A., Varese, G. C., Gnavi, E., Fusconi, A., Sampo, S., & Berta, G. (1996). Interactions between the soil borne pathogen *Phytophthora nicotianae var. parasitica* and the arbuscular mycorrhizal fungus *Glomus mosseae* in tomato plants. *Plant and Soil, 185*, 199–209.

van der Heijden, M. G. A., Martin, F. M., Selosse, M. A., & Sanders, I. R. (2015). Mycorrhizal ecology and evolution: The past, the present, and the future. *The New Phytologist, 205*, 1406–1423.

VanEtten, H. D., Sandrock, R., Wasmann, C., Soby, S., Mccluskey, K., & Wang, P. (1995). Detoxification of phytoanticipins and phytoalexins by phytopathogenic fungi. *Canadian Journal of Botany, 73*, 518–525.

Vierheilig, H., Steinkellner, S., & Khaosaad, T. (2008). The biocontrol effect of mycorrhization on soilborne fungal pathogens and the autoregulation of the AM symbiosis: One mechanism, two effects? In A. Varma (Ed.), *Mycorrhiza* (pp. 307–320). Berlin: Springer.

Vinayak, K., & Bagyaraj, D. J. (1990). Vesicular-arbuscular mycorrhizae screened for Troyer citrange. *Biology and Fertility of Soils, 9*, 311–314.

Vos, C. (2012). *Arbusculaire mycorrhizenschimmels in de biocontrole van plantenparasitaire nematoden*. Leuven: University of Leuven (KULeuven).

Vos, C. M., Yang, Y., DeConinck, B., & Cammue, B. P. A. (2014). Fungal(-like) biocontrol organisms in tomato disease control. *Biological Control, 74*, 65–81.

Waschkies, C., Schropp, A., & Marschner, H. (1994). Relations between grapevine replant disease and root colonization of grapevine (*Vitis* sp.) by fluorescent pseudomonads and endomycorrhizal fungi. *Plant and Soil, 162*, 219–227.

Wesemael, W., Viaene, N., & Moens, M. (2011). Root-knot nematodes (*Meloidogyne* spp.) in Europe. *Nematology, 13*, 3–16.

Wyss, P., Boller, T., & Wiemken, A. (1991). Phytoalexin response is elicited by a pathogen (*Rhizoctonia solani*) but not by a mycorrhizal fungus (*Glomus mosseae*) in soybean roots. *Cellular and Molecular Life Sciences, 47*(4), 395–399.

Xavier, L. J. C. (L. Johnny). (1999). *Effects of interactions between arbuscular mycorrhizal fungi and Rhizobium leguminosarum on pea and lentil*. PhD dissertation. Saskatoon: University of Saskatchewan

Xavier, L. J. C., & Germida, J. J. (2003). Bacteria associated with *Glomus clarum* spores influence mycorrhizal activity. *Soil Biology and Biochemistry, 35*, 471–478.

Yang, H., Zhang, Q., Dai, Y., Liu, Q., Tang, J., Bian, X., et al. (2014). Effects of arbuscular mycorrhizal fungi on plant growth depend on root system: A meta-analysis. *Plant and Soil, 389*, 361–374.

Zamioudis, C., & Pieterse, C. M. J. (2012). Modulation of host immunity by beneficial microbes. *Molecular Plant-Microbe Interactions, 25*, 139–150.

Mycoremediation of Environmental Pollutants from Contaminated Soil

15

Prem Chandra and Enespa

1 Introduction

The term "mycoremediation" refers to detoxificationof polluted site by using various fungal species coined by Paul Stamets. The fungi are the eukaryotic microorganisms, ubiquitous in nature and represent diversity of groups from various environments (Strong and Burgess 2008; Purohit et al. 2018). The sequestering toxic contaminants from soil by fungiusing this process. Soil health improves by the native and/or alien microflora of fungi (Stamets 2011; Andreoni and Gianfreda 2007). The precise augmentation of fungal cultures helps in speedy decay (Tahir et al. 2018). The fungi of saprophytic groups secrete diverse extracellular enzymes and those break down natural polymers like keratin, chitin, lignin, pectin, cellulose, and hemicellulose and play an important role in the decomposition of organic molecules (Walker and White 2017). The hazardous toxic heavy metals and organic pollutants as a significance of anthropogenic activities become a key apprehension in environmental and health problem by the soil and water contaminations (Vassilev et al. 2004). Several toxic metals such as Cd, Cu, Hg, Pb, Mn, As, Ni, Zn, etc. are released into the environment from industrial effluent and other human activities (Alloway 2013). These pollutants from heavy metals are not biodegradable and have capability to travel up the food chain via bioaccumulation, unlike organic contaminants. The several metals such as Cu, Fe, Mn and Zn are micronutrients and play a vital role in metalloenzymes for mostly organisms (Siddiquee et al. 2015). The cationic state usually increases the stability of membrane and play specific roles in the structure of nucleic acid, utilities and metabolisms. When the concentrations of beneficial metals increased in the environment e.g., mercury (Hg), lead (Pb), cadmium (Cd), they become more toxic (Lynes et al. 2010). The conventional

P. Chandra (✉) · Enespa
Department of Environmental Microbiology, School for Environmental Sciences,, Babasaheb Bhimrao Ambedkar (A Central) University, Lucknow, Uttar Pradesh, India

© Springer Nature Singapore Pte Ltd. 2019

A. Varma, D. K. Choudhary (eds.), *Mycorrhizosphere and Pedogenesis*,
https://doi.org/10.1007/978-981-13-6480-8_15

methods for the removal of heavy metals from the soil and the waste water including chemical precipitation and sludge separation, chemical oxidation or reduction, ion exchange, reverse osmosis, filtration, adsorption using activated charcoal, electrochemical treatment and evaporative recovery (Mohan and Pittman 2007). This physicochemical types techniques are very costly, and also their metal binding properties non-specific. The microbial process is more significant than the conventional process (Crini 2006). The naturally occurring fungi have adaptation capability in any ecological system and conditions, due to having a large range of extracellular proteins, organic acids and other metabolites. The species of filamentous fungi like *Trichoderma* sp. are reported in field soil mostly in the agriculture systems (Magan 2007). Commonly found in soil, root, foliar environments and other environmental conditions. Having the specific characters, it can grow faster, sporulation and biocontrol agent, cell wall degrading enzymes and eco-friendly in natures (Waghunde et al. 2016). It has high resistant capability towards the various heavy metal toxins and xenobiotic compounds, such as antibiotics, fungicides, herbicides and pesticides etc. (Imfeld and Vuilleumier 2012). Some *Trichoderma* sp. have potential and good antagonistic capabilities against some plant pathogenic fungi. Various heavy metals have significant role for the metabolic activities of fungi comparison to others (Bahn et al. 2007). The concentration of both essential and nonessential heavy metals increased in the soil it becomes toxic for all the fungal strains. The fruiting bodies of white-rot fungi get significant amounts of heavy metals from the environments (Baldrian 2003). The brown-rot fungi decomposed cellulose and hemicelluloses with the help of metal ions and copper and manganese directly participate in the process of lignin degradation in white-rot fungi (Dashtban et al. 2010). The participations of manganese in the reaction cycle of Mn-dependent peroxidase, and copper assists as a cofactor in the catalytic center of laccase. The role of Mn in lignin degradation has been the subject of various revisions (Cowan et al. 2016). The potential use of plants to remediate soils contaminated with metal and persistent organic pollutants (POPs) focused attention (Zhang et al. 2013). The rhizospheric remediation approaches is very cheap compare to other approaches and create low disturbance to decontaminating polluted soil (Tangahu et al. 2011). The phytoremediation is applied widely as a catch-all term for the use of plant life to remediate both metal- and POP-polluted soils. It is also known as hyper accumulation of metals by plants, since the plant tissues are the repository of the pollutants (Singh et al. 2003). Where the plants are used to remediate POPs, however, we have a preference the term "rhizospheric remediation", because the POP degrading activities will, in most circumstances, occur in the rhizosphere, rather than in the plant (Mueller 2005). The enhanced degradation or mineralization in the rhizosphere has been revealed for a range of polyaromatic hydrocarbons (PAHs), pesticides, oil (hydrocarbons), surfactants and chlorinated alkanes (Meharg and Cairney 2000). The plant-roused microbial activity in the rhizosphere, other biological (bacterial plasmid transfer) and physical (pollutants drawn into the rhizosphere by the transpiration stream, alteration of soil structure) and factors may also play a role. The rhizospheric microorganisms may not degrade POPs to yield energy; rather they may co-metabolize them as a consequence of utilizing plant-derived cyclic compounds (Meharg and Cairney 2000). For instance, that the plant phenolics, such as catechin and coumarin serve as co-metabolites fordegradation of polychlorinated biphenyls (PCBs) by bacteria (Bisht et al. 2015). The enzymes used to biodegrade plant-derived compounds by free-living rhizospheric microbial biomass may also degrade POPs (Juwarkar et al. 2010). A single microorganism may not be processed by the complete group of enzymatic procedures required to degrade a POP. The consortium of rhizospheric microorganisms degraded POPs and pesticide mecoprop (methylchlorophenoxypropionic acid) (Kvesitadze et al. 2006). Plants considered for remediation to POP contaminated sites have various challenges. These used plant species must be resistant to all contaminants present in such types of sites infrequently with only a single pollutant (Puglisi et al. 2012). The industrial sites have a series of pollutants so the plant growth cannot takes place easily at this site due to the complex soil structure and the low nutrients. The surface area of root soil is a crucial factor in the effectiveness and speed of rhizospheric remediation to be optimized (Cunningham et al. 1996). The enzymatic activities of the microorganisms have capability of degrading a wide range of POPs and set up to expedite remediation for contaminated sites with multiple pollutants (Teng et al. 2010). The attention has been given in this chapter to the mycoreremediation of inorganic and organic pollutants such as toxic heavy metals and pesticides, herbicides etc. (Alvarez et al. 2017).

2 Fungal Groups Involved in Remediation

The biodegradation and biotransformation of various hazardous and toxic compounds which interact to ecological diversity and its behavior is very important for understands (Van der Oost et al. 2003). Very few numbers of fungal species has been recognized to be connected with mycoremediation from diverse ecology from identified 69,000 fungal species worldwide (Pala et al. 2014). On the bases of utilizing the various species of fungi for bioremediation and it classified into various sections along with their mechanisms and functionary given in the (Table 15.1). The abiotic and biotic factors such as accessibility of carbon, nitrogen, phosphorus, metal ion concentration, temperature, aeration, moisture, and interspecific microbial competition are responsible for fungal growth and its presence (Lavelle and Spain 2001). Very little facts are obtainable on the ecology of fungi connected with mycoremediation, while the sufficient literature is availed on the ecology of fungal literature. These fungi grouped into following sections on the bases on ecology and functionality (Rillig and Mummey 2006).

2.1 Wood Rot Fungi

This group of fungi represents a diverse group with association of wood rotting. They have the capability to splitting of wood tissues to simpler form by using diversify group of enzymes to splitting of wood tissues to simpler form by using diversify

Table 15.1 A fungal diversity used for mycoremediation

Groups	Fungal spp.	Used in remediation of	Mechanisms	References
Wood-decaying white-rot fungi	*Pleurotus ostreatus*	Cadmium	Biosorption of heavy metals	Tay et al. (2011)
	Pleurotus ostreatus	Biodegradable plastic	Degradation of plastic	Kerem et al. (1992)
	Pleurotus sajorcaju	Heavy metal, Zn	Biosorption of heavy metals	Das (2005)
	Pleurotus tuberregium	Heavy metals	Biosorption of heavy metals	Oyetayo et al. (2012)
	Pleurotus pulmonarius	Crude oil	Degradation of crude oil	Adenipekun et al. (2015)
	Pleurotus tuberregium	Crude oil-polluted soil	Enzymatic degradation	Isikhuemhen et al. (2003)
	Bjerkandera adusta	PAHs, PCBs	Enzymatic modification of lignin	Yang et al. (2013)
Soil fungi	*Mucor* sp.	Heavy metals (Ni, Cd, Pb, Zn)	Bioadsorption of heavy metals	Wang and Chen (2009)
	Rhizopus sp.	Heavy metals (Ni, Cd, Pb, Zn)	Biosorption of heavy metals	Fu and Wang (2011)
	Cunninghamella sp.	Heavy metals (Ni, Cd, Pb, Zn)	Ions-sequestration	Purohit et al. (2018)
	Mortierella sp.	2,4-D (2,4-dichlorophenoxy acetic acid)	Hydroxylation and dechlorination	Nykiel-Szymańska et al. (2018)
	Aspergillus niger	Heavy metals (Cd, Zn, Ur, Ag, Cu)	Biosorption of heavy metals	Mudhoo et al. (2012)
	Aspergillus fumigates	Heavy metal (Ur)	Bioaccumulation of heavy metal	Alluri et al. (2007)
	Trichoderma viride	Heavy metal (Hg)	Biosorption of heavy metal	Siddiquee et al. (2015)
	Paecilomyces sp.,	PAH, Endosulfan	Hydroxylation	Nunes and Malmlöf (2018)
Leaf decomposing fungi	*Agrocybe praecox*	PAHs, TNT	Modification by laccase and MnP	Anasonye et al. (2015)
	Nematoloma frowardii	Radionuclide	Enzymatic degradation (MnP)	Ellouze and Sayadi (2016)
	Stropharia coronilla	Mineralization of ^{14}C-labeled synthetic lignin	Degradation by Ligninolytic enzymes	Joy et al. (2015)

(continued)

Table 15.1 (continued)

Groups	Fungal spp.	Used in remediation of	Mechanisms	References
Wood decaying brown-rot fungi	*Schizophyllum commune*,	Malachite green dye	Enzymatic degradation	Sudha et al. (2014)
	Polyporus sp.	Degradation of Polychlorophenol	Mineralization	Chiu et al. (1998)
	Gloeophyllum striatum	Polysaccharide decomposition	Mineralization	Anastasi et al. (2013)
	Flammulina velutipes	Copper	Biosorption	Qu et al. (2015)
	Fomes fasciatus	Copper (II)	Biosorption	Hamba and Tamiru (2016)
	Daedalea dickinsii, *Fomitopsis pinicola*,	DDT	Microbial biodegradation via Fenton reaction	Purnomo et al. (2017); Camacho-Morales et al. (2017)
Mycorrhizal fungi	*Glomus geosporum*	Zn	Enzymatic degradation	Lenoir et al. (2016)
	Suillus granulatus	Cresol, catechol	Biotransformation	Sardrood et al. (2013)
	Scutellospora heterogama	Cu	Enzymatic degradation	de Novais et al. (2014)
	Gigaspora gigantean	Zn, Cu, Pb, Ni, Cd	Enzymatic degradation	Cabral et al. (2015)
	Rhizopogon vinicolor	2,4-D	Mineralization	Ferreira-Guedes et al. (2012)
	Hymenoscyphus ericae, *Oidiodendron griseum*	2,4-D, atrazine	Mineralization via enzymatic degradation	Purohit et al. (2018)
Endophytic fungi	*Coriolopsis gallica*	PAH	Biotransformation	Couto and Herrera (2006)
	Ceratobasidum stevensii	Phenanthrene	Enzymatic degradation (MnP)	Tian et al. (2018)
	Phanerochaete chrysosporium	PAH degradation, Phenanthrene	Enzymatic degradation (LiP, MnP), oxidation	Eibes et al. (2006)
	Bjerkandera sp.	PAH degradation	Enzymatic degradation	Juhasz and Naidu (2000)
	Phomopsis sp.	Polymeric dyes	Enzymatic degradation	Xiong et al. (2014)

(continued)

Table 15.1 (continued)

Groups	Fungal spp.	Used in remediation of	Mechanisms	References
Aquatic fungi	*Nia vibrossa*, *Julella avicinnae*, *Lignincola laevis*	Polymeric dyes	Enzymatic degradation	Sarma (2018)
	Aspergillus sclerotiorum CBMAI 849, *Cladosporium cladosporioides CBMAI 857*, *Mucor racemosus* CBMAI 847	Lignin-based industrial pollutant	Mineralization	
	Phaeosphaeria spartinicola, *Sordaria fimicola*	Industrial pollutant	Mineralization	Raghukumar (2017)
	Penicillium raistrickii, *Trichoderma* sp.	Profenofos (organophosphate insecticide)	Enzymatic degradation	Kushwaha et al. (2016)

group of enzymes to degradation of the multifarious molecules such as cellulose, hemicellulose, lignin, etc. (Rajinipriya et al. 2018). These groups of fungi also play a substantial role in bioremediation of organic pollutants besides the wood rooting. They are not required any preconditioning prior to pollutants transformation (Adenipekun and Lawal 2012). Hence, they have widespread adaptability and capability to degradation of tissues. The wood-degradation fungal species expressively differ in their settlement capability and characterized as strong challengers such as *Pleurotus* sp., *Phanerochaete* sp., *T. versicolor* etc. and weak competitors such as *Dichomitus squalens* and *Ganoderma applanatum.* These fungi are classified in to white-rot fungi and brown-rot fungi on the bases of mode of action at the woody tissues (Martínez et al. 2005).

2.2 White Rot Fungi

Only some specified white-rot fungi have the unique capability to reduce the lignin along with cellulose and hemicellulose resulting decay and in white bleaching of woods ((Martínez et al. 2005). To the study of mycoremediation these fungi identified first in the group and produce some enzymes such as lignin peroxidase, manganese peroxidase, H_2O_2-generating enzymes, and laccase (Vara 2017). The extracellular oxidative ligninolytic enzymes have been studied in details in *Phanerochaete chrysosporium* for the biodegradation of complex compounds. The toxic or insoluble compounds degraded in to CO_2 and H_2O more efficiently by the *P. chrysosporium* than other fungi or microbes (Azmi et al. 1998). The degradation or biotransformation's of recalcitrant compounds make its presentation magnetic in environments by the diversified oxidative and reductive approaches. The broad spectrum of aromatic complex and the compounds of xenobiotics lies in contaminated soil easily remove due to the nonspecific and resilient nature of ligninolytic enzymes (Abuhussein 2018). Another white-rot fungus such as *Pleurotus ostreatus*, *Trametes versicolor*, *Bjerkandera adusta*, *Lentinula edodes*, and *Irpex lacteus* are also known for degradation of these compounds except *P. chrysosporium*. Currently, various studies revealed that the white rot fungi used at least 30% in the mycoremediation (Forgacs et al. 2004).

2.3 Brown Rot Fungi

The cellulose and hemicellulose present in wood also degraded by this group of fungi. The brown-rot fungi are distributed in *Agaricales*, *Hymenochaetales*, *Gloeophyllales*, and *Polyporales* in majority (Anastasi et al. 2013). The lignin modified partly by a non-enzymatic Fenton- type catalytic system due to the demethylation, oxidation, and depolymerization. In decaying wood a unique dark brown color produced by the partially modified lignin (Zavarzina et al. 2010). An oxidative process involved in the production of hydrogen peroxide degradation of cellulose and hemicellulose by the brown rot fungi, the also helps in the synthesis of free hydroxyl (OH) radicals and that in turn smooth the biodegradation and mineralization of artificial chemotherapeutants (Arantes et al. 2012). Additionally, the oxalic acid production and antimicrobial drug tolerance increases their capability to degrade the metals. So, in larger scale the potentiality can be exploited by brown rot fungi for bioremediation process (Dunwell et al. 2000).

2.4 Leaf Decomposing Fungi

In the forest ecology the leaf decomposing fungi is one of the chief components. The process of humification, mineralization, and decomposition of wood and litter of soil organic matter completed by these fungi actively (Coleman 2008). A rapid successional change takes during leaf litter decomposition by this community of fungi. During the initial stages of litter decay the phylum of Ascomycota fungal group are predominant, but during the later stages of decomposition their population gradually decreases with increase in fungi of the *Basidiomycota* phylum (Das et al. 2007). The lignocellulolytic materials of plant litters decomposed by these fungal groups vary actively. A plentiful ligninolytic enzyme produced by Basidiomycetous litter fungi is essential for biodegradation of plant materials such as cellulase, laccase, and oxidoreductases deposited on the floor of forest (Purahong et al. 2016). The persistent organic pollutants such as pesticides and herbicides in the soil also bio transformed by these enzymes. So, these fungi utilized for the bioremediation purposes will open up a new scope in soil contaminants (Harms et al. 2011).

2.5 Soil Fungi

A heterogeneous group represented by the soil fungi, especially *Ascomycota*, *Chytridiomycota*, and *Zygomycota*. These groups play an important role in carbon and nitrogen cycling and organic matter decomposition in soil are the chief components of soil ecology (Anastasi et al. 2013). Mostly, they are non-ligninolytic in nature known as saprophytes and have capability to decompose the cellulose (Fritsche et al. 2000). Various genera included in this group are *Acremonium*, *Allescheriella*, *Alternaria*, *Aspergillus*, *Beauveria*, *Cladosporium*, *Cunninghamella*, *Engyodontium*, *Fusarium*, *Geomyces*, *Microsporum*, *Mortierella*, *Paecilomyces*, *Penicillium*, *Phlebia*, *Rhizopus*, *Stachybotrys*, and *Trichoderma* (Purohit et al. 2018). The non-ligninolytic fungi produced extracellular enzymes monooxygenase which leads to degradation of PAHs via hydroxylation (Bamforth and Singleton 2005). And have capability tolerant to various pollutants like PAHs, polychlorobenzoic acids (PCBs), chlorobenzoic acids (CBA), and endosulfan, and indicated the potentiality as bioremediation mediators in soil. In the later stages of decomposition the fungi degraded the recalcitrant polymers. So, these are considered as xenobiotic-degrading fungi, and their consortia with diverse species certify a superior efficacy in the bioremediation of soil (Tan 2011).

2.6 Fungal Mycorrhiza

The symbiotic associations with plant roots established the mutualistic relationship by assisting the supply of nitrogen and phosphorus to the plants and in turning originate organic carbon from plants for the fungal metabolic activity (Lambers et al. 2008). There are numerous types of mycorrhizal associations with plant roots such as, ectomycorrhiza, ectendomycorrhiza, ericoid mycorrhiza, arbuscular mycorrhiza, arbutoid mycorrhiza, monotropoid mycorrhiza, and orchid mycorrhiza (Finlay 2008). They provide protection against various environmental pressures such as water and metal toxicity stress and also involved in nutrient supply to plant. The heavy metal toxicity ameliorated with in the plant system by reducing metal translocation (Sharma and Dietz 2006). Thereby, this translocation mechanism help to plants for adaptation and survivorship in heavy metal contaminated locations. By another way, the selective advantage accommodated by the host plant to fungus for retained a contaminated site and metabolized by enzymatic degradation of various petroleum, polycyclic aromatic hydrocarbons and chlorinated aromatic pesticides, for examples, 2,4-dichlorophenoxyacetic acid (2,4-D) and atrazine (John 2013). Thus, the mycorrhizal fungi will be very significant to biodegradation of hazardous components in soil (Lehmann et al. 2011).

2.7 Endophytic Fungi

The microbial groups of plants endophyte such as bacteria and fungi have the capability of colonization within the plants without any harmful effects at each other (Rout 2014). They secrete various metabolites for the nitrogen fixation and methane assimilation and are resides inside the specific plant tissues such as the cortical roots, vascular bundles, apoplastic space, young buds, and also in the dead bark cell (Roshchina and Roshchina 2012). The saprophytic capabilities are also found in them to sustain in the dead litter (Kubartová et al. 2009). They also produced a selection of enzyme group such as cellulase, lipase, peroxidase, and protease for the biodegradation of ecological chemicals such as pesticides, herbicides, insecticides, petrochemicals, polychlorobiphenyls, polyaromatic hydrocarbons, polyester polyurethane etc. and bio transform the heavy metal in their lower states (Purohit et al. 2018). Like this, they promote the capability and adaptability of tolerance to pollutants and other toxicity such as heavy metal. So, these groups are employed as valuable apparatuses for the bioremediation (Hossain et al. 2012).

3 Pollutants Reducing Fungal Mechanism

The microorganisms have wider delivery in various environmental conditions and known as ubiquitous (Schmidt and Schaechter 2012). In the middle of numerous microbes, fungi have extensive flexibility and rapid receptiveness to stress condition, environmental catastrophes, and extreme climatic circumstances due to its opportunistic nature (Ingram et al. 2015). Fungi can degrade a chain of hazardous molecules into simpler, nontoxic, and complex hydrocarbons to biodegradable form and clean up the environment (Xue et al. 2015). Various fungi also have outstanding capability to bind metal ions, which includes the efflux of metal ions outside the cell and formation of metal ion complex and accumulation inside the cell, and after transformation they decrease the toxic metal ions to a nonhazardous state (Gadd 2007). The metals can immobilize, mobilize, or transform rendering them inactive or tolerate the uptake of heavy metal ions by various mechanisms (Tak et al. 2013). The adopted mechanisms by fungi for bioremediation are given following;

(I) Exclusion- By the formation of a permeable barrier the metal ions kept away from the target sites.
(II) Extrusion- by the active transportation the metals are pushed out of cells.
(III) Fixation- forming the complex with metal-binding proteins or other cell components like detoxification due to the enzymes, extracellular and intracellular sequestration, acid production dissolution of metals, chelation, the production of organic bases precipitation, metal precipitation extracellular by fix metals (Jomova and Valko 2011).
(IV) Biotransformation - The process such as methylation, demethylation, volatilization, oxidation, and reduction involved in transformation of toxic metal to less toxic forms. Generally, The mycoremediation of hazardous substances in

agro ecosystem in order to avail good air and water quality for future generations by the various approaches such as immobilization, mobilization, biosorption, and biotransformation (Bolan et al. 2014).

3.1 Immobilization

In this approaches the microbes' utilizes to alter the physical or chemical properties of pollutants for decreasing of its mobility. The substantially constraining contact among the pollutants or by chemically changing the pollutants can be proficient by these methods (Kumar et al. 2015). Mostly the polluted sites used the methods such as solidification and stabilization for the immobilization of the toxic contaminants (Guo et al. 2006). The water and stabilizer are used with appropriate amounts to mixing the contaminated materials by these methods. A solidified matrix formed by this toxic waste in this mixture (Dermatas and Meng 2003). The formation of metal hydroxides by injecting chemicals to the contaminated soil and heavy metal precipitated. The chemical composition of the site, the amount of water present, and temperature are the key environmental factors for the efficacious use of this machinery (Crane and Scott 2012). In both in situ or ex situ processes the stabilization and solidification techniques occurred. Though, the volatile or semi-volatile organics are used in situ methods for the treatment of surface or shallow contamination of soil (Evanko and Dzombak 1997).

3.2 Mobilization

The metabolites and siderophores, alkylation, methylation, and redox transformations mobilized the pollutants through leaching and chelation. The free metal cations released due to the acidification of soil environment (Gadd 2004). The production of low-molecular-weight organic acids produced through leaching, which provides protons and metal-complexing organic acid anions by the most of the fungi. Metallic zinc, MnO_2, Fe_2O_3, and rock phosphate solubilized through the mechanism of chelation and reduction due to the potentiality of *Tricoderma harzianum* (Johnson et al. 2004). The iron-chelating legends with low-molecular weight have the capability to bind the other metals like magnesium, manganese, chromium, gallium, and radionuclide by the production of siderophores (Purohit et al. 2018). An alkyl group transferred from one molecule to another molecule known as alkylation, which can be moved as an alkyl carbocation, a free radical, a carbanion, or a carbine (Tsarevsky and Matyjaszewski 2007). The methyl groups that is enzymatically transferred to a metal, forming a number of different metalloids involved in methylation. The microorganisms to mobilize metals, metalloids, and organometallic compounds by reduction or oxidation processes known as redox transformations (Bolan et al. 2014). Additionally, numerous metal-mobilization techniques can also occur in nature.

3.3 Biosorption

A physicochemical method which involves in biosorption and uptake of toxic chemicals cell surface of dead or inactive biological sources using machineries like adsorption, chelation, precipitation, reduction, ion exchange, and coordination with active functional groups such as amine, hydroxyl, carboxyl, phosphate, and sulfhydryl (Sharma et al. 2018). The biosorption process involves a biosorbent and solvent having dissolved materials. Due to the high percentage of cell wall materials the fungal biomass acts as biosorbents (Crini 2006). The functional groups present at the surface of biomass involved in metal binding and their sequestration by the fungi such as *Mortierella ramannianc*, *Rhizopus sexualis*, *R. stolonifer*, *Zygorhynchusheterogamus*, *Z. moelleri*, *Aspergillus niger*, *Mucor racemosus*, *Penicillium chrysogenum*, and *Trichoderma viride* (Chandra and Singh 2014). So, in complementary manner the biosorption contributed in to the overall sequestration of toxic pollutant even from very small concentrations (Mkandawire and Dudel 2007).

3.4 Biotransformation

Metal/metalloids and radionuclide can be biotransform by exploiting the microorganisms known as biotransformation (Chan et al. 2016). The microenvironment near the microbial cell modify by the exploiting microorganisms through methylation or demethylation by catalysis, oxidation, and reduction of the solubility/mobility of metal (Bolan et al. 2013). The possible physicochemical mechanism of interaction with metals or metalions together with other metabolically mediated mechanisms such as bioprecipitation and bioreduction contributed by the microorganisms (Gavrilescu 2004). The elements of carbon, oxygen, nitrogen and sulphur act as methyl group acceptors in the secondary metabolite processes (Brosnan and Brosnan 2006). The conversion into methylarsonic acid or to dimethylarsinic acid is one of the possible mechanisms for the detoxification of arsenate. The water soluble arsenate converted by the multistep process into volatile trimethylarsine in which the reduction of arsenate into arsenite initially in the presence of arsenate reductase (Dembitsky and Rezanka 2003). The sequence of methylation and reoxidation process continued and monitored by the intermediates of organo-arsenical with S-adenosylmethionine as the usual methyl donor (Messens and Silver 2006). The microbial methylation of heavy metals resulting in volatilization used for contaminated sites of in situ bioremediation successfully. Various species of fungi converted As (V) or As (III) to their organic forms like monomethyl arsenate (MMA), dimethyl arsenate (DMA) or tri-methyl arsenate TMA (V) or tri-methyl arsine TMA (III) due to methylation are also reported (Messens and Silver 2006). Various fungal strains, like *Trichoderma* sp., *Neocosmospora* sp. and *Rhizopus* sp. have efficiency of As removal earlier reported in various growth medium (Zhao et al. 2013).

3.5 Bioprecipitation

Surrounding the microbial cell through the metabolic mediated process it helps in the modification of the environment (Dupraz et al. 2009). The microbes grow by the transfer of electrons available from the electron donor molecule to the oxygen under aerobic conditions (Pearce et al. 2003). The reduction of oxygen into the water and mineralization of organic carbon in to the carbon dioxide increase the alkalinity and pH of the cell microenvironment, and the formation of excess bicarbonate favors metal ions precipitation as hydroxides of metal Me(OH)x or carbonate $Me_2(CO_3)x$ (Konhauser 1998).

3.6 Biological Oxidation/Reduction

Various microorganisms reduced and catalyzed the heavy metals by the enzymatic process such as Fe (III) to Fe (II), Mn (VI) to Mn (II), Cr (VI) to Cr (III), Se (VI) to Se (IV), As (V) to As (III), Mo (VI) to Mo (IV), and U (VI) to U (IV). In alternative microbial respiration these reduced elements serve as electron acceptors or reduced without energy production by the enzymes (Rabaey and Verstraete 2005). The organic substances have potential for bioremediation of polluted sites with the use of native microorganism's especially fungal biomass by the mechanism of immobilization, mobilization, biosorption, and biotransformation of metals/metalloids, radionuclides (Gadd 2007).

4 Heavy Metal Toxicity

Various natural and anthropogenic activities such as volcanic activity, industrialization, urbanization have release and leaching of heavy metals into the soil, water and environment a cause of anxiety because of the hereditary traits of heavy metals (Ganeshamurthy et al. 2016). Various mechanisms of toxicity are found in various heavy metals such as oxidative stress due to free radical imbalance; the harmful thiol or methyl derivative formed due to the arsenic, chromium and mercury; metal ion replacement caused by aluminum and cadmium; cell membrane permeability and imbalances in ion channels and damages DNA and protein disturbed by Cr (VI); corrosion, saturation, organ penetration and lipid peroxidation leads by iron (Ercal et al. 2001). The quality of soil and water severely affected along with limited biodegradation of organic pollutants in the environment by additional accessibility of these heavy metals (Wuana and Okieimen 2011). Soil respiration, mineralization of nitrogen and nitrification inhibited by extra metals entering in the soil caused decreasing the cultivation of crops (Brookes 1995). The grown crop in polluted soil of heavy metal faces the difficulty of phytotoxicity and nutrient deficiency (Marques et al. 2009). These contaminations of metals transferred in the crop products through food chain via irrigation of industrial waste water and caused the human health risks (Chary et al. 2008).

4.1 Characteristics of Heavy Metal Contamination of Soils

(i) Strong latency

The contamination of heavy metal is dull and odorless, so, the analysis is very difficult that it does not destruct the environment in a very short period. However, the environmental tolerance or environmental conditions has changed, when the pollutants exceeded in the soil and cause serious ecological damage after activation. Therefore, the heavy metal contamination may be often looking upon as Chemical Time Bombs (CTBs) (Gossel 2018).

(ii) Irreversibility and remediation hardness

The problem of contamination can be overturned absolutely by the weakening and self-purification after blocking off the sources of pollution. However, the machinery of dilution or self-purification is very difficult to remove the heavy metal pollution and to get soils amended. Various pollutants of soil take a lot of times as hundred years to bioremediation. So, the pollution of heavy metal requires comparatively very high cost of remediation and the remediation cycle is relation longer (Lemery and Auerbach 2017).

4.2 Detoxification Mechanism of Heavy Metals

The mechanism of metal resistance in plants is polygenic and takes place via two ways, such as the mechanism of avoidance and tolerance (Hasanuzzaman et al. 2013). The first line of defence initiated the heavy metal exposure involving by plant structures such as the cuticles, cell wall, and trichomes which act as anobstruction and inhibits their entry in the plants (Doughari 2015). After entering the heavy metals in plants roots, shoots, leaf different cellular mechanisms like chelation and compartmentalization, interactions with proteins, dislocation of vital metal ions from their specific binding sites and ROS production are started in order to dissolve their negative effect on plants (Fig. 15.1) (Alam and Pantola 2016). Uptake of heavy metal depends on various environmental factors like as interaction of genotype and plant's nutritional requirement by internal sequestration and translocation in plant (Clemens 2006). The induction of stress molecules such as metallothionein, phytochelatins, glutathione, organic acids, cellular exudates such as flavonoid and phenolic compounds, heat-shock proteins, amino acids and hormones (Emamverdian et al. 2015). The superoxide dismutase (SOD) mechanisms, induction, activation and synthesis of antioxidant catalase protected to plants against metal-induced oxidative stress (Gratão et al. 2005). Various factors such as metal types, plant species and their tolerance level, environmental circumstances are involved in metal detoxification mechanisms in plants (Clemens 2001). Various tolerance mechanisms of plants are followed.

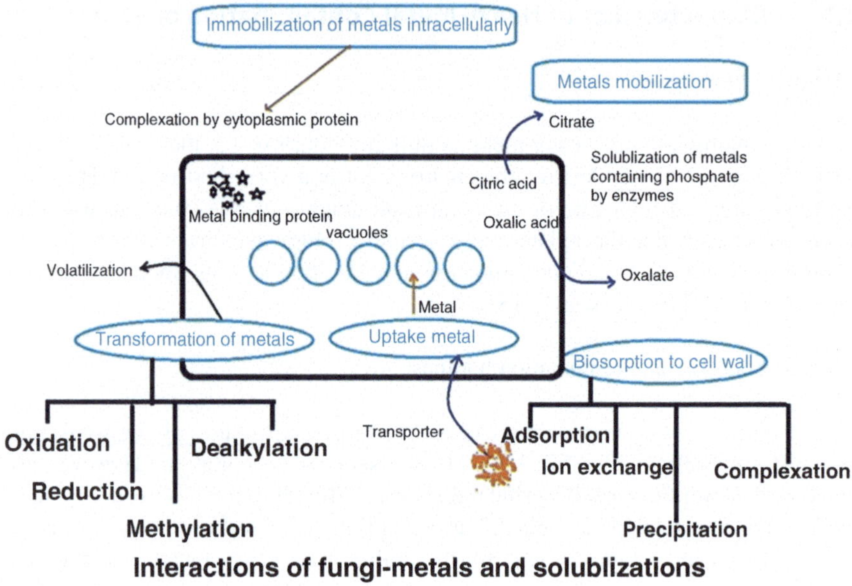

Fig. 15.1 Bioremediation mechanism of heavy metals by fungi

4.2.1 Phytochelatins

For the detection of heavy metal stress in plants the phytochelatins are testified as a biomarker (Sytar et al. 2013). And play an essential role in the detoxification of ions and the repairing of intracellular levels of essential metal ions (Hall 2002). The various metals like Cd, As, Cu, Zn, Hg, etc. initiated the phytochelatins productions in plants (Schutzendubel and Polle 2002). The synthesis of phytochelatin amounts depends on the plant species and the type of heavy metal after exposure (Rauser 1995). The three plant species *Onobrychis viciifolia*, *Lathyrus sativus* and *Medicago sativa* comparatively observed, in these observations under the Pb and Cu toxicity the more phytochelatins synthesized in *Onobrychis* as compared to the other two plant species (Zenk 1996). Additionally, the Pb was more effective in the synthesis of phytochelatin in all the three plants as compared to copper between the two heavy metals (Ha et al. 1999). Similar results were founds in plant root of *Solanum nigrum* L. was exposed to 200 μmol/L of Cu resulting in its accumulation in root and the production of phytochelatin was enhanced (Fidalgo et al. 2013). Thereby reducing its translocation from root to shoot found the over expression of phytochelatins in *Escherichia coli* for hyper accumulation of cadmium (Verbruggen et al. 2009). In cytosol from sulphur-rich and low-molecular-weight thiol glutathione (GSH) synthesized Phytochelatins in the presence of enzyme phytochelatin synthase (PCS) and are actively transported in vacuole in the form of metal-phytochelatin complex (Baghour 2017). The availability of GSH is another key factor to overcome and detoxify the oxidative stress caused by metal besides synthesis of phytochelatins (Jozefczak et al. 2012).

4.2.2 Glutathione

Three amino acids containing sulphur compound synthesized glutathione (Noctor et al. 2012). The formation of a peptide bond between γ-glutamate and cysteine in the presence of γ-glutamylcysteine synthetase 1 (GSH1) followed by addition of glycine by glutathione synthetase (GSH2) synthesized glutathione (Khullar and Reddy 2018). Stimulation of the sulphate uptake, reduction and its assimilation in order to fulfill the need of cysteine under the heavy metal stress produce high requirement of glutathione (Jozefczak et al. 2012). The contribution of cysteine residue to the antioxidant property of GSH and is also known as a substrate for rejuvenation of other antioxidants (Nordberg and Arner 2001). The high-affinity sulphate transporter gene ZmST1; 1 expression in maize unprotected to Cd, Zn and Cu also favors the role of glutathione. Consequently, various mechanisms involve in metal homeostasis, antioxidative defence and signal transduction to improve the heavy metal stress by GSH (Gill and Tuteja 2010).

4.2.3 Metallothioneins

Metallothioneins (MT) are low-molecular-weight cysteine-rich metal-binding protein molecules and gene-encoded polypeptides (Cobbett and Goldsbrough 2002). Biosynthesis of metallothionein is influenced by different factors such as cytotoxic agents, hormones, nutrient deprivation, heat shock and heavy metals such as Cd, Zn, Hg, Cu, Au, Ag, Co, Pb, Ni and Bi and regulated at the transcriptional level (Parmar et al. 2013). Under the different unfavorable conditions the expression of metallothionein genes is a part of general stress response and may be connected to heavy metal status indirectly (Ahsan et al. 2009). The physiological and pathophysiological processes such as homeostasis of essential metals (Zn and Cu), protection against oxidative stress (Matés et al. 1999), protecting cells against UV or ionic radiation and cytotoxic alkylating agents' reclamation of various heavy metals such as Cd, Pb, Cr (VI), Hg etc. and restraining oxygen free radicals, thus preventing the cells from apoptosis are finalized with the help of Metallothioneins (MT) (Mahmood et al. 2014). The greater appearance of MT1 in roots and MT2 and MT3 in seeds has been reported in various plants (Murphy et al. 1997). The MT is confined in cytosol and not transported to vacuoles like phytochelatins (Verbruggen et al. 2009). The growth and metabolism of microbes associated to the rhizospheric region of plants affecting the plants indirectly by the heavy metals and decrease the organic matter decomposition consequentially decline the nutrients in soil, and affects the growth of plants which sometimes results in death of plants (Spokas et al. 2012).

5 Bioremediation of Heavy Metal Contamination Through Fungi

The immobilization, evaporation, electroplating, toxicity reduction, physical separation, extraction, precipitation, ion exchange, etc. are the process of remediation of contaminated locations by the various approaches such as physical, chemical and biological (Mulligan et al. 2001). The physical and chemical approaches require high quantity of energy, very costly and non-eco-friendly has various demerits and merits (Chandra and Singh 2014). Currently, the various enzymes from the microorganisms are used for the elimination of toxic substances from the soil and water (Forgacs et al. 2004). The technologies based on microorganisms completely mineralized to the pollutants through iostimulation, bio-augmentation, bioaccumulation, biosorption, phytoremediation and rhizoremediation. So, the use of such microbial approaches for the elimination of heavy metals in a cheap to run and friendly with environments would be a vital solution for pollution abatement (Ayangbenro and Babalola 2017). The application of fungi at heavy metal-contaminated sites for reduction is pioneering machinery and it is known as mycoremediation. Due to the ubiquitous and dominant nature of fungi along with high surface area is a promising approach over conventional methods. In soil these fungi dominates because of their diversify nature and existence in risky conditions (Arnold 2007). The pollutants extracted and accumulated in the mycelia or fruiting bodies of fungi (Nasr and Arp 2011). The specific heavy metal pollutants such as arsenic (As), lead (Pb) and cadmium (Cd) targeted by the various fungal species known as mycoremediation (Pierart et al. 2015). Various fungal species have potential to bioremediation of heavy metal pollutants such as *Aspergillus niger*, *Aureobasidium pullulans*, *Cladosporium resinae*, *Funalia trogii*, *Pleurotus tuberregium*, *Ganoderma lucidum* and *Penicillium* sp. (Singh et al. 2018) (Table 15.2). Various types of mechanisms with contribution of numerous enzymes has capability to alleviation of heavy metal pollutants from the soil and waste water effluent, and are found in various groups of fungi (Jeffries et al. 2003) such as *Zygomycotina, Ascomycotina, Deuteromycotina* and *Basidiomycotina*shown in (Table 15.3). The isolation of various genera such as *Aspergillus, Penicillium* and *Cladosporium* from the oxidation ponds of sewage treatment plants has high level of resistance for various metal pollutants and offering them as attractive potential applicants for further analysis (Table 15.3) (Saranraj and Stella 2014). The harmful pollutants converted into harmless pollutant products by the enzymatically mechanisms, these processes of mycoremediation depends on fungi (Harms et al. 2011).

6 PAHs as Contaminants in Soil

6.1 Physicochemical and Toxicology Properties of PAHs

The physicochemical properties is controlled the fate of PAHs in the environment, also is well established (Kim et al. 2013). Generally the PAHs are nonpolar

Table 15.2 Fungi involved in mycoremediation of heavy metals

Fungi	Heavy metals	Mechanism Involved	References
Circinella sp.	Ni	Biosorption of Ni	Rashid et al. (2016)
Cunninghamella sp.	Pd	Biosorption	Sugasini et al. (2014)
Mucor sp.	Pb, Cd, Ni, Zn, Cu	Bioaccumulation	Deng et al. (2011)
Rhizopus sp.	Pb, Cd, Cr, Cu, Ni, Zn	Biosorption	Sud et al. (2008)
Trichoderma sp.	Cu, Pb, Cd, As, Zn	Bioaccumulation	Deng et al. (2014)
Penicillium sp.	Zn, Ni, Pb	Biosorption, bioaccumulation	Özdemir et al. (2009)
Paxillus involutus	Cd	Accumulation of Cd in vacuole	Hall (2002)
Phanerochaete chrysosporium	Pb, Cd	Bioaccumulation	Xu et al. (2016)
Aspergillus lentulus	Pb, Cu, Cr, Ni	Bioadsorption	Ahemad and Kibret (2013)
Pleurotus ostreatus	Pb, Cd	Bioaccumulation	Javaid et al. (2011)
Pisolithus sp.	Pb, Cu, Cr, Ni, Zn	Bioadsorption	Gonen et al. (2008)
Hymenoscyphus ericae	Zn, Cd and Fe	Absorption	Colpaert et al. (2011)
Agaricus bisporus, *Lactarius piperatus*	Cadmium (II) ions	Biosorption process	Nagy et al. (2014)
Fomes fasciatus	Copper (II)	Biosorption process	Sutherland and Venkobachar (2010)
Pleurotus platypus, *Agaricus bisporus*, *Calocybe indica*	Copper, zinc, iron, cadmium, lead, nickle	Biosorbent	Chatterjee and Abraham (2017).
Flammulina velutipes	Copper	Biosorbent	Encarnacion et al. (2012)
Pleurotus tuber- regium	Heavy metals	Biosorption	Nnorom et al. (2013)
Pleurotus ostreatus	Cadmium	Biosorption mechanism	Tay et al. (2011)
Pleurotus sajor-caju	Heavy metal Zn	Biosorption mechanism	Soden and Dobson (2001)

compounds with their physicochemical possessions such as lower water solubility, higher the hydrophobicity or lipophilicity, lower organic carbon partition coefficient and very high bioconcentration factor (Saichek and Reddy 2005; Duan et al. 2015). Normally, PAHs having lower molecular weight (LMW) are more volatile, more solubility in water, and less lipophilic comparison to high molecular weight (HMW) PAH. PAH persistence in the environment increased due to the hydrophobicity and electrochemical stability when the size and angularity of a PAH molecule increased (Lamichhane et al. 2016). PAHs are the first renowned environmental carcinogens, mutagen and toxic. The genotoxicity of PAH also increased with size, up to at least more than four fused benzene rings (Kim et al. 2013).PAHs are also caused

Table 15.3 Different class of fungi producing different heavy metal metabolizing enzymes

Enzymes occurrence in nature	Mode of mechanism	Taxon order	References
Laccases – extracellular	Oxidation of organic compounds(O_2-dependent)	*Ascomycota* and*Basidiomycota*	Castellet Rovira (2018)
Tyrosinases -mainly intracellular sometime extracellular	Cresolase activity and catechols	*Ascomycota*, and *Basidiomycota*	Flurkey et al. (2008)
Peroxidases – extracellular	Oxidation of organic compounds(H_2O_2-dependent); oxidation of Mn^{2+} to Mn^{3+}	*Basidiomycota*	Zhang et al. (2014)
Peroxidases – extracellular	Oxidation of aromatic compounds (H_2O_2-dependent)	*Basidiomycota*	Cameron et al. (2000)
Manganese, peroxidase-extracellular	Oxidation of Mn^{2+} to Mn^{3+}compounds (H_2O_2-dependent)	*Basidiomycota*	Alam et al. (2013)
Dye-decolorizingperoxidase – extracellular	Oxidation of organic compounds (H_2O_2-dependent) hydrolysing activity	*Basidiomycota*	Hong et al. (2017)
Caldariomyces fumago haem-thiolate chloroperoxidase-extracellular	Halogenation of organic compounds in the presence of halides (H_2O_2-dependent); oxidation of phenols andanilines in the absence of halides	*Ascomycota*	Singh et al. (2018)
Reductive dehalogenase – membrane bound	Membrane-bound glutathione S-transferase and glutathione conjugate reductase activity	*Ascomycota*, *Basidiomycota*	Stanic (2017)
Cytochrome P450 – membranebound	Reduction reaction	*Ascomycota*, *Basidiomycota*,	Moktali et al. (2012)
Haem thiolate peroxygenases – extracellular	Peroxigination of aromatic, aliphatic and alicyclic compounds (H_2O_2-dependent); bromination (H_2O_2-dependent); sulphoxidation	*Basidiomycota*	Singh et al. (2018)
Lignin peroxidizes – extracellular	Oxidation of aromatic compounds (H_2O_2-dependent)	*Basidiomycota*	Leonowicz et al. (2001)
Nitroreductases – cell bound	Reduction of nitroaromatic compounds (NADPH dependent)	*Ascomycota*, *Basidiomycota*,	Bugg et al. (2011)

potential risk to the soil ecosystems besides human and animal health risks. The US Environmental Protection Agency (EPA), the World Health Organization (WHO) classified 16 PAH as priority contaminants (Abrahams 2002).

6.2 Origen of PAHs Contamination Soil

During the thermal decomposition of organic molecules and their subsequent recombination the formation of PAHs takes place (Richter and Howard 2000). PAHs produced after incomplete combustion organic materials at high temperature (500–800 °C) or subjection of organic material at low temperature (100–300 °C) for a long period (Keiluweit et al. 2010) Various natural activities such as volcanoes, forest fires etc. and anthropogenic activities like tobacco smoke, burning of wood, fossil fuel, application of pesticides in agriculture or for wood protection, municipal solid waste incineration, and industrial activity related with petroleum refining and transport produced and released the PAHs in to the environment (Zhang and Tao 2009). So, the PAHs concentration varied in soils and its depend on the industrial development of sites. The PAHs in various municipal areas observed between 0.10 and 56.90 mg kg^{-1} (dry soil) of the upper soil layer of Ensenada City (Argentina) (Morelli et al. 2013).

6.3 Metabolic Pathways for Fungal Transformation of Pollutants

Due to the large array of inter unit linkages lignin is a complex aromatic hetero polymer having both chemical and biological transformation is one of the most recalcitrant compounds (Beckham et al. 2016). So, a key step in the carbon cycle represents its removal. Having the degradation and mineralization bases capability the fungi classified in mainly in to groups such as ligninolytic fungi i.e. white-rot fungi (WRF), growing on woods and leaf litter of forest, and another non-ligninolytic fungus, having various metabolic strategies to transforms PAHs and other pollutants (Rouches et al. 2016). The ligninolytic fungi have capability to degrade and mineralize the aromatic ring of lignin and PAHs (Pointing 2001). In the intracellular metabolism the enzymatic systems involved of PAHs by the fungi are cytochrome P450 monooxygenases and epoxide hydrolases including either ligninolytic or non-ligninolytic fungi (Bezalel et al. 1997). *Phanerochaete chrysosporium* and *Pleurotus ostreatus* are lignolytic fungi produce both non-ligninolytic and ligninolytic type enzymes, but the involvement of enzymes to the transformation of the PAH molecule is not clear (Kadri et al. 2017).

6.4 Ligninolytic System from White Rot Fungi (WRF)

Three p-hydroxycinnamyl alcohols, p-coumaryl, coniferyl, and sinapylalcohols, and their acylated forms are the result of three dimensional bulky polymer of lignin (Martínez et al. 2008). Lignin synthesized by the plants is the recalcitrant compounds found abundantly in trees and contributed mainly to wood strength of plant (Abril et al. 2011). Due to the heterogeneous structure and high molecular size of lignin, the fungal cell attacked outside the plant with the involvement of extracellular enzymes through the complex oxidative and unspecific system (Pérez et al. 2002). The WRF degrade the lignin by enzymatic incineration and the connected act of those extracellular mechanisms and the reductive reactions approved out by the cell-bound systems look like to regulate the efficiency of WRF to reduce and mineralize the lignin and other compounds like aromatic molecules, humic acids, xenobiotic such as PAHs (Chritian 2001) Similarly, after degradation the simple products entered in the hyphae of fungi and conformed into catabolic routes intra cellularly (Kirk and Farrell 1987). The ligninolytic peroxidases, laccases, oxidases are responsible for the production of extracellular hydrogen peroxide (H_2O_2), and reductases and the low substrate specificity have extracellular ligninolytic enzymatic systems (Dashtban et al. 2010). Various characteristic features of depends upon the fungi, strains, culture conditions and exhibited the enzyme systems, and regulated by various nutrients and chemical agents i.e. nitrogen and their concentration level (Sivaramakrishnan et al. 2006). WRF and ligninolytic enzyme synthesis deteriorates the various patterns of lignocellulose by involvement of a single enzyme orvarious other enzyme complexes due to the synergistic effects and equally participated in the degradation of lignin actively (Valášková 2010). For the catalysis of enzymes peroxidases require H_2O_2 as co-substrate.

(i) Lignin peroxidases (LiP, E.C.1.11.1.14) due to the high redox potential it has capability to oxidize the phenolic and non-phenolic lignin units directly.
(ii) Manganese peroxidases (MnP, E.C.1.11.1.13) specially act on phenolic units, and in the presence of mediators on non-phenolic units; generates MnP3 by this enzyme, and which acts on phenolic or nonphenolic lignin units as a diffusible oxidizer with the help of lipid peroxidation reactions.
(iii) Versatile peroxidases (VP, E.C.1.11.1.16), is a ligninolytic peroxidase and combines to catalytic properties of LiP and MnP, and microbial peroxidases oxidized phenolic compounds; and have capability to oxidized azo dyes instead of MnP or LiP.

P-diphenol: oxygen oxido reductase; EC 1.10.3.2 (Laccase), which participated in another processes of physiology in various fungal groups to be or not ligninolytic ones, and oxidize specially the phenolic units of lignin, likewise to the ligninolytic peroxidases, and in the presence of mediators it also acts on non-phenolic units (Zucca et al. 2014). Aromatic amines, phenolic compounds such as chlorophenols, anthraquinone dyes, and secondary aliphatic polyalcohals, anthracene, ferrocyanide, ferrocenes, and cytochrome c oxidized by Laccases or an electron transferred directly from an electrode. The peroxidases oxidized also these substrates. Though, the laccases, do not require hydrogen peroxide unlike peroxidases, rising the attention of biotech companies (Kirk and Farrell 1987). Moreover, the oxidation of azo and indigo dyes and in the production of active oxygen species the laccases also participated (Polak and Jarosz-Wilkolazka 2012). And the Laccases cannot oxidize other PAHs compounds. The laccase- mediator system (LMS) constituted by the laccase-substrate couple, and the free radicals generated by the laccases oxidizes the compounds (Morozova et al. 2007). The peroxidases are less active comparison to the laccase-mediator system (LMS). Consequently, the chlorine-free bleaching of paper pulp and the xenobiotic compounds degraded by LMS (Couto and Herrera 2006). The oxidation of lignin by one-electron as well as other aromatic compounds such as PAHs catalyzed by the laccases and peroxidases (Hofrichter 2002). Non-phenolic units and phenolic aromatic units generated phenoxy or cationic radicals, respectively (Crestini et al. 2010). The bond cleavage and the reaction with O_2 or water and active oxygen species evolved by the various non-enzymatic reactions to contribute hydroxy- or keto-derivatives, and produced a selection of acids and quinones (Tabibzadeh 2016). On another hands, glyoxal (GLOX, E.C.1.2.3.5) and aryl-alcohol (AAO, E.C.1.1.3.7) two extracellular oxidases, reported for the production of extracellularH_2O_2 (Morelli et al. 2013). The ligninolytic peroxidases required H_2O_2 but it can be involved as the precursor of hydroxyl radical (•OH) in the degradation of lignocellulose (Dashtban et al. 2010). And the fungi produced strongest oxidizing agent by the iron-catalyzed Haber–Weiss reaction. On the bases of theoretical performance previously proposed that in the initial stages of wood decay attack on lignin and initiated (Schiffer 1986). The cell wall prevents the penetration of ligninolytic enzymes due to the small size of pores in the still- intact. For the production of extracellular hydroxyl radical the quinones involved in redox cycling and oxygen activation, and peroxidases and laccases, and quinone reductases also reported for this reaction (Li and Jia 2008). Moreover, the cellobiose dehydrogenase (CDH, E.C.1.1.99.18) and cellobiose–quinone oxidoreductase (CQO, E.C.1.1.5.1) are extracellular reductases may catalyzed the phenol products reduction are proteolytic product of CDH, which is derived from lignin degradation, avoiding their subsequent repolymerizating, interconnected and act between lignolysis and cellulolysis (Morelli et al. 2013).

6.5 Mycoremediation of PAH-Contaminated Soil

The biological systems areattached to affect and the cleaning of environmental contaminants by the various technological approaches known as bioremediation. The microorganisms have capability to degrading the particular pollutants towards mineralization and the production of nonhazardous metabolites (Megharaj et al. 2011). The rate of diffusion and bioavailability, concentration, physical and chemical properties of PAH in soil caused effective biodegradation (Johnsen et al. 2005). The mycoremediation of pollutants depends upon the quantities and the categories of PAH degrading fungi and it depends upon the soil type, moisture content, aeration conditions, pH, temperature, and availability of carbon substrates, nitrogen and other nutrients also (Ennis et al. 2012). The cleanup of PAH- contaminated soil based on the fungal based remediation is a favorable performance of bioremediation (mycoremediation). Compared to other microbes the rates of PAHs degradation are slow and inefficient but the fungi have ability to degradation of HMW PAHs by various enzymatic activities (Harms et al. 2011). So, due to their capability to growing various environments and physiological versatility with less concentration of nutrients, less moisture contents, acidic pH and their filamentous nature of growing to allows to an ability to colonization and assessment of polluted soil, and the mycelia of fungi to constituted a bulky portion of soil biomass, arrange for to fungi a significant biological role, subsidizing significantly to the biotransformation of PAHs in soil contents (Leitão 2009). The microbial degradation denotes that the biological recovery of PAH contaminated sites responsible through major route. Though, the bioremediation projects have been success in limited by to remove HMW of PAHs. The exoenzymes of fungi initial attack on HMW PAHs in soil look like to be expected than attack by bacterial intracellular enzymes (Fulekar 2017).

7 Mycoremediation Technology in PAH-Contaminated Soil

7.1 Bioaugmentation

The bioremoval of pollutants and hazardous substances by the specific fungi or other microbes and improves the capability of removal by the technology defined as bioaugmentation (Tadkaew et al. 2011). The filamentous fungi particularly WRF used for the bioaugmentation techniques in mycoremediation (Wu et al. 2008). The unspecific enzymes synthesized by WRF can degrade HMW PAH due to their capability, and their growth of hyphae allows them to infiltrate and to diffuse into soil (Covino 2010). Heavy metal pollutants and PAH contaminated soil derived from the same sources as PAH (Saeedi et al. 2012).

7.2 Enzymatic Treatment

The enzymatic treatment of PAH-contaminated soil is an alternative of conventional bioremediation. The various other advantages present comparison to other microorganisms in the degradation of organic compounds using enzymatic activity (Mohan et al. 2006). Furthermore, the enzymes have capability to penetration of substrates in pores with the small dimensions, and approximately100 folds smaller than bacteria (Lynd et al. 2002). The enzymatic remediation of soil is limited, published in various reports and the price of these biocatalysts is too high for the implementation of commercially enzyme built managements in the field (Harms et al. 2011). The enzyme laccase have more efficiency and safe remediator for soil. Free laccase transforms various PAHs components very efficiently and decreased the genotoxicity and not affected the soil microbial community (Morelli et al. 2013).

7.3 Mycorrhizal Fungi and Xenobiotic Degradation

Most of the plant species have mutualistic associations with mycorrhizal fungi in the rhizosphere regions (Marschner 2012). In the majority of plants the symbiotic association of mycorrhiza uptake nutrients and approximately, 95%of land plants form another types of mycorrhizal associations (Brundrett 2009). The ectomycorrhizas, arbuscular mycorrhizas, ericaceous mycorrhizas, andorchid mycorrhizas known as mycorrhizal associations. The soil explored by the extra radical hyphae networks and mineralized and absorbs the nutrients and translocates them to the roots and benefitted to the plants (Smith and Read 2010). The chemical composition of root exudates changes by AM symbiotic status and influences the pH of soil, improved the soil structure and due to affecting the microbial populations in the rhizospheric regions (Johansson et al. 2004).AM extra radical mycelia produced exudates increasing the bacterial growth, vitality and change the composition of bacterial community and also influence the growth and development of bacterial communities (Toljander et al. 2007). Compared to the rhizosphere of non-mycorrhizal roots changed the bacterial communities in the mycorrhizosphere. The AMF *Glomus intraradices* and the plant growth promoting rhizobacteria *Paenibacillus polymyxa* in the mycorrhizosphere of *Cucumis sativus* interact among each other (Priyadharsini et al. 2016). The contaminated soil with heavy metals remediaton by AMF observed in maximum reports comparison to organic pollutants (Marques et al. 2009). *Acacia melanoxylum, Cytisus striatus, Allium cepa and Trifolium pretense* plant species and the fungi *Glomus deserticola* isolated from HCH- polluted soil inoculated in these plants enhanced the growth and colonization of root fungi, when these plants transferred to HCH- polluted soil comparison to other fungus *G. macrocarpum* or olonized by a consortium of indigenous AM fungal species (Azaizeh et al. 2011). Translocation of soil metals or uptake by plants influenced by Mycorrhizal fungus, in response to plant metal stress is variable by AMF. Metsulfuron-methyl in simulated *Triticum asetivum* L. in rhizospheric soil degraded and promoted by inoculation with *Penicillium* sp., the ectomycorrhizal fungi (ECMF) also degraded to organic contaminants such as PAHs (Duke et al. 2012). The fungal-bacterial co-cultures of *Penicillium janthine* and *Stenotrophomonas maltophilia* inoculated into the contaminated soil of PAH, it reduced and improved degradation of HMW PAHs significantly, mineralization of α- benzo pyrene and reduced the mutagenicity of organic soil extracts, comparison to other indigenous microbes and amend the soil with only axenic inoculum (Azaizeh et al. 2011). Four ectomycorrhizal fungi (ECMF) species, *Boletus edulis, Gomphidius viscidus, Laccaria bicolor*, and *Leccinum scabrum*, degraded to 1, 1, 1-trichloro-2, 2-bis (4-chlorophenyl) ethane (DDT) using a pathway similarly found in WRF (Read and Perez-Moreno 2003). Significantly, the ECMF play an important role in the nutrient cycling by degrading complex of minerals or organic substances present in soil and making them available to the host plant.

7.4 Disadvantages and Advantages of Mycoremediation

The biosorption, chelation on to polymers and forming insoluble oxalates of metals can easily immobilize the PTEs. The variety of contaminants in the soil mineralized by the nonspecific fungal enzymes such as laccase, lignin peroxidase, and Mn peroxidase and produced food or energy sources. Some fungi like mushroom degraded the organic pollutants such as PAHs, dyes, petroleum, hydrocarbon, PCBs, PTEs, pesticides and plastics by the mushroom fruiting bodies. The production of extracellular enzymes by the mushroom fungi reduces the toxicity in to the food chain and also reduces the risks of human health. The fruiting body of mushrooms acts as a storage pool of contaminants, these fruiting bodies used as food materials and for future research in bioremediation. On the bases of large scale availability the technology of mycoremediation can be succeed, the pollutants in soil to be absorbed with in the mycelium range and favorable conditions for the growth of mycelium. These physiological, biological and environmental aspects of mushroom cultivation covered all the characteristic features. Another methods currently in use for PTE abolition and has many advantages of mycoremediation. Generally, the growth of mycelia a well-engineered by the robust technique for the addressing PTE contaminants by mycoremediation. The mycoremediation are in situ or ex situ and eco-friendly which requires a smaller space low cost and managed in anywhere in the fields. The cultivation and harvesting of mushrooms on growing industrial effluents such as sludge, solid, fluidly waste and wastes generated from farmyards which entered through the food chain may be more toxic and pose risks to the human health. The contamination sites with organic and inorganic pollutants bioremediate very difficult and slow because of conflicting effects on the mobilization or immobilization of trace elements by the microorganisms. So, the nature of substrate should be measured when the mushrooms grown on polluted sites.

8 Conclusion and Future Perspective

The mycoremediation of environmental pollutants from the polluted soil utilizing the enzymes and biomass of various fungi. It is also known as ecofriendly or green technology and has more potential comparison to other conventional approaches. The symbiotic relationship of plant and fungi detoxify the hazardous substances, translocate and accumulate in the fruiting bodies of biomass, and then cultivated for the recovery of metals. It has extensively used on a large scale and not commercialized, there are various observations found for uptake, detoxification and accumulation of heavy metal/metalloid at the laboratory as well as field scale in model plants. In various studies the mycoremediation has been conducted on laboratory scale using various environmental factors. So, the further study for the measurement of taking capacities and their applicability at large-scale in contaminated fields. Additionally, the enzymatic activities required for the biodegradation of newly added contaminants due to the increasing of industry contamination by newly isolated fungus. The genetic engineering of newly isolated fungi with current biotechnologies utilizing the whole cell of fungi and their enzymatic activity for the mycoremediation will pave the way to future.

References

Abrahams, P. W. (2002). Soils: Their implications to human health. *Science of the Total Environment, 291*(1–3), 1–32.

Abril, N., Gion, J. M., Kerner, R., Müller-Starck, G., Cerrillo, R. M. N., Plomion, C., & Jorrin-Novo, J. V. (2011). Proteomics research on forest trees, the most recalcitrant and orphan plant species. *Phytochemistry, 72*(10), 1219–1242.

Abuhussein, A. (2018). Wastewater refining and reuse and city-level water decision making.. Electronic Thesis and Dissertation Repository. 5310. https://ir.lib.uwo.ca/etd/5310.

Adenipekun, C. O., & Lawal, R. (2012). Uses of mushrooms in bioremediation: A review. *Biotechnology and Molecular Biology Reviews, 7*(3), 62–68.

Adenipekun, C. O., Ipeaiyeda, A. R., Olayonwa, A. J., & Egbewale, S. O. (2015). Biodegradation of polycyclic aromatic hydrocarbons (PAHs) in spent and fresh cutting fluids contaminated soils by *Pleurotus pulmonarius* (Fries). Quelet and *Pleurotus ostreatus* (Jacq.) Fr. P. Kumm. *African Journal of Biotechnology, 14*(8), 661–667.

Ahemad, M., & Kibret, M. (2013). Recent trends in microbial biosorption of heavy metals: A review. *Biochemistry and Molecular Biology, 1*(1), 19–26.

Ahsan, N., Renaut, J., & Komatsu, S. (2009). Recent developments in the application of proteomics to the analysis of plant responses to heavy metals. *Proteomics, 9*(10), 2602–2621.

Alam, A., & Pantola, R. C. (2016). *Intracellular copper accumulation and biochemical changes in response to Cu induced oxidative stress in brassica species*. San Francisco: GRIN Publishing.

Alam, M. N., Bristi, N. J., & Rafiquzzaman, M. (2013). Review on in vivo and in vitro methods evaluation of antioxidant activity. *Saudi Pharmaceutical Journal, 21*(2), 143–152.

Alloway, B. J. (2013). Sources of heavy metals and metalloids in soils. In *Heavy metals in soils* (pp. 11–50). Dordrecht: Springer.

Alluri, H. K., Ronda, S. R., Settalluri, V. S., Bondili, J. S., Suryanarayana, V., & Venkateshwar, P. (2007). Biosorption: An eco-friendly alternative for heavy metal removal. *African Journal of Biotechnology, 6*(25), 2924–2931.

Alvarez, A., Saez, J. M., Costa, J. S. D., Colin, V. L., Fuentes, M. S., Cuozzo, S. A., Benimeli, C. S., Polti, M. A., & Amoroso, M. J. (2017). Actinobacteria: Current research and perspectives for bioremediation of pesticides and heavy metals. *Chemosphere, 166*, 41–62.

Anastasi, A., Tigini, V., & Varese, G. C. (2013). The bioremediation potential of different ecophysiological groups of fungi. In *Fungi as bioremediators* (pp. 29–49). Berlin/Heidelberg: Springer.

Anasonye, F., Winquist, E., Räsänen, M., Kontro, J., Björklöf, K., Vasilyeva, G., Jørgensen, K. S., Steffen, K. T., & Tuomela, M. (2015). Bioremediation of TNT contaminated soil with fungi under laboratory and pilot scale condition. *International Biodeterioration and Biodegradation., 105*, 7–12.

Andreoni, V., & Gianfreda, L. (2007). Bioremediation and monitoring of aromatic-polluted habitats. *Applied Microbiology and Biotechnology, 76*(2), 287–308.

Arantes, V., Jellison, J., & Goodell, B. (2012). Peculiarities of brown-rot fungi and biochemical Fenton reaction with regard to their potential as a model for bioprocessing biomass. *Applied Microbiology and Biotechnology, 94*(2), 323–338.

Arnold, A. E. (2007). Understanding the diversity of foliar endophytic fungi: Progress, challenges, and frontiers. *Fungal Biology Reviews, 21*(2–3), 51–66.

Ayangbenro, A. S., & Babalola, O. O. (2017). A new strategy for heavy metal polluted environments: A review of microbial biosorbents. *International Journal of Environmental Research and Public Health, 14*(1), 94.

Azaizeh, H., Castro, P. M., & Kidd, P. (2011). Biodegradation of organic xenobiotic pollutants in the rhizosphere. In *Organic xenobiotics and plants* (pp. 191–215). Dordrecht: Springer.

Azmi, W., Sani, R. K., & Banerjee, U. C. (1998). Biodegradation of triphenylmethane dyes. *Enzyme and Microbial Technology, 22*(3), 185–191.

Baghour, M. (2017). *Effect of seaweeds in phytoremediation. Biotechnological applications of seaweeds* (pp. 47–83). New York: Nova Science Publishers.

Bahn, Y. S., Xue, C., Idnurm, A., Rutherford, J. C., Heitman, J., & Cardenas, M. E. (2007). Sensing the environment: Lessons from fungi. *Nature Reviews Microbiology, 5*(1), 57.

Baldrian, P. (2003). Interactions of heavy metals with white-rot fungi. *Enzyme and Microbial Technology, 32*(1), 78–91.

Bamforth, S. M., & Singleton, I. (2005). Bioremediation of polycyclic aromatic hydrocarbons: Current knowledge and future directions. *Journal of Chemical Technology and Biotechnology, 80*(7), 723–736.

Beckham, G. T., Johnson, C. W., Karp, E. M., Salvachúa, D., & Vardon, D. R. (2016). Opportunities and challenges in biological lignin valorization. *Current Opinion in Biotechnology, 42*, 40–53.

Bezalel, L., Hadar, Y., & Cerniglia, C. E. (1997). Enzymatic mechanisms involved in phenanthrene degradation by the white rot fungus *Pleurotus ostreatus*. *Applied and Environmental Microbiology, 63*(7), 2495–2501.

Bisht, S., Pandey, P., Bhargava, B., Sharma, S., Kumar, V., & Sharma, K. D. (2015). Bioremediation of polyaromatic hydrocarbons (PAHs) using rhizosphere technology. *Brazilian Journal of Microbiology, 46*(1), 7–21.

Bolan, N. S., Choppala, G., Kunhikrishnan, A., Park, J., & Naidu, R. (2013). Microbial transformation of trace elements in soils in relation to bioavailability and remediation. In *Reviews of environmental contamination and toxicology* (pp. 1–56). New York, NY: Springer.

Bolan, N., Kunhikrishnan, A., Thangarajan, R., Kumpiene, J., Park, J., Makino, T., & Scheckel, K. (2014). Remediation of heavy metal(loid)s contaminated soils–to mobilize or to immobilize. *Journal of Hazardous Materials, 266*, 141–166.

Brookes, P. C. (1995). The use of microbial parameters in monitoring soil pollution by heavy metals. *Biology and Fertility of Soils, 19*(4), 269–279.

Brosnan, J. T., & Brosnan, M. E. (2006). The sulfur-containing amino acids: An overview. *The Journal of Nutrition, 136*(6), 1636S–1640S.

Brundrett, M. C. (2009). Mycorrhizal associations and other means of nutrition of vascular plants: Understanding the global diversity of host plants by resolving conflicting information and developing reliable means of diagnosis. *Plant and Soil, 320*(1–2), 37–77.

Bugg, T. D., Ahmad, M., Hardiman, E. M., & Rahmanpour, R. (2011). Pathways for degradation of lignin in bacteria and fungi. *Natural Product Reports, 28*(12), 1883–1896.

Cabral, L., Soares, C. R. F. S., Giachini, A. J., & Siqueira, J. O. (2015). Arbuscular mycorrhizal fungi in phytoremediation of contaminated areas by trace elements: Mechanisms and major benefits of their applications. *World Journal of Microbiology and Biotechnology, 31*(11), 1655–1664.

Camacho-Morales, R. L., Guillén-Navarro, K., & Sánchez, J. E. (2017). Degradation of the herbicide paraquat by macromycetes isolated from southeastern Mexico. *3 Biotech, 7*(5), 324.

Cameron, M. D., Timofeevski, S., & Aust, S. D. (2000). Enzymology of *Phanerochaetechrysosporium* with respect to the degradation of recalcitrant compounds and xenobiotics. *Applied Microbiology and Biotechnology, 54*(6), 751–758.

Castellet, R. F. (2018). Fungal biodegradation of pharmaceutical active compounds in wastewater. https://ddd.uab.cat/record/189612

Chan, W. K., Wildeboer, D., Garelick, H., & Purchase, D. (2016). Mycoremediation of heavy metal/metalloid-contaminated soil: Current understanding and future prospects. In *Fungal applications in sustainable environmental biotechnology* (pp. 249–272). Cham: Springer.

Chandra, P., & Singh, D. P. (2014). Removal of Cr (VI) by a halotolerant bacterium *Halomonas* sp. CSB 5 isolated from sāmbhar salt Lake Rajasthan (India). *Cellular and Molecular Biology, 60*(5), 64–72.

Chary, N. S., Kamala, C. T., & Raj, D. S. S. (2008). Assessing risk of heavy metals from consuming food grown on sewage irrigated soils and food chain transfer. *Ecotoxicology and Environmental Safety, 69*(3), 513–524.

Chatterjee, A., & Abraham, J. (2017). Efficient management of e-wastes. *International journal of Environmental Science and Technology, 14*(1), 211–222.

Chiu, S. W., ChingML, F. K. L., & Moore, D. (1998). Spent oyster mushroom substrate performs better than many mushroom mycelia in removing the biocide pentachlorophenol. *Mycological Research, 102*(12), 1553–1562.

Chritian, V. (2001). *Enzymes of lignin-degrading fungi: Degradation of xenobiotic compounds* (Doctoral dissertation). Saurashtra University.

Clemens, S. (2001). Molecular mechanisms of plant metal tolerance and homeostasis. *Planta, 212*(4), 475–486.

Clemens, S. (2006). Toxic metal accumulation, responses to exposure and mechanisms of tolerance in plants. *Biochimie, 88*(11), 1707–1719.

Cobbett, C., & Goldsbrough, P. (2002). Phytochelatins and metallothioneins: Roles in heavy metal detoxification and homeostasis. *Annual Review of Plant Biology, 53*(1), 159–182.

Coleman, D. C. (2008). From peds to paradoxes: Linkages between soil biota and their influences on ecological processes. *Soil Biology and Biochemistry, 40*(2), 271–289.

Colpaert, J. V., Wevers, J. H., Krznaric, E., & Adriaensen, K. (2011). How metal-tolerant ecotypes of ectomycorrhizal fungi protect plants from heavy metal pollution. *Annals of Forest Science, 68*(1), 17–24.

Couto, S. R., & Herrera, J. L. T. (2006). Industrial and biotechnological applications of laccases: A review. *Biotechnology Advances, 24*(5), 500–513.

Covino, S. (2010). In vivo and in vitro degradation of aromatic contaminants by white rot fungi. *A case study: Panus tigrinus CBS, 577*, 79.

Cowan, A. K., Lodewijks, H. M., Sekhohola, L. M., & Edeki, O. G. (2016). In situ bioremediation of South African coal discard dumps. In *Proceedings, mine closure-2016* (pp. 501–509). Perth: Australian Centre for Geomechanics.

Crane, R. A., & Scott, T. B. (2012). Nanoscale zero-valent iron: Future prospects for an emerging water treatment technology. *Journal of Hazardous Materials, 211*, 112–125.

Crestini, C., Crucianelli, M., Orlandi, M., & Saladino, R. (2010). Oxidative strategies in lignin chemistry: A new environmental friendly approach for the functionalization of lignin and lignocellulosic fibers. *Catalysis Today, 156*(1–2), 8–22.

Crini, G. (2006). Non-conventional low-cost adsorbents for dye removal: A review. *Bioresource Technology, 97*(9), 1061–1085.

Cunningham, S. D., Anderson, T. A., Schwab, A. P., & Hsu, F. C. (1996). Phytoremediation of soils contaminated with organic pollutants. *Advances in Agronomy, 56*(1), 55–114.

Das, N. (2005). Heavy metals biosorption by mushrooms. *Nilanjana Das Natural Product Radiance, 4*(6), 454–459.

Das, M., Royer, T. V., & Leff, L. G. (2007). Diversity of fungi, bacteria, and actinomycetes on leaves decomposing in a stream. *Applied and Environmental Microbiology, 73*(3), 756–767.

Dashtban, M., Schraft, H., Syed, T. A., & Qin, W. (2010). Fungal biodegradation and enzymatic modification of lignin. *International Journal of Biochemistry and Molecular Biology, 1*(1), 36.

de Novais, C. B., Borges, W. L., da Conceicão, J. E., Júnior, O. J. S., & Siqueira, J. O. (2014). Inter-and intraspecific functional variability of tropical arbuscular mycorrhizal fungi isolates colonizing corn plants. *Applied Soil Ecology, 76*, 78–86.

Dembitsky, V. M., & Rezanka, T. (2003). Natural occurrence of arseno compounds in plants, lichens, fungi, algal species, and microorganisms. *Plant Science, 165*(6), 1177–1192.

Deng, Z., Cao, L., Huang, H., Jiang, X., Wang, W., Shi, Y., & Zhang, R. (2011). Characterization of Cd-and Pb-resistant fungal endophyte *Mucor* sp. CBRF59 isolated from rapes (*Brassica chinensis*) in a metal-contaminated soil. *Journal of Hazardous Materials, 185*(2–3), 717–724.

Deng, Z., Zhang, R., Shi, Y., Tan, H., & Cao, L. (2014). Characterization of Cd-, Pb-, Zn-resistant endophytic *Lasiodiplodia* sp. MXSF31 from metal accumulating *Portulaca oleracea* and its potential in promoting the growth of rape in metal-contaminated soils. *Environmental Science and Pollution Research, 21*(3), 2346–2357.

Dermatas, D., & Meng, X. (2003). Utilization of fly ash for stabilization/solidification of heavy metal contaminated soils. *Engineering Geology, 70*(3–4), 377–394.

Doughari, J. (2015). An overview of plant immunity. *Journal of Plant Pathology and Microbiology, 6*(11), 10–4172.

Duan, L., Naidu, R., Thavamani, P., Meaklim, J., & Megharaj, M. (2015). Managing long-term polycyclic aromatic hydrocarbon contaminated soils: A risk-based approach. *Environmental Science and Pollution Research, 22*(12), 8927–8941.

Duke, S. O., Lydon, J., Koskinen, W. C., Moorman, T. B., Chaney, R. L., & Hammerschmidt, R. (2012). *Journal of Agricultural and Food Chemistry, 60*. ISSN: 1520-5118 ISO Abbreviation: J. Agric. Food Chem.

Dunwell, J. M., Khuri, S., & Gane, P. J. (2000). Microbial relatives of the seed storage proteins of higher plants: Conservation of structure and diversification of function during evolution of the cupin superfamily. *Microbiology and Molecular Biology Reviews, 64*(1), 153–179.

Dupraz, C., Reid, R. P., Braissant, O., Decho, A. W., Norman, R. S., & Visscher, P. T. (2009). Processes of carbonate precipitation in modern microbial mats. *Earth-Science Reviews, 96*(3), 141–162.

Eibes, G., Cajthaml, T., Moreira, M. T., Feijoo, G., & Lema, J. M. (2006). Enzymatic degradation of anthracene, dibenzothiophene and pyrene by manganese peroxidase in media containing acetone. *Chemosphere, 64*(3), 408–414.

Ellouze, M., & Sayadi, S. (2016). White-rot fungi and their enzymes as a biotechnological tool for xenobiotic bioremediation. In *Management of hazardous wastes*. Rijeka: InTech.

Emamverdian, A., Ding, Y., Mokhberdoran, F., & Xie, Y. (2015). Heavy metal stress and some mechanisms of plant defense response. *The Scientific World Journal, 2015*, 4.

Encarnacion, A. B., Fagutao, F., Jintasataporn, O., Worawattanamateekul, W., Hirono, I., & Ohshima, T. (2012). Applications of ergothioneine-rich extract from an edible mushroom *Flammulina velutipes* for melanosis prevention in shrimp, *Penaeus monodon* and *Litopenaeus vannamei*. *Food Research International, 45*(1), 232–237.

Ennis, C. J., Evans, A. G., Islam, M., Ralebitso-Senior, T. K., & Senior, E. (2012). Biochar: Carbon sequestration, land remediation, and impacts on soil microbiology. *Critical Reviews in Environmental Science and Technology, 42*(22), 2311–2364.

Ercal, N., Gurer-Orhan, H., & Aykin-Burns, N. (2001). Toxic metals and oxidative stress part I: Mechanisms involved in metal-induced oxidative damage. *Current Topics in Medicinal Chemistry, 1*(6), 529–539.

Evanko, C. R., & Dzombak, D. A. (1997). *Remediation of metals-contaminated soils and groundwater*. Pittsburg: Ground-Water Remediation Technologies Analysis Center.

Ferreira-Guedes, S., Mendes, B., & Leitão, A. L. (2012). Degradation of 2, 4-dichlorophenoxyacetic acid by a halotolerant strain of Penicillium chrysogenum: Antibiotic production. *Environmental Technology, 33*(6), 677–686.

Fidalgo, F., Azenha, M., Silva, A. F., de Sousa, A., Santiago, A., Ferraz, P., & Teixeira, J. (2013). Copper-induced stress in *Solanum nigrum* L. and antioxidant defense system responses. *Food and Energy Security, 2*(1), 70–80.

Finlay, R. D. (2008). Ecological aspects of mycorrhizal symbiosis: With special emphasis on the functional diversity of interactions involving the extraradical mycelium. *Journal of Experimental Botany, 59*(5), 1115–1126.

Flurkey, A., Cooksey, J., Reddy, A., Spoonmore, K., Rescigno, A., Inlow, J., & Flurkey, W. H. (2008). Enzyme, protein, carbohydrate, and phenolic contaminants in commercial tyrosinase preparations: Potential problems affecting tyrosinase activity and inhibition studies. *Journal of Agricultural and Food Chemistry, 56*(12), 4760–4768.

Forgacs, E., Cserhati, T., & Oros, G. (2004). Removal of synthetic dyes from wastewaters: A review. *Environment International, 30*(7), 953–971.

Fritsche, W., Scheibner, K., Herre, A., & Hofrichter, M. (2000). Fungal degradation of explosives: TNT and related nitroaromatic compounds. In *Biodegradation of nitroaromatic compounds and explosives* (pp. 213–237). Boca Raton: CRC Press.

Fu, F., & Wang, Q. (2011). Removal of heavy metal ions from wastewaters: A review. *Journal of Environmental Management, 92*(3), 407–418.

Fulekar, M. H. (2017). Microbial degradation of petrochemical waste-polycyclic aromatic hydrocarbons. *Bioresources and Bioprocessing, 4*(1), 28.

Gadd, G. M. (2004). Microbial influence on metal mobility and application for bioremediation. *Geoderma, 122*(2–4), 109–119.

Gadd, G. M. (2007). Geomycology: Biogeochemical transformations of rocks, minerals, metals and radionuclides by fungi, bioweathering and bioremediation. *Mycological Research, 111*(1), 3–49.

Ganeshamurthy, A. N., Varalakshmi, L. R., & Sumangala, H. P. (2016). Environmental risks associated with heavy metal contamination in soil, water and plants in urban and periurban agriculture. *Journal of Horticultural Science, 3*(1), 1–29.

Gavrilescu, M. (2004). Removal of heavy metals from the environment by biosorption. *Engineering in Life Sciences, 4*(3), 219–232.

Gill, S. S., & Tuteja, N. (2010). Reactive oxygen species and antioxidant machinery in abiotic stress tolerance in crop plants. *Plant Physiology and Biochemistry, 48*(12), 909–930.

Gonen Tasdemir, F., Yamac, M., Cabuk, A., & Yildiz, Z. (2008). Selection of newly isolated mushroom strains for tolerance and biosorption of zinc in vitro. *Journal of Microbiology and Biotechnology, 18*(3), 483–489.

Gossel, T. A. (2018). *Principles of clinical toxicology*. Boca Raton: CRC Press.

Gratão, P. L., Polle, A., Lea, P. J., & Azevedo, R. A. (2005). Making the life of heavy metal-stressed plants a little easier. *Functional Plant Biology, 32*(6), 481–494.

Guo, G., Zhou, Q., & Ma, L. Q. (2006). Availability and assessment of fixing additives for the in situ remediation of heavy metal contaminated soils: A review. *Environmental Monitoring and Assessment, 116*(1–3), 513–528.

Ha, S. B., Smith, A. P., Howden, R., Dietrich, W. M., Bugg, S., O'Connell, M. J., & Cobbett, C. S. (1999). Phytochelatin synthase genes from *Arabidopsis* and the yeast *Schizosaccharomycespombe*. *The Plant Cell, 11*(6), 1153–1163.

Hall, J. L. (2002). Cellular mechanisms for heavy metal detoxification and tolerance. *Journal of Experimental Botany, 53*(366), 1–11.

Hamba, Y., & Tamiru, M. (2016). Mycoremediation of heavy metals and hydrocarbons contaminated environment. *Asian Journal of Natural and Applied Sciences, 5*, 2.

Harms, H., Schlosser, D., & Wick, L. Y. (2011). Untapped potential: Exploiting fungi in bioremediation of hazardous chemicals. *Nature Reviews Microbiology, 9*(3), 177.

Hasanuzzaman, M., Nahar, K., Alam, M. M., Roychowdhury, R., & Fujita, M. (2013). Physiological, biochemical, and molecular mechanisms of heat stress tolerance in plants. *International Journal of Molecular Sciences, 14*(5), 9643–9684.

Hofrichter, M. (2002). Lignin conversion by manganese peroxidase (MnP). *Enzyme and Microbial Technology, 30*(4), 454–466.

Hong, C. Y., Ryu, S. H., Jeong, H., Lee, S. S., Kim, M., & Choi, I. G. (2017). *Phanerochaete chrysosporium* multienzyme catabolic system for in vivo modification of synthetic lignin to succinic acid. *ACS Chemical Biology, 12*(7), 1749–1759.

Hossain, M. A., Piyatida, P., da Silva, J. A. T., & Fujita, M. (2012). Molecular mechanism of heavy metal toxicity and tolerance in plants: Central role of glutathione in detoxification of reactive oxygen species and methylglyoxal and in heavy metal chelation. *Journal of Botany, 37*, 872875.

Imfeld, G., & Vuilleumier, S. (2012). Measuring the effects of pesticides on bacterial communities in soil: A critical review. *European Journal of Soil Biology, 49*, 22–30.

Ingram, D. S., Vince-Prue, D., & Gregory, P. J. (2015). *Science and the garden: The scientific basis of horticultural practice*. New York: Wiley.

Isikhuemhen, O. S., Anoliefo, G. O., & Oghale, O. I. (2003). Bioremediation of crude oil polluted soil by the white rot fungus, *Pleurotus tuberregium* (Fr.) Sing. *Environmental Science and Pollution Research, 10*(2), 108–112.

Javaid, A., Bajwa, R., Shafique, U., & Anwar, J. (2011). Removal of heavy metals by adsorption on *Pleurotus ostreatus*. *Biomass and Bioenergy, 35*(5), 1675–1682.

Jeffries, P., Gianinazzi, S., Perotto, S., Turnau, K., & Barea, J. M. (2003). The contribution of arbuscular mycorrhizal fungi in sustainable maintenance of plant health and soil fertility. *Biology and Fertility of Soils, 37*(1), 1–16.

Johansson, J. F., Paul, L. R., & Finlay, R. D. (2004). Microbial interactions in the mycorrhizosphere and their significance for sustainable agriculture. *FEMS Microbiology Ecology, 48*(1), 1–13.

John, J. (2013). Assessment of arbuscular mycorrhizal fungi in a green roof system. http://hdl.handle.net/10222/36259

Johnsen, A. R., Wick, L. Y., & Harms, H. (2005). Principles of microbial PAH-degradation in soil. *Environmental Pollution, 133*(1), 71–84.

Johnson, S. B., Yoon, T. H., Slowey, A. J., & Brown, G. E. (2004). Adsorption of organic matter at mineral/water interfaces: 3. Implications of surface dissolution for adsorption of oxalate. *Langmuir, 20*(26), 11480–11492.

Jomova, K., & Valko, M. (2011). Advances in metal-induced oxidative stress and human disease. *Toxicology, 283*(2–3), 65–87.

Joy, J. I. T. H. I. N., Jose, C. I. N. T. I. L., Mathew, P., Thomas, S. A. B. U., & Khalaf, M. N. (2015). Biological delignification of biomass. *Green Polymers and Environment Pollution Control, 2015*, 271.

Jozefczak, M., Remans, T., Vangronsveld, J., & Cuypers, A. (2012). Glutathione is a key player in metal-induced oxidative stress defenses. *International Journal of Molecular Sciences, 13*(3), 3145–3175.

Juhasz, A. L., & Naidu, R. (2000). Bioremediation of high molecular weight polycyclic aromatic hydrocarbons: A review of the microbial degradation of benzo [α] pyrene. *International Biodeterioration and Biodegradation, 45*(1–2), 57–88.

Juwarkar, A. A., Singh, S. K., & Mudhoo, A. (2010). A comprehensive overview of elements in bioremediation. *Reviews in Environmental Science and Bio/technology, 9*(3), 215–288.

Kadri, T., Rouissi, T., Brar, S. K., Cledon, M., Sarma, S., & Verma, M. (2017). Biodegradation of polycyclic aromatic hydrocarbons (PAHs) by fungal enzymes: A review. *Journal of Environmental Sciences, 51*, 52–74.

Keiluweit, M., Nico, P. S., Johnson, M. G., & Kleber, M. (2010). Dynamic molecular structure of plant biomass-derived black carbon (biochar). *Environmental Science and Technology, 44*(4), 1247–1253.

Kerem, Z., Friesem, D., & Hadar, Y. (1992). Lignocellulose degradation during solid-state fermentation: *Pleurotus ostreatus* versus *Phanerochaete chrysosporium*. *Applied and Environmental Microbiology, 58*(4), 1121–1127.

Khullar, S., & Reddy, M. S. (2018). Ectomycorrhizal fungi and its role in metal homeostasis through metallothionein and glutathione mechanisms. *Current Biotechnology, 7*(3), 231–241.

Kim, K. H., Jahan, S. A., Kabir, E., & Brown, R. J. (2013). A review of airborne polycyclic aromatic hydrocarbons (PAHs) and their human health effects. *Environment International, 60*, 71–80.

Kirk, T. K., & Farrell, R. L. (1987). Enzymatic "combustion": The microbial degradation of lignin. *Annual Reviews in Microbiology, 41*(1), 465–501.

Konhauser, K. O. (1998). Diversity of bacterial iron mineralization. *Earth-Science Reviews, 43*(3–4), 91–121.

Kubartová, A., Ranger, J., Berthelin, J., & Beguiristain, T. (2009). Diversity and decomposing ability of saprophytic fungi from temperate forest litter. *Microbial Ecology, 58*(1), 98–107.
Kumar, K. S., Dahms, H. U., Won, E. J., Lee, J. S., & Shin, K. H. (2015). Microalgae - a promising tool for heavy metal remediation. *Ecotoxicology and Environmental Safety, 113*, 329–352.
Kushwaha, M., Verma, S., & Chatterjee, S. (2016). Profenofos, an acetylcholinesterase-inhibiting organophosphorus pesticide: A short review of its usage, toxicity, and biodegradation. *Journal of Environmental Quality, 45*(5), 1478–1489.
Kvesitadze, G., Khatisashvili, G., Sadunishvili, T., & Ramsden, J. J. (2006). *Biochemical mechanisms of detoxification in higher plants: Basis of phytoremediation*. Berlin/Heidelberg: Springer.
Lambers, H., Raven, J. A., Shaver, G. R., & Smith, S. E. (2008). Plant nutrient-acquisition strategies change with soil age. *Trends in Ecology and Evolution, 23*(2), 95–103.
Lamichhane, S., Krishna, K. B., & Sarukkalige, R. (2016). Polycyclic aromatic hydrocarbons (PAHs) removal by sorption: A review. *Chemosphere, 148*, 336–353.
Lavelle, P., & Spain, A. V. (2001). *Soil ecology*. Dordrecht: Springer Science and Business Media.
Lehmann, J., Rillig, M. C., Thies, J., Masiello, C. A., Hockaday, W. C., & Crowley, D. (2011). Biochar effects on soil biota–a review. *Soil Biology and Biochemistry, 43*(9), 1812–1836.
Leitão, A. L. (2009). Potential of Penicillium species in the bioremediation field. *International Journal of Environmental Research and Public Health, 6*(4), 1393–1417.
Lemery, J., & Auerbach, P. (2017). *Enviromedics: The impact of climate change on human health*. Lanham: Rowman & Littlefield.
Lenoir, I., Fontaine, J., & Sahraoui, A. L. H. (2016). Arbuscular mycorrhizal fungal responses to abiotic stresses: A review. *Phytochemistry, 123*, 4–15.
Leonowicz, A., Cho, N., Luterek, J., Wilkolazka, A., Wojtas-Wasilewska, M., Matuszewska, A., & Rogalski, J. (2001). Fungal laccase: Properties and activity on lignin. *Journal of Basic Microbiology, 41*(3–4), 185–227.
Li, X., & Jia, R. (2008). Decolorization and biosorption for Congo red by system rice hull-*Schizophyllum* sp. F17 under solid-state condition in a continuous flow packed-bed bioreactor. *Bioresource Technology, 99*(15), 6885–6892.
Lynd, L. R., Weimer, P. J., Van Zyl, W. H., & Pretorius, I. S. (2002). Microbial cellulose utilization: Fundamentals and biotechnology. *Microbiology and Molecular Biology Reviews, 66*(3), 506–577.
Lynes, M. A., Pietrosimone, K., Marusov, G., Donaldson, D. V., Melchiorre, C., Yin, X., Lawrence, D. A., & McCabe, M. J. (2010). Metal influences on immune function. In *Cellular and molecular biology of metals* (p. 379). New York: CRC Press.
Magan, N. (2007). Fungi in extreme environments. *The Mycota, 4*, 85–103.
Mahmood, K., Jadoon, S., Mahmood, Q., Irshad, M., & Hussain, J. (2014). Synergistic effects of toxic elements on heat shock proteins. *BioMed Research International, 2014*, 1–17.
Marques, A. P., Rangel, A. O., & Castro, P. M. (2009). Remediation of heavy metal contaminated soils: Phytoremediation as a potentially promising clean-up technology. *Critical Reviews in Environmental Science and Technology, 39*(8), 622–654.
Marschner, P. (2012). Rhizosphere biology. In *Marschner's mineral nutrition of higher plants* (3rd ed., pp. 369–388). London: Academic.
Martínez, Á. T., Speranza, M., Ruiz-Dueñas, F. J., Ferreira, P., Camarero, S., Guillén, F., & Río Andrade, J. C. D. (2005). Biodegradation of lignocellulosics: Microbial, chemical, and enzymatic aspects of the fungal attack of lignin. *International Microbiology, 8*(3), 195–204.
Martínez, Á. T., Rencoret, J., Marques, G., Gutiérrez, A., Ibarra, D., Jiménez-Barbero, J., & José, C. (2008). Monolignol acylation and lignin structure in some non woody plants: A 2D NMR study. *Phytochemistry, 69*(16), 2831–2843.
Matés, J. M., Pérez-Gómez, C., & De Castro, I. N. (1999). Antioxidant enzymes and human diseases. *Clinical Biochemistry, 32*(8), 595–603.
Megharaj, M., Ramakrishnan, B., Venkateswarlu, K., Sethunathan, N., & Naidu, R. (2011). Bioremediation approaches for organic pollutants: A critical perspective. *Environment International, 37*(8), 1362–1375.

Meharg, A. A., & Cairney, J. W. (2000). Ectomycorrhizas-extending the capabilities of rhizosphere remediation. *Soil Biology and Biochemistry, 32*(11–12), 1475–1484.
Messens, J., & Silver, S. (2006). Arsenate reduction: Thiol cascade chemistry with convergent evolution. *Journal of Molecular Biology, 362*(1), 1–17.
Mkandawire, M., & Dudel, E. G. (2007). Are *Lemna* spp. effective phytoremediation agents. *Bioremediation, Biodiversity and Bioavailability, 1*(1), 56–71.
Mohan, D., & Pittman, C. U. (2007). Arsenic removal from water/wastewater using adsorbents – A critical review. *Journal of Hazardous Materials, 142*(1–2), 1–53.
Mohan, S. V., Kisa, T., Ohkuma, T., Kanaly, R. A., & Shimizu, Y. (2006). Bioremediation technologies for treatment of PAH-contaminated soil and strategies to enhance process efficiency. *Reviews in Environmental Science and Bio/Technology, 5*(4), 347–374.
Moktali, V., Park, J., Fedorova-Abrams, N. D., Park, B., Choi, J., Lee, Y. H., & Kang, S. (2012). Systematic and searchable classification of cytochrome P450 proteins encoded by fungal and oomycete genomes. *BMC Genomics, 13*(1), 525.
Morelli, I. S., Saparrat, M. C. N., Del Panno, M. T., Coppotelli, B. M., & Arrambari, A. (2013). Bioremediation of PAH-contaminated soil by fungi. In *Fungi as bioremediators* (pp. 159–179). Berlin/Heidelberg: Springer.
Morozova, O. V., Shumakovich, G. P., Shleev, S. V., & Yaropolov, Y. I. (2007). Laccase-mediator systems and their applications: A review. *Applied Biochemistry and Microbiology, 43*(5), 523–535.
Mudhoo, A., Garg, V. K., & Wang, S. (2012). Removal of heavy metals by biosorption. *Environmental Chemistry Letters, 10*(2), 109–117.
Mueller, K. E. (2005). *Investigations into the use of trees for phytoremediation of pah contaminated soils* (Doctoral dissertation). University of Cincinnati.
Mulligan, C. N., Yong, R. N., & Gibbs, B. F. (2001). Remediation technologies for metal-contaminated soils and groundwater: An evaluation. *Engineering Geology, 60*(1–4), 193–207.
Murphy, A., Zhou, J., Goldsbrough, P. B., & Taiz, L. (1997). Purification and immunological identification of Metallothioneins 1 and 2 from *Arabidopsis thaliana*. *Plant Physiology, 113*(4), 1293–1301.
Nagy, B., Măicăneanu, A., Indolean, C., Mânzatu, C., Silaghi-Dumitrescu, L., & Majdik, C. (2014). Comparative study of Cd (II) biosorption on cultivated Agaricus bisporus and wild Lactarius piperatus based biocomposites. Linear and nonlinear equilibrium modelling and kinetics. *Journal of the Taiwan Institute of Chemical Engineers, 45*(3), 921–929.
Nasr, M., & Arp, P. A. (2011). Hg concentrations and accumulations in fungal fruiting bodies, as influenced by forest soil substrates and moss carpets. *Applied Geochemistry, 26*(11), 1905–1917.
Nnorom, I. C., Jarzyńska, G., Drewnowska, M., Dryżałowska, A., Kojta, A., Pankavec, S., & Falandysz, J. (2013). Major and trace elements in sclerotium of Pleurotus tuber-regium (Ósū) mushroom – Dietary intake and risk in southeastern Nigeria. *Journal of Food Composition and Analysis, 29*(1), 73–81.
Noctor, G., Mhamdi, A., Chaouch, S., Han, Y. I., Neukermans, J., Marquez-Garcia, B. E. L. E. N., & Foyer, C. H. (2012). Glutathione in plants: An integrated overview. *Plant, Cell and Environment, 35*(2), 454–484.
Nordberg, J., & Arner, E. S. (2001). Reactive oxygen species, antioxidants, and the mammalian thioredoxin system1. *Free Radical Biology and Medicine, 31*(11), 1287–1312.
Nunes, C. S., & Malmlöf, K. (2018). Enzymatic decontamination of antimicrobials, phenols, heavy metals, pesticides, polycyclic aromatic hydrocarbons, dyes, and animal waste. In *Enzymes in human and animal nutrition* (pp. 331–359). New York: Academic.
Nykiel-Szymańska, J., Stolarek, P., & Bernat, P. (2018). Elimination and detoxification of 2, 4-D by Umbelopsis isabellina with the involvement of cytochrome P450. *Environmental Science and Pollution Research, 25*(3), 2738–2743.
Oyetayo, V. O., Adebayo, A. O., & Ibileye, A. (2012). Assessment of the biosorption potential of heavy metals by *Pleurotus tuberregium*. *International Journal of Advanced Biological Research, 2*, 293–297.

Özdemir, S., Kilinc, E., Poli, A., Nicolaus, B., & Güven, K. (2009). Biosorption of Cd, Cu, Ni, Mn and Zn from aqueous solutions by thermophilic bacteria, *Geobacillus toebii* sub. sp. *decanicus* and *Geobacillus thermoleovorans* sub. sp. *stromboliensis*: Equilibrium, kinetic and thermodynamic studies. *Chemical Engineering Journal, 152*(1), 195–206.

Pala, S. A., Wani, A. H., Boda, R. H., & Wani, B. A. (2014). Mushroom refinement endeavor auspicate non-green revolution in the offing. *Nusantara Bioscience, 6*(2), 173–185.

Parmar, P., Dave, B., Sudhir, A., Panchal, K., & Subramanian, R. B. (2013). Physiological, biochemical and molecular response of plants against heavy metals stress. *International Journal of Current Research, 5*(1), 80–89.

Pearce, C. I., Lloyd, J. R., & Guthrie, J. T. (2003). The removal of colour from textile wastewater using whole bacterial cells: A review. *Dyes and Pigments, 58*(3), 179–196.

Pérez, J., Munoz-Dorado, J., de la Rubia, T. D. L. R., & Martinez, J. (2002). Biodegradation and biological treatments of cellulose, hemicellulose and lignin: An overview. *International Microbiology, 5*(2), 53–63.

Pierart, A., Shahid, M., Séjalon-Delmas, N., & Dumat, C. (2015). Antimony bioavailability: Knowledge and research perspectives for sustainable agricultures. *Journal of Hazardous Materials, 289*, 219–234.

Pointing, S. (2001). Feasibility of bioremediation by white-rot fungi. *Applied Microbiology and Biotechnology, 57*(1–2), 20–33.

Polak, J., & Jarosz-Wilkolazka, A. (2012). Fungal laccases as green catalysts for dye synthesis. *Process Biochemistry, 47*(9), 1295–1307.

Priyadharsini, P., Rojamala, K., Ravi, R. K., Muthuraja, R., Nagaraj, K., & Muthukumar, T. (2016). Mycorrhizosphere: The extended rhizosphere and its significance. In *Plant-microbe interaction: An approach to sustainable agriculture* (pp. 97–124). Singapore: Springer.

Puglisi, E., Hamon, R., Vasileiadis, S., Coppolecchia, D., & Trevisan, M. (2012). Adaptation of soil microorganisms to trace element contamination: A review of mechanisms, methodologies, and consequences for risk assessment and remediation. *Critical Reviews in Environmental Science and Technology, 42*(22), 2435–2470.

Purahong, W., Wubet, T., Lentendu, G., Schloter, M., Pecyna, M. J., Kapturska, D., & Buscot, F. (2016). Life in leaf litter: Novel insights into community dynamics of bacteria and fungi during litter decomposition. *Molecular Ecology, 25*(16), 4059–4074.

Purnomo, A. S., Ashari, K., & Hermansyah, F. T. (2017). Evaluation of the synergistic effect of mixed cultures of white-rot fungus *Pleurotus ostreatus* and biosurfactant-producing bacteria on DDT biodegradation. *Journal of Microbiology and Biotechnology, 27*(7), 1306–1315.

Purohit, J., Anirudha, C., Mohan, K. B., & Singh, N. K. (2018). Mycoremediation of agricultural soil: Bioprospection for sustainable development. In *Mycoremediation and environmental sustainability* (pp. 91–120). Cham: Springer.

Qu, J., Zang, T., Gu, H., Li, K., Hu, Y., Ren, G., & Jin, Y. (2015). Biosorption of copper ions from aqueous solution by *Flammulina velutipes* spent substrate. *Bio Resources, 10*(4), 8058–8075.

Rabaey, K., & Verstraete, W. (2005). Microbial fuel cells: Novel biotechnology for energy generation. *Trends in Biotechnology, 23*(6), 291–298.

Raghukumar, S. (2017). Physiology, biochemistry, and biotechnology. In *Fungi in coastal and oceanic marine ecosystems* (pp. 265–306). Cham: Springer.

Rajinipriya, M., Nagalakshmaiah, M., Robert, M., & Elkoun, S. (2018). Importance of agricultural and industrial waste in the field of nanocellulose and recent industrial developments of wood based nanocellulose: A review. *ACS Sustainable Chemistry and Engineering, 6*(3), 2807–2828.

Rashid, A., Bhatti, H. N., Iqbal, M., & Noreen, S. (2016). Fungal biomass composite with bentonite efficiency for nickel and zinc adsorption: A mechanistic study. *Ecological Engineering, 91*, 459–471.

Rauser, W. E. (1995). Phytochelatins and related peptides. Structure, biosynthesis, and function. *Plant Physiology, 109*(4), 1141.

Read, D. J., & Perez-Moreno, J. (2003). Mycorrhizas and nutrient cycling in ecosystems–A journey towards relevance. *New Phytologist, 157*(3), 475–492.

Richter, H., & Howard, J. B. (2000). Formation of polycyclic aromatic hydrocarbons and their growth to soot a review of chemical reaction pathways. *Progress in Energy and Combustion Science, 26*(4–6), 565–608.
Rillig, M. C., & Mummey, D. L. (2006). Mycorrhizas and soil structure. *New Phytologist, 171*(1), 41–53.
Roshchina, V. V., & Roshchina, V. D. (2012). *The excretory function of higher plants*. Berlin/Heidelberg: Springer.
Rouches, E., Herpoël-Gimbert, I., Steyer, J. P., & Carrere, H. (2016). Improvement of anaerobic degradation by white-rot fungi pretreatment of lignocellulosic biomass: A review. *Renewable and Sustainable Energy Reviews, 59*, 179–198.
Rout, M. E. (2014). The plant microbiome. In *Advances in botanical research* (Vol. 69, pp. 279–309). Waltham: Academic.
Saeedi, M., Li, L. Y., & Salmanzadeh, M. (2012). Heavy metals and polycyclic aromatic hydrocarbons: Pollution and ecological risk assessment in street dust of Tehran. *Journal of Hazardous Materials, 227*, 9–17.
Saichek, R. E., & Reddy, K. R. (2005). Electrokinetically enhanced remediation of hydrophobic organic compounds in soils: A review. *Critical Reviews in Environmental Science and Technology, 35*(2), 115–192.
Saranraj, P., & Stella, D. (2014). Impact of sugar mill effluent to environment and bioremediation: A review. *World Applied Sciences Journal, 30*(3), 299–316.
Sardrood, B. P., Goltapeh, E. M., & Varma, A. (2013). An introduction to bioremediation. In *Fungi as bioremediators* (pp. 3–27). Berlin/Heidelberg: Springer.
Sarma, V. V. (2018). Obligate marine fungi and bioremediation. In *Mycoremediation and environmental sustainability* (pp. 307–323). Cham: Springer.
Schiffer, M. B. (1986). Radiocarbon dating and the "old wood" problem: The case of the Hohokam chronology. *Journal of Archaeological Science, 13*(1), 13–30.
Schmidt, T., & Schaechter, M. (2012). *Topics in ecological and environmental microbiology*. Burlington: Elsevier.
Schutzendubel, A., & Polle, A. (2002). Plant responses to abiotic stresses: Heavy metal-induced oxidative stress and protection by mycorrhization. *Journal of Experimental Botany, 53*(372), 1351–1365.
Sharma, S. S., & Dietz, K. J. (2006). The significance of amino acids and amino acid-derived molecules in plant responses and adaptation to heavy metal stress. *Journal of Experimental Botany, 57*(4), 711–726.
Sharma, S., Tiwari, S., Hasan, A., Saxena, V., & Pandey, L. M. (2018). Recent advances in conventional and contemporary methods for remediation of heavy metal-contaminated soils. *3 Biotech, 8*(4), 216.
Siddiquee, S., Rovina, K., Azad, S. A., Naher, L., Suryani, S., & Chaikaew, P. (2015). Heavy metal contaminants removal from wastewater using the potential filamentous fungi biomass: A review. *Journal of Microbial and Biochemical Technology, 7*(6), 384–393.
Singh, O. V., Labana, S., Pandey, G., Budhiraja, R., & Jain, R. K. (2003). Phytoremediation: An overview of metallic ion decontamination from soil. *Applied Microbiology and Biotechnology, 61*(5–6), 405–412.
Singh, P. C., Srivastava, S., Shukla, D., Bist, V., Tripathi, P., Anand, V., & Srivastava, S. (2018). Mycoremediation mechanisms for heavy metal resistance/tolerance in plants. In *Mycoremediation and environmental sustainability* (pp. 351–381). Cham: Springer.
Sivaramakrishnan, S., Gangadharan, D., Nampoothiri, K. M., Soccol, C. R., & Pandey, A. (2006). α-Amylases from microbial sources–an overview on recent developments. *Food Technology and Biotechnology, 44*(2), 173–184.
Smith, S. E., & Read, D. J. (2010). *Mycorrhizal symbiosis*. New York: Academic.
Soden, D. M., & Dobson, A. D. (2001). Differential regulation of laccase gene expression in *Pleurotus sajor-caju*. *Microbiology, 147*(7), 1755–1763.
Spokas, K. A., Cantrell, K. B., Novak, J. M., Archer, D. W., Ippolito, J. A., Collins, H. P., & Lentz, R. D. (2012). Biochar: A synthesis of its agronomic impact beyond carbon sequestration. *Journal of Environmental Quality, 41*(4), 973–989.

Stamets, P. (2011). *Growing gourmet and medicinal mushrooms*. Berkeley: Ten Speed Press.
Stanic, A. (2017). Preparation of Thiol Conjugates of the Mycotoxin Deoxynivalenol and their Occurrence in Nature. http://urn.nb.no/URN:NBN:no-58380
Strong, P. J., & Burgess, J. E. (2008). Treatment methods for wine-related and distillery wastewaters: A review. *Bioremediation Journal, 12*(2), 70–87.
Sud, D., Mahajan, G., & Kaur, M. P. (2008). Agricultural waste material as potential adsorbent for sequestering heavy metal ions from aqueous solutions – A review. *Bioresource Technology, 99*(14), 6017–6027.
Sudha, M., Saranya, A., Selvakumar, G., & Sivakumar, N. (2014). Microbial degradation of azo dyes: A review. *International Journal of Current Microbiology and Applied Sciences, 3*(2), 670–690.
Sugasini, A., Rajagopal, K., & Banu, N. (2014). A study on biosorption potential of *Aspergillus* sp. of tannery effluent. *Advances in Bioscience and Biotechnology, 5*(10), 853.
Sutherland, C., & Venkobachar, C. (2010). A diffusion-chemisorption kinetic model for simulating biosorption using forest macro-fungus, fomes fasciatus. *International Research Journal of Plant Science, 1*(4), 107–117.
Sytar, O., Kumar, A., Latowski, D., Kuczynska, P., Strzałka, K., & Prasad, M. N. V. (2013). Heavy metal-induced oxidative damage, defense reactions, and detoxification mechanisms in plants. *Acta Physiologiae Plantarum, 35*(4), 985–999.
Tabibzadeh, S. (2016). Nature creates, adapts, protects and sustains life using hydrogen sulfide. *Frontiers in Bioscience, 21*, 528–560.
Tadkaew, N., Hai, F. I., McDonald, J. A., Khan, S. J., & Nghiem, L. D. (2011). Removal of trace organics by MBR treatment: The role of molecular properties. *Water Research, 45*(8), 2439–2451.
Tahir, M. W., Zaidi, N. A., Rao, A. A., Blank, R., Vellekoop, M. J., & Lang, W. (2018). A fungus spores dataset and a convolutional neural networks based approach for fungus detection. *IEEE Transactions on Nano Bioscience, 17*(3), 5–22.
Tak, H. I., Ahmad, F., & Babalola, O. O. (2013). Advances in the application of plant growth-promoting rhizobacteria in phytoremediation of heavy metals. In *Reviews of environmental contamination and toxicology* (Vol. 223, pp. 33–52). New York: Springer.
Tan, Y. H. (2011) *Behavioral properties of locally isolated Acinetobacter species in degrading hydrocarbon chain in crude oil and used cooking oil* (Doctoral dissertation). UTAR.
Tangahu, B. V., Abdullah, S., Rozaimah, S., Basri, H., Idris, M., Anuar, N., & Mukhlisin, M. (2011). A review on heavy metals (As, Pb, and Hg) uptake by plants through phytoremediation. *International Journal of Chemical Engineering, 2011*, 1–31.
Tay, C. C., Liew, H. H., Yin, C. Y., Abdul-Talib, S., Surif, S., Suhaimi, A. A., & Yong, S. K. (2011). Biosorption of cadmium ions using *Pleurotus ostreatus*: Growth kinetics, isotherm study and biosorption mechanism. *Korean Journal of Chemical Engineering, 28*(3), 825–830.
Teng, Y., Luo, Y., Sun, M., Liu, Z., Li, Z., & Christie, P. (2010). Effect of bioaugmentation by *Paracoccus* sp. strain HPD-2 on the soil microbial community and removal of polycyclic aromatic hydrocarbons from an aged contaminated soil. *Bioresource Technology, 101*(10), 3437–3443.
Tian, H., Ma, Y. J., Li, W. Y., & Wang, J. W. (2018). Efficient degradation of triclosan by an endophytic fungus *Penicillium oxalicum* B4. *Environmental Science and Pollution Research, 25*(9), 8963–8975.
Toljander, J. F., Lindahl, B. D., Paul, L. R., Elfstrand, M., & Finlay, R. D. (2007). Influence of arbuscular mycorrhizal mycelial exudates on soil bacterial growth and community structure. *FEMS Microbiology Ecology, 61*(2), 295–304.
Tsarevsky, N. V., & Matyjaszewski, K. (2007). Green atom transfer radical polymerization: From process design to preparation of well-defined environmentally friendly polymeric materials. *Chemical Reviews, 107*(6), 2270–2299.
Valášková, V. (2010). Physiology and ecology of saprotrophic basidiomycetes degrading dead plant biomass. http://hdl.handle.net/20.500.11956/24782

Vander Oost, R., Beyer, J., & Vermeulen, N. P. (2003). Fish bioaccumulation and biomarkers in environmental risk assessment: A review. *Environmental Toxicology and Pharmacology, 13*(2), 57–149.

Vara, S. (2017). Mycoremediation of lignocelluloses. In *Handbook of research on inventive bioremediation techniques* (pp. 264–286). New York: IGI Global.

Vassilev, A., Schwitzguébel, J. P., Thewys, T., Van Der Lelie, D., & Vangronsveld, J. (2004). The use of plants for remediation of metal-contaminated soils. *The Scientific World Journal, 4*, 9–34.

Verbruggen, N., Hermans, C., & Schat, H. (2009). Mechanisms to cope with arsenic or cadmium excess in plants. *Current Opinion in Plant Biology, 12*(3), 364–372.

Waghunde, R. R., Shelake, R. M., & Sabalpara, A. N. (2016). *Trichoderma*: A significant fungus for agriculture and environment. *African Journal of Agricultural Research, 11*(22), 1952–1965.

Walker, G. M., & White, N. A. (2017). Introduction to fungal physiology. In *Fungi: Biology and applications* (pp. 1–35). Hoboken: Wiley.

Wang, J., & Chen, C. (2009). Biosorbents for heavy metals removal and their future. *Biotechnology Advances, 27*(2), 195–226.

Wu, Y., Luo, Y., Zou, D., Ni, J., Liu, W., Teng, Y., & Li, Z. (2008). Bioremediation of polycyclic aromatic hydrocarbons contaminated soil with *Monilinia* sp.: Degradation and microbial community analysis. *Biodegradation, 19*(2), 247–257.

Wuana, R. A., & Okieimen, F. E. (2011). Heavy metals in contaminated soils: A review of sources, chemistry, risks and best available strategies for remediation. *Isrn Ecology, 11*, 1–19.

Xiong, X. Q., Liao, H. D., Ma, J. S., Liu, X. M., Zhang, L. Y., Shi, X. W., & Zhu, Y. H. (2014). Isolation of a rice endophytic bacterium, *Pantoea* sp. Sd-1, with ligninolytic activity and characterization of its rice straw degradation ability. *Letters in Applied Microbiology, 58*(2), 123–129.

Xu, P., Zeng, G., Huang, D., Liu, L., Zhao, M., Lai, C., & Zhang, C. (2016). Metal bioaccumulation, oxidative stress and antioxidant defenses in *Phanerochaete chrysosporium* response to Cd exposure. *Ecological Engineering, 87*, 150–156.

Xue, J., Yu, Y., Bai, Y., Wang, L., & Wu, Y. (2015). Marine oil-degrading microorganisms and biodegradation process of petroleum hydrocarbon in marine environments: A review. *Current Microbiology, 71*(2), 220–228.

Yang, S., Hai, F. I., Nghiem, L. D., Price, W. E., Roddick, F., Moreira, M. T., & Magram, S. F. (2013). Understanding the factors controlling the removal of trace organic contaminants by white-rot fungi and their lignin modifying enzymes: A critical review. *Bioresource Technology, 141*, 97–108.

Zavarzina, A. G., Lisov, A. A., Zavarzin, A. A., & Leontievsky, A. A. (2010). Fungal oxidoreductases and humification in forest soils. In *Soil enzymology* (pp. 207–228). Berlin/Heidelberg: Springer.

Zenk, M. H. (1996). Heavy metal detoxification in higher plants-A review. *Gene, 179*(1), 21–30.

Zhang, Y., & Tao, S. (2009). Global atmospheric emission inventory of polycyclic aromatic hydrocarbons (PAHs) for 2004. *Atmospheric Environment, 43*(4), 812–819.

Zhang, X., Wang, H., He, L., Lu, K., Sarmah, A., Li, J., Bolan, N. S., Pei, J., & Huang, H. (2013). Using biochar for remediation of soils contaminated with heavy metals and organic pollutants. *Environmental Science and Pollution Research, 20*(12), 8472–8483.

Zhang, C., Chen, W., & Alvarez, P. J. (2014). Manganese peroxidase degrades pristine but not surface-oxidized (carboxylated) single-walled carbon nanotubes. *Environmental Science and Technology, 48*(14), 7918–7923.

Zhao, F. J., Zhu, Y. G., & Meharg, A. A. (2013). Methylated arsenic species in rice: Geographical variation, origin, and uptake mechanisms. *Environmental Science and Technology, 47*(9), 3957–3966.

Zucca, P., Rescigno, A., Rinaldi, A. C., & Sanjust, E. (2014). Biomimetic metalloporphines and metalloporphyrins as potential tools for delignification: Molecular mechanisms and application perspectives. *Journal of Molecular Catalysis A: Chemical, 388*, 2–34.

The Mycorrhizoshpere Effect on Pedogenesis and Terrestrial Biomes

16

Sanjukta Dey and Rabindranath Bhattacharyya

1 Introduction

Microorganisms are found in plentiful in soil. It is very fascinating to know that there are more microbes in 1 g of soil than there are human beings on this earth! And as they co-exists with every other members of the terrestrial ecosystem they respond to and influence each other's living. Bacteria, fungi, pseudomonads, blue green algae, actinomycetes are often encountered with living as mixed population in soil. And as they respond to each other's presence they enter into different types of community interactions among themselves and with other members of the biome; some of which are beneficial, while others are harmful and yet some other that have no net effect. Of special interest is the symbiotic mutualism between land plants and members of kingdom Fungi. This association is known as mycorrhizae meaning "fungal root" and around more than three-fourth of extant vascular land plants form mycorrhizal association with members of fungal class like Glomeromycota, Ascomycota, Zygomycota and Basidiomycota. The benefit of this symbiosis is the Carbon-for-nutrient exchange between vascular plants and mycorrhizal fungi and plants invest anywhere between 5% and 15% of C assimilated by photosynthesis, in supporting their fungal symbiont and the fungus in return help in efficient nutrient and water uptake by the host. A global estimate made of mycorrhizal hyphae present in the top 10 cm of soil layer extends over a distance of approximately 45×10^{16} km which is almost half the diameter of our galaxy (95×10^{16} km). It has been calculated that surface area of mycorrhizal hyphae is approximately 3 times the area of

S. Dey
Department of Botany, School of Life Science and Biotechnology, Adamas University, Barasat, West Bengal, India
e-mail: sanjukta.dey@outlook.com

R. Bhattacharyya (✉)
Department of Life Sciences (Plant Science Section), Presidency University, Kolkata, West Bengal, India

© Springer Nature Singapore Pte Ltd. 2019

A. Varma, D. K. Choudhary (eds.), *Mycorrhizosphere and Pedogenesis*,
https://doi.org/10.1007/978-981-13-6480-8_16

continental land masses. These astronomical figures indicated the extensivity of mycorrhizal fungi in soil (Leake et al. 2004; Leake and Read 2017).

There are six distinct groups of mycorrhizal association (Smith and Reed 1997; Brundrett 2002). They have been classified as arbuscular, ecto-, ericoid, arbutoid, monotropoid and orchid.

1.1 Arbuscular Mycorrhizae

They are the most abundant type and members of monophyletic phylum Glomeromycota form this type of mycorrhizal association with bryophytes, pteridophytes, gymnosperms and angiosperms. Arbuscular mycorrhizae are type of endomycorrhizae because they form intraradical (i.e. inside of plant root) structures. The fungus interacts with the host plant with specialised highly coiled hyphae called arbuscules. The coiled structure of arbuscules greatly increases surface area for exchange of C and nutrients. Often times the fungus produce sac like storage organs known as vesicles; thus arbuscular mycorrhizal fungi are also known as vesicular-arbuscular mycorrhizal (VAM) fungi.

1.2 Ectomycorrhizae

The fungal classes that predominates in ectomycorrhizal association are Basidiomycota, Ascomycota and Zygomycota. Host plant ranges from woody gymnosperms like members of Pinaceae and angiosperms like species of Dipterocarpaceae and Betulaceae. Ectomycorrhizal association is of extreme importance in boreal and temperate forests which are dominated mostly by woody gymnosperms and tree like angiosperms.

Structurally, ectomycorrhizae (EcM) are characterised by presence of fungal mantel and rhizomorphs (fused, coenocytic fungal hyphae). Number of extraradical hyphae is higher in EcM and they extend a greater distance relative to arbuscular mycorrhizae. Thus mycorrhizosphere effect of EcM is more extensive than arbuscular mycorrhizae (AM).

1.3 Mycorrhizae in Ericales

Angiosperms belonging to the Ericales contain a worldwide distribution of closely related members. Ericaceous plants form three distinct class of mycorrhizae viz. ericoid, arbutoid and monotropoid. Ascomycota (Leotiales) form ericoid type of association with three families of Ericales (Ericaceae, Epacridaceae and Empetraceae). Basidiomycotan fungi dominates arbutoid type. Plant partner includes members of Ericales *Arbutus, Arctostaphylos* and Pyrolaceae. It even enters in mycorrhization with bryophytes. Monotropoid mycorrhizae are partnership between certain members of Basidiomycota and members of non-photosynthetic angiosperms Monotropaceae. In monotropoid association the fungus transfers food from a neighbouring photosynthetic plant to achlorophyllous (myco-heterotrophic) host plant (Johnson and Gehring 2007). These fungus produce additional hyphal peg inside plant root, apart from hyphal Hartig net and hyphal mantel (Smith and Reed 1997; Johnson and Gehring 2007).

1.4 Orchid Mycorrhizae

Orchids form mycorrhizal association with members of Basidiomycota. Orchids differ from other plants in having an extended protocorm stage during which the plant cannot photosynthesise and rely on their fungal partners for fixed carbon from other plants. Thus they are myco-heterotrophic in nature. While most adult orchid plants become chlorophyllous some 200 species have been reported to remain achlorophyllous for their entire life stay myco-heterotrophic. Structurally, orchid mycorrhizae can be distinguished from other groups of mycorrhizae by the presence of intracellular hyphae which is termed a peloton.

1.5 Evolution of Mycorrhizae

Considering evolution it is seen that ectomycorrhizae have evolved until recently and arbuscular mycorrhizae are more primitive type. The earliest evidence of mycorrhiza like symbiosis comes from Devonian Rhynie-Chert (411.5 million years ago) where the early vascular land plants have been so well preserved that it has enabled detailed characterization of plant-fungus interaction even to the cellular level; studies show presence of highly branched fungal structure and fungal vesicles in cells resembling arbuscules. Fungal spores preserved at the Rhynie-Chert bed are that of Glomeromycota genera *Aculospora* and *Scutellospora*. Both molecular and fossil evidence indicate the mycorrhizal nature of early land plants (Redeckeret al. 2000). The early bryophytes were devoid of roots and had stem-like rhizome colonized by fungus shared resemblances to present day Glomeromycotan fungi (Stubblefield et al. 1987). Comparing these spores and other fungal structures with that of present day members of Glomeromycota shows that these have undergone little change over more than 400 million years. It is in fact true that plants successfully colonized land with aid from its fungal partner who were capable of acquiring nutrients from under developed soil that existed during Silurian and Devonian period (Pirozynski and Malloch 1975).

1.6 Benefits from Mycorrhizal Symbioses

Mycorrhiza is a mutualistic symbiosis where the fungal partner derives carbon from host plant photosynthate and in return help plants to acquire mineral nutrients from soil, especially immobile elements like phosphorous, zinc and copper but also more mobile ions such as sulphur, calcium, potassium, iron, magnesium, manganese, chlorine, bromine and nitrogen. Mycorrhizal association enhance plant growth by aiding in efficient mineral nutrient uptake, in soils which are deficient in the above elements. They also increase water uptake by plants and alter plant's physiology to cope better with drought stress. Furthermore, mycorrhizal associations can reduce plants' response to stresses like high salinity, mineral and heavy metal toxicities or toxicity due to minor element imbalance like Mn toxicity. Mycorrhizal plants show less susceptibility to pathogens because of altered membrane permeability and system physiology. Some mycorrhizal fungi produce metabolites that alter plant's ability to produce adventitious roots from cuttings and change root regeneration capacity and morphology; it greatly increases absorptive surface area and feeder root longevity. They are known to alter soil texture by promoting soil particle aggregation and thus stability. Some mycorrhizal fungi produce edible fruit bodies and is eaten by people worldwide.

2 Concept of Rhizosphere

Plant roots provide structural support, act in the absorption of mineral nutrients and water, are site of synthesis of growth regulator like cytokinin and gibberellin, site for storage of starch and provide nutrient supply for a wide range soil inhabiting micro-organisms. The area under the influence of nutrient released from roots is termed as rhizosphere (Hiltner 1904). It is a site of significantly increased microbial activities and is a dynamic zone in the soil. Nutrients released from roots are organic non metabolic exudates or metabolic secretions. The root exudates and secretions are water soluble sugars, organic acids, antibiotics and volatile compounds. The quality and quantity of organic root exudation and release is determined by host factors such as species, stage of development, age; soil factors like pH, physical properties and moisture content; environmental factors such as temperature and light; cultural properties which includes use of pesticides and insecticides. And their nature and amount determine the rhizosphere microbial population (Linderman 1988).

3 Influence of Soil Microflora on Mycorrhization

As other soil-borne and root inhabiting fungus mycorrhizal fungal propagule exists in soil as spores and vegetative propagule in root fragments. There are two factors responsible for initiation of colonization of root tissue by mycorrhizal fungus. Firstly, the root exudates stimulates both the plant root and mycorrhizal fungi and the latter penetrate epidermal root cells by protruding hyphae or germ tube. Depending on the host symbiont combination, morphology of both host and fungus changes and results in endo-, ectendo-, ericoid or ecto- mycorrhizal relationship.

Secondly, study reports from ecto-mycorrhizal type of association suggests that if certain associative bacteria and fungi, like *Azotobacter*, *Trichoderma*, and certain fluorescent pseudomonads were present at the time of inoculation then it showed a marked increase in the development of mycorrhizae. The reason postulated is production of some growth promoting substances like thiamine which stimulates both plant roots and mycorrhizal fungi. Contrary to this, certain bacteria and fungi may be detrimental to development of mycorrhizae. The above mentioned factors help in both external and internal colonization of the host tissue by the fungus (Fig. 16.1).

Root exudates around non-mycorrhizal and mycorrhizal root vary considerably. Plant roots infected with mycorrhizal fungus show decreased root permeability which results in qualitative and quantitative changes in root exudates. Thus a marked shift in microbiota population in and around mycorrhizal root is observed. These changes along with changes in physical and chemical composition of surrounding soil is very different from rhizosphere, so much so that Rambelli (1973) termed the soil surrounding and under the influence of mycorrhizae as "mycorrhizosphere". Mycorrhization significantly changes the root morphology and nature of root exudates and secretions; consequently shifting the diversity of original microbiota and establishing a new equilibrium of soil micro flora and micro fauna.

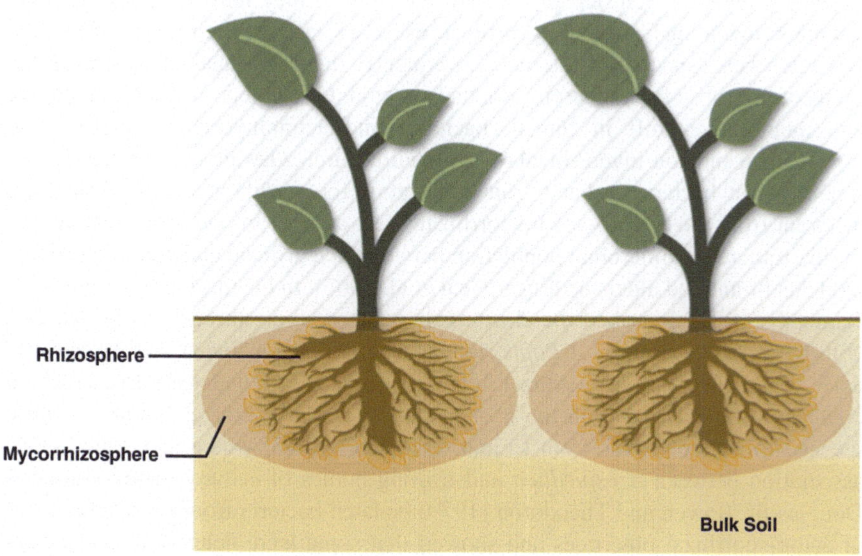

Fig. 16.1 Diagrammatic representation of rhizosphere and mycorrhizosphere. Hyphosphere (not shown) is the soil immediately surrounding hyphal strands of mycorrhizal fungus. Rhizosphere is the soil under the influence of plant roots. Mycorrhizosphere encompasses both rhizosphere and hyphosphere and its limits are extended much beyond the limits of rhizosphere

4 Plant-Mycorrhizae-Soil Microbiota: A Tripartite Relationship

4.1 Ericoid Mycorrhizal Interactions with Soil Protist

Interactions of ericoid mycorrhizal fungi with organisms inhabiting the surrounding soil have been studied to a limited extent until recently where it has been reported that there exists a tripartite relationship between ericoid mycorrhizal fungi, their host plant and soil protist testate amoebae (TA). They studied TA spectra of three different European *Rhododendron* species naturally occurring in the rhizoplane. Interestingly they found that shells of testate amoebae in inhabited by some dark septate endophytic (DSE)fungi. From their field observations they concluded that ericoid mycorrhizal fungi and DSE exploit the TA shells as a source of nutrition. They calculated that some DSE species have colonized more than 45% of *Trigonopyxis* shell. They supported their study by performing in vitro experiments with ericoid mycorrhizal fungi Rhizoscyphus ericae and dark septate endophytic fungi *Phialocephala fortinii* which utilize TA shells for nutrition (Vohník et al. 2009).

4.2 Ectomycorrhizal Interactions with Soil Bacteria

In experiments by Schisler and Linderman (1989), they have demonstrated that volatiles emanating from ectomycorrhizae exerts a selective pressure on surrounding soil microbes, especially bacteria. Consequently there is a qualitative shift in the kinds of microbes found therein. Katznelson et al. (1962) reported both qualitative and quantitative shift in fungal, bacterial and actinomycetes in and around mycorrhizal and non-mycorrhizal roots of yellow birch. Oswald and Ferchau (1968) reported that bacteria isolated from coniferous roots without mycorrhizae were different from those that were mycorrhizal. Neal et al. (1968) reported differences in rhizosphere soil microbial population between mycorrhizal and non-mycorrhizal Douglas fir and red alder seedlings (Neal et al. 1967). In his most noteworthy work Rambelli (1973) described the close association of some nitrogen fixing bacteria with the ectomycorrhizae of *Pinus radiata* wherein he concluded that the bacteria derived nutrition from mycorrhizal fungus in exchange for its fixed nitrogen which could now be used by both the host plant and ectomycorrhizal fungus. A more recent similar report was from Li and Castellano (1985) where they found close association between *Azospirillum* and fruiting bodies of ectomycorrhiza fungi of Douglas fir. Bowen and Theodorou (1979) isolated bacteria from mycosphere soil of ectomycorrhizal pine trees and showed that some were deleterious and others were beneficial for development of mycorrhizae. Duponnois and Garbaye (1991) in their work demonstrated a phenomena which is known as dual inoculation. This is part of management strategy in which two bare root forest nurseries of Douglas fir seedlings were inoculated with Calcium alginate seed encapsulation containing ectomycorrhizal fungi *Laccaria laccata* and some Mycorrhiza Helper Bacterium (MHB). The MHB was later reported to be *Pseudomonas fluorescens* strain BBc6R8, while working with Douglas fir and ectomycorrhizal fungus *Laccaria bicolor* strain S238N (Frey-Klett et al. 1999). In all dual inoculation studies it has been found that interaction between the ectomycorrhizal fungus and the bacteria has had a synergistic effect on plant growth by increasing the percentage of mycorrhizal short roots. It has been reported by Duponnois and Garbaye (1991) that when compared with the control plants with no bacterial inoculation, the percent of mycorrhizal short root increased from 60% to 90% and from 80% to 100% depending on the nursery. It has also been stated in a study that effective population density of *Pseudomonas fluorescens* strain BBc6R8, in dual inoculation is as low as 102 colony forming unit per gram soil, which is in contrast to most Plant Growth Promoting Rhizobacteria (PGPR) where to obtain beneficial effect a population density of 105 colony forming unit per gram soil is required (Frey-Klett et al. 1999). A second type of dual inoculation has been reported by Founoune et al. (2002) where they have inoculated *Acacia holosericea* with *Glomus aggregatum* strain IR27 (AM fungi) and *Pisolithnus tinctorius* strain COI024 (ectomycorrhizal fungus). They also performed single inoculation with *G. aggregatum* strain IR27 or *P. tinctorius* strain COI024 and observed greater plant height and shoot biomass from dual inoculation. They concluded that this may be due to nodule formation in hosts subjected to co-inoculation. They also observed higher ectomycorrhizal colonization in dually inoculated hosts.

4.3 Dual Inoculation with AM Fungi and Beneficial Bacteria

Arbuscular Mycorrhiza (AM) or Vesicular Arbuscular Mycorrhiza (VAM) is the predominant and most primitive of mycorrhizal fungi. Most of the reports on microbial interactions in the rhizosphere of arbuscular mycorrhiza is of dual inoculation kind. The co-inoculants are selected groups of bacteria and mycorrhizal fungi who work in synergy to enhance plant growth. Bagyaraj and Menge (1978) reported increased activity of rhizosphere microflora (bacteria and actinomycetes) upon inoculating plants with *Azotobacter* and vesicular arbuscular mycorrhizae. Maximum activity was reported in co-inoculation than inoculating the host singly with either *Azotobacter* or VA fungi. Meyer and Linderman (1986) studied the shift in the microflora diversity in rhizoplane and rhizosphere soil. They observed qualitative shift as facultative anaerobes (nitrogen fixers and ethylene producers) increased and fluorescent pseudomonads decreased. Fluorescent pseudomonads were high in number in the rhizoplane and also the total number of bacteria. Their study also showed that compositional shift of microflora in the mycorrhizosphere adversely affected the microbial induction of sporangia of root pathogen *Phytophthora cinnamomi*. Ames et al. (1984) reported that some bacteria were preferentially selected over the others in mycorrhizal soil. And it is certain that some plant responses are due to the selective microbial shift in the rhizosphere soil of AM fungi. In most cases dual inoculation is done with beneficial bacteria like nitrogen fixers, phosphate solubilizers, or PGPR and besides increasing host short root percent they have been found, to increase the colonization percentage on the host plant, of the mycorrhizal fungi. And it is certain that enhanced plant short root percent is due to higher colonization. Meyer and Linderman (1986) demonstrated that when subclover (*Trifolium subterraneum*) was co-inoculated with VA fungi and growth promoting pseudomonad they significantly enhanced the growth rate of host plant. Uptake of minor elements increased with dual inoculation possibly due to increased rate of colonization. Nodulation by *Rhizobium* also increased when hosts were co inoculated with VA mycorrhizal fungi and PGPR.

Some bacteria may also have suppressive effect on AM fungi and that is why often a time plants and their mycorrhizal fungal partners fail to establish the symbiosis. Work on inhibition of mycorrhizae formation by bacteria have been demonstrated by Krishna et al. (1982) with *Streptomyces* and VAM. In cases of dual inoculation, failure to derive complete benefit from this may be due to suppression and competition from other bacteria present in soil.

4.4 Mycorrhizal Parasitism

A variety of plants and fungus enter into mycorrhizal symbiosis. Mostly they are of mutualistic benefit type. Mycorrhizae are often used as a management system to increase crop productivity. But sometimes in man-managed ecosystems, net cost of symbiosis for mycorrhization exceeds net cost of productivity and the mycorrhizal fungus becomes parasitic on the host. Induction of mycorrhizal parasitism can be genetic, developmental and even environmental (Johnson et al. 1997).

5 Mycorrhizosphere Population as Drivers of Pedogenesis and Biogeochemical Cycles

The process of formation of soil is known as pedogenesis and mycorrhizae are important drivers of the same. Jenny (1941, 1980) was the first to propose a model which emphasize on the importance of biotic components in pedogenesis. The model describes soil as a function of five factors viz. climate, relief/topography, organism, time and source material even considering stochastic factors such as flood and fire. According to this models organisms are both drivers of pedogenesis as well as depend and are responsive to environmental context in which they helped to create the soil. According to Jenny (1941) soil is an open system to which substances may be added or removed from, and that soil and environment is a coupled system which changes when the functions that had created them changed.

According to Soudzilovskaia et al. (2015) climate and soil type preferences explain the present global pattern of mycorrhization intensity by AM fungi, and more independently and recently evolved EcM fungi. However there exist a reciprocal interaction where the hostplant-mycorrhizal fungi influence soil properties in rhizoplane and around rhizosphere.

One of the major feature of ecosystem and a key to survival of every biotic components on earth is the biogeochemical cycle where elements like carbon oxygen, nitrogen, phosphorous and water move through biotic (biosphere) and abiotic (lithosphere, hydrosphere, atmosphere) compartments of earth. It is a type of mass and energy transfer. Plant and their symbiotic mycorrhizal partner play a major role in biogeochemical cycling of carbon, nitrogen and phosphorous. It has been estimated that three-fourth of the organic carbon stock in soil of the contemporary ecosystem is the accumulation of plant photosynthate carbon which have been stabilized by plant-microbial interactions in soil organic matter pool, possibly by mycorrhizae (Hiederer and Köchy 2011; Scharlemann et al. 2014). And in fact, co-evolution of plant and mycorrhizae influenced terrestrial ecosystem, biogeochemical cycle and pedogenesis.

One of the most important function of mycorrhiza is their role in physically structuring soils. Mycorrhization has transformative effect on soil chemistry and physical properties. They increase soil fertility through their input of organic matter. They aid the host plant to achieve deeper root penetration for selective uptake of elements and water. Mycorrhizal fungal cells secrete organic acids and chelators which disintegrates mineral ores and from rocks and consequently accelerate chemical and physical weathering process (Leake et al. 2008; Taylor et al. 2009).

Mycorrhizae can affect processes of pedogenesis either directly or indirectly by improving plant nutrition, plant health and biomass production. Their effects are intrinsically scale dependent i.e. large plants with deeper roots and more extensive mycorrhizal colonization have larger effects on mineral weathering, pedogenic clay formation, accumulation of soil organic matter and developing structures such as soil aggregates and pores; than small plants with shallower roots. In turn, these factors control the core soil functions including carbon, water, nutrient storage capacity and major pathways for movement of some particle fluid and gases.

Stable soil aggregates are generated by mycorrhizae. According to Miller and Jastrow (2000) mycorrhizae physically and chemically bind soil particles into stable macro aggregates. Furthermore, mycorrhizal hyphae arrange macro and micro aggregates of in soil matrix in hierarchical fashion. The contributions of mycorrhizae to soil structure vary with soil type and also with plant and fungal phenotype. Studies from experiments show that AM fungi generally influence aggregate formation more in sandy soil than in clayey soil (Miller and Jastrow 2000). Although both saprotrophic and mycorrhizal fungi facilitate the formation of soil aggregates, mycorrhizal fungi stabilize soil much more effectively than saprotrophic fungi, three reasons for more effective stabilization of soil aggregates by mycorrhizal fungi over saprotrophic fungi are: (1) direct access to plant photosynthate, and consequently less carbon limitation than saprotrophic fungi; (2) hyphae care often more persistent than hyphae of saprotrophic fungi; and (3) hyphae of ericoid, EcM, and AM fungi exude sticky glycoproteinaceous slimes that help to bind soil particles within the mycorrhizal "string bags". Glomalin is a glycoprotein that has been linked with stability of soil aggregates (Miller and Jastrow 2000).

Mycorrhizae play a very important role in C sequestration. Plants convert atmospheric Carbon dioxide into organic C by photosynthesis; some of the photosynthate is used to support mycorrhizal fungal partner and induces the fungus to secret organic acids and enzymes such as reductases whereby they dissolute mineral elements in soil. This changes soil properties (Taylor et al. 2009). The "C energy hypothesis" links below-ground allocation of photosynthate to mineral weathering and element mass transfers by mycorrhizae (Leake et al. 2008; Quirk et al. 2014). This vast mycorrhizal hyphosphere interacts with soil minerals, organic matter, and other soil microorganisms in ways that inevitably affect the chemical, physical, and biological properties of soils. C flows thorough plant to the mycorrhizal fungus. Some of the allocated carbon is used by the fungal symbiont to support other microbes in the mycorrhizosphere; their presence may act synergistically and enhance plant growth. Typically AM fungi mycelium supply C which support phosphate solubilising bacteria and PGPR like *Pseudomonas* and *Burkholderia* that ultimately influence soil processes and soil functions such as mineralisation of organic phosphate and weathering of mineral phosphate, such as calcium phosphate apatite by local acidification. Brantley et al. (2011) conceptualised the effects of mycorrhizal plants on pedogenesis that "solar to-chemical conversion of energy by plants regulates flows of carbon, water, and nutrients through plant-microbe soil networks, thereby controlling the location and extent of biological weathering". This is consistent with the fact that photosynthate C allocation affects landscape evolution and of the findings of Phillips (2009), as he states that present day potential energy of net primary production by trees is 3–7 times greater than kinetic energy generated from tectonic plate uplift and exogenic denudation that are traditionally been considered to control changes in topography over time.

5.1 Carbon Sequestration in Ocean and Ca, P and Si Biogeochemical Cycles

Mycorrhizae have been in association with even the earliest forms of amphibians of the plant kingdom but whether the association was of mutualistic symbiosis type has not yet been determined as there are very poor and only few fossil records of the latter. But it is certain that they have been on this planet ever since the evolution of early land plants. And it seems reasonable to hypothesise that if only a small percentage of photosynthate carbon is allocated to the roots and mycorrhizal hyphae and it directly impacts the processes involved in geomorphology, such as physical and chemical weathering of minerals and rocks; then considering the period they have been on earth their effects on soil and landscape would be extremely important over geological time. The co-evolution of plant-mycorrhiza have had an enhanced effect on weathering of continental silicate rocks and is a key to long term pedogenic and biogeochemical cycles. Release of calcium and magnesium to the ocean by weathering of terrestrial silicate ore such as basalt, ultimately removes CO_2 from the atmosphere over millions of years in geochemical carbon cycle. According to Berner (2006), calcium precipitates marine carbonates such as limestone and chalk. These rocks are uplifted onto land or sub ducted and $CO_2$is released by volcanic degassing. In addition, dissolution of continental silicate and its fluvial export of dissolved and particulate Si into ocean supports the productivity and C sequestration by marine animals such as *Hydra*, sponges and even diatoms that build Si skeletons.

Conley and Carey (2015) reports that the rise of the diatoms over the past 65 million years contributed to half of marine net carbon fixation and was coincident with the rise and expansion of grassy biomes that dominate the present day terrestrial Si cycle. Olssen et al. (2011) hypothesised that AM fungi may be involved in global silicone cycle based on the strong linear relationship obtained phosphorous and silicone content of AM fungal vesicles. Quirk et al. (2014, 2015) in an experiment demonstrated that AM fungi when grow along with members of Hepatics dissolute and trench primary silicate minerals. These observations are suggestive of the role of mycorrhizal associations in the global biogeochemical cycling of silicone as well as calcium and phosphorous. Marine and terrestrial productivity become influenced through this activity and the effects are important from the early Ordovician period when liverworts like plants first appeared and evolved having been selected by nature because they developed symbiotic association with soil fungi.

Taylor et al. (2009) reports that mycorrhization contribute different of processes which affects weathering and pedogenesis. Roots and mycorrhizal hyphae secrete protons, which facilitate in H+ associated cation uptake, and exude organic chelators having molecular weight. The secretions accelerate dissolution of minerals, elemental leaching and ion exchange in soil. Zhang et al. (2014, 2016) reports that some mycorrhizosphere bacteria like *Pseudomonas* are involved in mineral dissolution and help promote plant growth. These bacteria are supported by plant roots and exudates from mycorrhizal fungal hyphae that provide carbon source.

Plant roots with the hyphae of mycorrhizae stabilises soil particles by enmeshment of the soil; thereby it intensifies chemical changes by checking erosion and reburial in less active soil. Glomeromycotan members of mycorrhizae are known to have strong influence in direct contribution to soil organic matter through enmeshment by hyphae which generates water soluble macro aggregates that store organic carbon, reduce surface runoff and erosion as well as improve soil drainage. Clemmensen et al. (2013) reports that EcM fungi increase the recalcitrance of soil organic matter by preferential uptake of most labile forms organic phosphorous and nitrogen. Reports state that EcM fungi also excrete siderophores and organic acids which help in dissolution of complex mineral ores (Haselwandter 2008). EcM fungal necrotic mass have high in polyphenols viz. tannin which play a vital role in C sequestration. Additionally, mycorrhization produce more plant biomass with productivity and hydrological fluxes via increased evapotranspiration and contributes to overall effect of plants on pedogenesis (Taylor et al. 2009).

5.2 Phosphorous Rarity Is the Cause for Mycorrhizal Symbiosis

Beerling (2007) opines that Earth become evolved, diversified and greened over the last 500 million years. For this, plants developed some strategies for its own. For example: increase of biomass, root and shoot length, nutrient demand, developments of its complete structure and water transport regulation to check water loss through stomata etc. Plants uptake phosphorus from soil in the form of phosphate ion.

Mycorrhizae help in this P uptake from the soil for the plant. As plant roots uptake P it a creates a depletion zone around rhizoids and roots. Soon the area of the soil accessible to plant roots, root hair and rhizoids become P exhausted. But the fine distal mycorrhizal hyphae whose average diameter is 2–7 μm can access to even the finest pores of soil, which are often abundant in nutrients. Data by Leake et al. 2008 on average diameter of fine roots and fine distal hyphae of AM fungi explains how the fungal distal hyphae can access to the smallest of pores in soil. According to their calculations, there is average pore diameter in 2–7 μm for the distal absorptive hypha of AM fungi which is far less than the pores of fine roots (100–500 μm) and root hairs (10–15 μm).

5.3 Weathering of Phosphorous by Mycorrhizae

Calcium and phosphorous are the main composition of apatite and earth crust have 95% P. For the formation of soil, weathering of apatite was the main source of P. Weathering, by physical and chemical means, of apatite and other phosphorous containing ore over the course of millions of year have resulted in progressive limitation on the element from Earth's crust. Roots, mycorrhizae and mycorrhizosphere microorganisms transforms phosphorous, through biological and chemical processes, into secondary iron, aluminium phosphates, organic phosphate in soil organic matter and (Walker and Syers 1976). In two models of 120,000 years chronosequences in New Zealand (Turner et al. 2013) and Australia (Albornoz et al. 2016) it was observed that there is a change to EcM mycorrhization from AM due to increase of P limitation. This is due to the fact that EcM fungi have even finer distal hyphae than AM fungi. From the above mentioned model it has been seen that increasing phosphorous limitation resulted in decline of and consequent shift in symbiosis from AM to EcM (Albornoz et al. 2016).

Filippelli (2008) states that the loss of phosphorous from soil by runoff and leaching is balanced by weathering input of apatite fund. But once the mineral is depleted, over the course of time, the pool of phosphorous is depleted to the ocean primarily by fluvial export. Deposition of the element in ocean affects marine primary productivity.

Members of Marchantiopsida diversified about 370 million years ago and they lacked leaves, stomata and roots but had rhizoids that interacted with shallow depths of soil. During the Ordovician era CO_2 concentration in the atmosphere was high. In an experimental study simulating high atmospheric CO_2 concentration it has been seen that mycorrhiza like symbiosis in the early liverwort plants had been extremely beneficial in phosphorous nutrition and in the overall establishment of liverworts class. The then members of liverworts had a symbiotic association with Glomeromycotan fungi and this symbiosis helped the plant to access phosphorous pools by extending extra rhizoidal hyphae which penetrated anywhere between 100–400 m per plant. As phosphorous was growth limiting, acquiring greater amount of this element helped the plants gain biomass. It was found that higher rate of photosynthesis and reproduction help for the development of early land plants diversification. It was suggested that the prevailing environmental conditions have positively favoured mycorrhization of rhizoids of the gametophytes of early liverworts (Humphreys et al. 2010). Quirk et al. (2015) hypothesised that liverwort rhizoids and associated AM mycelia had significant impact on total weather flux of pre historic soil minerals and consequent establishment of land plants. The extra rhizoidal hyphae (0.1 m) were approximately 1/8th the diameter of extra radical hyphae of tree roots and associated mycorrhizae (0.75 m). Therefore they could access finer pores of densely packed pre historic rocks and perform disintegration and consequent pedogenesis and establishment of successive seral plant communities.

Mycorrhization of roots help its host to access nutrient pools generally inaccessible by roots. EcM fungi secrete organic acids in rhizosphere helping weathering of minerals in the soil (van Breemen et al. 2000). Ericoid mycorrhiza are also known to derive nitrogen, phosphorous and sulphur by performing degradation of complex organic matter in soil. However AM fungi have limitation in degrading organic molecule which are complex in nature.

6 Determination of Ecosystem Response and Functioning by the Members of Mycorrhizal Fungi

The stability and functionality of a terrestrial ecosystem relies on its plant biodiversity and species composition. However, biome diversity and richness, as seen from a study conducted, is dependent on the diversity and composition of mycorrhizal fungus in the soil. Two independent but complementary ecological experiments were conducted to show that below ground diversity of particularly AM fungi is a major contributing factor to maintain plant biodiversity and ecosystem functioning. In the study van der Heijden et al. (1998) simulated European calcareous grassland and North American old-fields. They found that at low AM fungi diversity, the plant population and overall structure or microcosms fluctuate greatly when there is a shift in AM fungi taxa, in European calcareous grasslands. And in North American old-fields increase in AM fungal species richness increases plant biodiversity, nutrient capture and productivity in macrocosms. Their results emphasize the need to protect and conserve AM fungi and to consider them in future management practices in order to maintain diverse ecosystem.

6.1 How Nitrogen in Soil Affects Mycorrhizal Community and Determines Ecosystem of the Area

One of the major determinant of biome diversity in a geographical area is the composition of inorganic N in soil. Heavy nitrogen deposition have significantly changed different aspect of functioning of ecosystem (Vitousek et al. 1997; Aber et al. 2003). EcM and AM respond differently to increasing N concentration in soils. EcM being integral to plant nitrogen uptake is more sensitive to increase in inorganic N in soil (Read 1991). Increased nitrogen availability have no net effect on AM fungal abundance (Lilleskov et al. 2001). A recent review by Prado et al. (2011) viewed that for the changes in the diversity of EcM, N addition thresholds were 5–10 kg N per hectare per year but that were 7.8–12 kg N per hectare per year for compositional shifts in diversity of AM members. Reay et al. (2008) predicts that Nitrogen deposition rate, is likely to double by 2030 thus exposing part of the world's ecosystems to threshold nitrogen deposition rate for mycorrhizae.

Eom et al. (1999) in a study reported that nitrogen fertilization of 1 quintal N ha^{-1} $year^{-1}$in large grass prairies increased extra radical hyphal length and root colonization in AM fungi. However, in a more recent study no net effect on colonization was observed but significant shift in mycorrhizal diversity was detected (Jumpponen et al. 2005). Contrary to this addition of nitrogen fertilization to perennial grassland ecosystem, AM abundances is lowered along with lowering of richness of mycorrhizal member in P rich sites and higher AM fungal abundance in P deficit regions of soils. Thus in presence of P addition of inorganic N to soil reduces the abundance of AM fungi and so it seems reasonable to hypothesise that P and N acts antagonistically to induce colonization by AM fungi.

To know the cause for shifting in mycorrhizae that affect the ecosystem, an inoculation experiment was carried out. In a RBD experiment, plots were treated with different nitrogen fertilization and consequently inoculated seedlings of shrub land species (*Artemisia californica*) and grassland species (*Bromusmadritensis*) were sown. It was observed that nitrogen fertilization suppressed growth of the shrubland species while enhancing growth of grassland species. This Inoculations with AM fungi, collected from varying nitrogen fertilization plots reduced the growth of A. californica, however, opposite picture is found in B. madritensis which is a grassland member (Siguenza et al. 2006a, b). However, as molecular techniques continue to advance we now have a clear idea in getting shifts of mycorrhizal diversity in composition. We know that community composition of mycorrhizal fungus affects ecosystem processes viz. change in major plant responses (biomass, biomass distribution, demography and physiology), CO_2 efflux from soil, decomposition, and modified nature of nutrient distribution and ecological cycling and its impact on different aspects of soil nitrogen cycle.

6.2 Mycorrhizae Response to Elevated Carbon Dioxide Concentration

From 1995 to 2005 atmospheric CO_2 concentration have increased at the rate of approximately 2 ppm per year and it is likely to further increase to 2.8–4.2 ppm thorough 2030 (IPCC 2007). Ecosystem responses of mycorrhizae to elevated CO_2 concentration in atmosphere varies with the type of association. Because soil concentration of CO_2 is high therefore direct effects of elevated CO_2 concentration in atmosphere has is less profound on mycorrhizal fungi (Mohan et al. 2007. But indirect effects are enhanced nutritive demand of plants (Finzi et al. 2007). In higher carbon dioxide exposure there are variation within associated species for the effects of AM on ecosystem response (Allen et al. 2005; Clark et al. 2009) and may enhance or inhibit plant growth, root colonization, and spore and extraradical hyphal production with increasing atmospheric CO_2. On the contrary, plant growth is higher enhanced CO_2 concentration is exhibited by EcM fungi (Kasurinen et al. 1999; Garcia et al. 2008).

A study in desert environment by Clark et al. (2009) reported that AM fungal hyphae were unaffected under higher carbon dioxide concentration of atmosphere. However, Allen et al. (2005) on the contrary reported that amount of hyphae in *Scutellospora* and *Acaulospora* and fungal glomalin present inside soil micro aggregates increased with increasing CO_2 concentration in the chaparral. Rillig et al. (1999) observed that elevated CO_2 concentration in the atmosphere promoted AM fungi hyphal length in sandstone grassland whereas AMF mycorrhizae of serpentine grassland remained unaffected; however percentage of root colonization increase in both cases. Results experimentally obtained showed that in warm temperate forest, higher CO_2 concentration failed to affect AMF hyphal length or glomalin stocks. Gracia et al. (2008) observed higher root colonization by EcM fungi in elevated CO_2. However, results obtained from some studies indicated no impact of CO_2on biomass or root colonization by EcM (Kasurinen et al. 1999). More recently Chung et al. (2006) reported that overall fungal abundance of AM and EcM fungi remains unaffected by elevated CO_2 concentration.

Elevated CO_2 level in atmosphere profoundly affects ecosystem processes by changing responses of mycorrhizal fungi. Hyphal length, root colonization and nutrient uptake by fungal members of mycorrhiza are affected by higher CO_2 concentration. With increased availability of CO_2 rate of metabolic processes of the host increases. Photosynthetic rate is heightened significantly. Consequently, greater amount of sugar (synthesized by host) is allocated to the fungus. This results in increased extra-radical hyphal growth, greater glomalin production and increased sequestration of organic C to the soil. Cheng et al. (2012) reports that metabolic processes that uses energy stored in the chemical bonds of reduced C (sugars), like respiration, by both host and mycorrhizal fungi may balance the potential soil organic C accumulation. Furthermore, result of higher activity viz. respiration, extra-radical hyphal growth and plant nutrient transfer may be variable among AM and EcM symbionts (Johnson et al. 2005, 2013).

Increasing CO_2 concentration in atmosphere results in compositional shift in AM community in the soil. Host plant and fungi both respond to changing community structure of AM fungi (Compant et al. 2010). Parrent and Vilgalys (2007) reported similar shifts in EcM communities. However, the cause for shift in mycorrhizal fungal community and differential response of AM and EcM fungi is still unclear and can be a potential area for future investigation.

6.3 Effect of Rising Temperature on Mycorrhizal Species Composition

Rising temperature can effect mycorrhizal colonization of host plant. The effects can be direct or indirect. Direct effect of increasing temperature on mycorrhizal fungi have not been studied extensively and reports suggest that some increased temperature enhances functionality of some AM fungus (measured as a response of increasing extra-radical hyphal length). Therefore in the predicted 3 °C rise of global temperature by next century will increase mycorrhization by some fungus (Heinemeyer and Fitter 2004) which will help its host to acquire more nutrients which will be the demand of higher rate of photosynthesis in increasing temperature. Indirect effects include heat induced change in physiology and metabolism of host plant, altered resource demand of host and changing rates of soil nutrient biogeochemistry. One of the best studied indirect effect is where warming changes rates of viz. nitrogen mineralization (Peterjohn et al. 1994) and nitrification (Butler et al. 2012) thus causing a warming-induced indirect "fertilization effect."Olsrud et al. (2010) on a study on arctic grass reported that warming resulted in lowered mycorrhization by AM members and elevated foliar N percentage. However, no net change in above ground grass cover was observed. This indicates that higher rates of N mineralization inhibited mycorrhization. Majority of the field studies conducted to see the effect of temperature has been from arctic regions of the world and other parts, particularly warm temperate and tropical regions need to be studied. By this we can get rid of the inconsistencies and bias in our results and gain a clear understanding of effect of temperature on mycorrhizal fungi.

One consequence of rising temperature is altered precipitation having decreased moisture in sub-tropical and tropical areas and lower precipitation at low latitudes (Trenberth et al. 2007). Longer dry periods and increased episodic rainfall change the quantity, frequency and degree of precipitation for terrestrial ecosystem; which will lead to either flood or drought and put the plant under water stress (Trenberth 2011). EcM fungus are known to alleviate drought stress in plants. Results drawn from field studies suggest that seedlings pre inoculated with EcM fungus have higher survival rate upon being subsequently subjected to drought stress. This survivorship is due to favourable hydraulic conductance for higher connectivity to EcM network and greater diversity in mycorrhizal mycoflora in plants surviving extreme drought (Mohan et al. 2014).

6.4 Effect of UV Radiation on Mycorrhization

Stratospheric ozone depletion is a concern of recent years that help to conduct number of investigations to check effect UV radiation on plant (Laing 1991; Reboredo and Lidon 2012). Most of these studies focuses on above ground processes viz. photosynthesis and biomass production, however less work was carried out on effect of UV on root and soil changes and only a few studies examine effects on mycorrhizal colonization and growth (De La Rosa et al. 2003; Kristian et al. 2008).

Results obtained from these studies shows inconsistencies as some reports that enhanced UV leads to decrease in AM colonization (van de Staaij et al. 2001), while others report no net effect (Klironomos and Allen 1995). This is suggestive of additive ecosystem variables like community members, latitude and soil nutrient levels etc. which function in conjugation with UV radiation to play an important part changing mycorrhizal response.

7 Conclusion

From the foregoing discussion it seems very obvious that mycorrhizosphere has profound effect in pedogenesis, biogeochemical cycles and in determining the diversity and functionality of terrestrial ecosystem. In a study on pedogenesis determining plant diversity of an ecosystem it was found that soils that are older and more weathered are richer in plant species and they have been under the influence of mycorrhizal fungi for longer period of time. It is also evident from other studies that mineral weathering is in majority done by mycorrhizal fungi and mycorrhizosphere microbiota. And studies report that tropics and sub-tropics which are the most biome rich places on the globe have the most old and weathered rocks than less diverse temperate and arctic regions.

Development of mycorrhizae is influenced by the rhizosphere microflora and diversity of microbial population in the mycorrhizosphere is controlled by the mycorrhizal fungal type, where some microbes are selected over the other. Host plant plays a pivotal role in determining the mycorrhizal fungal type and indirectly controls the rhizosphere population around mycorrhizal fungal hyphae. Thus it can be said that this host-mycorrhizae-soil microbiota is a tripartite relationship and each component equally and universally determines the diversity and number of every other component.

Mycorrhizal symbiosis have been playing significant role since pre-historic pedogenesis and evolution of terrestrial ecosystem. It has helped early vascular land plants to transit from aquatic to terrestrial habitat and 'conquer' the land by facilitating plant access to spread out nutrient pools and water as well as better anchorage to the substratum. Systemic acquired resistance that developed in mycorrhizal plants have helped them survive selection pressure while the non-mycorrhizal trees remained vulnerable to diseases and extinction. Pre-historic pedogenesis have been initiated by mycorrhizal fungi and they continue to drive soil formation to present day, especially by organic matter inputs and effects on composition of soil structure. The symbiosis has long helped in biogeochemical cycling.

Using mycorrhizae as a part of agricultural field management strategies may help increase productivity and mitigate problems of scarcity in food. They have been reported to alleviate drought stress and also help host plant cope with infections from exopathogens.

A considerable amount of work has been done on mycorrhizae but there still remains a lot unexplored. With the advancement of sophisticated molecular techniques we hope to solve more and understand the underlying molecular mechanism by which mycorrhizae drives pedogenesis and respond to microflora and fauna present in the mycorrhizosphere ultimately leading to ecosystem development of an area.

References

Aber, J. D., Goodale, C. L., Ollinger, S. V., Smith, M. L., Magill, A. H., Martin, M. E., Hallett, R. A., & Stoddard, J. L. (2003). Is nitrogen deposition altering the nitrogen status of northeastern forests? *Bioscience, 53*, 375–389.

Albornoz, F. E., Lambers, H., Turner, B. L., Teste, F. P., & Laliberté, E. (2016). Shifts in symbiotic associations in plants capable of forming root symbioses across a long-term soil chronosequence. *Ecology and Evolution, 6*, 2368–2377.

Allen, M. F., Klironomos, J. N., Treseder, K. K., & Oechel, W. C. (2005). Responses of soil biota to elevated CO2 in a chaparral ecosystem. *Ecological Applications, 15*, 1701–1711.

Ames, R. N., Reid, C. P. P., & Ingham, E. R. (1984). Rhizosphere bacterial population responses to root colonization by vesicular-arbuscular mycorrhizal fungus. *The New Phytologist, 96*, 555–563.

Bagyaraj, D. J., & Menge, J. A. (1978). Interaction between VA Mycorrhiza and Azotobacter and their effects on rhizosphere microflora and plant growth. *The New Phytologist, 80*, 567–573.

Beerling, D. (2007). *The Emerald planet. How plants changed earth's history* (288 pages). Oxford: Oxford University Press.

Berner, R. A. (2006). GEOCARBSULF: A combined model for Phanerozoic atmospheric O2 and CO2. *Geochimica et Cosmochimica Acta, 70*, 5653–5664.

Bowen, G. D., & Theodorou, C. T. (1979). Interactions between bacteria and ectomycorrhizal fungi. *Soil Biology and Biochemistry, 11*, 119–126.

Brantley, S. L., Megonigal, J. P., Scatena, F. N., Balogh-Brunstad, Z., Barnes, R. T., Bruns, M. A., Van Cappellen, P., Dontsova, K., Hartnett, H. E., Hartshorn, A. S., Heimsath, A., Herndon, E., Jin, L., Keller, C. K., Leake, J. R., McDowell, W. H., Meinzer, F. C., Mozdzer, T. J., Petsch, S., Pett-Ridge, J., Pregitzer, K. S., Raymond, P. A., Riebe, C. S., Shumaker, K., Sutton-Grier, A., Walter, R., & Yoo, K. (2011). Twelve testable hypotheses on the geobiology of weathering. *Geobiology, 9*, 140–165.

Brundrett, M. C. (2002). Coevolution of roots and mycorrhizae of land plants. *The New Phytologist, 154*, 275–304.

Butler, S. M., Melillo, J. M., Johnson, J., Mohan, J., Steudler, P. A., Lux, H., Burrows, E., Smith, R., Vario, C., & Scott, L. (2012). Soil warming alters nitrogen cycling in a New England forest: Implications for ecosystem function and structure. *Oecologia, 168*, 819–828.

Cheng, L., Booker, F. L., Tu, C., Burkey, K. O., Zhou, L., Shew, H. D., Rufty, T. W., & Hu, S. (2012). Arbuscular mycorrhizal fungi increase organic carbon decomposition under elevated CO_2. *Science, 337*, 1084–1087.

Chung, H., Zak, D. R., & Lilleskov, E. A. (2006). Fungal community composition and metabolism under elevated CO_2 and O_3. *Oecologia, 147*, 143.

Clark, N. M., Rillig, M. C., & Nowak, R. S. (2009). Arbuscular mycorrhizal fungal abundance in the Mojave Desert: Seasonal dynamics and impacts of elevated CO_2. *Journal of Arid Environments, 73*, 834–843.

Clemmensen, K. E., Bahr, A., Ovaskainen, O., Dahlberg, A., Ekblad, A., Wallander, H., Stenlid, J., Finlay, R. D., Wardle, D. A., & Lindahl, B. D. (2013). Roots and associated fungi drive long-term carbon sequestration in boreal forest. *Science, 339*, 1615–1618.

Compant, S., Van Der Heijden, M. G., & Sessitsch, A. (2010). Climate change effects on beneficial plant-microorganism interactions. *FEMS Microbiology Ecology, 73*, 197–214.

Conley, D. J., & Carey, J. C. (2015). Silica cycling over geologic time. *Nature Geoscience, 8*, 431–432.

De La Rosa, T. M., Aphalo, P. J., & Lehto, T. (2003). Effects of ultraviolet-B radiation on growth, mycorrhizae and mineral nutrition of silver birch (*Betula pendula* Roth) seedlings grown in low-nutrient conditions. *Global Change Biology, 9*, 65–73.

Duponnois, R., & Garbaye, J. (1991). Effect of dual inoculation of Douglas fir with the ectomycorrhizal fungus *Laccaria laccata* and mycorrhization helper bacteria (MHB) in two bare-root forest nurseries. *Plant and Soil, 138*, 169–176.

Eom, A. H., Hartnett, D. C., Wilson, G. W., & Figge, D. A. (1999). The effect of fire, mowing and fertilizer amendment on arbuscular mycorrhizae in tallgrass prairie. *The American Midland Naturalist, 142*, 55–70.

Filippelli, G. (2008). The global phosphorus cycle: Past, present, and future. *Elements, 4*, 89–95.

Finzi, A. C., Norby, R. J., Calfapietra, C., Gallet-Budynek, A., Gielen, B., Holmes, W. E., Hoosbeek, M. R., Iversen, C. M., Jackson, R. B., Kubiske, M. E., Ledford, J., Liberloo, M., Oren, R., Polle, A., Pritchard, S., Zak, D. R., Schlesinger, W. H., & Ceulemans, R. (2007). Increases in nitrogen uptake rather than nitrogen-use efficiency support higher rates of temperate forest productivity under elevated CO_2. *Proceedings of the National Academy of Sciences USA, 104*, 14014–14019.

Founoune, H., Duponnois, R., Bâ, A. M., & El Bouami, F. (2002). Influence of the dual arbuscular endomycorrhizal/ectomycorrhizal symbiosis on the growth of Acacia holosericea (A. Cunn. ex G. Don) in glasshouse conditions. *Annals of Forest Science, 59*, 93–98.

Frey-Klett, P., et al. (1999). Dose effect in the dual inoculation of an ectomycorrhizal fungus and a mycorrhiza helper bacterium in two forest nurseries. *Soil Biology and Biochemistry, 31*, 1555–1562.

Garcia, M. O., Ovasapyan, T., Greas, M., & Treseder, K. K. (2008). Mycorrhizal dynamics under elevated CO2 and nitrogen fertilization in a warm temperate forest. *Plant and Soil, 303*, 301–310.

Haselwandter, K. (2008). Structure and function of siderophores produced by mycorrhizal fungi. *Mineralogical Magazine, 72*, 61–64.

Heinemeyer, A., & Fitter, A. H. (2004). Impact of temperature on the arbuscular mycorrhizal (AM) symbiosis: Growth responses of the host plant and its AM fungal partner. *Journal of Experimental Botany, 55*(396), 525–534.

Hiederer, R., & Köchy, M. (2011). *Global soil organic carbon estimates and the harmonized world soil database* (EUR 25225 EN). Luxembourg: Publications Office of the EU.

Hiltner, L. (1904). Überneuere Erfahrungen und Probleme auf dem Gebiete der Bodenbakteriol ogieunterbesonderer Berücksichtigung der Gründüngung und Brache. *Arbeiten der DLG, 98*, 59–78.

Humphreys, C. P., Franks, P. J., Rees, M., Bidartondo, M. I., Leake, J. R., & Beerling, D. J. (2010). Mutualistic mycorrhiza-like symbiosis in the most ancient group of land plants. *Nature Communications, 1*, 7.

IPCC. (2007). In S. Solomon, D. Qin, M. Manning, Z. Chen, M. Marquis, K. B. Averyt, M. Tignor, & H. L. Miller (Eds.), *Climate change 2007 – The physical science basis: Working Group I contribution to the fourth assessment report of the IPCC*. Cambridge: Cambridge University Press.

Jenny, H. (1941). *Factors of soil formation – A system of quantitative pedology*. New York: McGraw-Hill.

Jenny, H. (1980). *The soil resource, origin and behavior*. New York/Heidelberg/Berlin: Springer.

Johnson, N. C., & Gehring, C. A. (2007). Chapter-4 mycorrhizae: Symbiotic mediators of rhizosphere and ecosystem processes. In Z. G. Cardon & J. L. Whitbeck (Eds.), *The rhizosphere: An ecological perspective*. Amsterdam: Elsevier Academic Press.

Johnson, N. C., Graham, J. H., & Smith, F. A. (1997). Functioning of mycorrhizal associations along the mutualism-parasitism continuum. *The New Phytologist, 135*, 575–585.

Johnson, N. C., Wolf, J., Reyes, M. A., Panter, A., Koch, G. W., & Redman, A. (2005). Species of plants and associated arbuscular mycorrhizal fungi mediate mycorrhizal responses to CO_2 enrichment. *Global Change Biology, 11*, 1156–1166.

Johnson, N. C., Caroline Angelard, I. R. S., & Kiers, E. T. (2013). Predicting community and ecosystem outcomes of mycorrhizal responses to global change. *Ecology Letters, 16*, 140–153.

Jumpponen, A., Trowbridge, J., Mandyam, K., & Johnson, L. (2005). Nitrogen enrichment causes minimal changes in arbuscular mycorrhizal colonization but shifts community composition-evidence from rDNA data. *Biology and Fertility of Soils, 41*, 217–224.

Kasurinen, A., Helmisaari, H. S., & Holopainen, T. (1999). The influence of elevated CO_2 and O_3 on fine roots and mycorrhizae of naturally growing young Scots pine trees during three exposure years. *Global Change Biology, 5*, 771–780.

Katznelson, H., Rouatt, J. W., & Peterson, E. A. (1962). The rhizosphere effect of mycorrhizal and non mycorrhizal roots of yellow birch seedlings. *Canadian Journal of Botany, 40*, 377–382.

Klironomos, J. N., & Allen, M. F. (1995). UV-B-mediated changes on below-ground communities associated with the roots of *Acer saccharum*. *Functional Ecology, 9*, 923–930.

Krishna, K. R., Balakrishna, A. N., & Bagyaraj, D. J. (1982). Interaction between a vesicular-arbuscular mycorrhizal fungus and *Streptomyces cinnamomeous* and their effects on finger millet. *The New Phytologist, 92*, 401–405.

Kristian, R. A., Riikka, R., Helge, R. P., Teis, N. M., Kirsten, B. H., Marie, F. A., & Anders, M. (2008). Solar Ultraviolet-B radiation at Zackenberg: The impact on higher plants and soil microbial communities. *Advances in Ecological Research, 40*, 421–440.

Laing, W. A. (1991). The consequences of increased ultraviolet-B radiation for plants. *DSIR Fruit and Trees Internal Report, 206*.

Leake, J. R., & Read, D. J. (2017). Chapter-2 Mycorrhizal symbioses and pedogenesis throughout earth's history. In N. C. Johnson, C. Gehring, & J. Jansa (Eds.), *Mycorrhizal mediation of soil*. Amsterdam: Elsevier.

Leake, J. R., Johnson, D., Donnelly, D., Muckle, G. E., Boddy, L., & Read, D. J. (2004). Networks of power and influence: The role of mycorrhizal mycelium in controlling plant communities and agro-ecosystem functioning. *Canadian Journal of Botany, 82*, 1016–1045.

Leake, J. R., Duran, A. L., Hardy, K. E., Johnson, I., Beerling, D. J., Banwart, S. A., & Smits, M. M. (2008). Biological weathering in soil: The role of symbiotic root-associated fungi biosensing minerals and directing photosynthate-energy into grain-scale mineral weathering. *Mineralogical Magazine, 72*, 85–89.

Li, C. Y., & Castellano, M. A. (1985). Nitrogen-fixing bacteria isolated from within sporocarps of three ectomycorrhizal fungi. In: *Proceedings of the 6th North American coference on mycorrhizae* (p. 264), June 25–29, 1984, Bend.

Lilleskov, E. A., Fahey, T. J., & Lovett, G. M. (2001). Ectomycorrhizal fungal aboveground community change over an atmospheric nitrogen deposition gradient. *Ecological Applications, 11*, 397–410.

Linderman, R. G. (1988). *Mycorrhizal interactions with the rhizosphere microflora: The mycorrhizosphere effect*. Paper presented at symposium: Interaction of Mycorrhizal Fungi, Aps Symp Ser.

Meyer, J. R., & Linderman, R. G. (1986). Response of subterranean clover to dual inoculation with vesicular-arbuscular mycorrhizal fungi and a plant growth-promoting bacterium, Pseudomonas putida. *Soil Biology and Biochemistry, 18*(2), 185–190.

Miller, R. M., & Jastrow, J. D. (2000). Mycorrhizal fungi influence soil structure. In Y. Kapulnik & D. D. Douds Jr. (Eds.), *Arbuscular mycorrhizae: Physiology and function* (pp. 3–18). London: Kluwer Academic Publishers.

Mohan, J. E., Clark, J. S., & Schlesinger, W. H. (2007). Long-term CO2 enrichment of a forest ecosystem: Implications for forest regeneration and succession. *Ecological Applications, 17*, 1198–1212.

Mohan, J. E., Cowden, C. C., Baas, P., Dawadi, A., Frankson, P. T., Helmick, K., Hughes, E., Khan, S., Lang, A., Machmuller, M., Taylor, M., & Witt, C. A. (2014). Mycorrhizal fungi mediation of terrestrial Ecosystem responses to global change: Mini-review. *Fungal Ecology, 10*, 3–19. Science Direct. Elsevier.

Neal, J. L. Jr., Lu, K. C., Bollen, W. B., & Trappe, J. M. (1968). A comparison of rhizosphere microfloras associated with mycorrhizae of red alder and Douglas-fir. Pager 57–71 In: J. M. Trappe, J. F. Franklin, R. F. Tarrant & G. M. Hansen (Eds.), *Biology of Alder. USDA Forest Service, Pacific Northwest Forest and Range Experiment Station* (292 pp).

Neal Jr, J. L., Lu, K. C., Bollen, W. B., & Trappe, J. M. (1967, April). A comparison of rhizosphere microfloras associated with mycorrhizae of red alder and Douglas-fir. In Biology of Alder, Proceedings of Northwest Scientific Association Annual Meeting.

Olsrud, M., Carlsson, B. A., Svensson, B. M., Michelsen, A., & Melillo, J. M. (2010). Responses of fungal root colonization, plant cover and leaf nutrients to long-term exposure to elevated atmospheric CO_2 and warming in a subarctic birch forest understory. *Global Change Biology, 16*, 1820–1829.

Olssen, P. A., Hammer, E. C., Pallon, J., & Van Aarle, I. M. (2011). Elemental composition in vesicles of an arbuscular mycorrhizal fungus, as revealed by PIXE analysis. *Fungal Biology-UK, 115*, 643–648.

Oswald, E. T., & Ferchau, H. A. (1968). Bacterial association of coniferous mycorrhizae. *Plant and Soil, 28*, 187–192.

Pardo, L. H., Fenn, M. E., Goodale, C. L., Geiser, L. H., Driscoll, C. T., Allen, E. B., Baron, J. S., Bobbink, R., Bowman, W. D., Clark, C. M., Emmett, B., Gilliam, F. S., Greaver, T. L., Hall, S. J., Lilleskov, E. A., Liu, L. L., Lynch, J. A., Nadelhoffer, K. J., Perakis, S. S., Robin-Abbott, M. J., Stoddard, J. L., Weathers, K. C., & Dennis, R. L. (2011). Effects of nitrogen deposition and empirical nitrogen critical loads for ecoregions of the United States. *Ecological Applications, 21*, 3049–3082.

Parrent, J. L., & Vilgalys, R. (2007). Biomass and compositional responses of ectomycorrhizal fungal hyphae to elevated CO_2 and nitrogen fertilization. *The New Phytologist, 176*, 164–174.

Peterjohn, W. T., Melillo, J. M., Steudler, P. A., Newkirk, K. M., Bowles, F. P., & Aber, J. D. (1994). Responses of trace gas fluxes and N availability to experimentally elevated soil temperatures. *Ecological Applications, 4*, 617–625.

Phillips, J. D. (2009). Biological energy in landscape evolution. *American Journal of Science, 309*, 271–290.

Pirozynski, K. A., & Malloch, D. W. (1975). The origin of land plants: A matter of mycotrophism. *Biosystems, 6*, 153–164.

Quirk, J., Leake, J. R., Banwart, S. A., Taylor, L. L., & Beerling, D. J. (2014). Weathering by tree-root-associating fungi diminishes under simulated Cenozoic atmospheric CO2 decline. *Biogeosciences, 11*, 321–331.

Quirk, J., Leake, J. R., Johnson, D. A., Taylor, L. L., Saccone, L., & Beerling, D. J. (2015). Constraining the role of early land plants in Palaeozoic weathering and global cooling. *Proceedings of the Royal Society B, 282*, 20151115. https://doi.org/10.1098/rspb.2015.1115.

Rambelli, A. (1973). The rhizosphere of mycorrhizae. In G. L. Marks & T. T. Koslowski (Eds.), *Ectomycorrhizae* (pp. 299–343). New York: Academic.

Read, D. J. (1991). Mycorrhizae in ecosystems. *Experientia, 47*, 376–391.

Reay, D. S., Dentener, F., Smith, P., Grace, J., & Feely, R. A. (2008). Global nitrogen deposition and carbon sinks. *Nature Geoscience, 1*, 430–437.

Reboredo, F., & Lidon, F. J. C. (2012). UV-B radiation effects on terrestrial plants – A perspective. *Emirates Journal of Food and Agriculture, 24*, 502–509.

Redecker, D., Kodner, R., & Graham, L. E. (2000). Glomalean fungi from the Ordovician. *Science, 289*, 1920–1921.

Rillig, M. C., Field, C. B., & Allen, M. F. (1999). Soil biota responses to long-term atmospheric CO_2 enrichment in two California annual grasslands. *Oecologia, 119*, 572–577.

Scharlemann, J. P. W., Tanner, E. V. J., Hiederer, R., & Kapos, V. (2014). Global soil carbon: Understanding and managing the largest terrestrial carbon pool. *Carbon Management, 5*, 81–91.

Schisler, D. A., & Linderman, R. G. (1989). Influence of humic-rich organic amendments to coniferous nursery soils on Douglas-fir growth, damping-off and associated soil microorganisms. *Soil Biology and Biochemistry, 21*(3), 403–408.

Siguenza, C., Corkidi, L., & Allen, E. B. (2006a). Feedbacks of soil inoculum of mycorrhizal fungi altered by N deposition on the growth of a native shrub and an invasive annual grass. *Plant and Soil, 286*, 153–165.

Siguenza, C., Crowley, D. E., & Allen, E. B. (2006b). Soil microorganisms of a native shrub and exotic grasses along a nitrogen deposition gradient in southern California. *Applied Soil Ecology, 32*, 13–26.

Smith, S. E., & Read, D. J. (1997). *Mycorrhizal symbiosis*. New York: Academic.

Soudzilovskaia, N. A., Douma, J. C., Akhmetzhanova, A. A., Bodegom, P. M., Cornwell, W. K., Moens, E. J., Treseder, K. K., Tibbett, M., Wang, Y. P., & Cornelissen, J. H. C. (2015). Global patterns of plant root colonization intensity by mycorrhizal fungi explained by climate and soil chemistry. *Global Ecology and Biogeography, 24*, 371–382.

Stubblefield, S. P., Taylor, T. N., & Trappe, J. M. (1987). Fossil mycorrhizae: A case for symbiosis. *Science, 237*, 59–60.

Taylor, L. L., Leake, J. R., Quirk, J., Hardy, K., Banwart, S. A., & Beerling, D. J. (2009). Biological weathering and the longterm carbon cycle: Integrating mycorrhizal evolution and function into the current paradigm. *Geobiology, 7*, 171–191.

Trenberth, K. E. (2011). Changes in precipitation with climate change. *Climate Research, 47*, 123–138.

Trenberth, K. E., Smith, L., Qian, T., Dai, A., & Fasullo, J. (2007). Estimates of the global water budget and its annual cycle using observational and model data. *Journal of Hydrometeorology, 8*, 758–769.

Turner, B. L., Lambers, H., Condron, L. M., Cramer, M. D., Leake, J. R., Richardson, A. E., & Smith, S. E. (2013). Soil microbial biomass and the fate of phosphorus during long-term ecosystem development. *Plant and Soil, 367*, 225–234.

van Breemen, N., Finlay, R., Lundström, U., Jongmans, A. G., Giesler, R., & Olsson, M. (2000). Mycorrhizal weathering: A true case of mineral plant nutrition? *Biogeochemistry, 49*(1), 53–67.

van de Staaij, J., Rozema, J., van Beem, A., & Aerts, R. (2001). Increased solar UV-B radiation may reduce infection by arbuscular mycorrhizal fungi (AMF) in dune grassland plants: Evidence from five years of field exposure. *Plant Ecology, 154*(1–2), 169.

van der Heijden, M. G. A., Klironomos John, N., Margot, U., Peter, M., Ruth, S.-E., Thomas, B., Andres, W., & Sanders, I. R. (1998). Mycorrhizal fungal diversity determines plant biodiversity, ecosystem variability and productivity. *Nature, 396*, 69–72.

Vitousek, P. M., Mooney, H. A., Lubchenco, J., & Melillo, J. M. (1997). Human domination of Earth's ecosystems. *Science, 277*, 494–499.

Vohník, M., Burdíková, Z., Albrechtová, J., & Vosátka, M. (2009). Testate Amoebae (Arcellinida and Euglyphida) vs. Ericoid Mycorrhizal and DSE fungi: A possible novel interaction in the Mycorrhizosphere of Ericaceous plants? *Microbial Ecology, 57*, 203–214.

Walker, T. W., & Syers, J. K. (1976). The fate of phosphorus during pedogenesis. *Geoderma, 15*, 1–19.

Zhang, L., Fan, J., Ding, X., He, X., Zhang, F., & Feng, G. (2014). Hyphosphere interactions between an arbuscular mycorrhizal fungus and a phosphate solubilizing bacterium promote phytate mineralization in soil. *Soil Biology and Biochemistry, 74*, 177–183.

Zhang, L., Xu, M., Liu, Y., Zhang, F., Hodge, A., & Feng, G. (2016). Carbon and phosphorus exchange may enable cooperation between an arbuscular mycorrhizal fungus and a phosphate-solubilizing bacterium. *The New Phytologist, 210*, 1022–1032.

Mycorrhiza Based Approaches for Soil Remediation and Abiotic Stress Management

17

Ratul Moni Ram, Prakash Jyoti Kalita, Rahul Singh Rajput, and H. B. Singh

1 Introduction

Soil contains various forms of metals and trace elements although trace elements are required for the normal growth of the plant but heavy metals are very harmful for the plants. Trace elements like Cu, Fe, Mn or Zn in small amount is beneficial for the plants but higher concentration again makes it unsuitable for the proper development of the plants (Ruotsalainen et al. 2007). The main reason for the spread of these elements is the unscrupulous human mining activities, smelting, petroleum industries, fertilizer industries and many more. All these sources in turn adds a huge amount of such wastes in the nearby areas and which in turn makes the agricultural land unsuitable for the crop production. These heavy metals can significantly affect the photosynthetic rates of the crops and thereby can hamper the overall productivity (Leyval and Binet 1998). Further these heavy metals when gets absorbed by the roots they affects the metabolism by impairing the normal enzymatic activities as they can distort the protein folding and its overall activities by replacing itself with the essential elements from the enzymes. These heavy metals also contaminate the water resources and thereby affect the aquatic organisms. As these heavy metals are very difficult to degrade so sometimes they even enters inside the food chain *via* several sources and thereby severely affects the human health. Apart from these inorganic elements soils are also gets contaminated with organic chemicals which in turn are highly toxic in nature. Organic chemical mostly includes the chlorinated aromatic herbicides, polycyclic aromatic hydrocarbons (PAHs) and several antibiotics. Among these the contamination of aromatic herbicides occurs during

R. M. Ram (✉) · R. S. Rajput · H. B. Singh
Department of Mycology and Plant Pathology, Institute of Agricultural Sciences, Banaras Hindu University, Varanasi, India

P. J. Kalita
Myongji University, Yongin, Gyeonggi-Do, South Korea

© Springer Nature Singapore Pte Ltd. 2019

A. Varma, D. K. Choudhary (eds.), *Mycorrhizosphere and Pedogenesis*,
https://doi.org/10.1007/978-981-13-6480-8_17

the excessive use of these herbicides in the agricultural land for the control of weeds. As these herbicides are not completely absorbed by the plants so a huge amount of residues remain in the soil which in turn contaminates the soil (Chaudhry and Chapalamadugu 1991). Spread of PAH is mainly attributed to the chemical industries and through aerosols (Braun-Lullemann et al. 1999). Apart from this, in the livestock farming for the control of several diseases many antibiotics are being used, but these antibiotics are not at all completely metabolized inside the body of the animal and hence a huge portion of the un-metabolized antibiotics are excreted out. These livestock excreta are generally used for the production of manures and this way these antibiotics come in the direct contact of the soil microorganisms and plants. These antibiotics by several means affect the normal growth and development of the plants and the soil microorganisms (Tang et al. 2015). In order to tackle these problems several approaches has been used but the use of arbuscular mycorrhizal fungi (AMF) is more reliable as it is cost effective and environmental friendly.

AMF are the most common mycorrhizal association and reported to form mutualistic association with nearly 74% of flowering plants (Brundrett 2009) and with over 80% of vascular plants (Peterson et al. 2004). AMF are characterized by presence of arbuscles (modified hyphae for resource exchange), intraradical hyphae, extraradical mycelium and large spores. The AMF plant hosts range from nonvascular plants to gymnosperms alongwith majority of angiosperm families. (Peterson et al. 2004). Most AMF species are found to be associated with the economically important families *viz.* Gramineae, Compositae, Fabacae, Rosaceae and Labiaceae (Newman and Reddell 1987). The beneficial role of AMF has been depicted in Fig. 17.1.

2 Reclamation of Soil Polluted with Inorganic Chemicals by Using AMF

It has been reported previously that mycorrhiza have considerable roles in imparting tolerance against various abiotic stresses, this in turn has lead the researchers to find out it's possible role in the degradation of the organic and inorganic pollutants (Donnelly and Fletcher 1994; Meharg and Cairney 2000). Major part of the research in this field has been carried out to find out the possible role of the mycorrhiza in the phytoremediation of heavy metals which are highly phytotoxic in nature (Meharg and Cairney 2000; Leyval et al. 2002). Reports suggest that AMF have considerable potential in the removal of these toxic organic pollutants from the soil, thereby helping in the reclamation of the contaminated soil (Leyval and Binet 1998; Joner and Leyval 2001, 2003). This section of the chapter will briefly focus on such type of research work carried out in the recent past for the reclamation of contaminated soils by using AMF. Unscrupulous human mining activities have resulted into the deterioration of both soil and water resources (Dong et al. 2008). These activities brings the highly hazardous elements *viz.* arsenic into the direct contact of human beings which in turn can impose some serious threat to human health as well as to the environment (Toujaguez et al. 2013; Bundschuh et al. 2013). Apart from affecting human health they can significantly affect the productivity of the crops in

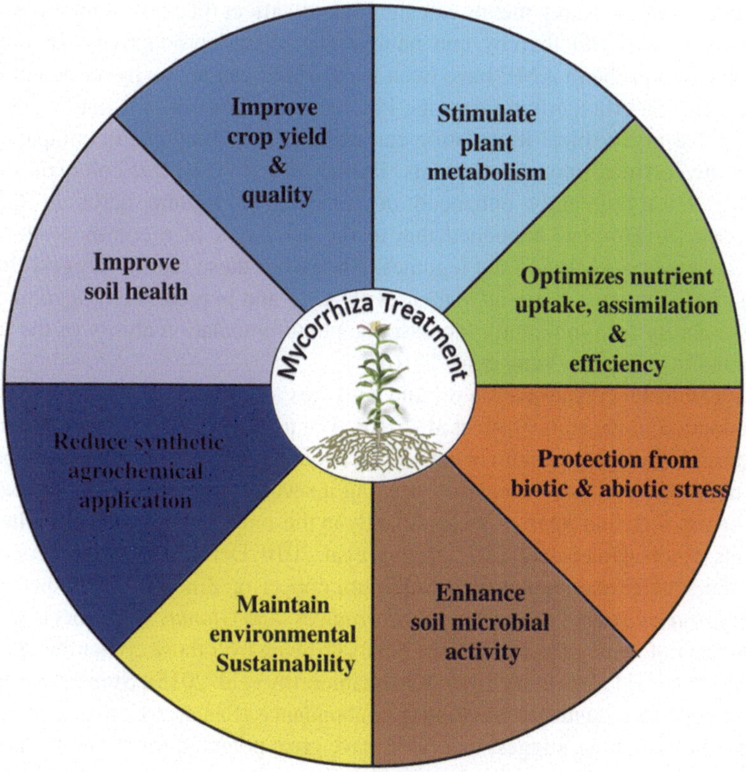

Fig. 17.1 Multifarous applications of Arbuscular Mycorrhiza fungi (AMF)

the affected regions. Upon absorption by the plants these metals can get reduced from its As(V) to As(III) form which in turn affects several enzymes and also leads to the generation of reactive oxygen species (ROS) which in turn causes some serious damage to the plant cell (Meharg and Hartley-Whitaker 2002). However, plants also show some defense mechanisms against these processes by inducing its phytochelatin synthesis and by increasing its antioxidant activities (Tiwari and Sarangi 2017; Bhattacharya and Bhattacharya 2007). But the plants defense mechanism alone is not sufficient to deal with these problems hence many methods have been developed to deal with the problem of arsenic stress and among which use of AMF is showing some promising effects (Gohre and Paszkowski 2006; Vangronsveld et al. 2009; Harms et al. 2011; Rangel et al. 2014; Cabral et al. 2015). It was reported that AMF can reduce the phytotoxicity caused by arsenic by stimulating the plant to increase its phosphate uptake ability (Chen et al. 2007a, b; Xu et al. 2008; Gomes et al. 2013; Schneider et al. 2013).

The mining activities also results into the contamination of the soil and water resources with various types of toxic elements among which lead (Pb) contamination is most abundant (Sharma and Dubey 2005). High level of Pb contamination can impose some serious threats to environment and to the crop species grown in the affected soil. These heavy metals can significantly affect the photosynthetic rates of the crops species and thereby can hamper the overall productivity. In order to overcome this problem AMF have been used which can act as hyper accumulator and thereby can efficiently remove the Pb contamination from the soil. It has been reported that mycorrhizal inoculation can enhance the Pb uptake in comparison to the non-mycorrhizal inoculated plants. Further, the mycorrhizal colonization was found to be significantly enhanced in presence of legume herbs in *Robinia pseudoacacia*. This has happened due to the activation of a common symbiosis pathway between AMF and the legumes. Altogether these reports suggest that the inoculation of the tree *R. pseudoacacia* with AMF and in presence of legume herbs have the ability to significantly increase the phytoremediation ability of the tree in the contaminated soils (Yang et al. 2016).

Apart from Pb contamination mining activities also results in the trace elements contamination in the soils (Toth et al. 2016). The trace elements are very difficult to degrade and hence they remain in the soil for a long time. The trace elements not only contaminate the soil and the environment but it severely affects the human health. It has been reported that AMF grow abundantly in the soils contaminated with the trace elements (Ruotsalainen et al. 2007; Regvar et al. 2010; Deram et al. 2011). For example in the soils contaminated with cadmium, copper or Zinc the AMF species *viz. Funneliformis mosseae*, *Rhizophagus intraradices*, and *Glomus sp* has been reported in abundant colonizing the roots of the plants growing over those contaminated fields (Hassan et al. 2011; Ban et al. 2015; Krishnamoorthy et al. 2015). Similarly in the Pb affected soils *R. irregularis* was found in abundance (Sanchez-Castro et al. 2017). Thus all these findings suggest that AMF have great potential for the reclamation of the soils polluted with trace elements and hence must be used effectively (Gohre and Paszkowski 2006). AMF generally interact with the plant and thereby prevent the metal translocation in the shoots by accumulating it in the roots itself. This type of immobilization in turn prevents the absorption of trace elements by the plants (White et al. 1997; Gohre and Paszkowski 2006).

It has been reported that even the industrial activities have also resulted into the contamination of nearby sites with the trace elements *viz.* copper. In general copper is required by for the normal growth and development of the plant and deficiency of this element may result into severe growth defects in plants. Copper is required for several metabolic processes of plants *viz.* photosynthesis, respiration, CO_2 fixation and for cell wall formation (Mocquot et al. 1996, Burkhead et al. 2009). But on the other hand excessive dose of Cu can induce toxicity in the plants and thereby reduces its overall growth. In leaves the most visual symptom of Cu toxicity is chlorosis of leaves followed by reduction in the photosynthetic rate. These are observed mainly because of the formation of ROS which is triggered by Cu which in turn disrupts the photosynthetic apparatus (Kuper et al. 2004, Bernal et al. 2006). Hence in order to tackle this problem, AMF has been used which can not only help the plant to grow under nutrient depleted condition but can also help its growth under soil contaminated with trace elements like Cu (Chen et al. 2007a, b, Novoa et al. 2009, Cicatelli et al. 2010, Lin et al. 2014). AMF can act as a suitable phytostabilizer which can regulate the plant response under Cu stress and thereby help in improving the overall growth and development of the plants.

In today's world the major concern is to increase the global crop productivity from the limited available land resources. In order to achieve this target, several approaches were used and among these, use of inorganic fertilizers for increasing the crop productivity is the most prevalent and widely used approach. It has been observed that increased use of fertilizers leads to the rapid increase in the Cadmium level in the soil. Cadmium is among one of the most hazardous pollutant which severely deteriorates the environment and affects the crop growth (Douchiche et al. 2012). In order to tackle this problem recently AMF have been used which generally helps the plants in increasing its phytoremediation ability (Gao et al. 2010a, b; Wu et al. 2010; Ali et al. 2015). Potential use of AMF increases the capability of plant's heavy metal uptake which significantly increases their ability to promote growth in Cd or other heavy metal polluted soils (Leyval et al. 2001; Wang et al. 2007a, b; Li et al. 2011; Shahabivand et al. 2012). It has been reported that under greenhouse condition inoculation of *Phragmites australis* with *Rhizophagus irregularis* has improved its tolerance against Cd stress. It was found that the inoculated plants have significant tolerance against Cd stress when it was treated with 0–20 mg L^{-1} of Cd for 21 days. Further, the inoculated plants have shown better uptake ability of Mn and P along with providing significant tolerance against Cd stress (Wang et al. 2017a).

The soils contaminated with heavy metals severely affect the growth of the plants hence it is gaining considerable attention of the scientific communities for developing suitable method to overcome this problem. For removal of these heavy metals from soil it is essential that it should be first extracted followed by its concentration and finally its disposal. For carrying out this task several techniques are used in general which includes thermal processes, physical separation method and electrochemical methods, but these techniques are very costly and they are not environmental friendly as well (McGrath et al. 2015; Mulligan et al. 2001). Hence in order to tackle this problem biological approach has been used which includes the use of several species of AMF for the reclamation of polluted soil. It has been observed that AMF generally immobilizes the heavy metals by either forming polyphosphate granule which in turn gets precipitated in the soil or it absorbs the metal inside the fungal cell wall where it is chelated (Li and Christie 2000; Zhu et al. 2001; Huang et al. 2002; Malcova and Gryndler 2003; Andrade et al. 2009) and in this way it removes the toxic heavy metals from the soil and allow the growth of the plants. Several plant associated fungi has been found to produce siderophores, they can form a complex with iron and thereby helps in chelating iron. But apart from iron, siderophores form complex with the metals like aluminum, copper, cadmium, zinc and lead (Glick and Bashan 1997). Certain ectomycorrhizal fungi (EMF) *viz. Scleroderma verrucosum*, *Suillus luteus* and *Rhizopogon luteolus* were reported to produce catecholates and hydroxamates siderophores which in turn help in the chelation of iron in the soil (Machuca et al. 2007). The various sources of soil contamination with AMF reclamation has been represented in Fig. 17.2.

Like siderophores, organic acids are also very efficient in the removal of heavy metals from the soil. Microorganisms produce several types of organic acids which

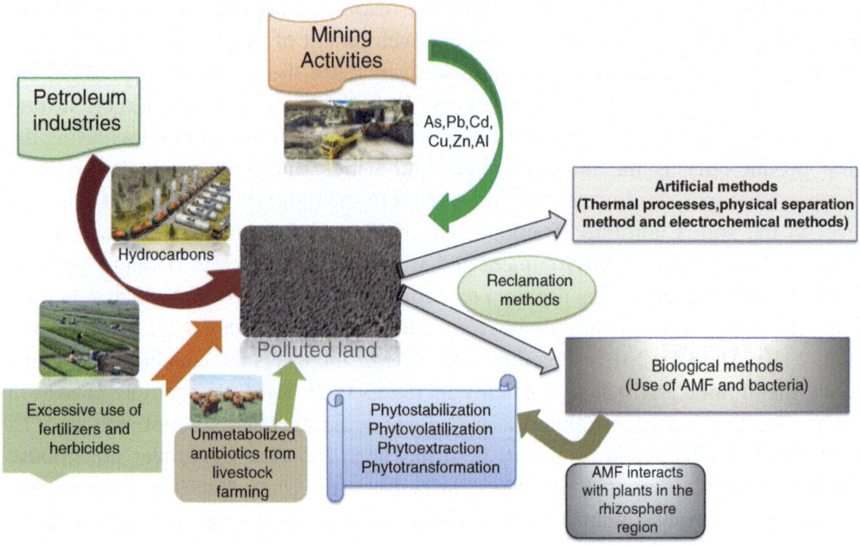

Fig. 17.2 Sources of soil contamination and its reclamation with AMF

include gluconic acid, oxalic acid and citric acid. These organic acids in turn form a strong complex with the metals and thereby help in the removal of such metals from the rhizosphere region (Ryan et al. 2001). It has been reported that the ericoid mycorrhizal fungi *Oidiodendron maius* carries out the release of Zn from its insoluble form ZnO and Zn_3 $(PO4)_2^-$ through citric acid and malic acid which in turn chelates Zn out of the soil (Martino et al. 2003). Gonzalez-Chavez et al. (2004) reported that the glycoprotein's glomalin produced by the AMF have the tendency to form complex with the heavy metals and thereby contributes in the extraction of heavy metals from soil.

3 Reclamation of Soil Polluted with Organic Chemicals by Using AMF

Chlorinated aromatic herbicides are widely used across the globe for controlling the unwanted growth of the weeds. But the major problem is that these organic chemicals are highly toxic and they are very resistant against degradation (Chaudhry and Chapalamadugu 1991). Another problem is that the entire amount of herbicide applied on the plants is not absorbed by them and hence the residual amount remains in the soil for a considerable period of time thereby making it difficult for the cultivation of several crop species (Beste 1983). Amongst the various chlorinated chemicals, atrazine and 2,4-dichlorophenoxyacetic acid (2,4-D) were commonly used for the agricultural purpose. The herbicide mainly inhibits the photosynthetic machinery in the leaves once it is absorbed by the roots (Beste 1983). Atrazine is strongly absorbed by the clay soil, so its leaching tendency is very low further its degradation rate is very slow so it remains on the soil for a considerable period of the time. Another herbicide 2,4-D is used against the broadleaf vegetation and it is also absorbed by the root and thereby it spread systemically inside the plant. It has been reported that AMF are quite efficient in degradation of these harmful chemicals. It has been found that *Phanerochaete chrysosporium* can degrade 2,4-D by incorporating the herbicide carbon into tissues but it can't mineralize Atrazine. In case of Atrazine, ericoid mycorrhizal fungi *Hymenoscyphus ericae* was found to be most efficient in its degradation (Donnelly et al. 1993).

Another organic compound known as polycyclic aromatic hydrocarbons (PAHs) are highly toxic in nature and they acts as an important source of environmental pollution. The major source of spread of PAH pollution is through chemical industries, aerosols etc. Ectomycorrhizal fungi have been used potentially for the degradation of PAHs. Several species like *Amanita excelsa*, *Leccinum versipelle*, *Suillus grevillei*, *S. luteus*, and *S. variegatus* has been reported to be efficient in the degradation of PAHs *viz*. phenanthrene, chrysene, pyrene and benzo[a]pyrene (Braun-Lullemann et al. 1999). In pot experiment it was observed that the soils inoculated with *Glomus mosseae* (BEG69) shows significantly higher rates of PAH degradation then the non-inoculated soils (Binet et al. 2000). Further it has been reported that PAHs with low molecular weight and higher water solubility can be easily translocated through the roots which is generally mediated by the fungal hyphae. In this way AMF contributes significantly in the removal of these PAHs from the soil (Gao et al. 2010a, b). Some Mycobacterium species has been isolated from the soils contaminated with PAHs and were found to be efficient in at least 60% degradation of the compound pyrene within a period of 8 days (Rehmann et al. 1998). In case of soils contaminated with the compounds viz. PHE, PYR and fluoranthene (Flu), it was found that inoculation of such type of soils with *Burkholderia sp.* can efficiently degrade at least 97% of PHE within 30 days and around 74% of PYR and 88% of Flu after a period of 60 days (Somtrakoon et al. 2008).

Although it is quite simple to degrade the low molecular weight PAHs but however it is very difficult to degrade the high molecular weight PAHs (Somtrakoon et al. 2008). Microorganisms catalyze the degradation reaction through some oxidative enzymes *viz*. oxygenase, dehydrogenase and lignolytic enzymes (Baldrian et al. 2000). Further it has been reported that the root exudates from the plants can significantly increase the process of PAHs degradation. Hence, by keeping this in view several groups has used Ryegrass for this purpose in the contaminated soils because of its higher number of fibrous root system which in turn provides huge amount of root surface area for the colonization of these microorganisms (Kang et al. 2010). It has been postulated that in the contaminated soil there occurs a significant amount of interaction between the PAHs degrading microorganisms, AMFs and plants but the mechanism is not well understood till date thus more research work should be carried out in this field in order to get a deeper insight into the underlying mechanism.

In poultry farming and animal husbandry, the use of antibiotics is very common and from here itself these antibiotics are passed into the excreta which in turn are used for the production of manures. These manures are subsequently applied on the agricultural fields and hereby they get spread over the entire field (Bartikova et al. 2016). The main reason for the transfer of these antibiotics into the manure is the incomplete metabolism of the injected antibiotics. Agricultural fields affected by these antibiotics may affect the microbial population flourishing there and in many cases they affect the normal growth and development of the plants (Tang et al. 2015). One such antibiotic is oxytetracycline (OTC), whose contamination is widespread; this antibiotic is mostly used for animals, poultry or aquaculture (Ma et al. 2016). Apart from these OTC various other metabolized forms of OTC antibiotics has been detected from the urine and animal feces *viz.* 4-epi-OTC (EOTC) and -apo-OTC which still bear around 30% and 10% of OTC activity respectively (Lykkeberg et al. 2004). Hence, better understanding of these OTC metabolized products is required to develop some suitable methods for complete degradation of these antibiotics. In accordance to this the use of AMF has been proposed for carrying out the degradation of these antibiotics. It has been found that the exudates of AMF contains certain organic acids, sugars and certain polymeric compounds which have the ability to induce the growth of certain soil bacteria's which can carry out the degradation of these OTC (Toljander et al. 2007). Certain studies show that the inoculation of *Rhizophagus intraradices* can significantly stimulate the degradation of these OTC (Cao et al. 2015). In this way several studies shows that AMF shows considerable ability to carry out the degradation of these antibiotics either alone or by stimulating other microorganisms. Apart from potential use of all above mentioned AMF some more information about the use of AMF for the reclamation of contaminated soils Table 17.1.

4 Role of AMF in Reclamation of Abiotic Stresses

Drought, soil salinity and metal toxicity etc. are the most potent environmental stresses that affects a huge range of crop species annually (Kramer and Boyer 1997; Cattivelli et al. 2008; Lambers et al. 2008; Trenberth et al. 2014) although it affects the crops to different levels depending upon the availability of water but its impact is more severe in the arid and semi-arid regions (Knapp et al. 2001; Seki et al. 2003; Fischlin et al. 2007). The main reason for these types of stresses is the rapid decline in the ground water table followed by high rate of evapo-transpiration which ultimately results into decrease in the soil moisture beyond the field capacity level thereby leads to the wilting of plants (Wery et al. 1994; Rapti-Caputo 2010). Now a days, in order to increase the crop productivity and to meet the demand of plant nutrient supply a huge amount of chemical fertilizers are being used globally. In recent past an unprecedented hike in the consumption of these fertilizers has been reported which ultimately increase the soil salinization. Apart from these chemical fertilizers, even the use of irrigation water which is generally higher in ion content also contributes to the rise in soil salinity and thereby contributes to the overall salinity stress in agriculturally important crops (Cantrell and Linderman 2001; Al-Karaki 2006; Kapoor et al. 2008; Abdel Latef and Chaoxing 2014).

Table 17.1 List of some tolerant AMF species used for reclamation of contaminated soils

S. No.	Tolerant species	Contaminants present in soil	References
1	*Phanerochaete chrysosporium*	2,4-D	Donnelly et al. (1993)
2	*Glomus mosseae*	Cu, Cd and Pb	Gonzalez-Chavez et al. (2004)
3	*Suillus luteus*	Cu	Adriaensen et al. (2005)
4	*Glomus mosseae*	Cd	Repetto et al. (2007)
5	*Glomus aggregatum, Glomus claroideum, Glomus constrictum, Glomus etunicatum, Glomus mosseae, Glomus rubiforme, and Glomus tortuosum*	Ni	Doherty et al. (2008)
6	*Suillus luteus*	Zn and Cd	Krznaric et al. (2010)
7	*Glomus mosseae*	Al, Mn, Cu, Cd, Mo, Zn, As, and Ni	Azcon et al. (2010)
8	*Pisolithus albus*	Ni	Majorel et al. (2014)
9	*Glomus mosseae* and *G. intraradices*	Pb and Cd	Sheikh- Assadi et al. (2015)
10	*Rhizophagus irregularis*	Fe	Rodrigues and Rodrigues (2015)
11	*Glomus intraradices*	Petroleum products	Xun et al. (2015)
12	*Rhizophagus intraradices*	Oxytetracycline	Cao et al. (2016)
13	*Rhizophagus irregularis* and *Funneliformis mosseae*	Hg	Cozzolino et al. (2016)
14	*Rhizophagus intraradices*	Arsenic	Li et al. (2016)
15	*Rhizophagus intraradices*	Pb	Yang et al. (2016)
16	*Scutelospora* sp., *Glomus sp., Claroideoglomus* sp., *Acaulospora* sp.	Ag, Au	Kasiamdari et al. (2016)
17	*Acaulospora* sp.	Cd and Pb	Gonzalez-Chavez et al. (2017)
18	*Pisolithus tinctorius* and *Cenococcum geophilum*	Cu	Wen et al. (2017)
19	*Funneliformis mosseae, Rhizophagus intraradices* and *Claroideoglomus etunicatum*	Hydrocarbons	Sut et al. (2016)
20	*Rhizophagus irregularis*	Benzo[a]pyrene	Calonne-Salmon et al. (2018)

Further, these stresses results into the change in the osmotic potential inside the root cells, which ultimately results into the rapid decline in the absorption of water by the roots (Zhu et al. 1997; Seki et al. 2003; Aroca et al. 2012). All these phenomenon altogether leads to the osmotic stress in plants. These stresses altogether leads to the decrease in the rate of photosynthesis, increase in photorespiration and subsequent decline in the sugar and protein metabolism in the plant which results in decline of the crop productivity (Sircelj et al. 2005; Ashraf and Foolad 2007; Anjum et al. 2011). Now, in order to tackle these problems several approaches has been used which range from the use of drought or salinity tolerant varieties which can be obtained by introgressing the corresponding QTLs from the resistant varieties, or it deals with the use of genetic engineering approaches to transfer such gene from different sources. Apart from these approaches recently some authors have reported that even the use of some beneficial soil microorganism's *viz.* AMF can helps the plant to cope up against such abiotic stresses (Barea et al. 2005; Hidri et al. 2016). These AMF not only increase the nutrient uptake capability of the plants but also helps in the accumulation of osmoprotectants, antioxidant enzymes and even it rejuvenates the rhizosphere environment (Barzana et al. 2015; Calvo-Polanco et al. 2016; Yin et al. 2016). All these activities in turn results in the significant rise in the ability of the plants to cope up against these abiotic stresses. The benefits of AMF on plants have been depicted in Fig. 17.3.

AMF plays a potent role in ameliorating the abovementioned abiotic stresses (Al-Karaki 2013; Soares and Siqueira 2008; Amir et al. 2013). The mechanism by which AMF bestow tolerance against such abiotic stresses is primarily nutritional (Soares and Siqueira 2008; Birhane et al. 2012; Al-Karaki 2013; Navarro et al.

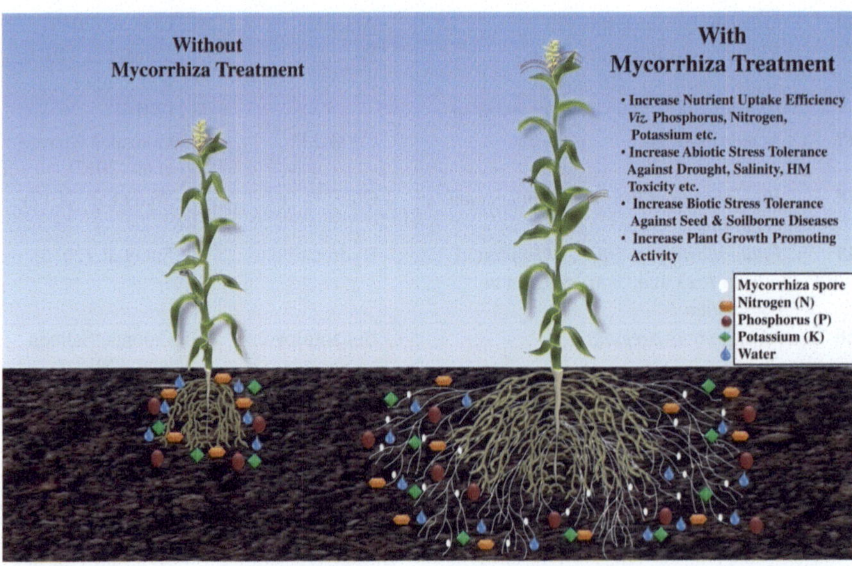

Fig. 17.3 Schemtatic representation on benefits of Mycorrhiza tratment on plants vs control

2013) Similarly, Navarro et al. (2013) reported that *Citrus* rootstocks inoculated with AMF showed significantly enhanced growth compared to non-inoculated ones, despite irrigating inoculated individuals with saline water and vice versa.

5 Drought

Drought is regarded as one of the most important environmental stresses limiting the productivity of crop plants around the world (Bohnert et al. 1995). The major feature of lands in arid zones is the low precipitation. In addition, they are characterized by fluctuations of temperature, soil and climate diversity with extreme patchiness of soils (Skujins and Allen 1986). Drought stress decreases the rate of photosynthesis (Kawamitsu *et al.* 2000; Flexas et al. 2004). Plants grown under drought condition have a lower stomatal conductance in order to conserve water. Thus, CO_2 fixation is reduced and photosynthetic rate declines, resulting in less assimilate production for growth and development of plant (Cornic 2000). However, severe drought stress also inhibits the plant photosynthetic ability through alteration in chlorophyll content, or by affecting cholorophyll components and by damaging the photosynthetic apparatus. In addition, it also leads to formation of toxic free radicals (Chaves et al. 2002; Johnson et al. 2003).

The functioning of AM fungi in arid communities is based on interaction of several physiological mechanisms to overcome the extreme variability of environmental conditions (Allen 1984). AMF have been reported to enhance water uptake and increase drought tolerance of several plant species (Al-Karaki et al. 2004; Safir and Nelsen 1985). It has been estimated that approximately 42% reduction in plants water necessity could be made upon inoculation of plants with drought tolerant AMF (Gianinazzi et al. 2010). Increased water uptake by mycorrhizal plants has been governed by many factors like increased P uptake and stomatal responses (Auge 2001). Fungal hyphae facilitate in transportation of water to the deeper roots of the plant (Allen 1984, 2007).

6 Salinity

Salinity is one of the most brutal abiotic constraints in arid and semiarid zones which limit plant growth and yield. Soil salinization is a major threat in those regions where saline water is used for irrigating crops. Salinity not only decreases the crop yield of most crops, in turn also affects physicochemical properties of soil and ecological balance of the region. The effect of salinity are the results of complex interactions among various morphological, physiological and biochemical processes *viz.* seed germination, nutrient and water uptake and ultimately plant growth (Akbarimoghaddam et al. 2011; Singh and Chatrath 2001). In addition to it, salinity also alters photosynthesic rate mainly through a reduction in leaf area, stomatal conductance and chlorophyll content (Netondo et al. 2004). Thus, owing to such harsh effect of salinity on plants, its reclamation is one of the major challenges in agriculture. AMF have been reported to play a crucial role in mitigating salinity stress to a considerable extent (Al-Karaki 2000; Al-Karaki et al. 2004; Al-Karaki and Hammad 2001). Mechanisms underlying the role of AM fungi in plant salinity tolerance may be attributed to nutritional, physiological and biochemical effects. These factors basically include: enhanced nutrient uptake (Al-Karaki 2006, 2000), changes in plant hormones, accumulation of osmoregulators (Kaya et al. 2009; Sharifi et al. 2007), increased water use efficiency and photosynthetic rate (Danneberg et al. 1992), increased activity of enzymes involved in antioxidant defense (Ruiz-Lozano et al. 1996), elevated leaf gas exchange and photosynthetic rate (Ruiz-Lozano et al. 1996), enhanced water uptake through increasing leaf conductance and photosynthetic activity (Dell'Amico et al. 2002), maintaining osmotic potential (Al-Garni 2006), alteration in cell-wall elasticity (Auge et al. 1987) and stability of cell membranes (Kaya et al. 2009). Several scientists have reported that inoculation with AMF aid in plant growth promotion through increased root and shoot growth under saline conditions. These growth promotion effect have been reported in many crop plants viz. tomato (Al-Karaki 2006), bell pepper (Kaya et al. 2009).

In case of salinity reclamation by VAM, enhanced water uptake by plants results in dilution of high salt concentration within the plants cells (Larcher 1995). However, the non-nutritional mechanisms attributed to AMF plant's salinity tolerance include: exclusion of salt from plant cells by accumulating the salt within fungal hyphae, production of antioxidants and various enzymes which facilitate changes in cell wall elasticity and membrane stability.

7 Heavy Metal Toxicity

Heavy metals (HMs) adversely affect the morphological, physiological and biochemical functions of plants, most common ones are inhibition of growth rate, altered stomatal action, decreased water potential, leaf rolling, chlorosis, necrosis, changes in functions of various membranes, inhibition of photosynthesis, respiration, altered metabolism and activities of several key enzymes (Dubey 2011; Hossain et al. 2010; Sharma and Dubey 2007). HMs are highly toxic in nature, if the cytosolic concentration in plant turns out of control, phytotoxicity occur leading to inhibition of photosynthesis, cell respiration and nitrogen metabolism and induction of oxidative stress, which ultimately hamper the overall plant growth and development (Romanowska et al. 2006; Maksymiec et al. 2007).

AMF have also been reported to mitigate HMs stress in agricutural soil. In an experiment, Soares and Siqueira (2008) demonstrated that both AMF inoculation and P fertilization of plants significantly elevated plant growth on heavy metal contaminated soils. Thus according to the obtained results, the workers stated that AMF increases metal stress tolerance of plants through P nutrition. Plant tolerance by AMF through non-nutritional mechanisms include: hormonal change, increased plant photosynthetic rate, hyphal soil improvement, hyphal ability to quench water from tiny soil pores and accumulation of compatible osmolites (Birhane et al. 2012;

Al-Karaki 2013). Similarly, the non-nutritional mechanism includes immobilizing heavy metals in their biomass mainly involving cell wall, vesicles and in the glomaline (Hildebrandt et al. 2007).

8 Ozone Stress

Besides drought salinity and HM toxicity, ozone stress is also very detrimental for the normal growth and development of the crops. O_3 acts as a very strong phytotoxic pollutant which affects huge range of crops (Ashmore 2005; Van Dingenen et al. 2009). It's level has been significantly increased in some of the Asian countries including India and China (Wang et al. 2007a, b; IPCC 2013). Rise in O_3 level can significantly affect the productivity by either decreasing the photosynthetic rates of the crops, or by causing significant leaf injury and through inducing early senescence (Ismail et al. 2014; Wang et al. 2015). It was reported that the colonization rates of AMF were either decreased in the crops affected with O_3 stress (McCool and Menge 1984; Wang et al. 2015) or remain unaffected (Duckmanton and Widden 1994; Wang et al. 2011) or may get increased under O_3 stress (Brewer and Heagle 1983).

The AMF generally acts as an obligate biotrophs and they depends completely upon the host plant for procuring its food supply and this may be the reason for decrease in the colonization rate under O_3 stress. Further it was found that O_3 acts as a strong oxidant, so it generally acts on the photosynthetic machinery of the plants and thereby decreases its photosynthetic rate which finally reduces the food supply to the AMF (McCool and Menge 1984; Morgan et al. 2003; Feng et al. 2008; Parniske 2008; Wang et al. 2011). Apart from AMF susceptibility to O_3 stress, many authors have observed that AMF even shows positive effects on the plant growth upon O_3 stress. In some studies it was reported that at O_3 concentration above 80 ppb, AMF symbiosis has resulted into 68% of increase in the shoot biomass and around 131% increase in the root biomass (Wang et al. 2017b). As a whole it can be stated that AMF can be used efficiently to improve the overall growth under O_3 stress condition.

9 Effect of AMF on Soil Structure

AMF provides a direct physical link between the host and soil resource through its external hyphal network. The hyphal networks facilitate the uptake of mineral ions of its host but also represent a large carbon sink within the soil (Jakobsen and Rosendahl 1990; Miller et al. 1995). Thus the external hyphae may be regarded as a stabilizing agent in formation and maintainence of soil structure. It depends upon various factors like soil properties, vegetation, management practices used as well as characteristics of the associated mycorrhiza. AMF can persist in soil for longer periods due to their filamentous nature, branching pattern and large diameter. The AMF appear to be the most important mediator of soil aggregation (Rilling et al.

2000). External hyphae of AMF bind the small soil particles into aggregates by producing a glycoprotein (glomalin) which itself can account for 35–60% of C in undisturbed soils (Treseder and Allen 2000). The association of microaggregates leads to formation of macroaggregates that finally leads to improved structure and aggregation stability in soils. This phenomenon can be seen in wide texture of soils ranging from sandy, loamy, clayey and many intermediate textures (Bearden and Petersen 2000).

10 Drawbacks of Mycorrhiza Assisted Remediation (MAR)

MAR is relatively a slow method of bioremediation. It may take months for a particular mycorrhizal species to colonise and express its potential. However, there are few species of mycorrhizal fungi which are pollutant specific. Thus, wrong selection of a species may not provide the desired results and it would be wastage of time and economy. Efficiency depends on the type of plant used. Some plants do not form mycorrhizal association; hence, remediation may not be accomplished when these plants are used.

11 Conclusion

Bioremediation has emerged as a low cost, alternative technology for alleviation of biotic and abiotic stress. AMF provide an attractive system to advance plant-based clean-up of the ecosystem. Mycorrhizal fungi are potential symbionts that are at the interface between plant roots and soil. They are partners of plant in remediation of ill effects of soil pollution, through various mechanisms. Although the role of AMF in phytoremediation is quite evident, still there is a need to completely understand the complexities of the plant-microbe soil interactions and their better exploitation in remediation strategies of the polluted and degraded lands. Multi-disciplinary investigations using physiological, molecular and biochemical techniques could facilitate in understanding the complex phenomenon. MAR is gaining worldwide popularity and is tremendously used as a xenobiotic tool. The benefits derived from mycorrhizal fungi make MAR a suitable method for the clean-up of soils whose intended use is required. MAR effectively detoxifies both organic and inorganic pollutants and facilitate in better crop production. However, the efficiency of MAR depends on the type of fungal association, type and nature and concentration of the pollutants. Despite having innumerable beneficial effect, MAR still have some drawbacks. A consortium of MAR with other bioremediation methods could be useful and effective in improving its efficiency. Finally, it can be stated that MAR is a valuable strategy for phytoremediation of and should be further studied to explore more of its beneficial attributes to humankind.

Acknowledgement Rahul Singh Rajput is grateful to UGC for providing RET Fellowship as a source of financial assistance. HB Singh is grateful to DST for providing funding under grant (BT/PR5990/AGR/5/587/2012).

12 References

Abdel-Latef, A., & Chaoxing, H. (2014). Does the inoculation with *Glomus mosseae* improve salt tolerance in pepper plants? *Journal of Plant Growth Regulation, 33*(3), 644–653.

Adriaensen, K., Vralstad, T., Noben, J. P., Vangronsveld, J., & Colpaert, J. V. (2005). Copper-adapted Suillus luteus, a symbiotic solution for pines colonizing Cu mine spoils. *Applied and Environmental Microbiology, 71*(11), 7279–7284.

Akbarimoghaddam, H., Galavi, M., Ghanbari, A., & Panjehkeh, N. (2011). Salinity effects on seed germination and seedling growth of bread wheat cultivars. *Trakia Journal of Sciences, 9*(1), 43–50.

Al-Garni, S. M. S. (2006). Increasing NaCl – Salt tolerance of a halophytic plant *Phragmites australis* by mycorrhizal symbiosis. *American-Eurasian Journal of Agricultural & Environmental Sciences, 1*, 119–126.

Ali, N., Masood, S., Mukhtar, T., Kamran, M. A., Rafique, M., Munis, M. F. H., & Chaudhary, H. J. (2015). Differential effects of cadmium and chromium on growth, photosynthetic activity, and metal uptake of *Linum usitatissimum* in association with *Glomus intraradices*. *Environmental Monitoring and Assessment, 187*, 1–11.

Al-Karaki, G. N. (2000). Growth of mycorrhizal tomato and mineral acquisition under salt stress. *Mycorrhiza, 10*, 51–54.

Al-Karaki, G. N. (2006). Nursery inoculation of tomato with arbuscular mycorrhizal fungi and subsequent performance under irrigation with saline water. *Scientia Horticulturae, 109*, 1–7.

Al-Karaki, G. N. (2013). The role of mycorrhiza in the reclamation of degraded lands in arid environments. In S. A. Shahid, F. K. Taha, & M. A. Abdelfattah (Eds.), *Developments in soil classification, land use planning and policy implications: Innovative thinking of soil inventory for land use planning and management of land resources* (pp. 823–836). Dordrecht: Springer.

Al-Karaki, G. N., & Hammad, R. (2001). Mycorrhizal in fl uence on fruit yield and mineral contents of tomato grown under salt stress. *Journal of Plant Nutrition, 24*, 1311–1323.

Al-Karaki, G. N., McMichael, B., & Zak, J. (2004). Field response of wheat to arbuscular mycorrhizal fungi and drought stress. *Mycorrhiza, 14*, 263–269.

Al-Karaki, G. N., Othman, Y., & Al-Ajmi, A. (2007). Effects of mycorrhizal fungi inoculation on landscape turf establishment under Arabian Gulf region conditions. *Arab Gulf Journal of Scientific Research, 25*(3), 147–152.

Allen EB (1984) The role of mycorrhiza in mined land diversity. In: Proceedings of the third biennial symposium surface mine reclamation on the great plains, Montana, 19–21 March 1984, pp 273–295.

Allen, M. F. (2007). Mycorrhizal fungi: Highways for water and nutrients in arid soils. *Vadose Zone Journal, 6*, 291–297.

Amir, H., Lagrange, A., Hassaine, N., & Cavaloc, Y. (2013). Arbuscular mycorrhizal fungi from New Caledonian ultramafic soils improve tolerance to nickel of endemic plant species. *Mycorrhiza, 23*, 585–595.

Andrade, S. A. L., Gratao, P. L., Schiavinato, M. A., Silveira, A. P. D., Azevedo, R. A., & Mazzafera, P. (2009). Zn uptake, physiological response and stress attenuation in mycorrhizal jack bean growing in soil with increasing Zn concentrations. *Chemosphere, 75*, 1363–1370.

Anjum, S. A., Wang, L. C., Farooq, M., Hussain, M., Xue, L. L., & Zou, C. M. (2011). Brassinolide application improves the drought tolerance in maize through modulation of enzymatic antioxidants and leaf gas exchange. *Journal of Agronomy and Crop Science, 197*, 177–185.

Aroca, R., Porcel, R., & Ruiz-Lozano, M. J. (2012). Regulation of root water uptake under abiotic stress conditions. *Journal of Experimental Botany, 63*, 43–57.

Ashmore, M. R. (2005). Assessing the future global impacts of ozone on vegetation. *Plant, Cell & Environment, 28*, 949–964.

Ashraf, M., & Foolad, M. R. (2007). Roles of glycine betaine and proline in improving plant abiotic stress resistance. *Environmental and Experimental Botany, 59*, 206–216.

Auge, R. M. (2001). Water relations, drought and vesicular-arbuscular mycorrhizal symbiosis. *Mycorrhiza, 11*, 3–42.

Auge, R. M., Schekel, K. A., & Wample, R. L. (1987). Leaf water and carbohydrate status of VA mycorrhizal rose exposed to drought stress. *Plant and Soil, 99*, 291–302.

Azcón, R., Perálvarez Mdel, C., Roldán, A., & Barea, J. M. (2010). Arbuscular mycorrhizal fungi, *Bacillus cereus*, and *Candida parapsilosis* from a multi contaminated soil alleviate metal toxicity in plants. *Microbial Ecology, 59*(4), 668–677.

Baldrian, P., Wiesche, C., Gabriel, J., Nerud, F., & Zadrazi, F. (2000). Influence of cadmium and mercury on activities of ligninolytic enzymes and degradation of polycyclic aromatic hydrocarbons by *Pleurotus ostreatus* in soil. *Journal ofApplied & Environmental Microbiology, 66*, 2471–2478.

Ban, Y., Xu, Z., Zhang, H., Chen, H., & Tang, M. (2015). Soil chemistry properties, translocation of heavy metals, and mycorrhizal fungi associated with six plant species growing on lead-zinc mine tailings. *Annales de Microbiologie, 65*, 503–515.

Barea, J. M., Pozo, M. J., Azcon, R., & Azcon-Aguilar, C. (2005). Microbial cooperation in the rhizosphere. *Journal of Experimental Botany, 56*, 1761–1778.

Bartikova, H., Podlipna, R., & Skalova, L. (2016). Veterinary drugs in the environment and their toxicity to plants. *Chemosphere, 144*, 2290–2301.

Bárzana, G., Aroca, R., & Ruiz-Lozano, J. M. (2015). Localized and nonlocalized effects of arbuscular mycorrhizal symbiosis on accumulation of osmolytes and aquaporins and on antioxidant systems in maize plants subjected to total or partial root drying. *Plant, Cell & Environment, 38*, 1613–1627.

Bearden, B. N., & Petersen, L. (2000). Influence of arbuscular mycorrhizal fungi on soil structure and aggregate stability of a vertisol. *Plant and Soil, 218*, 173–183.

Bernal, M., Ramiro, M. V., Cases, R., Picorel, R., & Yruela, I. (2006). Excess copper effect on growth, chloroplast ultrastructure, oxygen-evolution activity and chlorophyll fluorescence in Glycine max cell suspensions. *Physiologia Plantarum, 127*, 312–325.

Beste, C. E. (Ed.). (1983). *Herbicide handbook of the weed science Society of America* (5th ed., p. 515). Champaign: Weed Science Society of America.

Bhattacharya, A., & Bhattacharya, S. (2007). Induction of stress by arsenic in *Clarius batrachus*: Involvement of peroxisomes. *Ecotoxicology and Environmental Safety, 66*(2), 178–187.

Binet, P., Portal, J. M., & Leyval, C. (2000). Dissipation of 3-6-ring polycyclic aromatic hydrocarbons in the rhizosphere of rye grass. *Soil Biology and Biochemistry, 32*, 2011–2017.

Birhane, E., Sterck, F. J., Fetene, M., Bongers, F., & Kuyper, T. W. (2012). Arbuscular mycorrhizal fungi enhance photosynthesis, water use efficiency, and growth of frankincense seedlings under pulsed water availability conditions. *Oecologia, 169*, 895–904.

Bohnert, H. J., Nelson, D. E., & Jensen, R. G. (1995). Adaptations to environmental stress. *Plant Cell, 7*, 1099–1111.

Braun-Lullemann, A., Huttermann, A., & Majcherczyk, A. (1999). Screening of ectomycorrhizal fungi for degradation of polycyclic aromatic hydrocarbons. *Applied Microbiology and Biotechnology, 53*, 127–132.

Brewer, P. F., & Heagle, A. S. (1983). Interactions between *Glomus geosporum* and exposure of soybeans to ozone or simulated acid rain in the field. *Phytopathology, 73*, 1035–1040.

Brundrett, M. C. (2009). Mycorrhizal associations and other means of nutrition of vascular plants: Understanding the global diversity of host plants by resolving conflicting information and developing reliable means of diagnosis. *Plant and Soil, 320*(1), 37–77.

Bundschuh, J., Bhattacharya, P., Nath, B., Naidu, R., Jack, N. G., Guilherme, L. R. G., Ma, L. Q., Kim, K. W., & Jean, J. S. (2013). Arsenic ecotoxicology: The interface between geosphere, hydrosphere and biosphere. *Journal of Hazardous Materials, 262*, 883–886.

Burkhead, J. L., Gogolin Reynolds, K. A., Abdel-Ghany, S. E., Cohu, C. M., & Pilon, M. (2009). Copper homeostasis. *The New Phytologist, 182*, 799–816.

Cabral, L., Soares, C. R. F. S., Giachini, A. J., & Siqueira, J. O. (2015). Arbuscular mycorrhizal fungi in phytoremediation of contaminated areas by trace elements: Mechanisms and major benefits of their applications. *World Journal of Microbiology and Biotechnology, 31*(11), 1655–1664.

Calonne-Salmon, M., Plouznikoff, K., & Declerck, S. (2018). The arbuscular mycorrhizal fungus *Rhizophagus irregularis* MUCL 41833 increases the phosphorus uptake and biomass of *Medicago truncatula*, a benzo[a]pyrene-tolerant plant species. *Mycorrhiza, 18*, 861–869.

Calvo-Polanco, M., Sánchez-Romera, B., Aroca, R., Asins, M. J., Declerck, S., Dodd, I. C., Martínez-Andújar, C., Albacete, A., & Ruiz-Lozano, J. M. (2016). Exploring the use of recombinant inbred lines in combination with beneficial microbial inoculants (AM fungus and PGPR) to improve drought stress tolerance in tomato. *Environmental and Experimental Botany, 131*, 47–57.

Cantrell, I. C., & Linderman, R. G. (2001). Preinoculation of lettuce and onion with VA mycorrhizal fungi reduces deleterious effects of soil salinity. *Plant and Soil, 233*, 269–281.

Cao, J., Ji, D. G., & Wang, C. (2015). Interaction between earthworms and arbuscular mycorrhizal fungi on the degradation of oxytetracycline in soils. *Soil Biology and Biochemistry, 90*, 283–292.

Cao, J., Wang, C., & Ji, D. (2016). Improvement of the soil nitrogen content and maize growth by earthworms and arbuscular mycorrhizal fungi in soils polluted by oxytetracycline. *Science of the Total Environment, 571*, 926–934.

Cattivelli, L., Rizza, F., Badeck, F. W., Mazzucotelli, E., & Mastrangelo, A. M. (2008). Drought tolerance improvement in crop plants: An integrated view from breeding to genomics. *Field Crops Research, 105*, 1–14.

Chaudhry, G. R., & Chapalamadugu, S. (1991). Biodegradation of halogenated organic compounds. *Microbiological Reviews, 55*, 59–79.

Chaves, M. M., Pereira, J. S., Maroco, J., Rodrigues, M. L., Ricardo, C. P. P., Osorio, M. L., Carvalho, I., Faria, T., & Pinheiro, C. (2002). How plants cope with water stress in the field. *Annals of Botany, 89*, 907–916.

Chen, B., Xiao, X., Zhu, Y., Smith, F. A., Xie, Z. M., & Smith, S. E. (2007a). The arbuscular mycorrhizal fungus *Glomus mosseae* gives contradictory effects on phosphorus and arsenic acquisition by *Medicago sativa* Linn. *Science of the Total Environment, 379*, 226–234.

Chen, B. D., Zhu, Y. G., Duan, J., Xiao, X. Y., & Smith, S. E. (2007b). Effects of the arbuscular mycorrhizal fungus *Glomus mosseae* on growth and metal uptake by four plant species in copper mine tailings. *Environmental Pollution, 147*, 374–380.

Cicatelli, A., Lingua, G., Todeschini, V., Biondi, S., Torrigiani, P., & Castiglione, S. (2010). Arbuscular mycorrhizal fungi restore normal growth in a white poplar clone grown on heavy metal-contaminated soil, and this is associated with upregulation of foliar metallothionein and polyamine biosynthetic gene expression. *Annals of Botany, 106*, 791–802.

Cornic, G. (2000). Drought stress inhibits photosynthesis by decreasing stomatal aperture – Not by affecting ATP synthesis. *Trends in Plant Science, 5*, 187–188.

Cozzolino, V., De Martino, A., Nebbioso, A., Di Meo, V., Salluzzo, A., & Piccolo, A. (2016). Plant tolerance to mercury in a contaminated soil is enhanced by the combined effects of humic matter addition and inoculation with arbuscular mycorrhizal fungi. *Environmental Science and Pollution Research International, 23*(11), 11312–11322.

Danneberg, G., Latus, C., Zimmer, W., Hundeshagen, B., Schneider-Poetsch, H. J., & Bothe, H. (1992). Influence of vesicular-arbuscular mycorrhiza on phytohormone balances in maize (Zea mays L.). *Journal of Plant Physiology, 141*, 33–39.

Dell'Amico, J., Torrecillas, A., Rodriguez, P., Morte, A., & Sanchez-Blanco, M. (2002). Responses of tomato plants associated with the arbuscular mycorrhizal fungus Glomus clarum during drought and recovery. *The Journal of Agricultural Science, 138*, 387–393.

Deram, A., Languereau, F., & Haluwyn, C. (2011). Mycorrhizal and endophytic fungal colonization in *Arrhenatherum elatius* L. roots according to the soil contamination in heavy metals. *Soil and Sediment Contamination, 20*, 114–127.

Doherty, J. H., Ji, B., & Casper, B. B. (2008). Testing nickel tolerance of *Sorghastrum nutans* and its associated soil microbial community from serpentine and prairie soils. *Environmental Pollution, 151*(3), 593–598.

Dong, Y., Zhu, Y. G., Smith, F. A., Wang, Y., & Chen, B. (2008). Arbuscular mycorrhiza enhanced arsenic resistance of both white clover (*Trifolium repens* Linn.) and ryegrass (*Lolium perenne* L.) plants in an arsenic-contaminated soil. *Environmental Pollution, 155*, 174–181.

Donnelly, P. K., Entry, J. A., & Crawford, D. L. (1993). Degradation of atrazine and 2,4-dichlorophenoxyacetic acid by mycorrhizal fungi at three nitrogen concentrations in vitro. *Applied and Environmental Microbiology, 59*, 2642–2647.

Donnelly, P. K., & Fletcher, J. S. (1994). Potential use of mycorrhizal fungi as bioremediation agents. In T. A. Anderson & J. R. Coats (Eds.), *Bioremediation through rhizosphere technology* (pp. 93–99). Washington: American Chemical Society.

Douchiche, O., Chaïbi, W., & Morvan, C. (2012). Cadmium tolerance and accumulation characteristics of mature flax, cv. Hermes: Contribution of the basal stem compared to the root. *Journal of Hazardous Materials, 235*, 101–107.

Dubey, R. S. (2011). Metal toxicity, oxidative stress and antioxidative defense system in plants. In S. D. Gupta (Ed.), *Reactive oxygen species andantioxidants in higher plants* (pp. 177–203). Boca Raton: CRC Press.

Duckmanton, L., & Widden, P. (1994). Effect of ozone on the development of vesicular arbuscular mycorrhiza in sugar maple saplings. *Mycologia, 86*, 181–186.

Feng, Z. Z., Kobayashi, K., & Ainsworth, E. A. (2008). Impact of elevated ozone concentration on growth, physiology and yield of wheat (*Triticum aestivum* L.): A meta analysis. *Global Change Biology, 14*, 2696–2708.

Fischlin, A., Midgley, G. F., Price, J. T., Leemans, R., Gopal, B., Turley, C., Rounsevell, M. D. A., Dube, O. P., Tarazona, J., & Velichko, A. A. (2007). Ecosystems, their properties, goods and services. In M. L. Parry, O. F. Canziani, J. P. Palutikof, P. J. Van der Linden, & C. E. Hanson (Eds.), *Climate change 2007: impacts, adaptation and vulnerability. Contribution of working group II to the fourth assessment report of the intergovernmental panel of climate change (IPCC)* (pp. 211–272). Cambridge: Cambridge University Press.

Flexas, J., Bota, J., Loreto, F., Cornic, G., & Sharkey, T. D. (2004). Diffusive and metabolic limitations to photosynthesis under drought and salinity in C3 plants. *Plant Biology, 6*, 1–11.

Gao, Y., Cheng, Z., Ling, W., & Huang, J. (2010a). Arbuscular mycorrhizal fungal hyphae contribute to the uptake of polycyclic aromatic hydrocarbons by plant roots. *Bioresource Technology, 101*, 6895–6901.

Gao, Y. Z., Li, Q. L., Ling, W. T., & Zhu, X. Z. (2010b). Arbuscular mycorrhizal phytoremediation of soils contaminated with phenanthrene and pyrene. *Journal of Hazardous Materials, 185*(2–3), 703–709.

Gianinazzi, S., Gollotte, A., Binet, M. N., van Tuinen, D., Redecker, D., & Wipf, D. (2010). Agroecology: The key role of arbuscular mycorrhizas in ecosystem services. *Mycorrhiza, 20*(8), 519–530.

Glick, B. R., & Bashan, Y. (1997). Genetic manipulation of plant growth-promoting bacteria to enhance biocontrol of phytopathogens. *Biotechnology Advances, 15*, 353–378.

Gohre, V., & Paszkowski, U. (2006). Contribution of the arbuscular mycorrhizal symbiosis to heavy metal phytoremediation. *Planta, 223*, 1115–1123.

Gomes, M. P., Carvalho, M., Carvalho, G. S., Marques, T. C. L. L. S. M., Garcia, Q. S., Guilherme, L. R. G., & Soares, A. M. (2013). Phosphorus improves arsenic phytoremediation by *Anadenanthera peregrina* by alleviating induced oxidative stress. *International Journal of Phytoremediation, 15*(7), 633–646.

Gonzalez-Chavez, M. C., Carrillo-Gonzalez, R., Wright, S. F., & Nichols, K. A. (2004). The role of glomalin, a protein produced by arbuscular mycorrhizal fungi, in sequestering potentially toxic elements. *Environmental Pollution, 130*, 317–323.

González-Chávez, M. D., Carrillo-González, R., Hernández Godínez, M. I., & Evangelista Lozano, S. (2017). *Jatropha curcas* and assisted phytoremediation of a mine tailing with biochar and a mycorrhizal fungus. *International Journal of Phytoremediation, 19*(2), 174–182.

Harms, H., Schlosser, D., & Wick, L. Y. (2011). Untapped potential: Exploiting fungi in bioremediation of hazardous chemicals. *Nature Reviews. Microbiology, 9*, 177–192.

Hassan, S. E. D., Boon, E., St-Arnaud, M., & Hijri, M. (2011). Molecular biodiversity of arbuscular mycorrhizal fungi in trace metal-polluted soils. *Molecular Ecology, 20*, 3469–3483.

Hidri, R., Barea, J. M., Metoui-Ben Mahmoud, O., Abdelly, C., & Azcon, R. (2016). Impact of microbial inoculation on biomass accumulation by *Sulla carnosa* provenances, and in regulating nutrition, physiological andantioxidant activities of this species under non-saline and saline conditions. *Journal of Plant Physiology, 201*, 28–41.

Hildebrandt, U., Regvar, M., & Bothe, H. (2007). Arbuscular mycorrhiza and heavy metal tolerance. *Phytochemistry, 68*, 139–146.

Hossain, M. A., Hasanuzzaman, M., & Fujita, M. (2010). Up-regulation of antioxidant and glyoxalase systems by exogenous glycinebetaine and proline in mung bean confer tolerance to cadmium stress. *Physiology and Molecular Biology of Plants, 16*, 259–272.

Huang, Y., Chen, Y. J., & Tao, C. (2002). Uptake and distribution of Cu, Zn, Pb and Cd in maize related to metals speciation change in rhizosphere. *Chinese Journal of Applied Ecology, 13*, 860–862.

IPCC. (2013). *Climate change 2013: The physical science basis*. New York: Cambridge University Press.

Ismail, I. M., Basahi, J. M., & Hassan, I. A. (2014). Gas exchange and chlorophyll fluorescence of pea (*Pisum sativum* L.) plants in response to ambient ozone at a rural site in Egypt. *Science of the Total Environment, 497*, 585–593.

Jakobsen, I., & Rosendahl, L. (1990). Carbon flow into soil and external hyphae from roots of mycorrhizal cucumber plants. *New Phytologist, 115*(1), 77–83.

Johnson, S. M., Doherty, S. J., & Croy, R. R. D. (2003). Biphasic superoxide generation in potato tubers. A self amplifying response to stress. *Plant Physiology, 13*, 1440–1449.

Joner, E. J., & Leyval, C. (2001). Influence of arbuscular mycorrhiza on clover and ryegrass grown together in a soil spiked with polycyclic aromatic hydrocarbons. *Mycorrhiza, 10*, 155–159.

Joner, E. J., & Leyval, C. (2003). Rhizosphere gradients of polycyclic aromatic hydrocarbon (PAH) dissipation in two industrial soils, and the impact of arbuscular mycorrhiza. *Environmental Science & Technology, 37*, 2371–2375.

Kang, F. X., Chen, D. S., Gao, Y. Z., & Zhang, Y. (2010). Distribution of polycyclic aromatic hydrocarbons in subcellular root tissues of ryegrass '*Lolium multiflorum* Lam. *BMC Plant Biology, 10*, 210–216.

Kapoor, R., Sharma, D., & Bhatnagar, A. (2008). Arbuscular mycorrhizae in micropropagation systems and their potential applications. *Scientia Horticulturae, 116*, 227–239.

Kasiamdari, R. S., Soetarto, E. S., & Sancayaningsih, R. P. (2016). Presence of arbuscular mycorrhizal fungi on fern from tailing deposition area of gold mine in Timika, Indonesia. *International Journal of Environmental Bioremediation & Biodegradation, 4*, 1–7.

Kawamitsu, Y., Driscoll, T., & Boyer, J. S. (2000). Photosynthesis during desiccation in an intertidal alga and a land plant. *Plant & Cell Physiology, 41*(3), 344–353.

Kaya, C., Ashraf, M., Sonmez, O., Aydemir, S., Tuna, A. L., & Cullu, M. A. (2009). The influence of arbuscular mycorrhizal colonisation on key growth parameters and fruit yield of pepper plants grown at high salinity. *Scientia Horticulturae, 121*, 1–6.

Knapp, A. K., Briggs, J. M., & Koelliker, J. K. (2001). Frequency and extent of water limitation to primary production in a mesic temperate grassland. *Ecosystems, 4*, 19–28.

Kramer, P. J., & Boyer, J. S. (1997). *Water relations of plants and soils* (p. 495). London: Academic.

Krishnamoorthy, R., Kim, C. G., Subramanian, P., Kim, K. Y., Selvakumar, G., & Sa, T. M. (2015). Arbuscular mycorrhizal fungi community structure, abundance and species richness changes in soil by different levels of heavy metal and metalloid concentration. *PLoS One, 10*, 0128784.

Krznaric, E., Wevers, J. H., Cloquet, C., Vangronsveld, J., Vanhaecke, F., & Colpaert, J. V. (2010). Zn pollution counteracts Cd toxicity in metal-tolerant ectomycorrhizal fungi and their host plant, *Pinus sylvestris*. *Environmental Microbiology, 12*(8), 2133–2141.

Kuper, J., Llamas, A., Hecht, H. J., Mendel, R. R., & Schwarz, G. (2004). Structure of the molybdopterin-bound Cnx1G domain links molybdenum and copper metabolism. *Nature, 430*, 803–806.

Lambers, H., Chapin, F. S., & Pons, T. L. (2008). *Plant physiological ecology* (2nd ed.). New York: Springer.

Larcher, W. (1995). *Physiological plant ecology*. Berlin: Springer.

Leyval, C., & Binet, P. (1998). Effect of polyaromatic hydrocarbons in soil on arbuscular mycorrhizal plants. *Journal of Environmental Quality, 27*, 402–407.

Leyval, C., Joner, E., Del Val, C., & Haselwandter, K. (2001). Potential of arbuscular mycorrhiza for bioremediation. *Mycorrhiza, 7*, 308–317.

Leyval, C., Joner, E. J., del Val, C., & Haselwandter, K. (2002). Potential of arbuscular mycorrhizal fungi for bioremediation. In S. Gianinazzi, H. Schuepp, J. M. Barea, & K. Haselwandter (Eds.), *Mycorrhizal technology in agriculture: From genes to bioproducts* (pp. 175–186). Basel: Birkhauser.

Li, H., Chen, X. W., & Wong, M. H. (2016). Arbuscular mycorrhizal fungi reduced the ratios of inorganic/organic arsenic in rice grains. *Chemosphere, 145*, 224–230.

Li, H., Ye, Z., Chan, W., Chen, X., Wu, F., Wu, S., & Wong, M. (2011). Can arbuscular mycorrhizal fungi improve grain yield, as uptake and tolerance of rice grown under aerobic conditions? *Environmental Pollution, 159*, 2537–2545.

Li, X. L., & Christie, P. (2000). Changes in soil solution Zn and pH and uptake of Zn by arbuscular mycorrhizal red clover in Zn-contaminated soil. *Chemosphere, 42*, 201–207.

Lin, A., Zhang, X., & Yang, X. (2014). *Glomus mosseae* enhances root growth and Cu and Pb acquisition of upland rice (*Oryza sativa* L.) in contaminated soils. *Ecotoxicology, 23*, 2053–2061.

Lykkeberg, A. K., Sengelov, G., Cornett, C., Tjornelund, J., Hansen, S. H., & Halling-Sorensen, B. (2004). Isolation, structural elucidation and in vitro activity of 2-acetyl-2-decarboxamido-oxytetracycline against environmental relevant bacteria, including tetracycline-resistant bacteria. *Journal of Pharmaceutical and Biomedical Analysis, 34*, 559–567.

Ma, T., Pan, X., Liu, W., Christie, P., Luo, Y., & Wu, L. (2016). Effects of different concentrations and application frequencies of oxytetracycline on soil enzyme activities and microbial community diversity. *European Journal of Soil Biology, 76*, 53–60.

Machuca, A., Pereira, G., Aguiar, A., & Milagres, A. M. F. (2007). Metal-chelating compounds produced by ectomycorrhizal fungi collected from pine plantations. *Letters in Applied Microbiology, 44*, 7–12.

Majorel, C., Hannibal, L., Ducousso, M., Lebrun, M., & Jourand, P. (2014). Evidence of nickel (Ni) efflux in Ni-tolerant ectomycorhizal *Pisolithus albus* isolated from ultramafic soil. *Environmental Microbiology Reports, 6*(5), 510–518.

Maksymiec, W., Wojcik, M., & Krupa, Z. (2007). Variation in oxidative stress and photochemical activity in *Arabidopsis thaliana* leaves subjected to cadmium and excess copper in the presence or absence of jasmonate and ascorbate. *Chemosphere, 66*, 421–427.

Malcova, R., & Gryndler, M. (2003). Amelioration of Pb and Mn toxicity to arbuscular mycorrhizal fungus *Glomus intraradices* by maize root exudates. *Biologia Plantarum, 47*, 297–299.

Martino, E., Perotto, S., Parsons, R., & Gadd, G. M. (2003). Solubilization of insoluble inorganic zinc compounds by ericoid mycorrhizal fungi derived from heavy metal polluted sites. *Soil Biology and Biochemistry, 35*, 133–141.

McCool, P. M., & Menge, J. A. (1984). Interaction of ozone and mycorrhizal fungi on tomato as influenced by gungal species and host variety. *Soil Biology and Biochemistry, 16*, 425–427.

McGrath, S. P., Chaudri, A. M., & Giller, K. E. (2015). Long-term effects of metals in sewage sludge on soils, microorganisms and plants. *Journal of Industrial Microbiology & Biotechnology, 14*, 94–104.

Meharg, A. A., & Cairney, J. W. G. (2000). Ectomycorrhizas extending the capacities of rhizosphere remediation? *Soil Biology and Biochemistry, 32*, 1475–1484.
Meharg, A. A., & Hartley-Whitaker, J. (2002). Arsenic uptake and metabolism in arsenic resistant and nonresistant plant species. *The New Phytologist, 154*(1), 29–43.
Miller, R. M., Jastrow, J. D., & Reinhardt, D. R. (1995). External hyphal production of vesicular-arbuscular mycorrhizal fungi in pasture and tallgrass prairie communities. *Oecologia, 103*(1), 17–23.
Mocquot, B., Vangronsveld, J., Clijsters, H., & Mench, M. (1996). Copper toxicity in young maize (*Zea mays* L.) plants: Effects on growth, mineral and chlorophyll contents, and enzyme activities. *Plant and Soil, 182*, 287–300.
Morgan, P. B., Ainsworth, E. A., & Long, S. P. (2003). How does elevated ozone impact soybean? A meta-analysis of photosynthesis, growth and yield. *Plant, Cell & Environment, 26*, 1317–1328.
Mulligan, C. N., Yong, R. N., & Gibbs, B. F. (2001). Remediation technologies for metal-contaminated soils and groundwater: An evaluation. *Engineering Geology, 60*, 193–207.
Navarro, J. M., Perez-Tornero, O., & Morte, A. (2013). Alleviation of salt stress in citrus seedlings inoculated with arbuscular mycorrhizal fungi depends on the rootstock salt tolerance. *Journal of Plant Physiology, 171*, 76–85.
Netondo, G. W., Onyango, J. C., & Beck, E. (2004). Sorghum and salinity: II. Gas exchange and chlorophyll fluorescence of sorghum under salt stress. *Crop Science, 44*, 806–811.
Newman, E. I., & Reddell, P. (1987). The distribution of mycorrhizas among families of vascular plants. *The New Phytologist, 106*, 747.
Novoa, D., Palma, S., & Gaete, H. (2009). Effect of arbuscular mycorrhizal fungi *Glomus* spp. inoculation on alfalfa growth in soils with copper. *Chilean Journal of Agricultural Research, 70*, 259–265.
Parniske, M. (2008). Arbuscular mycorrhiza: The mother of plant root endosymbioses. *Nature Reviews. Microbiology, 6*, 763–775.
Peterson, R. L., Massicotte, H. B., & Melville, L. H. (2004). *Mycorrhizas: Anatomy and cell biology*. Ottawa: NRC Research Press.
Rangel, W. M., Schneider, J., Costa, E. T. S., Soares, C. R. F. S., Guilherme, L. R. G., & Moreira, F. M. S. (2014). Phytoprotective effect of arbuscular mycorrhizal fungi species against arsenic toxicity in tropical leguminous species. *International Journal of Phytoremediation, 16*, 840–858.
Rapti-Caputo, D. (2010). Influence of climatic changes and human activities on the salinization process of coastal aquifer systems. *Italian Journal of Agronomy, 5*(3), 67–79.
Regvar, M., Likar, M., Piltaver, A., Kugonic, N., & Smith, J. E. (2010). Fungal community structure under goat willows (*Salix caprea* L.) growing at metal polluted site: The potential of screening in a model phytostabilisation study. *Plant and Soil, 330*, 345–356.
Rehmann, K., Noll, H. P., Steinberg, C. E. W., & Kettrup, A. A. (1998). Pyrene degradation by *Mycobacterium* sp. strain KR2. *Chemosphere, 36*, 2977–2992.
Repetto, O., Massa, N., Gianinazzi-Pearson, V., Dumas-Gaudot, E., & Berta, G. (2007). Cadmium effects on populations of root nuclei in two pea genotypes inoculated or not with the arbuscular mycorrhizal fungus *Glomus mosseae*. *Mycorrhiza, 17*(2), 111–120.
Rilling, M. C., Hernandez, G. Y., & Newton, P. C. D. (2000). Arbuscular mycorrhizae respond to elevated atmospheric CO2 after long-term exposure: Evidence from a CO2 spring in New Zealand supports the resource balance model. *Ecology Letters, 3*, 475–478.
Rodrigues, C. R., & Rodrigues, B. F. (2015). Use of arbuscular mycorrhiza and organic amendments to enhance growth of *Macaranga peltata* (Roxb.) Müll. Arg. in iron ore mine wastelands. *International Journal of Phytoremediation, 17*(1–6), 485–492.
Romanowska, E., Wroblewska, B., Drozak, A., & Siedlecka, M. (2006). High light intensity protects photosynthetic apparatus of pea plants against exposure to lead. *Plant Physiology, 44*, 387–394.
Ruiz-Lozano, J. M., Azcon, R., & Gomez, M. (1996). Alleviation of salt stress by arbuscular mycorrhizal *Glomus* species in *Lactuca sativa* plants. *Physiologia Plantarum, 98*, 767–772.

Ruotsalainen, A. L., Markkola, A., & Kozlov, M. V. (2007). Root fungal colonisation in *Deschampsia flexuosa:* Effects of pollution and neighbouring trees. *Environmental Pollution, 147*, 723–728.

Ryan, P. R., Delhaize, E., & Jones, D. L. (2001). Function and mechanism of organic anion exudation from plant roots. *Annual Review of Plant Physiology and Plant Molecular Biology, 52*, 527–560.

Safir, G. R., & Nelsen, C. E. (1985). VA mycorrhizas: Plant and fungal water relations. In R. Molina (Ed.), *Proceedings of the sixth North American conference on mycorrhizae* (pp. 161–164). Corvallis: Forest Research Laboratory.

Sanchez-Castro, I., Gianinazzi-Pearson, V., Cleyet-Marel, J. C., Baudoin, E., & van Tuinen, D. (2017). Glomeromycota communities survive extreme levels of metal toxicity in an orphan mining site. *Science of the Total Environment, 598*, 121–128.

Schneider, J., Sturmer, S. L., Guilherme, L. R. G., Moreira, F. M. S., & Soares, C. R. F. S. (2013). Arbuscular mycorrhizal fungi in arsenic-contaminated areas in Brazil. *Journal of Hazardous Materials, 262*, 1105–1115.

Seki, M., Kamei, A., Yamaguchi-Shinozaki, K., & Shinozaki, K. (2003). Molecular responses to drought, salinity and frost: Common and different paths for plant protection. *Current Opinion in Biotechnology, 14*, 194–199.

Shahabivand, S., Maivan, H. Z., Goltapeh, E. M., Sharifi, M., & Aliloo, A. A. (2012). The effects of root endophyte and arbuscular mycorrhizal fungi on growth and cadmium accumulation in wheat under cadmium toxicity. *Plant Physiology and Biochemistry, 60*, 53–58.

Sharifi, M., Ghorbanli, M., & Ebrahimzadeh, H. (2007). Improved growth of salinity stressed soybean after inoculation with salt pre-treated mycorrhizal fungi. *Journal of Plant Physiology, 164*, 1144–1151.

Sharma, P., & Dubey, R. S. (2005). Lead toxicity in plants. *Brazilian Journal of Plant Physiology, 17*, 35–52.

Sharma, P., & Dubey, R. S. (2007). Involvement of oxidative stress and role of antioxidative defense system in growing rice seedlings exposed to toxic levels of aluminium. *Plant Cell Reports, 26*, 2027–2038.

Sheikh-Assadi, M., Khandan-Mirkohi, A., Alemardan, A., & Moreno-Jiménez, E. (2015). Mycorrhizal limonium sinuatum (L.) mill. Enhances accumulation of lead and cadmium. *International Journal of Phytoremediation, 17*(1–6), 556–562.

Singh, K. N., & Chatrath, R. (2001). Salinity tolerance. In M. P. Reynolds, M. Jio, & A. McNab (Eds.), *Application of Physiology in Wheat Breeding* (pp. 101–110). Mexico, DF: CIMMYT.

Sircelj, H. M., Tausz, M., Grill, D., & Batic, F. (2005). Biochemical responses in leaves of two apple tree cultivars subjected to progressing drought. *Journal of Plant Physiology, 162*, 1308–1318.

Skujins, J., & Allen, M. F. (1986). Use of mycorrhizae for land rehabilitation. *Mircen Journal, 2*, 161–176.

Soares, C. R. F. S., & Siqueira, J. O. (2008). Mycorrhiza and phosphate protection of tropical grass species against heavy metal toxicity in multi-contaminated soil. *Biology and Fertility of Soils, 44*, 833–841.

Somtrakoon, K., Suanjit, S., Pokethitiyook, P., Kruatrachue, M., Lee, H., & Upatham, S. (2008). Phenanthrene stimulates the degradation of pyrene and fluoranthene by *Burkholderia* sp. VUN10013. *World Journal of Microbiology and Biotechnology, 24*, 523–531.

Sut, M., Boldt-Burisch, K., & Raab, T. (2016). Possible evidence for contribution of arbuscular mycorrhizal fungi (AMF) in phytoremediation of iron–cyanide (Fe–CN) complexes. *Ecotoxicology, 25*(6), 1260–1269.

Tang, X., Lou, C., Wang, S., Lu, Y., Liu, M., Hashmi, M. Z., & Fan, F. (2015). Effects of long-term manure applications on the occurrence of antibiotics and antibiotic resistance genes (ARGs) in paddy soils: Evidence from four field experiments in south of China. *Soil Biology and Biochemistry, 90*, 179–187.

Tiwari, S., & Sarangi, B. K. (2017). Comparative analysis of antioxidant response by *Pteris vittata* and *Vetiveria zizanioides* towards arsenic stress. *Ecological Engineering, 100*, 211–218.

Toljander, J. F., Lindahl, B. D., Paul, L. R., Elfstrand, M., & Finlay, R. D. (2007). Influence of arbuscular mycorrhizal mycelial exudates on soil bacterial growth and community structure. *FEMS Microbiology Ecology, 61*, 295–304.

Toth, G., Hermann, T., Da Silva, M. R., & Montanarella, L. (2016). Heavy metals in agricultural soils of the European Union with implications for food safety. *Environment International, 88*, 299–309.

Toujaguez, R., Ono, F. B., Martins, V., Cabrera, P. P., Blanco, A. V., Bundschuh, J., & Guilherme, L. R. G. (2013). Arsenic bioaccessibility in gold mine tailings of Delita, Cuba. *Journal of Hazardous Materials, 262*, 1004–1013.

Trenberth, K. E., Dai, A., Van Der Schrier, G., Jones, P. D., Barichivich, J., Briffa, K. R., & Sheffield, J. (2014). Global warming and changes in drought. *Nature Climate Change, 4*, 17–22.

Treseder, K. K., & Allen, M. F. (2000). Mycorrhizal fungi have a potential role in soil carbon storage under elevated CO_2 and nitrogen deposition. *The New Phytologist, 147*, 189–200.

Van Dingenen, R., Raes, F., Krol, M. C., Emberson, L., & Cofala, J. (2009). The global impact of O_3 on agricultural crop yields under current and future air quality legislation. *Atmospheric Environment, 43*, 604–618.

Vangronsveld, J., Herzig, R., Weyens, N., Boulet, J., Adriaensen, K., Ruttens, A., Thewys, T., Vassilev, A., Meers, E., Nehnevajova, E., van der Lelie, D., & Mench, M. (2009). Phytoremediation of contaminated soils and groundwater: Lessons from the field. *Environmental Science and Pollution Research International, 16*, 765–794.

Wang, F. Y., Lin, X. G., & Yin, R. (2007a). Effect of arbuscular mycorrhizal fungal inoculation on heavy metal accumulation of maize grown in a naturally contaminated soil. *International Journal of Phytoremediation, 9*, 345–353.

Wang, L., Huang, X., Ma, F., Ho, S. H., Wu, J., & Zhu, S. (2017a). Role of Rhizophagus irregularis in alleviating cadmium toxicity via improving the growth, micro-and macroelements uptake in *Phragmites australis*. *Environmental Science and Pollution Research, 24*(4), 3593–3607.

Wang, S., Augé, R. M., & Toler, H. D. (2017b). Arbuscular mycorrhiza formation and its function under elevated atmospheric O3: A meta-analysis. *Environmental Pollution, 226*, 104–117.

Wang, S. G., Diao, X. J., Li, Y. W., & Ma, L. M. (2015). Effect of *Glomus aggregatum* on photosynthetic function of snap bean in response to elevated ozone. *The Journal of Agricultural Science, 153*, 837–852.

Wang, S. G., Feng, Z. Z., Wang, X. K., & Gong, W. L. (2011). Arbuscular mycorrhizal fungi alter the response of growth and nutrient uptake of snap bean (*Phaseolus vulgaris* L.) to O_3. *Journal of Environmental Sciences, 23*, 968–974.

Wang, X. K., Manning, W. J., Feng, Z. W., & Zhu, Y. G. (2007b). Ground-level ozone in China: Distribution and effects on crop yields. *Environmental Pollution, 147*, 394–400.

Wen, Z., Shi, L., Tang, Y., Shen, Z., Xia, Y., & Chen, Y. (2017). Effects of *Pisolithus tinctorius* and *Cenococcum geophilum* inoculation on pine in copper-contaminated soil to enhance phytoremediation. *International Journal of Phytoremediation, 19*(4), 387–394.

Wery, J., Silim, S. N., Knights, E. J., Malhotra, R. S., & Cousin, R. (1994). Screening techniques and sources and tolerance to extremes of moisture and air temperature in cool season food legumes. *Euphytica, 73*, 73–83.

White, C., Sayer, J. A., & Gadd, G. M. (1997). Microbial solubilization and immobilization of toxic metals: Key biogeochemical processes for treatment of contamination. *FEMS Microbiology Reviews, 20*, 503–516.

Wu, Q. S., Zou, Y. N., & He, X. H. (2010). Contributions of arbuscular mycorrhizal fungi to growth, photosynthesis, root morphology and ionic balance of citrus seedlings under salt stress. *Acta Physiologiae Plantarum, 32*, 297–304.

Xu, P., Christie, P., Liu, Y., Zhang, J., & Li, X. (2008). The arbuscular mycorrhizal fungus Glomus mosseae can enhance arsenic tolerance in *Medicago truncatula* by increasing plant phosphorus status and restricting arsenate uptake. *Environmental Pollution, 156*, 215–220.

Xun, F., Xie, B., Liu, S., & Guo, C. (2015). Effect of plant growth-promoting bacteria (PGPR) and arbuscular mycorrhizal fungi (AMF) inoculation on oats in saline-alkali soil contaminated by petroleum to enhance phytoremediation. *Environmental Science and Pollution Research International, 22*(1), 598–608.

Yang, Y., Liang, Y., Han, X., Chiu, T. Y., Ghosh, A., Chen, H., & Tang, M. (2016). The roles of arbuscular mycorrhizal fungi (AMF) in phytoremediation and tree-herb interactions in Pb contaminated soil. *Scientific Reports, 6*, 20469.

Yin, N., Zhang, Z., Wang, L., & Qian, K. (2016). Variations in organic carbon, aggregation, and enzyme activities of gangue-fly ash-reconstructed soils with sludge and arbuscular mycorrhizal fungi during 6-year reclamation. *Environmental Science and Pollution Research International, 23*(17), 17840–17849.

Zhu, J. K., Hasegawa, P. M., & Bressan, R. A. (1997). Molecular aspects of osmotic stress in plants. *CRC Critical Reviews in Plant Sciences, 16*, 253–277.

Zhu, Y. G., Christie, P., & Laidlaw, A. S. (2001). Uptake of Zn by arbuscular mycorrhizal white clover from Zn-contaminated soil. *Chemosphere, 42*, 193–199.

Mycorrhizosphere: Microbial Interactions for Sustainable Agricultural Production

18

Biplab Dash, Ravindra Soni, Vinay Kumar,
Deep Chandra Suyal, Diptimayee Dash, and Reeta Goel

1 Mycorrhizosphere

In the year 1904, the concept of rhizosphere was first projected by Lorenz Hiltner (a German scientist) at a meeting of German Agricultural Society while explaining the role of bacteria in fixation of nitrogen in soil (Hartmann et al. 2007). It is well known that rhizosphere is surrounded by a number of microbes (Mendes et al. 2013) but often their diversity differs from that present around roots colonised by mycorrhiza (Duponnois et al. 2008) and thus gave birth to a term called "mycorrhizosphere" (Rambelli 1973; Linderman 1988). Mycorrhizosphere is often called as an extension of rhizosphere consisting of hyposphere (a zone of soil surrounded by fungal hyphae) where interaction occurs between these two (Priyadharsini et al. 2016). Broadly, mycorrhizosphere can further be divided into inner mycorrhizosphere (mostly consisting of rhizosphere) and outer mycorrhizosphere (these consist of hyposphere along with external mycelium and associated microbes) (Timonen and Marschner 2006).

It has been reported that the number and diversity of bacterial species can be greatly influenced by mycorrhizal presence in rhizosphere, hyposphere and bulk soil (Andrade et al. 1997; Mansfeld-Giese et al. 2002; Frey et al. 1997). Due to presence of varied number of bacterial population in mycorrhizosphere, a series of interaction takes place between them; thus ultimately affecting the plant growth and subsequently improving soil quality (Barea et al. 2002). Due formation of symbiotic

B. Dash · R. Soni · D. Dash
Department of Agricultural Microbiology, College of Agriculture,
Indira Gandhi Krishi Vishwavidyalaya, Raipur, Chhatisgarh, India

V. Kumar
National Institute of Biotic Stress Management, Raipur, India

D. C. Suyal · R. Goel (✉)
Department of Microbiology, CBSH, G.B. Pant University of Agriculture and Technology,
Pantnagar, Uttarakhand, India

© Springer Nature Singapore Pte Ltd. 2019

A. Varma, D. K. Choudhary (eds.), *Mycorrhizosphere and Pedogenesis*,
https://doi.org/10.1007/978-981-13-6480-8_18

relationship of mycorrhiza with roots of plant, certain varied physical, physiological, morphological and biochemical changes takes place, thereby establishing a new norm for creation of microbial equilibrium in mycorrhizosphere often called as "The Mycorrhizosphere Effect" (Linderman 1988; Frey-Klett et al. 2005).

1.1 Microbial Interactions in Mycorrhizosphere

Symbionts and saprophytes are the two groups of micoorganisms that mainly interact with mycorhizal fungi in mycorrhizosphere (Barea et al. 2002) comprising of both fungi and bacteria having antagonistic, synergistic or neutral effect on plant (Bianciotto et al. 2002). Furthermore, mycorrhiza seems to interact with several other soil fauna like protozoa, collembolan, nematodes, and earthworms in rhizosphere (Hodge 2014; Balaes and Catalin 2011; Fitter and Garbay 1994). Of all these, its interaction with rhizobacteria is of significant importance as it influence a number of crucial processes like nutrient cycling (Azcón-Aguilar and Barea 2015), carbon sequestration (Churchland and Grayston 2014), sulphur supply to plants (Gahan and Schmalenberger 2014), by providing stability to soil due to soil aggregation (Andrade et al. 1998) as well as promotes mineral weathering in forest soils (Uroz et al. 2007). These were certain positive microbial interactions going on in mycorrhizosphere. But there are also some examples of negative interactions like reduction in zinc (Zn) uptake by plants from mycorrhizosphere as compared to rhizosphere in AM plants (Audet and Charest 2010). This is followed by the suppression of extra radical mycelium of mycorrhizal fungus by several bacterial groups including Acidobacteria and Actinobacteria (Svenningsen et al. 2018). Sometimes, mycorhizal fungus seems to be antagonistic on some biocontrol agents like *Trichoderma harzianum* by suppressing its growth and thus adversely affects its biocontrol activity (Green et al. 1999). On other occasions, multiple interactions may be noticed among mycorrhizal fungi and saprophytic fungi (Fracchia et al. 1998) having the examples of all three forms of interactions (neutral, antagonistic and synergistic) which symbolises that mycorrhizosphere is a place filled with complexcity not only on front of microbial diversity but also on ground depicting interconnecting associations with them.

One of the most important interaction going on in mycorrhizosphere is that of Mycorrhization Helper Bacteria (MHB) with mycorrhizal fungi. They not only help in receptivity and recognition of plant root but also promote germination of fungal spores (Rigamonte et al. 2010). The role mycorrhiza with that of N_2 fixing bacteria cannot be ignored which suggest that their interaction is quiet beneficial in nitrogen fixation (Siviero et al. 2008; Matsumoto et al. 2005; Tian et al. 2003). In another such instance, genetically modified nitrogen fixing bacteria *Rhizobium* proved quiet successful in increasing the AM colonisation in host plant (Tobar et al. 1996). Beside all these, mycorrhiza interacts with several other group of bacteria like plant growth promoting rhizobacteria (Larsen et al. 2009; Jaderlund et al. 2008), endocellular bacteria (Bianciotto and Bonfante 2002), deleterious bacteria (Miransari 2011); thereby opening several new dimensions for understanding mycorrhizal interaction.

2 Mycorrhiza for Sustainable Agriculture

A potentially concealed problem largely looming over agriculture is its vicious battle with sustainability when agriculture is consider as a profession for a poor farmer. In order to tackle this problem, a root cause analysis is imperative which clearly advocates a better understanding of soil microbe interactions with their surrounding (rhizosphere). A much genuine and eclectic microbial association yet sluggish in harnessing its full potential which can relatively change the very narrative of agricultural sustainability is non-other than mycorrhiza. In it, we can find a perfect example of harmonious symbiotic association between plant and fungi depicting as a symbol of nature's creation showcasing the importance of mutualistic interactions. Mycorrhizal association is found in 336 plant families of which 99% are flowering plants (Brundrett 2009), which if brought into use can have a daunting impact on plant growth considering its invaluable resources in form of a biocontrol agent and a biofertilizer. However, the type of mycorrhiza that is found more in agricultural and horticultural crops are the Arbuscular mycorrhiza which finds their name perfectly fit when the topic of sustainable agriculture comes into play (Bagyaraj 2014). It is well known that mycorrhiza's are excellent absorbers of phosphorous but beside that it is equally fit in providing micronutrients to plants; thus befittingly establishing itself as a Natural biofertilizer (Berruti et al. 2016). Having said that, it is noteworthy to take into account its proficiency for shielding plants against soil borne pathogens and thus clinching the tittle of bio-control agent (Tahat et al. 2010). Not only for controlling of pathogens but also mycorrhiza is known to have kept a check on weed growth which was successfully demonstrated by Bethlenfalvay et al. 1996. Nevertheless, the role Mycorrhiza in alleviation of water stress in drought conditions (Auge et al. 2015) and effective mitigation of heavy metal contamination (Tamayo et al. 2014) can never be underestimated (Farahani et al. 2008). Enhancement in photosynthesis level with increased water use efficiency under the shadow of mycorrhizal application was noted in *Boswellia papyrifera* seedlings (Brihane et al. 2012) were as the same result was observed by Sheng et al. (2008) in Maize plant under salt stress condition. Adding some more teeth to its resources, it is also an effective soil aggregator and thus has an invisible hand behind the improvement of soil structure as well as preventing from soil erosion which is yet to be unchartered fully (Schreiner and Bethlenfalvay 1995). Zhu and Miller (2003) have concluded that arbuscular mycorrhizal fungi takes part in carbon cycling by modulating the quantity of carbon fluxes to and fro between atmosphere and biosphere. Realistically, a more diverse interaction occurs in Mycorhizosphere with a galaxy of microbes virtually having its own world of life and thereby influencing the plant growth (Paulilz and Linderman 1991). Arbuscular mycorrhizal is vital for sustainable agriculture production as it primarily colonies the root of crop plants where as in the same note, forest plants are colonized by Ectomycorrhizal fungi discharging their role for phytoremediation. Peeping into a much broader perspective, mycorrhiza can address the hot boiling issue of climate change through carbon sequestration (Staddon et al. 2002). A very profound and in-depth hypothesis regarding the probable impact of mycorrhizal diversity due to ongoing climate change is given by

Bellgard and Williams 2011. Several other aspects of mycorrhiza, like its role in food chain, bio-geochemical cycling and the inter or intra specific interactions of plants with it, gives us an insight about its influential role on natural ecosystems (Jha and Kumar 2011). Coming into tropics, their role is inevitable and indispensable because of the fact that 75–80% of the phosphatic fertilizer applied here gets fixed immediately which is not seen in case of temperate region (Bagyaraj 2014). This reality even gives more substance to their immense importance and stature, knowing by the fact that Mycorrhiza can simply change very narrative of sustainable agriculture, especially in tropics. After a brief description of Mycorrhiza and knowing its multifarious usefulness, throwing some more light on its two most potent natural hideous forms (as a Biofertilizer and as a Biocontrol agent) surely comes into play in order to establish itself as a champion player in the sphere of sustainable agriculture.

2.1 Biofertilizer

Among several good attributes of Mycorrihza, its applicability as a natural biofertilizer is of paramount importance and will always and every time finds its name at the pole position in its list of qualities. Mycorrhiza has its inherent property of colonising roots of plants by which it forms a symbiotic relationship with it, thereby cementing its role by providing essential nutrients at tough times and conditions (Syafruddin et al. 2016). From an array of mycorrhizal types, AM (Arbuscular Mycorrhiza) fungi, a form of endomycorrhiza is best suited for biofertiliser purpose due its ubiquitous nature, particularly in context of agricultural and horticultural crops. On the basis of its diversity at morphological, molecular and ecological level from its other fungal counterparts, it was placed in a new phylum called *Glomeromycota* after being separated from Zygomycota, in which three families are constituted, namely *Archaeosporales*, *Paraglomerales*, *Diversisporales* with over 150 species described (Schubler et al. 2001). AM fungi can exponentially increase the effectiveness of absorption capacity of host plant root by as much as ten times and at the same time can also help their hyphe in spreading infection to other surrounding plants (Sadhana 2014).

The most fundamental objective of AM fungi is absorption of phosphorous from soil, which is immobile and then finally translocating it into the host plant. As per the report of Bagyaraj et al. (2015) the productive efficiency of crops can be increased by simultaneous inoculation of phosphorous solubilising microorganisms (PSM) along with AM fungi. This is because PSM solubilises unavailable phosphorous into available form which is there upon taken up by AM fungi. Moreover, they have reported in a reduction of 50% less phosphatic fertiliser applicability in field condition with AM fungi inoculation which is a very promising sign. Besides phosphorous, Hodge et al. (2001) have reported an AM fungus *Glomus hoi*, which is said to have enhanced the decomposition rate of organic material and showed its proficiency in absorbing more amount of nitrogen from organic matter. Both macro and micro nutrients were enhanced by Mycorrhiza like Ca, K, S and Zn (Abbasi et al. 2015) and Ca, Si, Ni, Co (Mirzakhani et al. 2009). Arbuscular mycorrhiza was seen to have a favourable effect on plant nutrition by significantly increasing the concentrations of Iron (Fe) and Copper (Cu) in plant tissues but having a limiting effect on Mn concentration in plants (Lehmann and Rillig 2015). Significance of Mycorrihza as a biofertilizer is enlisted in Table 18.1.

Table 18.1 Effect of Mycorrihza as a biofertilizer

Crop	Mycorrihza	Significance as a biofertiliser	References
Rice	*Rhizophagus intraradices*	Decreased the cost of production by a factor of 18.5–16.3%	Orona-Castro et al. (2013)
Maize	*Glomus intraradices*	Improvement in relative water content and cell membrane stability	Naghashzadeh (2014)
Safflower	*Glomus intraradices*	Increase in grain yield	Mirzakhani et al. (2009)
Tobacco	*Glomus intraradices*	Combination of AMF and potassium solublising bacteria enhanced P and K availability with increase in leaf quality	Subhashini (2016)
Cowpea	*Glomus etunicatum* and *Gigasporaalbida*	Increased in grain and shoot length	Andrade et al. (2013)
Durian, Cashew and Longan	*Glomus mosseae* and *Glomus manihotis*	Increase in overall growth	Thamsurakul and Charoensook (2006)
Black gram	*Glomus mossae*	Increment in seed germination along with leghaemoglobin content	Bharti and Kumar (2016)
Tomato	*Acaulospora laevis* and *Gigaspora margarita*	Significant increase in plant height and number of leaves along with early onset of flower and fruit emergence	Osillos and Nagpala (2014)
Sesame	VAM (Vascular Arbuscular Mycorrhiza)	Positive response on most of the growth traits of plant	Alsamowal et al. (2016)
Potato	*Glomus* sp.	Reduces the use of chemical fertilizer	Nurbaity et al. (2016)
Maize	*Glomus mossea*	Substantial increase in seed yield as well as number of seeds per ear	Mobasser and Moradgholi (2012)
Soybean	*Glomus fasciculatum*	Mycorhiza in combination with rhizobium and diammonium phosphate boosted the yield.	Salih et al. (2015)
Barley	*Glomus intraradices*	Enhancement in phosphorous uptake	Zhu et al. (2003)
Capsicum annuum	*Glomus mosseae, Glomus intraradices*	Total chlorophyll and carotenoid content	Cekic et al. (2012)
Tea	*Glomus etunicatum, Glomus intraradices, Glomus versiforme*	*G. versiforme* resulted in highest phosphorous in leaf while inoculation with *G. intraradices* showed a marked increase in Fe, Zn and Cu in plant.	Kahneh et al. (2006)

2.2 Biocontrol Agent

Failure of several strategies including both chemical and biological methods for controlling soil borne diseases lead to the call for some unconventional way to deal with and here comes the usage of mycorrhiza as a possible biocontrol agent. Besides being eco-friendly and cost effective, VAM (Vesicular Arbuscular Mycorrhizal) fungi have several other facets for protecting plants against pathogens; thus bring sustainability and stability in multidimensional agriculture system (Dar and Reshi 2017). Due to its intensive power of root colonisation coupled with its symbiotic nature that covers most part of rhizosphere, mycorrhiza was seen as an alternate. Caron (2009) highlighted the usefulness of vesicular arbuscular mycorrhiza (VAM) in controlling soil borne diseases, especially caused by nematodes and fungi while Waschkies et al. (1994) studied the anti-bacterial action of *Glomus mosseae* on fluorescent pseudomonads, thereby reducing the incidence of replant disease in grapevine. Pinochet et al. (1996) showed the efficiency of mycorrhizal fungi in successfully controlling of the nematode *Pratylenchus vulnus* when being inoculated in early stages of plant development. In contrast, the use of mycorrhizal fungus as a potential biocontrol agent against viruses are not yet quiet fruitful (Xavier and Boyetchko 2004). Decrease in root colonisation by fungus *Phytophthora nicotianae var. parasitica* on tomato plant which were pre-treated with mycorrhizal fungus *Glomus mosseae* was highlighted by Cordier et al. (1996) in their experiment. Moreover, it is also imperative to note down the role of rhizobacteria and their tandem association with mycorrhizal fungi for fighting against the soil borne pathogens. It the efficacy of AM fungi which selectively leads to the establishment of rhizobacteria in and around mycorrhizosphere which ultimately has an antagonistic effect on soil borne pathogens (Lioussanne 2013). The beneficial interaction among several group of microorganisms like Mycorrhizal fungus with that of Actinomycetes and Plant growth promoting rhizobacteria for alleviation of plant diseases through improvement in plant defence mechanism was reported by Kamal et al. 2014. There are several defence mechanisms being employed by mycorrhizal fungus towards soil borne pathogens like (a) increase in nutrient uptake, (b) several changes in anatomical and morphological structure of root, (c) compensating the loss of biomass, (d) changes in soil microbial interaction, (e) competition shown for colonisation and photosynthesis in host, (f) alteration in chemical constituents in host tissues (Aguilar and Barea 1996; Tahat et al. 2010). Schouteden et al. (2015) extensively described the mechanisms that are involved in controlling plant pathogenic nematodes by mycorrhiza fungi. It has been noted that plant immune system gets activated systemically throughout the plant during establishment of mycorrhiza (Jung et al. 2012). Application of Mycorrihzae against fungal pathogen and Nematodes are summarized in Tables 18.2 and 18.3, respectively.

Table 18.2 Mycorrhiza as a biocontrol agent against Fungal pathogen

Mycorrhiza	Fungi	Crop	References
Glomus etunicatum *Glomus leptotichum* *Rhizophagus intraradices*	*Fusarium oxysporum* f. sp. *lycopersic*	Tomato	Muhsen et al. (2015)
Glomus hoi *Glomus fasciculatum*	*Fusarium oxysporum* f. sp. *ciceris*	Chickpea	Singh et al. (2010)
Piriformospora indica *Sebacina vermifera*	*Gaeumannomyces graminis* var. *tritici*	Wheat	Ghahfarokhy et al. (2011)
Glomus mosseae *Glomus intraradices* *Glomus clarum* *Glomus gigantean* *Glomus margarita*	*Fusarium solani*	Bean plant (Phaseolus vulgaris)	Askar and Rashad (2010)
Glomus intraradice *Glomus constrictum* *Glomus claroideum*	*Cochliobolus sativus*	Barley	Arabi et al. (2013)
Glomus intraradice	*Fusarium oxysporum* f. sp. *asparagi*	Asparagus	Arriola et al. (2000)
Glomus aggregatum	*Sclerotium cepivorum*	Onion	Leta and Selvaraj (2013)
Glomus spp.	*Fusarium oxysporum* f. sp. *sesame* *Macrophomina phaseolina*	Sesame	Ziedan et al. (2011)
Glomus fasiculatum *Glomus mossae* *Acaulispora laevis*	*Cephalosporium acremonium*	Maize	Veerabhadraswamy and Rajkumar (2011)
Glomus etunicatum *Glomus caledonium*	*Sclerotium rolfsii*	Peanut	Ozgonen et al. (2010)

3 Molecular Techniques to Explore Mycorrhizosphere

Morphological identification of AMF is time-consuming and requires considerable expertise, whereas DNA-based methods are less time-consuming and more reliable. For application of AMF as biological fertilizer for agricultural and environmental uses it is prerequisite to perform strict quality control of the inoculum. The quality control measure consists of identifying and quantifying species of AMF present in the inoculum and determining the absence of pathogens. A quick, accurate taxonomic identification of AMF isolates is necessary for culture collections, research and large-scale AMF usage.

Molecular understanding of AM fungi is limited, due to unavailable *in vitro* multiplication system to get spores without microbial contamination. Molecular techniques are highly useful for identification of mycorrhizal fungi in absence of morphological characters. With the advent of polymerase chain reaction (PCR) based molecular techniques has provided a valuable and alternative approach to morphology, which represents the primary criterion to define the taxonomic position of fungi (Kohn 1992). Molecular characterization using DNA sequences was

Table 18.3 Biocontrol agent against Nematodes

Mycorrhiza	Nematode	Crop	References
Glomus mossae	*Meloidogyne incognita*	Cowpea	Odeyemi et al. (2010)
Glomus intraradices	*Nacobbus aberrans*	Tomato	Marro et al. (2014) and Lax et al. (2011)
Glomus intraradices	*Meloidogyne incognita*	Tomato	Sharma and Sharma (2015)
Glomus fasciculatum	*Meloidogyne incognita*	Tomato	Shreenivasa et al. (2007)
AMF	*Meloidogyne exigua*	Coffee	Alban et al. (2013)
Glomus intraradices	*Xiphinema index*	Grape	Hao et al. (2012)
Glomus versiforme	*Meloidogyne incognita*	Grape	Li et al. (2006)
Glomus coronatum and AMF consortium	*Meloidogyne incognita*	*Impatiens balsamina* L.	Banuelos et al. (2014)
Scutellospora heterogama	*Meloidogyne incognita*	Sweet Passion Fruit (*Passiflora alata*)	Anjos et al. (2010)
Scutellospora castanea and *Glomus* spp.	*Pratylenchus penetrans*	Dune grass (*Ammophila arenaria*)	de la Pena et al. (2006)
Glomus mossae *Glomus versiforme*	*Meloidogyne incognita*	Cucumber	Zhang et al. (2008)

introduced in the 1990s to detect, identify and quantify AMF in the roots of plants (Simon et al. 1992a, b). For molecular identification of fungi, (i) the entire internal transcribed spacer (ITS) region is often between 600 and 800 bp, and can be readily amplified with universal primers, complementary to sequences within the rRNA genes (White et al. 1990), (ii) the multicopy nature of rDNA repeat makes the ITS region easy to amplify from spore samples, and (iii) the ITS region is often highly variable among morphologically distinct fungal species (Gardes et al. 1991). Spores of AM fungi contain thousands of nuclei (Kohn 1992).

The major challenges faced in AMF research includes power of molecular markers used for identification and quantification of AMF at intra and inter-specific levels. Large intra-isolate variation of nuclear ribosomal genes complicates assignment of a single marker gene sequence to a fungal strain and assessment of the diversity of AMF in molecular field studies. (Hijri and Sanders 2005 and Croll et al. 2009). An often-used fragment for resolving closely related species comprises the SSU rRNA gene, the whole internal transcribed spacer (ITS) rDNA region, including the 5.8S rRNA gene, and the partial large subunit (LSU) rRNA gene, herein referred to as SSU–ITS–LSU18 (Kruger et al. 2012). The long sequences allow robust phylogenetic analyses and species level resolution by inclusion of the variable ITS and LSU rDNA region (Walker et al. 2007; Gamper et al. 2009), whereas formerly used primers mainly amplified rDNA fragments of up to 800 bp (Lee et al. 2008). The search for new marker genes has additionally been hampered by the great difficulty encountered in assembling the nuclear genome of Rhizophagus irregularis (Martin et al. 2008). Therefore some alternate techniques have been utilized for identification and quantification of AMF. An alternative approach to DNA-based isolation and characterization applied to AMF, is proteomic-based chemotaxonomic biotyping using MALDI-TOF mass spectrometry.

Matrix-Assisted Laser Desorption Ionisation Time of Flight Mass Spectrometry (MALDI-TOF-MS) based identification of fungi provided more accurate results than morphology-based analyses (Gautier et al. 2014). Furthermore, this technology is less expensive, easier and faster than current DNA based-identification (Tran et al. 2015). Arbuscular mycorrhizal fungi (AMF, Glomeromycota) are mutualistic symbionts associated with majority of land plants. These fungi play an important role in plant growth, but their taxonomic identification remains a challenge for academic research, culture collections and inoculum producers. Identification of these fungi was traditionally performed based on their spore morphology and DNA sequence data to study the evolutionary relationships. MALDI-TOF-MS proteomic-based biotyping is a highly efficient approach for AMF identification. Crossay et al. (2017) differentiated 19 isolates belonging to 14 species, seven genera and five families using MALDI biotyping at the species level, and intraspecific differentiation was achieved for the majority. AMF identification using MALDI biotyping could be highly useful for research as well as agricultural and environmental applications. MALDI-TOF-MS may be used as an alternative to conventional morphological and molecular methods for AMF identification. Some PCR based approaches to identified Mycorrihzae are enlisted in Table 18.4.

Table 18.4 Name of the molecular markers used for the characterization of AMF

Molecular Technique	Molecular markers	Specific Primer/s	Name of the target organism	References
PCR	SSU rDNA	*VANS1*	*Glomales*	Simon et al. (1992a, b)
PCR–RAPD	Genomic DNA	OPA-02, OPA-04 OPA-18, P124	*Glomus versiforme, Gl. mosseae* *Gl. caledonium, Acaulospora laevis,*	Wyss and Bonfante (1993)
PCR	SSU rDNA	VANS1 and NS21	*G. intraradices*	Di Bonito et al. (1995)
Competitive PCR	Genomic DNA	PO and M3	*G. mosseae*	Edwards et al. (1997)
PCR–RFLP	ITS	ITS1 and ITS4	*Glomus* sp., *Scutellospora* sp., *Gigaspora* sp.	Redecker et al. (1997)
PCR	SSU 1492′	NS71 and SSU1492′	*Gigaspora* sp.	Bago et al. (1998)
PCR-partial	Partial rDNA	SS38, VANS1 VANS1	Roots and spores of AM, *Scutellospora* and *Glomus*	Clapp et al. (1999)
PCR	ITS1 and ITS2	ITS1 and ITS2	*G. margarita*	Lanfranco et al. (1999)
PCR	ITS	ITS1 and ITS4	*G. mosseae* and *Gigaspora margarita*	Antoniolli et al. (2000)

(continued)

Table 18.4 (continued)

Molecular Technique	Molecular markers	Specific Primer/s	Name of the target organism	References
PCR–RFLP	SSU rDNA	LR1, FLR2	Subgroups of Glomales	Jacquot et al. (2000)
PCR–SSCPs	28S rDNA	LSU-Primers	*Glomus* sp.	Kjφller and Rosendahl (2000)
PCR	SSU rDNA	NS31, AM1	*Glomus* sp.	Daniell et al. (2001)
PCR–SSCP	SSU rDNA	VANS1	Subgroups of Glomales	Redecker (2002)
Nested–PCR		ITS, AM1		
PCR–RFLP	ITS	ITS1, ITS4	*G. mosseae*	Giovannetti et al. (2003)
Nested PCR–SSCP	ITS	Eukaryotic universal primer *Glomus*-specific ITS primer	*Glomus* sp.	Kjφller and Rosendahl (2003)
Nested-PCR	ITS	ITS 5, ITS4	Glomeromycota (except Archaeosporaceae)	Renker et al. (2003)
PCR	ITS	SSU-Glom/ LSU-Glom1 ITS5 and ITS4	Major groups within Glomeromycota	Renker et al. (2003)

References

Abbasi, H., Akhtar, A., & Sharf, R. (2015). Vesicular Arbuscular mycorrhizal (VAM) Fungi: A tool for sustainable Agriculture. *American Journal of Plant Nutrition and Fertilization Technology, 5*, 40–49.

Alban, R., Guerrero, R., & Toro, M. (2013). Interactions between a root knot nematode (Meloidogyne exigua) and Arbuscular Mycorrhizae in Coffee Plant Development (Coffea arabica). *American Journal of Plant Sciences, 4*(7B), 19–23.

Alsamowal, M. M., Hadad, M. A., & Sharif, Z. (2016). Response of sesame (Sesasum indicum L.) to Vesicular Arbuscular Mycorrhiza (VAM) and Mineral phosphorus Additions at different moisture regimes under greenhouse conditions in Sudan. *International Journal of Scientific and Research Publications (IJSRP), 6*(8), 82–83.

Andrade, G., Mihara, K., Linderman, R., & Bethlenfalvay, G. J. (1997). Bacteria from rhizosphere and hyphosphere soils of different arbuscular-mycorrhizal fungi. *Plant and Soil, 192*, 71. https://doi.org/10.1023/A:1004249629643.

Andrade, G., Mihara, K., Linderman, R., & Bethlenfalvay, G. J. (1998). Bacterial associations with the mycorrhizosphere and hyphosphere of the arbuscular mycorrhizal fungus Glomus mosseae. *Plant and Soil, 202*, 79–87.

Andrade, M. M. M., Stamford, N. P., Santos, C. E. R. S., Freitas, A. D. S., Sousa, C. A., & Junior, M. A. L. (2013). Effects of biofertilizer with diazotrophic bacteria and mycorrhizal fungi in soil attribute, cowpea nodulation yield and nutrient uptake in field conditions. *Scientia Horticulturae, 162*, 374–379.

Anjos, E. C. T. D., Cavalcante, U. M. T., Gonçalves, D. M. C., Pedrosa, E. M. R., Santos, V. F. D., & Maia, L. C. (2010). Interactions between an Arbuscular mycorrhizal Fungus (*Scutellospora heterogama*) and the Root-knot nematode (*Meloidogyne incognita*) on Sweet passion fruit (*Passiflora alata*). *Brazilian Archives of Biology and Technology, 53*(4), 801–809.

Antoniolli, Z. I., Schachtman, D. P., Ophel, K. K., & Smith, S. E. (2000). Variation in rDNA its sequences in Glomus mosseae and Gigaspora margarita spores from a permanent pasture. *Mycological Research, 104*, 708–715.

Arabi, M. I. E., Ayoubi, S. K. Z., & Jawhar, M. (2013). Mycorrhizal application as a biocontrol agent against common root rot of barley. *Research in Biotechnology, 4*(4), 07–12.

Arriola, L. L., Hausbeck, M. K., Rogers, J., & Safir, G. R. (2000). The effect of *Trichoderma harzianum* and Arbuscular mycorrhizae on Fusarium root rot in Asparagus. *Hort Technology, 10*(1), 141–144.

Askar, A. A. A., & Rashad, Y. M. (2010). Arbuscular mycorrhizal fungi: A biocontrol agent against common bean Fusarium rot disease. *Plant Pathology Journal, 9*(1), 31–38.

Audet, P., & Charest, C. (2010). Determining the impact of the AM-Mycorrhizosphere on "Dwarf" sunflower Zn uptake and soil-Zn bioavailability. *Journal of Botany*, Article ID 268540, 1–11. https://doi.org/10.1155/2010/268540.

Augé, R. M., Toler, H. D., & Saxton, A. M. (2015). Arbuscular mycorrhizal symbiosis alters stomatal conductance of host plants more under drought than under amply watered conditions: A meta-analysis. *Mycorrhiza, 25*, 13. https://doi.org/10.1007/s00572-014-0585-4.

Azcon-Aguilar, C., & Barea, J. (1996). Arbuscular mycorrhizas and biological control of soil-borne plant pathogens – An overview of the mechanisms involved. *Mycorrhiza, 6*, 457–464. https://doi.org/10.1007/s005720050147.

Azcon-Aguilar, C., & Barea, J. M. (2015). Nutrient cycling in the mycorrhizosphere. *Journal of Soil Science and Plant Nutrition, 15*(2), 372–396.

Bago, B., Bentivenga, S. P., Brenec, V., Dodd, J. C., Piche, Y., & Simon, L. (1998). Molecular analysis of Gigaspora, Glomales, Gigasporaceae. *The New Phytologist, 139*, 581–588.

Bagyaraj, D. J. (2014). Mycorrhizal fungi. *Proceedings of the Indian National Science Academy, 80*(2), 415–428.

Bagyaraj, D. J., Sharma, M. P., & Maiti, D. (2015). Phosphorus nutrition of crops through arbuscular mycorrhizal fungi. *Current Science, 108*(7), 1288–1293.

Balaes, T., & Catalin, T. (2011). Interrelations between the mycorrhizal systems and soil organisms. *Journal of Plant Development, 18*, 55–69.

Banuelos, J., Alarcón, A., Larsen, J., Cruz-Sánchez, S., & Trejo, D. (2014). Interactions between arbuscular mycorrhizal fungi and *Meloidogyne incognita* in the ornamental plant *Impatiens balsamina*. *Journal of Soil Science and Plant Nutrition, 14*(1), 63–74. https://doi.org/10.4067/S0718-95162014005000005.

Barea, J. M., Azcón, R., & Azcón-Aguilar, C. (2002). Mycorrhizosphere interactions to improve plant fitness and soil quality. *Antonie Van Leeuwenhoek, 81*, 343. https://doi.org/10.1023/A:1020588701325.

Bellgard, S. E., & Williams, S. E. (2011). Response of mycorrhizal diversity to current climatic changes. *Diversity, 3*, 8–90.

Berruti, A., Lumini, E., Balestrini, R., & Bianciotto, V. (2016). Arbuscular mycorrhizal fungi as natural biofertilizers: Let's benefit from past successes. *Frontiers in Microbiology, 6*(1559), 1–13. https://doi.org/10.3389/fmicb.2015.01559.

Bethlenfalvay, G. J., Mihara, K., Schreiner, R. P., & McDaniel, H. (1996). Mycorrhizae, biocides and biocontrol. 1. Herbicide – Mycorrhiza interaction in soybean and cocklebur treated with bentazon. *Applied Soil Ecology, 3*, 197–204.

Bharti, N., & Kumar, A. (2016). Response of mycorrhiza on physiological and biochemical parameters of black gram Vigna mungo (l.) hepper. *IJPRBS, 5*(2), 143–157.

Bianciotto, V., & Bonfante, P. (2002). Arbuscular mycorrhizal fungi: A specialised niche for rhizospheric and endocellular bacteria. *Antonie Van Leeuwenhoek, 81*, 365. https://doi.org/10.1023/A:1020544919072.

Bianciotto, V., Perotto, S., Ruiz-Lozano, J. M., & Bonfante, P. (2002). Arbuscular mycorrhizal fungi and soil bacteria: From cellular investigations to biotechnological perspectives. In

S. Gianinazzi, H. Schüepp, J. M. Barea, & K. Haselwandter (Eds.), *Mycorrhizal technology in agriculture*. Basel: Birkhäuser.

Birhane, E., Sterck, F. J., Fetene, M., et al. (2012). Arbuscular mycorrhizal fungi enhance photosynthesis, water use efficiency, and growth of frankincense seedlings under pulsed water availability conditions. *Oecologia, 169*, 895–904.

Brundrett, M. C. (2009). Mycorrhizal associations and other means of nutrition of vascular plants: Understanding the global diversity of host plants by resolving conflicting information and developing reliable means of diagnosis. *Plant and Soil, 320*, 37. https://doi.org/10.1007/s11104-008-9877-9.

Caron, M. (2009). Potential use of mycorrhizae in control of soil-borne diseases. *Canadian Journal of Plant Pathology, 11*(2), 177–179. https://doi.org/10.1080/07060668909501135.

Cekic, O. F., Unyayar, S., & Ortas, I. (2012). Effects of arbuscular mycorrhizal inoculation on biochemical parameters in *Capsicum annuum* grown under long term salt stress. *Turkish Journal of Botany, 36*, 63–72.

Churchland, C., & Grayston, S. J. (2014). Specificity of plant-microbe interactions in the tree mycorrhizosphere biome and consequences for soil C cycling. *Frontiers in Microbiology, 5*(261), 1–20.

Clapp, J. P., Fitter, A. H., & Young, J. P. W. (1999). Ribosomal small subunit sequence variation within spores of an arbuscular mycorrhizal fungus Scutellospora sp. *Molecular Ecology, 8*, 915–921.

Cordier, C., Gianinazzi, S., & Gianinazzi-Pearson, V. (1996). Colonisation patterns of root tissues by *Phytophthora nicotianae var. parasitica* related to reduced disease in mycorrhizal tomato. *Plant and Soil, 185*(2), 223–232.

Croll, D., et al. (2009). Nonself vegetative fusion and genetic exchange in the arbuscular mycorrhizal fungus Glomus intraradices. *New Phytologist, 181*, 924–937.

Crossay, T., Antheaume, C., Redecker, D., Bon, L., Chedri, N., Richert, C., Guentas, L., Cavaloc, Y., & Amir, H. (2017). New method for the identification of arbuscular mycorrhizal fungi by proteomic-based biotyping of spores using MALDI-TOF-MS. *Scientific Reports, 7*, 14306. https://doi.org/10.1038/s41598-017-14487-6.

Daniell, T. J., Husband, R., Fitter, A. H., & Young, J. P. W. (2001). Molecular diversity of arbuscular mycorrhizal fungi colonising arable crops. *FEMS Microbiology Ecology, 36*, 203–209.

Dar, M. H., & Reshi, Z. A. (2017). Vesicular Arbuscular Mycorrhizal (VAM) Fungi- as a major biocontrol agent in modern sustainable agriculture system. *Russian Agricultural Sciences, 43*(2), 138–143.

de la Pena, E., Rodriguez-Echeverria, S., van der Putten, W. H., Freitas, H., & Moens, M. (2006). Mechanism of control of root-feeding nematodes by mycorrhizal fungi in the dune grass *Ammophila arenaria*. *New Phytologist, 169*, 829–840.

Di Bonito, R., Elliott, M. L., & Des Jardin, E. A. (1995). Detection of an arbuscular mycorrhizal fungus in roots of different plant species with the PCR. *Applied and Environmental Microbiology, 61*, 2809–2810.

Duponnois, R., Galiana, A., & Prin, Y. (2008). The Mycorrhizosphere effect: A multitrophic interaction complex improves mycorrhizal symbiosis and plant growth. In Z. A. Siddiqui, M. S. Akhtar, & K. Futai (Eds.), *Mycorrhizae: Sustainable agriculture and forestry*. Dordrecht: Springer.

Edwards, S. G., Fitter, A. H., & Young, J. P. W. (1997). Quantification of an arbuscular mycorrhizal fungus, Glomus mosseae within plant roots by competitive polymerase chain reaction. *Mycological Research, 10*, 1440–1444.

Farahani, A., et al. (2008). Effects of arbuscular mycorrhizal fungi, different levels of phosphorus and drought stress on water use efficiency, relative water content and proline accumulation rate of Coriander (*Coriandrum sativum* L.). *Journal of Medicinal Plant Research, 2*(6), 125–131.

Fitter, A. H., & Garbaye, J. (1994). Interactions between mycorrhizal fungi and other soil organisms. *Plant Soil, 159*, 123–132. https://doi.org/10.1007/BF00000101.

Fracchia, S., Mujica, M., García-Romera, I., et al. (1998). Interactions between *Glomus mosseae* and arbuscular mycorrhizal sporocarp-associated saprophytic fungi. *Plant and Soil, 200*, 131. https://doi.org/10.1023/A:1004349426315.

Frey, P., Frey-Klett, P., Garbaye, J., Berge, O., & Heulin, T. (1997). Metabolic and genotypic fingerprinting of fluorescent pseudomonads associated with the douglas Fir-*Laccaria bicolor* Mycorrhizosphere. *Applied and Environmental Microbiology, 63*(5), 1852–1860.

Frey-Klett, P., Chavatte, M., Clausse, M. L., Courrier, S., Le Roux, C., Raaijmakers, J., Martinotti, M. G., Pierrat, J. C., & Garbaye, J. (2005). Ectomycorrhizal symbiosis affects functional diversity of rhizosphere fluorescent pseudomonads. *New Phytologist, 165*, 317–328.

Gahan, J., & Schmalenberger, A. (2014). The role of bacteria and mycorrhiza in plant sulfur supply. *Frontiers in Plant Science, 5*(723), 1–7.

Gamper, H., Walker, C., & Schüßler, A. (2009). Diversispora celata sp. nov.: Molecular ecology and phylotaxonomy of an inconspicuous arbuscular mycorrhizal fungus. *New Phytologist, 182*, 495–506.

Gardes, M., White, T. J., Fortin, J. A., Bruns, T. D., & Taylor, J. W. (1991). Identification of indigenous and introduced symbiotic fungi in endomycorrhizae by amplification of nuclear and mitochondrial ribosomal DNA. *Canadian Journal of Botany, 69*, 180.

Gautier, M., et al. (2014). Matrix-assisted laser desorption ionization time-of-fight mass spectrometry: Revolutionizing clinical laboratory diagnosis of mould infections. *Clinical Microbiology and Infection, 20*, 1366–1371.

Ghahfarokhy, M. R., Goltapeh, E. M., Purjam, E., Pakdaman, B. S., Modarres Sanavy, S. A. M., & Varma, A. (2011). Potential of mycorrhiza-like fungi and Trichoderma species in biocontrol of take-all disease of wheat under greenhouse condition. *Journal of Agricultural Technology, 7*, 185–195.

Giovannetti, M., Sbrana, C., Strani, P., Agnolucci, M., Rinaudo, V., & Avio, L. (2003). Genetic diversity of isolates of Glomus mosseae from different geographic areas detected by vegetative compatibility testing and biochemical and molecular analysis. *Applied and Environmental Microbiology, 69*, 615–624.

Green, H., Larsen, J., Olsson, P. A., Jensen, D. F., & Jakobsen, I. (1999). Suppression of the biocontrol agent *Trichoderma harzianum* by Mycelium of the Arbuscular mycorrhizal fungus *Glomus intraradices* in Root-free soil. *Applied and Environmental Microbiology, 65*(4), 1428–1434.

Hao, Z., Fayolle, L., van Tuinen, D., Chatagnier, O., Li, X., Gianinazzi, S., & Gianinazzi-Pearson, V. (2012). Local and systemic mycorrhiza-induced protection against the ectoparasitic nematode *Xiphinema index* involves priming of defence gene responses in grapevine. *Journal of Experimental Botany, 63*(10), 3657–3672.

Hartmann, A., Rothballer, M., & Schmid, M. (2007). Lorenz Hiltner, a pioneer in rhizosphere microbial ecology and soil bacteriology research. *Plant and Soil, 312*, 7–14.

Hijri, M., & Sanders, I. R. (2005). Low gene copy number shows that arbuscular mycorrhizal fungi inherit genetically diferent nuclei. *Nature, 433*, 160–163.

Hodge, A. (2014). Chapter two – interactions between Arbuscular mycorrhizal fungi and organic material substrates. In S. Sariaslani & G. M. Gadd (Eds.), *Advances in applied microbiology* (Vol. 89, pp. 47–99). Saint Louis: Academic.

Hodge, A., Campbell, C. D., & Fitter, A. H. (2001). An arbuscular mycorrhizal fungus accelerates decomposition and acquires nitrogen directly from organic material. *Nature*, 297–299. ISSN 1476-4687.

Jacquot, E., Van Tuinen, D., Gianinazzi, S., & Gianinazzi Pearson, V. (2000). Monitoring species of arbuscular mycorrhizal fungi in planta and in soil by nested PCR: Amplification to the study of the impact of sewage sludge. *Plant and Soil, 226*, 179–188.

Jaderlund, L., Arthurson, V., Granhall, U., & Jansson, J. K. (2008). Specific interactions between arbuscular mycorrhizal fungi and plant growth-promoting bacteria: As revealed by different combinations. *FEMS Microbiology Letters, 287*, 174–180.

Jha, S. K., & Kumar, N. (2011). Potential of mycorrhizal fungi in Ecosystem: A Review. *International Journal of Research in Botany, 1*(1), 1–7.

Jung, S. C., Martinez-Medina, A., Lopez-Raez, J. A., & Pozo, M. J. (2012). Mycorrhiza-induced resistance and priming of plant defenses. *Journal of Chemical Ecology, 38*, 651–664. https://doi.org/10.1007/s10886-012-0134-6.

Kahneh, E., Ramezan Pour, H., Haghparast, M., & Shirinfekr, A. (2006). Effects of Arbuscular mycorrhizal fungi and phosphorus supplement on Leaf P, Zn, Cu and Fe concentrations of tea seedlings. *Caspian Journal of Environmental Sciences, 4*(1), 53–58.

Kamal, R., Gusain, Y. S., & Kumar, V. (2014). Interaction and symbiosis of AM fungi, actinomycetes and plant growth promoting Rhizobacteria with plants: Strategies for the improvement of plants health and defense system. *International Journal of Current Microbiology and Applied Sciences, 3*(7), 564–585.

Kjφller, R., & Rosendahl, S. (2000). Detection of arbuscular mycorrhizal fungi (Glomales) in roots by nested PCR and SSCP (single stranded conformation polymorphism). *Plant and Soil, 226*, 189–196.

Kjφller, R., & Rosendahl, S. (2003). Molecular diversity of Glomalean (arbuscular mycorrhizal) fungi determined as distinct Glomus specific DNA sequences from roots of field-grown peas. *Mycological Research, 105*, 1027–1032.

Kohn, L. M. (1992). Developing new characters for fungal systematics: An experimental approach for determining the rank of resolution. *Mycologia, 84*, 139.

Krüger, M., Krüger, C., Walker, C., Stockinger, H., & Schüßler, A. (2012). Phylogenetic reference data for systematics and phylotaxonomy of arbuscular mycorrhizal fungi from phylum to species level. *New Phytologist, 193*, 970–984.

Lanfranco, L., Delpero, M., & Bonfante, P. (1999). Intrasporal variability of ribosomal sequences in the endomycorrhizal fungus Gigaspora margarita. *Molecular Ecology, 8*, 37–45.

Larsen, J., Cornejo, P., & Barea, J. M. (2009). Interactions between the arbuscular mycorrhizal fungus Glomus intraradices and the plant growth promoting rhizobacteria *Paenibacillus polymyxa* and *P. macerans* in the mycorrhizosphere of *Cucumis sativus*. *Soil Biology and Biochemistry, 41*(2), 286–292.

Lax, P., Becerra, A. G., Soteras, F., Cabello, M., & Doucet, M. E. (2011). Effect of the arbuscular mycorrhizal fungus *Glomus intraradices* on the false Root-knot nematode Nacobbus aberrans in tomato plants. *Biology and Fertility of Soils, 47*, 591–597.

Lee, J., Lee, S., & Young, J. P. W. (2008). Improved PCR primers for the detection and identification of arbuscular mycorrhizal fungi. *FEMS Microbiology Ecology, 65*, 339–349.

Lehmann, A., & Rillig, M. (2015). Arbuscular mycorrhizal contribution to copper, manganese and iron nutrient concentrations in crops – A meta-analysis. *Soil Biology and Biochemistry, 81*, 147–158. https://doi.org/10.1016/j.soilbio.2014.11.013.

Leta, A., & Selvaraj, T. (2013). Evaluation of Arbuscular mycorrhizal fungi and *Trichoderma* species for the control of onion white rot (*Sclerotium cepivorum* Berk). *Journal of Plant Pathology & Microbiology, 4*(1), 1–6. https://doi.org/10.4172/2157-7471.1000159.

Li, H., Nishida, I., Shu, H., Yang, G., Yang, Y., Ye, B., & Zheng, C. (2006). Colonization by the arbuscular mycorrhizal fungus *Glomus versiforme* induces a defense response against the Root-knot nematode Meloidogyne incognita in the grapevine (*Vitis amurensis* Rupr.), which includes transcriptional activation of the class III chitinase gene VCH3. *Plant & Cell Physiology, 47*(1), 154–163.

Linderman, R. G. (1988). Mycorrhizal interaction with the rhizosphere microflora: The Mycorrhizosphere Effect. *Phytopathology, 78*, 366–371.

Lioussanne, L. (2013). The role of the arbuscular mycorrhiza-associated rhizobacteria in the biocontrol of soilborne phytopathogens: A review. *Spanish Journal of Agricultural Research, 8*(S1), 51–61.

Mansfeld-Giese, K., Larsen, J., & Bodker, L. (2002). Bacterial populations associated with mycelium of the arbuscular mycorrhizal fungus *Glomus intraradices*. *FEMS Microbiology Ecology, 41*, 133–140.

Marro, N., Lax, P., Cabello, M., Doucet, M. E., & Becerra, A. G. (2014). Use of the arbuscular mycorrhizal fungus Glomus intraradices as biological control agent of the nematode *Nacobbus*

aberrans parasitizing tomato. *Brazilian Archives of Biology and Technology, 57*(5), 668–674. https://doi.org/10.1590/S1516-8913201402200.
Martin, F., et al. (2008). Te long hard road to a completed Glomus intraradices genome. *New Phytologist, 180*, 747–750.
Matsumoto, L. S., Martines, A. M., Avanzi, M. A., et al. (2005). Interactions among functional groups in the cycling of, carbon, nitrogen and phosphorus in the rhizosphere of three successional species of tropical woody trees. *Applied Soil Ecology, 28*(1), 57–65.
Mendes, R., Garbeva, P., & Raaijmakers, J. M. (2013). The rhizosphere microbiome: Significance of plant beneficial, plant pathogenic, and human pathogenic microorganisms. *FEMS Microbiology Reviews, 37*(5), 634–663.
Miransari, M. (2011). Interactions between arbuscular mycorrhizal fungi and soil bacteria. *Applied Microbiology and Biotechnology, 89*, 917. https://doi.org/10.1007/s00253-010-3004-6.
Mirzakhani, M., Ardakani, M. R., Band, A. A., Rad, A. H. S., & Rejali, F. (2009). Effects of dual inoculation of azotobacter and mycorrhiza with nitrogen and phosphorus fertilizer rates on grain yield and some of characteristics of spring safflower. *International Journal of Civil and Environmental Engineering, 1*(1), 39–43.
Mobasser, H. R., & Moradgholi, A. (2012). Mycorrhizal bio-fertilizer applications on yield seed corn varieties in Iran. *Annals of Biological Research, 3*(2), 1109–1116.
Muhsen, T. A. A., Al-Attabi, M. S. Y., & Ali, B. Z. (2015). Effect of arbuscular mycorrhizal fungi as a biocontrol agent and organic matter against fusarium wilt in tomato. *Journal of Genetic and Environmental Resources Conservation, 3*(3), 237–245.
Naghashzadeh, M. (2014). Response of relative water content and cell membrane stability to mycorrhizal biofertilizer in maize. *Electronic Journal of Biology, 10*(3), 68–72.
Nurbaity, A., Sofyan, E. T., & Hamdani, J. S. (2016). Application of Glomus sp. and Pseudomonas diminuta reduce the use of chemical fertilizers in production of potato grown on different soil types. *IOP Conference Series: Earth and Environmental Science, 41*, 012004. https://doi.org/10.1088/1755-1315/41/1/012004.
Odeyemi, I., Afolami, S., & Sosanya, O. (2010). Effect of *Glomus mosseae* (Arbuscular Mycorrhizal Fungus) On host – parasite relationship of *Meloidogyne incognita* (Southern Root-knot nematode) on four improved Cowpea varieties. *Journal of Plant Protection Research, 50*(3), 321–325.
Orona-Castro, F., Lozano-Contreras, M., Tucuch-Cauich, M., Grageda-Cabrera, O., Medina-Mendez, J., Díaz-Franco, A., Ruiz-Sánchez, E., & Soto-Rocha, J. (2013). Response of rice cultivation to biofertilizers in Campeche, Mexico. *Agricultural Sciences, 4*, 715–720. https://doi.org/10.4236/as.2013.412097.
Osillos, P. L., & Nagpal, A. L. (2014). The effects of Arbuscular Mycorrhizal Fungi (AMF) as biofertilizer on the growth, yield and nutrient uptake of tomato (Lycopersicon esculentum Mill). *International Journal of Scientific and Engineering Research, 3*(11), 49–65.
Ozgonen, H., Akgul, D. S., & Erkilic, A. (2010). The effects of arbuscular mycorrhizal fungi on yield and stem rot caused by *Sclerotium rolfsii Sacc.* in peanut. *African Journal of Agricultural Research, 5*(2), 128–132.
Paulilz, T. C., & Lindennan, R. G. (1991). Mycorrhizal inieractions wilh soil organisms. In D. K. Aurora, B. Rai, K. G. Mukerji, & G. Knudsen (Eds.), *Handbook of applied mycology. Vol. 1: Soils and plants* (pp. 77–129). New York: Marcel Dekker.
Pinochet, J., Calvet, C., Camprubí, A., & Fernández, C. (1996). Interactions between migratory endoparasitic nematodes and arbuscular mycorrhizal fungi in perennial crops: A review. *Plant and Soil, 185*(2), 183–190.
Priyadharsini, P., Rojamala, K., Ravi, R. K., Muthuraja, R., Nagaraj, K., & Muthukumar, T. (2016). Mycorrhizosphere: The extended rhizosphere and its significance. In D. Choudhary, A. Varma, & N. Tuteja (Eds.), *Plant-microbe interaction: An approach to sustainable agriculture*. Singapore: Springer.
Rambelli, A. (1973). The rhizosphere of mycorrhizae. In A. C. Marks & T. T. Kozlowski (Eds.), *Ectomycorrhizae: Their ecology and physiology* (pp. 229–249). London: Academic.

Redecker, D. (2002). Molecular identification and phylogeny of arbuscular mycorrhizal fungi. *Plant and Soil, 244*, 67–73.

Redecker, D., Thierfelder, H., Walker, C., & Werner, D. (1997). Restriction analysis of PCR-amplified internal transcribed spacer of ribosomal DNA as a tool for species identification in different genera of the order Glomales. *Applied and Environmental Microbiology, 63*, 1756–1761.

Renker, C., Heinrichs, J., Kaldorf, M., & Buscot, F. (2003). Combining nested PCR and restriction digest of the internal transcribed spacer region to characterize arbuscular mycorrhizal fungi on roots from the field. *Mycorrhiza, 13*, 191–198.

Rigamonte, T. A., Pylro, V. S., & Duarte, G. F. (2010). The role of mycorrhization helper bacteria in the establishment and action of ectomycorrhizae associations. *Brazilian Journal of Microbiology, 41*, 832–840.

Sadhana, B. (2014). Arbuscular mycorrhizal fungi (AMF) as a biofertilizer – A review. *International Journal of Current Microbiology and Applied Sciences, 3*(4), 384–400.

Salih, S. H., Hamd, S. A. M., & Dagash, Y. M. I. (2015). The Effects of rhizobium, mycorrhizal inoculations and Diammonium Phosphate (DAP) on nodulation, growth, and yield of soybean. *Universal Journal of Agricultural Research, 3*(1), 11–14.

Schouteden, N., Waele, D. D., Panis, B., & Vos, C. M. (2015). Arbuscular mycorrhizal fungi for the biocontrol of plant-parasitic nematodes: A review of the mechanisms involved. *Frontiers in Microbiology, 6*(1280), 1–12.

Schreiner, R. P., & Bethlenfalvay, G. J. (1995). Mycorrhizal interactions in sustainable agriculture. *Critical Reviews in Biotechnology, 15*(3/4), 271–285.

Schubler, A., Schwarzott, D., & Walker, C. (2001). A new fungal phylum, the Glomeromycota: Phylogeny and evolution. *Mycological Research, 105*(12), 1413–1421.

Sharma, I. P., & Sharma, A. K. (2015). Root–knot Nematodes (*Meloidogyne incognita*) suppression through Pre-colonized Arbuscular Mycorrhiza (*Glomus intraradices*) in Tomato-PT3. *Science in Agriculture, 12*(1), 52–57.

Sheng, M., Tang, M., Chen, H., et al. (2008). Influence of arbuscular mycorrhizae on photosynthesis and water status of maize plants under salt stress. *Mycorrhiza, 18*, 287. https://doi.org/10.1007/s00572-008-0180-7.

Shreenivasa, K. R., Krishnappa, K., & Ravichandra, N. G. (2007). Interaction effects of Arbuscular mycorrhizal fungus *Glomus fasciculatum* and Root –knot nematode, *Meloidogyne incognita* on growth and phosphorous uptake of tomato. *Karnataka Journal of Agricultural Sciences, 20*(1), 57–61.

Simon, L., Lalonde, M., & Bruns, T. D. (1992a). Specific amplification of 18S fungal ribosomal genes from vesicular-arbuscular endomycorrhizal fungi colonizing roots. *Applied and Environmental Microbiology, 58*, 291–295.

Simon, L., Levesque, R. C., & Lalonde, M. (1992b). Identification of endomycorrhizal fungi colonizing roots by fluorescent single-strand conformation polymorphism–polymerase chain reaction. *Applied and Environmental Microbiology, 59*, 4211–4215.

Singh, S. R., Singh, U., Chaubey, A. K., & Bhat, M. I. (2010). Mycorrhizal fungi For sustainable agriculture-a review. *Agricultural Reviews, 31*(2), 93–104.

Siviero, M. A., Motta, A. M., Lima, D. S., Birolli, R. R., Huh, S. Y., Santinoni, I. A., Murate, L. S., Castro, C. M. A., Miyauchi, M. Y. H., Zangaro, W., Nogueira, M. A., & Andrade, G. (2008). Interaction among N-fixing bacteria and AM fungi in Amazonian legume tree (Schizolobium amazonicum) in field conditions. *Applied Soil Ecology, 39*, 144–152.

Staddon, P., Heinemeyer, A., & Fitter, A. (2002). Mycorrhizas and global environmental change: Research at different scales. *Plant and Soil, 244*(1/2), 253–261.

Subhashini, D. V. (2016). Effect of NPK fertilizers and Co-inoculation with phosphate-solubilizing Arbuscular mycorrhizal fungus and potassium-mobilizing bacteria on growth, yield, nutrient acquisition, and quality of tobacco (*Nicotiana tabacum* L.). *Communications in Soil Science and Plant Analysis, 47*(3), 328–337. https://doi.org/10.1080/00103624.2015.1123724.

Svenningsen, N. B., Watts-Williams, S. J., Joner, E. J., Battini, F., Efthymiou, A., Cruz-Paredes, C., Nybroe, O., & Jakobsen, I. (2018). Suppression of the activity of arbuscular mycorrhizal fungi by the soil microbiota. *The ISME Journal, 12*, 1–12.

Syafruddin, S., Syakur, S., & Arabia, T. (2016). Propagation techniques of mycorrhizal bio-fertilizer with different types of mycorrhiza inoculant and host plant in Entisol Aceh. *International Journal of Agricultural Research, 11*, 69–76.

Tahat, M. M., Kamaruzaman, S., & Othman, R. (2010). Mycorrhizal fungi as a biocontrol agent. *Plant Pathology Journal, 9*(4), 198–207.

Tamayo, E., Gómez-Gallego, T., Azcón-Aguilar, C., & Ferrol, N. (2014). Genome-wide analysis of copper, iron and zinc transporters in the arbuscular mycorrhizal fungus *Rhizophagus irregularis*. *Frontiers in Plant Science, 5*(547), 1–13.

Thamsurakul, S., & Charoensook, S. (2006, October 16–20). *Mycorrhizal fungi as biofertilizer for fruit tree production in Thailand*. Paper presented at international workshop on sustained management of the soil-Rhizosphere system for efficient crop production and fertilizer use.

Tian, C., He, X., Zhong, Y., et al. (2003). Effect of inoculation with ecto- and arbuscular mycorrhizae and Rhizobium on the growth and nitrogen fixation by black locust, *Robinia pseudoacacia*. *New Forests, 25*, 125.

Timonen, S., & Marschner, P. (2006). Mycorrhizosphere concept. In K. G. Mukerji, C. Manoharachary, & J. Singh (Eds.), *Microbial activity in the rhizoshere* (Soil biology) (Vol. 7). Berlin/Heidelberg: Springer.

Tobar, R. M., Azcon-Aguilar, C., Sanjuan, J., & Barea, J. M. (1996). Impact of a genetically modified Rhizobium strain with improved nodulation competitiveness on the early stages of arbuscular mycorrhiza formation. *Applied Soil Ecology, 4*, 15–21.

Tran, A., Alby, K., Kerr, A., Jones, M., & Gilligan, P. H. (2015). Cost savings realized by implementation of routine microbiological identification by matrix-assisted laser desorption ionization–time of flight mass spectrometry. *Journal of Clinical Microbiology, 53*, 2473–2479.

Uroz, S., Calvaruso, C., Turpault, M. P., Pierrat, J. C., Mustin, C., & Frey-Klett, P. (2007). Effect of the mycorrhizosphere on the genotypic and metabolic diversity of the bacterial communities involved in mineral weathering in a forest soil. *Applied and Environmental Microbiology, 73*(9), 3019–3027. https://doi.org/10.1128/AEM.00121-07.

Veerabhadraswamy, A. L., & Garampalli, R. H. (2011). Effect of Arbuscular mycorrhizal fungi in the management of black bundle disease of maize caused by *Cephalosporium acremonium*. *Science Research Reporter, 1*(2), 96–100.

Walker, C., Vestberg, M., Demircik, F., Stockinger, H., Saito, M., Sawaki, H., Nishmura, I., & Schüßler, A. (2007). Molecular phylogeny and new taxa in the Archaeosporales (Glomeromycota): Ambispora fennica gen. sp nov., Ambisporaceae fam. nov., and emendation of Archaeospora and Archaeosporaceae. *Mycological Research, 111*, 137–153.

Waschkies, C., Schropp, A., & Marschner, H. (1994). Relations between grapevine replant disease and root colonization of grapevine (*Vitis* sp.) by fluorescent pseudomonads and endomycorrhizal fungi. *Plant and Soil, 162*(2), 219–227.

White, T. J., Bruns, T., Lee, S., & Taylor, J. (1990). Amplification and direct sequencing of ribosomal RNA genes for phylogenetics. In M. A. Innis, D. H. Gelfand, J. J. Sninsky, & T. J. White (Eds.), *PCR protocol – A guide to methods and applications* (p. 315). New York: Academic.

Wyss, P., & Bonfante, P. (1993). Amplification of genomic DNA of arbuscular mycorrhizal (AM) fungi by PCR using short arbitrary primers. *Mycological Research, 97*, 1351–1357.

Xavier, L. J. C., & Boyetchko, S. M. (2004). Arbuscular mycorrhizal fungi in plant disease control. In D. K. Arora (Ed.), *Fungal biotechnology in agricultural, food, and environmental applications* (pp. 183–194). New York: Dekker.

Zhang, L. D., Zhang, J. L., Christie, P., & Li, X. L. (2008). Pre-inoculation with arbuscular mycorrhizal fungi suppresses root knot nematode (*Meloidogyne incognita*) on cucumber (*Cucumis sativus*). *Biology and Fertility of Soils, 45*(2), 205–211. https://doi.org/10.1007/s00374-008-0329-8.

Zhu, Y.-G., & Miller, R. M. (2003). Carbon cycling by arbuscular mycorrhizal fungi in soil–plant systems. *Trends in Plant Science, 8*(9), 407–409.

Zhu, Y. G., Smith, A. F., & Smith, S. E. (2003). Phosphorus efficiencies and responses of barley (*Hordeum vulgare* L.) to arbuscular mycorrhizal fungi grown in highly calcareous soil. *Mycorrhiza, 13*, 93. https://doi.org/10.1007/s00572-002-0205-6.

Ziedan, E., Elewa, I., Mostafa, M., et al. (2011). Application of mycorrhizae for controlling root diseases of sesame. *Journal of Plant Protection Research, 51*(4), 355–361.

Mycorrhizal Fungi: Potential Candidate for Sustainable Agriculture

19

Monika, S. Devi, S. S. Arya, N. Kumar, and S. Kumar

1 Introduction

The word 'sustainability' is very broad term; it cannot define by single line. Sustainability is combination of all agriculture practices maintain the environment, plant, human and animal health with low input and labor. Microbial inoculants have potential to increase sustainability of agriculture crop production through more efficient utilization of nitrogen and phosphorus. They have potential to increase the production of crop by improving yield and quantity. Such improvements includes production efficiency at given level of inputs and consequently reduce input level to achieve same yield. Reducing input level can help in resolving some core issues of sustainability like eutrophication of water resources caused by excessive use of phosphorus (P) and nitrogen fertilizers and also overcome the depletion of nonrenewable resources such as rock phosphate. To effective in addressing any of these concern microbial inoculant can use as biofertilizer in crop production. Mycorrhizal fungi are one of the suitable microbial inoculant in terms of maintaining sustainability.

Frank (1885) first used the term 'mycorrhiza' to describe the modified root structures of forest trees, and has a mutualistic, symbiotic associations between fungi and plant roots. The word mycorrhiza originated from Greek word mushroom and

Monika
Department of Microbiology, CCS Haryana Agricultural University, Hisar, Haryana, India

S. Devi
Department of Botany and Plant Physiology, CCS HAU, Hisar, Haryana, India

S. S. Arya (✉) · N. Kumar
Department of Botany, Maharshi Dayanand University, Rohtak, Haryana, India

S. Kumar
Department of Environmental Science, Maharshi Dayanand University, Rohtak, Haryana, India

© Springer Nature Singapore Pte Ltd. 2019

A. Varma, D. K. Choudhary (eds.), *Mycorrhizosphere and Pedogenesis*,
https://doi.org/10.1007/978-981-13-6480-8_19

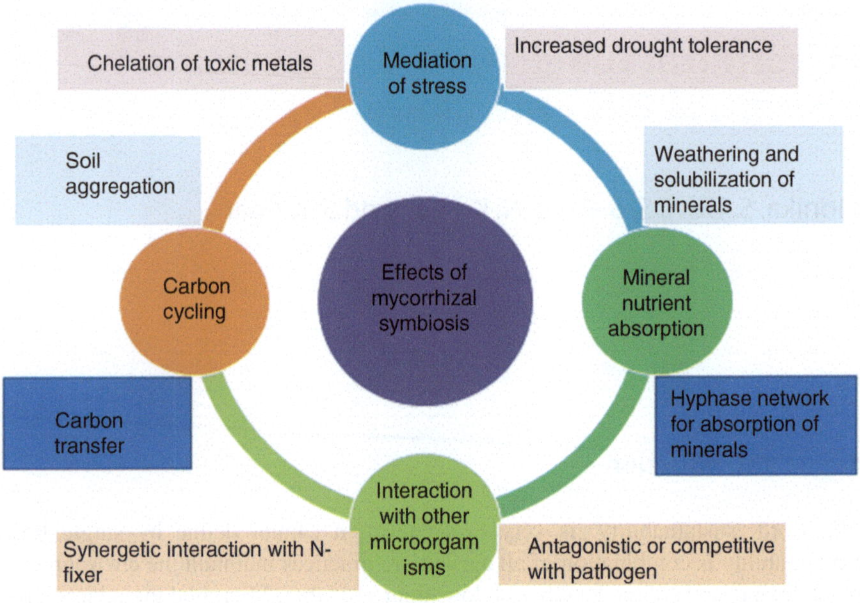

Fig. 19.1 Effects of mycorrhizal fungi on plant and soil health

root. In mycorrhizal-plant association underground fungal mycelium are in contact with plant root but without causing any harm to the plant (Smith and Read 2008). Among small portion of all plant species examined 95% of those plant species predominately by mycorrhizal fungi. In addition to supply of macronutrient like N, P, K, mycorrhiza increase the supply and availability of micronutrient like Zinc (Zn), Iron (Fe), Molybdenum (Mo) and Copper (Cu) to the host plant (Alori et al. 2017). Mycorrhiza plays an important role in disease protection, mitigation of soil stresses and increasing grain production; indirect contributions of AM fungi and soil aggregation to plant growth. Mycorrhiza also makes contribution in the restoration of native ecosystems by symbiotic association with other important soil microflora. Improving the quality of crops by improving nutrient status of crops also improves the sustainability of commercial crop is less tangible but it is equally important because main aim of sustainable agriculture is well being of human population (Table 19.2, Fig. 19.1).

2 Functional Diversity and Classification of Arbuscular Mycorrhizal Fungi

Arbuscular mycorrhiza fungus (AMF) is the most ancient and widespread form, therefore AMF more emphasized. Paleobotanical and molecular sequence data suggest that the first land plants formed associations with Glomalean fungi from the Glomeromycota about 460 million years ago (Redecker et al. 2000) Arbuscular mycorrhizal (AM) symbioses can be formed with a large number (250,000) of plant species. On the basis morphology, only 150–200 species of AM fungi have so far been distinguished, while the molecular (DNA-based) studies revealed the true diversity of these symbionts may be very much higher (Fitter et al. 2000; Santos-Gonzalez et al. 2007). The symbiosis is characterized by highly branched fungal structures, arbuscules, which grow intracellularly without penetrating the host plasmalemma. Today, we accept three classes (Archaeosporomycetes, Glomeromycetes, and Paraglomeromycetes) of mycorrhizal fungi. Mycorrhizal fungi widely distributed but the arbuscular mycorrhiza is most widespread association.

Arbuscular mycorrhiza can classified on the basis of molecular, morphological (spore formation) and biochemical characteristics. In addition, it can classify on the basis of hyphae and mycelial structure (staining process) and genetic features (β-tubulin structure). In this system of classification fungi which constitute arbuscular mycorrhizal comes under phylum Glomeromycota having class Glomeromycetes. Glomeromycetes class consists of five order *i.e.* Glomerales, Diversisporales, Paraglomerales, Archeosporales and Gigasporales which in turn constitute of 14 families. These families consist of 29 genera and 230 species as depicted in Table 19.1.

Arbuscular mycorrhizae are mutualistic association among plant and mycorrhizal fungi that play an central role in plant growth, soil quality and plant protection. The AM fungi spread their filaments in soil and plant roots. This fungal filamentous network promotes bi-directional movement of plant mineral nutrient and water to the plants and plant photosynthetic movement to the fungal hyphae. AM fungi colonize the cortical tissue of plant roots of most plant species and thus increase root surface area. The AM fungi are ubiquitous in nature and can form mutualistic association with most of the terrestrial plants including horticulture plants, most of the

Table 19.1 Arbuscular mycorrhizal classification (Oehl 2011a, b, 2015)

Class	Orders	Families	Genera
Glomeromycetes	Glomerales	Glomeraceae	*Glomus, Septoglomus, Simiglomus, Funnelformies*
	Diversisporales	Acaulosporaceae	*Acaulospora, Kuklospora*
		Entrophosporaceae	*Entrophospora*
		Pacisporaceae	*Pacispora*
		Diversisporaceae	*Diversispora, Redeckera, Otospora, Tricispora*
	Gigasporales	Gigasporaceae	*Gigaspora, Scutellospora*
		Sacculosporaeae	*Sacculospora, Orbispora*
		Dentiscutataceae	*Dentiscutata, Quatunica, Fuscutata*
		Racocetraceae	*Cetrospora, Racocetra*
Aracheosporomycetes	Archeosporales	Geosiphonaceae	*Geosiphon*
		Arthaeosporaceae	*Archaeospora, Inraospora*
		Ambisporaceae	*Ambispora*
Paraglomeromycetes	Paraglomerales	Paraglomeraceae	*Paraglomus*

crops, vegetables and cereals. Cultivation of commercial crop like soyabean which is one of the most promising oil seed crop in reclaimed soil is one of the ways to achieved agriculture sustainability. Enhancement in yield and production of soybean on interaction with mycorrhizal fungi has been linked to improved assimilation of phosphorus.

3 Mycorrhizal Fungi and Sustainable Agriculture

3.1 Acquisition of Mineral Nutrients

Mycorrhizal fungi provide nutrient to the host plant in exchange of photosynthetic material or sugar. Fungal mycelium penetrates deeper and wider into the soil as compared to host plant root. Fungal hyphae network make available inaccessible mineral nutrient to plant due to their thinner network which penetrate into very small pore of soil. Arbuscules acts as a functional site for exchange of mineral nutrients (N, P, K, Fe and Zn etc.) and carbohydrates between host and mycorrhiza (Balestrini et al. 2015). Therefore, AMF can overcome the limitation in plant nutrition due to insufficient nutrient availability (Nouri et al. 2014). Acquisition of mineral nutrient depends on plant identity, the extent to which AMF colonization benefit plant also depends on traits of physiology and morphology of roots. Maximum crop production using AMF association with plant rely on root characteristics of host plant like morphogenesis, physiology, nutrient availability, plant pathogen influence and genetic make up of plant.

Mycorrhizal infection to the host plant causes physiological changes in host plant like number of mitochondria increases, number of nucleus increases, number of plastid increase in size which move towards arbuscule and forms Hertig net like structure over fungus. These physiological changes cause some metabolic changes in host plant. Increased numbers of mitochondria and plastids triggers production of metabolites like fatty acids, amino acids and carotenoids. As these metabolic changes occur, phosphorus is transferred from the mycorrhiza to the host cell in exchange for fatty acids, amino acids, and sugars (Fig. 19.2, French 2017). Since the characterization of high affinity phosphate transporter in mycorrhiza, it is clearly demonstrated that plant uptake phosphorus through mycorrhizal fungi in low nutrient conditions (Smith and Smith 2011). Xie et al. (2013) observed the arbuscular mycorrhizal fungi role in induction and expression of plant transporter genes. It was also reported that there is direct correlation between phosphorus acquisition and arbuscular morphogenesis (Walder et al. 2015). Volpe et al. (2015) demonstrated the role of arbuscular mycorrhiza fungi (AMF) in expression of PT4 genes which is part of pi sensing machinery in root tip of *Medicago truncatula* and *Lotus japonicus*. Their studies revels that AMF inoculation upon early stage inoculation, induced pi transporters named as MtPT4 and LjPT4 found in root tip of *Medicago truncatula* and *Lotus japonicas* respectively. Symanczik et al. (2017) assessed root trap culture of Narnajila using next generation sequencing (NGS) to find out the Pi transporter and it was found that on inoculation of AMF strain in narnajila phosphorus

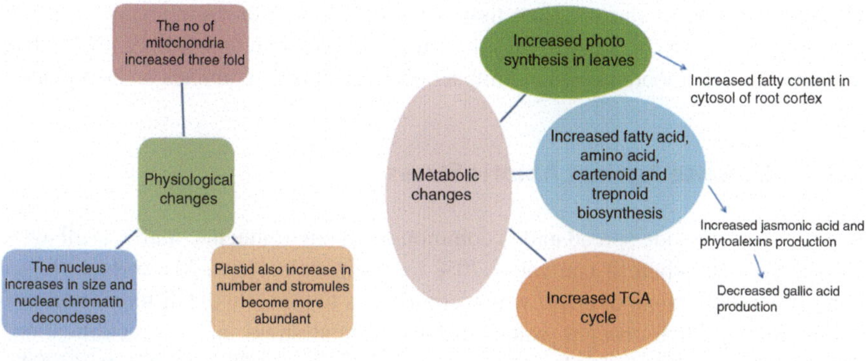

Fig. 19.2 Physiological and metabolic changes during mycorrhizal symbiosis of host plant for mineral nutrient exchange

acquisition increased to 104% as compared to uninoculated control. Presence of *Piriformospora indica* as revealed by NGS confirms the role of mycorrhizal fungi in phosphorus acquisition.

Pi transporters are well known in mycorrhiza, besides pi transporters mycorrhiza inducible transporters have also been recognized (Koegel et al. 2013). Periarbuscular membrane that envelops arbuscule is the main site of for mineral nutrient exchange, plat transporter located in membrane transfer nutrient from periarbuscular apoplast to cortical cells. Breullin-sessoms et al. (2015) suggested role of ammonium transporter in mineral transfer due to their location in periarbuscular membrane. It has been also reported that phosphorous transporter and ammonium transporter not only involved in a mineral nutrient transfer but also maintain arbuscule between plant root and mycorrhizal. Recently, Giovannetti et al. 2014 speculated the role of a sulphate transporter in uptake of mineral nutrient from mycorrhizal arbuscule. Potassium (K^+) play an important role in plant cell machinery, although the involvement of AM symbiosis rarely studied (Garcia and Zimmermann 2014) but the presence of K^+ reported in arbuscule by Olsson et al. (2011) speculated the role of association in accessibility of K^+ during unavailability. Fungal arbsucule maintain balance between nutrients transfer, excessive amount absorbed by arbuscule network, only required quantity is transfer through transporter inside the cortex cell of host plant. Interestingly, it has been proposed that the nutrient dependent regulation of AM colonization provides an important feedback mechanism for plants to promote or limit fungal colonization according to their needs (Nouri et al. 2014). It has already been established that phosphorous availability represents an environmental factor that can disturb the symbiotic interaction of AM. In fact, the suppression of AM colonization due to high Pi levels has been reported in several experiments. Recently metagenomic studies show that role of mycorrhizal symbiosis in micronutrients (Zn, Fe, Cu, Mo and B) transfer and it is depicted from reports mycorrhiza could use as a sustainable tool to improve micronutrient pool in crops (Lehmann et al. 2012; Lehmann and Rillig 2015). Moving attention on Zn nutrition in plant, Lehmann et al. (2012) speculated that mycorrhizal symbiosis positively affected the

Zn concentration in various plant tissues under different environmental condition and time. Focus in Cu, Fe and Mn, the study by Lehmann and Rilling (2015) has demonstrated that there is positive impact of mycorrhizal symbiosis on host plants.

3.2 Defense Against Abiotic Stress

Abiotic stress is wide spread and is common in all environments, it affects all agriculture systems where it can cause 70% losses in field crops. Mycorrhizal fungi comprise a wide range of effect which contributes to mitigate the different types of abiotic stresses like drought, salts, oxidative, metal stress and soil acidification. These have recently been demonstrated by Colpaert (2008) and Finlay (2004). Drought introduced production of excessive reactive oxygen species. AMF provide protection against these reactive oxygen species via production of antioxidant or induction of regulatory gene. Meanwhile water stress or drought causes generation reactive oxygen species (ROS) such as superoxides, singlet oxygen hydrogen peroxides and hydroxyl which either leads to cell death or mutation in functional genes. Several studies reveal that oxidative stress offered by ROS could overcome by activation of regulatory gene by mycorrhizal fungi. It was shown that production of peroxidase, superoxide dismutase and catalase increased in AMF inoculated or associated plant during ROS production. Several scientists reported activation of antioxidants during ROS production in plant. Functional CuZn superoxide dismutase stimulated by AMF provides defense against ROS in inoculated plant. In *P. indica* colonized the roots of Chinese cabbage and promotes root and shoot growth and lateral root formation. When Chinese cabbage plant further exposed to polyethylene glycol (comparable to drought stress), the activities of peroxidase and dismutase in the leaves is increased (Lanfranco et al. 2005). Similarly Sanchez et al. (2016) observed that AMF association overcome drought stress by improving photosynthetic performance. AMF increased accumulation of antioxidant compounds like glutathione which is capable in reduction of oxidative damage caused through ROS generation during drought stress in plant. Further they also reported that level of glutathione is found only in mycorrhizal inoculated plant as compared to non mycorrhizal inoculated. These studied supports that mycorrhizal mediated production antioxidant and regulatory gene against drought stress may be one of the important mechanism through which AM symbiosis offers resistance against drought to the plants. Other studies suggest that effects of mycorrhizal fungi on tolerance of water stress are difficult to study since the supply of poorly diffusible nutrients such as P in dry soil will become limited by the increasing tortuosity of the diffusion path and mycorrhizal hyphae will make an increasingly important contribution to P uptake as the soil dries, confounding the effects of water and nutrients. However, it was observed that increased tolerance of AM plants to drought may possess modification of drought induced gene (Ruiz-Lozano et al. 2006). Down regulation of some gene may play important role in drought tolerance mechanism of AM inoculated plant. Porcel et al. 2006 reported downregulation of gene that encodes membrane aquaporins.

Many recent studies have indicated that AM fungi could enhance the ability of plants to cope with salt stress by improving mineral nutrient absorption, maintaining ion balance, protecting enzyme activities and increasing water use efficiency. Karaki (2017) investigated that pre-inoculation effect of *Glomus mosseae* fungus on green paper under high saline condition. Their reports concluded that AM inoculated plants had root dry matter and greater shoot, fruit yield and plant height than non AM inoculated plants at same salinity level. Shoot concentrations and contents of P and K were higher and Na concentration and content were lower in AM compared with non AM plants at pre-flowering stage. The High nutrient uptake in mineral absorption in AM inoculated plant is due to improved soil extraradical hyphae by mycorrhizal fungi as compared to non-inoculated plant. Improved mineral uptake by AMF is one of the big reason for better yield in AMF inoculated plant (Balliu et al. 2015). Mycorrhizal inoculation reduce the effect of salt stress by lowering Na^+ absorption in root and thus prevents the ion from disturbing in growth metabolic pathways (Ronco et al. 2008).

3.3 Defense Against Biotic Stress

Mycorrhizal symbiosis plays an important role on interaction of plant with pathogens and pest. It was attributed that AMF inoculated plant shows improved disease resistance mechanism than non-inoculated plant due to improved nutrient status. However, it may be contribute, it seems to unlikely to be single reason involved in some cases it was observed that AMF inoculation not necessarily correlated with nutrient uptake. It has also been suggested that AMF colonization induced direct competition for space and resources which leads to reduction in pathogen population nearby host plant. Direct competition for space and other resources one of the leading mechanism which can reduce root pathogen in plant. Highest AMF diversity was reported in healthy roots as compared to *M. incognita* infected roots and galls, and that the composition of the AMF community was different between Infected and uninfected roots. They also demonstrated that galls produced by *M. incognita* in *Prunus persica* roots could be colonized by AMF.AM-colonized plants show increased production of defense compounds such as phenolics-1,3-glucanase and chitinolytic enzymes (Del Maria et al. 2011).

It was shown that AMF inoculation or indigenous AMF induced priming in host plant. Priming stage to plant is alert state which contributes active or faster response against pathogens attack than non-inoculated plant. Thus, priming stage is an important state for plant for better survival and defense against attacking pathogens (Dos-Anjos et al. 2010). Competition for space and nutrients also a limiting factor in AMF colonization which decided the degree of AMF colonization and AMF mediated bio control level. Alban et al. (2013) when co-inoculated native AMF together with *Meloidogyne exigua* in coffee plants observed better biocontrol effect as compared exogenous strain.

The dynamics and mechanisms of AM-induced resistance have been studied. Some signal exchange takes place between host plant and associated AM fungus during the plant-mycorrhizal symbiosis. This signal may bring activation of some defense related gene in host gene. During pathogen attack may be there is upregulation of some defense related gene which provide resistant to host plant. Pieterse et al. (2012) suggest that host plants initially treat AMF as potential invaders, activating defense gene that are later downregulated to allow colonization. Lopez-Raez et al (2010) carried out comparative study on the transcriptional response of tomato colonized by *F. mosseae* and *R. irregularis.* Comparison of transcriptional profiles in tomato suggest the rate of induction of defense related gene and JA biosynthesis gene was more in AMF inoculated tomato plant as compared to control. Different AM fungus induces different defense related gene, differ at species level in all fungus. Li et al. (2007) studied the transcriptional response of *Medicago truncatula* to different AM fungi. It was suggested that a core set of genes (similar for all AMF) activated against different AMF. A core of set gene activated is related to defense mechanisms were induced against to AMF reveals that gene induction is not specific for single AM fungus. A protein named dafens in activated due to antifungal activity was upregulated in AMF inoculated plant. Inoculation of non mycorrhizal colonized plant and different mycorrhizal colonized plant either by *G. mosseae* and *G. intraradices* suggest that on application of these fungi different defense related gene stimulated in host plant and it was found that stronger induction of defence related gene found in plant inoculated with *G. mosseae.* These studies discussed here show differences in regulating different defense regulated gene. Assemblage of different consortium of AM fungus may regulate defense related gene more efficiently than single species (Pozo et al. 2002).

4 Ecological Significance

Soil structure is central point which ensuring plant health, better crop yield and sustainability in agriculture. AMF play an important role in maintains of soils structure and soil health. Extra radical hyphae present in AMF have been suggested to hold soil aggregates and particles together. This can be happen due to proteinaceous secretion of AMF which overcome mineral nutrient leaching and greater reduction of soil erosion. This is further indirect potential benefit of AMF in sustainable agriculture.

Moreover, AMF increase the ability of phytoremediation and environment health. Phytoremediation is removal of toxic pollutant and harmful chemicals or compounds from the soil through the help of plants. AMF due to their developed hyphae allows host plant to explore and mitigate pollutant from larger volume of soil. Besides this, AMF also protect plant from excessive absorption of harmful chemicals, ions and metals. It has been suggested that heavy metals bound to carbonyl group of hemicellulose between fungus and host cells. It has been suggested the plant survival in mine contaminated soil (Table 19.2).

There are many findings suggest that ability of AMF to overcome the adsorption of heavy metals above ground part and ensure the edibility of food crops. Fomina et al. (2005) studied solubilization of toxic metals and metal tolerance by

Table 19.2 Contribution of AMF in sustainable agriculture

Mycorrhizal fungi	Benefits	References
Glomus intraradices+*Funneliformis mosseae*	AM symbiosis can improve the harmful effects of salt stress on the growth of durum wheat plants	Fileccia et al. (2017)
Necrotrophic mycorrhizal species	Augment the nutrient status and secure plant against herbivore (such as aphids) attack	Gilbert and Johnson (2015)
Rhizophagus intraradices	Higher leaf water potential in mycorrhizal plants and mycorrhiza protected the plants against oxidative stress	Meddich et al. (2015)
Glomus etunicatum	Improved vegetative growth, chlorophyll and nutrient level in maize more than non mycorrhizal maize plant	Sadhana (2014)
AMF	Improved nutrient uptake capability of soybean and consequently increased yield	Liu et al. (2012) and Tian et al. (2013)
Acaulospora lacunosa and *Glomus constrictum*	Improved foliar nutrient status of onions (*Allium porrum*)	Hart and Forsythe (2012)
Funneliformis mosseae	Inoculated plants produced more dry matter, heavier seeds and greater seed and oil yields with *F. mosseae*. Despite of reduction in N percentage due to drought, N percentage was higher in inoculated plants compared to control	Gholamhoseini et al. (2013)
Vigna radiate	The fungus had positive impact on N, K, P and protein content of the green grain	Bhat et al. (2010)
Gomus intraradices	Mycoremediate zinc from *Medicago truncatula* environment with the resultant effect of improved growth and development	Hildebrandt et al. (2007)
Glomus mosseae	Improved wheat yield even under more efficient in enhancing crop yield under stress	Daei et al. (2009)
Glomus fascicualtum	Improved stress tolerance in *Capsicum annuum*	Bagyaraj and Sreeramulu (1982)
Glomus coronatum, G. intraradices, Glomus caledonium, G. versiforme	Tolerance to drought, increase shoot biomass, increase plant height and fruit yield	Hu et al. (2013) Marulanda et al. (2003) Ortas et al. (2013)
Glomus spp.	Help in early establishment of host plant *Citrullus lanatus,* provide resistance against nematode *Meloidogyne incognita*	Westphal et al. (2008)
Glomus mosseae	Provide defence against *Meloidogyne incognita* nematode in host plant *Solanum lycopersicum* cv. Marmande	Vos et al.(2012)
Soil indigenous AMF population	Increase shoot biomass, water use efficiency	Watts-Williams et al. (2014)

mycorrhizal fungi. It was reported that metal dissolution by mycorrhizal fungi takes place proton promoted or ligand process and organic acid provides both a source of proton for solubilization and metal chelating anions to complex metal cations. Recently applied AMF have been discovered to increase food crop quality or for fortification of food crops. Use of applied AMF increased the amount of sugar, elements (useful and trace elements) and antioxidants in plant have been reported.

5 Constraints in Use

Major constraint in use of AMF as inoculum in crop plants as AMF are obligate mutualistic partner and therefore cannot grown in culture media in the absence of host plants. Soil from the native colonized AMF population is the main source or form of inocula. The multiplied populations of AMF inocula consist of AMF propagules and mixture of soil and thus it is impossible to established pure culture of fungal inoculum that is free from other harmless microorganisms. Moreover, transportation cost of AMF is enormously high in developing nations because of high weight of soil containing AMF inocula (Ceballos et al. 2013, Rodriguez and Sanders 2015).

Unless so many advantages, soil inocula may harmful for plant health due to possibilities of transferring disease causing microbes and seeds of weeds. Crude inocula could be obtained by propagates AMF with the host plant under standardized conditions that can mediate AMF proliferation, and this is the most frequently utilized kind of inoculum for massive inoculation. In addition, root fragments detached from the inoculated or AMF associated host plant could also serve as inoculum for future use. Large scale production of AMF inoculum is still completely baffling despite recent advanced and seed coating method for large scale production of inoculum (Vosatka et al. 2012, Heijden et al. 2015).

6 Strategies to Improve AMF Inocula

From the last 20 years, researchers focused on improvement of AMF diversity to improve crop health. Benefit of mycorrhiza in sustainable agriculture could be intensified by developing mycorrhizal technology. Mycorrhizal technology term refers to some sets of measures to optimize and intensify mycorrhizal population for gaining sustainability in agriculture production. Some key elements of mycorrhizal technology listed here: monitoring, management, database tools, plant breeding and mycorrhizal engineering. Monitoring represents assessment of mycorrhizal diversity in soil. Monitoring of abundance and diversity of AMF overcome deleterious measures used at farm level. Management constitute of certain set of measures in agriculture, agronomic practices which influence and maintain mycorrhizal diversity in soil. Database acts as a store house of information for abundance and diversity of mycorrhizal population. Mycorrhizal engineering is an approach to maintain and produce AMF inoculum of desirable traits. A number of steps have taken to create optimized AMF inoculum for plant health improvement. AMF inoculum generally consists of fungal rich soil that can be transferred from one location to other, one plant to other plant species as AMF readily form symbiosis. Researchers established novel strain of AMF from undistributed sagebrush grasslands increased the phosphorus content of crop planted on coal pits in Wyoming (Stahl et al. 1988). In future we can developed inoculum of AMF have more beneficial traits for example some hyphae like *Acaulospora laevis* (10.55), *Glomus calospora* (12.3) and *Glomus tanue* (14.2) have 10× long hyphae than usual mycorrhizal. Longer hyphae have benefit over small hyphae that they could increase phosphorus uptake to the host plants due to more developed hyphae network. In addition, it improved plant health by developing better plant resistance due to high phytoalexins production. Moreover, metagenomics research on mycorrhizal population of extreme environments claims that there are many other AMF species which could be propagated in other harsh environmental condition and provide abiotic resistance to the host plant. For example, *Diversispora omaniana*, *Septoglomus nakheelum*, and *Rhizophagus arabicus* spp. nov. three new species were identified from Arabian desert and could be propagated as inoculum in arid region (Symanczik et al. 2014). Till date there is negligible research on use of mycorrhizal in synthetic biology, either to improve plant health or to perform better and advanced biological function. Synthetic biology can increase expression of host gene by altering transcription rates or inserting new genes from other species (Khalil and Collins 2010). As mycorrhizal do not sexually reproduce, there is lesser chances of genetic alteration in native gene pool of AMF (Pawlowska 2005). Therefore, specific traits of some AMF species could be chosen and selected and can be expressed in modified fungal inoculum. Tools and techniques used in synthetic biology have enormous potential in modification of AMF inoculum with better traits like phosphorus uptake, abiotic stress, phytoalexin production in host plants. Moreover, nitrogen fixation in host plant can be improved using synthetic biology by modifying metabolism or by engineering N-fixing bacteria engage in symbiosis with AMF strain (Manchanda and Garg 2007).

7 Conclusions

It is necessity to enhance the productivity of AM fungi contribute in maintaining plant health through absorption of scarce nutrient, soil bioremediation, water absorption and that help plants to resist various biotic and abiotic stress to sustain and enhance worldwide food production and preserving environmental health. AMF get less attention to despite their multipurpose role in plant health maintenance. Fresh insight on mycorrhizal fungi diversity in rhizosphere could lead to detection of new bioinoculant cum improved performance of valuable mycorrhizal fungi on different plant. Developments in metagenomics sequencing could allow us to enhance native AMF diversity cum boosting crop fitness. Technique present in synthetic biology may also lead to new revaluations in how these symbiosis function benefits provided to the host plant. Future research should focus on identifying new mycorrhizal genes that effect plant growth and starts experimenting with genetic alteration of potential benefit. Environmental and ecological disturbance due to climate change may harm the functioning and disturbance of AMF. Method of conservation of AMF should be extended below the soil if we want to ensure the preservation of this resource for generations. Moreover, indigenous AMF was found to be better performing than non- indigenous or commercial AMF isolates. Therefore, it is necessities to encourage farmers for production and maintenance of indigenous AMF inocula. This makes the biofertilization technology more affordable to our poor farmers and acts as milestone towards the progress of sustainable agriculture.

References

Alban, R., Guerrero, R., & Toro, M. (2013). Interactions between a root- knot nematode (*Meloidogyne exigua*) and arbuscular mycorrhizae in coffee plant development (*Coffea arabica*). *American Journal of Plant Sciences, 4*, 19–23.

Alori, E. T., Dare, M. O., & Babalola, O. O. (2017). Microbial inoculants for soil quality and plant health. In *Sustainable agriculture review* (pp. 281–230). Cham: Springer International Publishing.

Bagyaraj, D. J., & Sreeramulu, K. R. (1982). Preinoculation with VA mycorrhiza improves growth and yield of chilli transplanted in the field and saves phosphatic fertilizer. *Plant and Soil, 69*, 375–381.

Balestrini, R., Lumini, E., Borriello, R., & Bianciotto, V. (2015). Plant-soil biota interactions. In E. A. Paul (Ed.), *Soil microbiology, ecology and biochemistry* (pp. 311–338). London: Academic/Elsevier.

Balliu, A., Sallaku, G., & Rewald, B. (2015). AMF inoculation enhances growth and improves the nutrient uptake rates of transplanted, salt-stressed tomato seedlings. *Sustainability, 7*(12), 15967–15981.

Bhat, M., Yadav, S., Ali, T., & Bangroo, S. (2010). Combined effects of *Rhizobium* and vesicular arbuscular fungi on green gram (*Vigna radiata* L.) under temperate conditions. *Indian Journal of Ecology, 37*, 157–161.

Breuillin Sessoms, F., Floss, D. S., Gomez, S. K., Pumplin, N., Ding, Y., & Levesque Tremblay, V. (2015). Suppression of arbuscule degeneration in *Medicago truncatula* phosphate transporter 4 mutants is dependent on the ammonium transporter 2 family protein AMT2. *Plant Cell, 27*, 1352–1366. https://doi.org/10.1105/tpc.114.131144.

Ceballos, I., Ruiz, M., Fernández, C., Peria, R., Rodriguez, A., & Sanders, I. R. (2013). The in vitro mass-produced model mycorrhizal fungus, Rhizophagus irregularis, significantly increases yields of the globally important food security crop cassava. *PLoS One, 8*, e70633.

Colpaert, J. V. (2008). Heavy metal pollution and genetic adaptations in ectomycorrhizal fungi. In S. Avery, M. Stratford, & P. van West (Eds.), *Stress in yeasts and filamentous fungi* (pp. 157–173). Amsterdam: Elsevier.

Daei, G., Ardekani, M., Rejali, F., Teimuri, S., & Miransari, M. (2009). Alleviation of salinity stress on wheat yield, yield components, and nutrient uptake using arbuscular mycorrhizal fungi under field conditions. *Journal of Plant Physiology, 166*, 617–625.

Del Maria Alguacil, M., Torrecillas, E., Lozano, Z., & Roldan, A. (2011). Evidence of differences between the communities of arbuscular mycorrhizal fungi colonizing galls and roots of *Prunus persica* infected by the root-knot nematode *Meloidogyne incognita*. *Applied and Environmental Microbiology, 77*, 8656–8661. https://doi.org/10.1128/AEM.05577-11.

DosAnjos, É. C. T., Cavalcante, U. M. T., Gonçalves, D. M. C., Pedrosa, E. M. R., dosSantos, V. F., & Maia, L. C. (2010). Interactions between an arbuscular mycorrhizal fungus (*Scutellospora heterogama*) and the root-knot nematode (*Meloidogyne incognita*) on sweet passion fruit

(*Passiflora alata*). *Brazilian Archives of Biology and Technology, 53*, 801–809. https://doi.org/10.1590/S1516-89132010000400008.
Fileccia, V., Ruisi, P., Ingraffia, R., Giambalvo, D., Frenda, A. S., & Martinelli, F. (2017). Arbuscular mycorrhizal symbiosis mitigates the negative effects of salinity on durum wheat. *PLoS One, 12*(9), e0184158. https://doi.org/10.1371/journal.pone.0184158.
Finlay, R. D. (2004). Mycorrhizal fungi and their multifunctional roles. *Mycologist, 18*, 91–96.
Fitter, A. H., Heinemeyer, A., & Staddon, P. L. (2000). The impact of elevated CO2 and global climate change on arbuscular mycorrhizas: Amycocentric approach. *New Phytologist, 147*, 179–187.
Fomina, M. A., Alexander, I. J., Colpaert, J. V., & Gadd, G. M. (2005). Solubilization of toxic metal minerals and metal tolerance of mycorrhizal fungi. *Soil Biology and Biochemistry, 37*, 851–866.
Frank, A. B. (1885). Uber die auf Wurzelsymbiose beruhende Emahrung gewisser Baume durch Uniterirdische Pilze. *Ber.Dent. Bot. Gesell., 3*, 128–145.
French, K. E. (2017). Engineering mycorrhizal symbioses to alter plant metabolism and improve crop health. *Frontiers in Microbiology, 8*, 1–8. https://doi.org/10.3389/fmicb.2017.01403.
Garcia, K., & Zimmermann, S. D. (2014). The role of mycorrhizal associations in plant potassium nutrition. *Frontiers in Plant Science, 5*, 337. https://doi.org/10.3389/fpls.2014.00337.
Gholamhoseini, M., Ghalavand, A., Dolatabadian, A., & Jamshidi, E. (2013). Effects of arbuscular mycorrhizal inoculation on growth, yield, nutrient uptake and irrigation water productivity of sunflowers grown under drought stress. *Agricultural Water Management, 117*, 106–114.
Gilbert, L., & Johnson, D. (2015). Plant-mediated 'apparent effects' between mycorrhiza and insect herbivores. *Current Opinion in Plant Biology, 26*, 100–105.
Giovannetti, M., Tolosano, M., Volpe, V., Kopriva, S., & Bonfante, P. (2014). Identification and functional characterization of a sulfate transporter induced by both sulfur starvation and mycorrhizal formation in *Lotus japonicus*. *The New Phytologist, 204*, 609–619.
Hart, M. M., & Forsythe, J. A. (2012). Using arbuscular mycorrhizal fungi to improve the nutrient quality of crops; nutritional benefits in addition to phosphorus. *Scientia Horticulturae, 148*, 206–214.
Heijden, M. G., Martin, F. M., Selosse, M. A., & Sanders, I. R. (2015). Mycorrhizal ecology and evolution: The past, the present, and the future. *The New Phytologist, 205*, 1406–1423.
Hildebrandt, U., Regvar, M., & Bothe, H. (2007). Arbuscular mycorrhiza and heavy metal tolerance. *Phytochemistry, 68*, 139–146.
Hu, J., Chan, P. T., Wu, F., Wu, S., Zang, J., Lin, X., & Wong, M. H. (2013). Arbuscular mycorrhizal fungi induce differential Cd and P acquisation by Alfred stonecrop (*Sedum alfredii* Hance) and upland kangkong (*Ipomoea aquactica* Forsk.) in an intercropping system. *Applied Soil Ecology, 63*, 29–35.
Karaki, G. N. (2017). Effects of mycorrhizal fungi inoculation on green pepper yield and mineral uptake under irrigation with saline water. *Advances in Plants & Agriculture Research, 6*(5), 00231. https://doi.org/10.15406/apar.2017.06.00231.
Khalil, A. S., & Collins, J. J. (2010). Synthetic biology: Applications come of age. *Nature Reviews. Genetics, 11*, 367–379.
Koegel, S., AitLahmidi, N., Arnould, C., Chatagnier, O., Walder, F., & Ineichen, K. (2013). The family of ammonium transporters (AMT) in Sorghum bicolor: Two AMT members are induced locally, but not systemically in roots colonized by arbuscular mycorrhizal fungi. *The New Phytologist, 198*, 853–865.
Lanfranco, L., Novero, M., & Bonfante, P. (2005). The mycorrhizal fungus Gigaspora margarita possesses a CuZn superoxide dismutase that is up-regulated during symbiosis with legume-hosts. *Plant Physiology, 137*, 1319–1330.
Lehmann, A., & Rillig, M. C. (2015). Arbuscular mycorrhizal contribution to copper, manganese and iron nutrient concentrations in crops – A metaanalysis. *Soil Biology and Biochemistry, 81*, 147–158.
Lehmann, R. M., Taheri, W. I., Osborne, S. L., Buyer, J. S., & Douds, D. D., Jr. (2012). Fall cover cropping can increase arbuscular mycorrhizal in soils supporting intensive agricultural production. *Agriculture, Ecosystems & Environment, Applied Soil Ecology, 61*, 300–304.

Li, J., Maldonado-Mendoza, I., Lopez-Meyer, M., Cheung, F., Town, C. D., & Harrison, M. J. (2007). Arbuscular mycorrhizal symbiosis is accompanied by local and systemic alterations in gene expression and an increase in disease resistance in the shoots. *The Plant Journal, 50*, 529–544.

Liu, X., Li, Y., Han, B., Zhang, Q., Zhou, K., Zhang, X., & Hashemi, M. (2012). Yield response of continuous soybean to one-season crop disturbance in a previous continuous soybean field in Northeast China. *Field Crops Research, 138*, 52–56.

Lopez-Raez, J. A., Verhage, A., Fernandez, I., Garcia, J. M., Azcon-Aguilar, C., & Flors, V. (2010). Hormonal and transcriptional profiles highlight common and differential host responses to arbuscular mycorrhizal fungi and the regulation of the oxylipin pathway. *Journal of Experimental Botany, 61*, 2589–2601.

Manchanda, G., & Garg, N. (2007). Endomycorrhizal and rhizobial symbiosis: How much do they share? *Journal of Plant Interactions, 2*, 79–88.

Marulanda, A., Azcón, R., & Ruiz-Lozano, J. M. (2003). Contribution of six arbuscular mycorrhizal fungal isolates to water uptake by Lactuca sativa L. plants under drought stress. *Physiologia Plantarum, 119*, 526–533.

Meddich, A., Jaiti, F., Bourzik, W., Asli, A. E., & Hafidi, M. (2015). Use of mycorrhizal fungi as a strategy for improving the drought tolerance in date palm (Phoenix dactylifera). *Scientia Horticulturae, 192*, 468–471.

Nouri, E., Breuillin-Sessoms, F., Feller, U., & Reinhardt, D. (2014). Phosphorus and nitrogen regulate arbuscular mycorrhizal symbiosis in Petunia hybrida. *PLoS One, 9*, e90841. https://doi.org/10.1371/journal.pone.0090841.

Oehl, F., Da Silva, G. A., Goto, B. T., Maia, L. C., & Sieverding, E. (2011a). Glomeromycota: Two new classes and a new order. *Mycotaxon, 116*, 365–379.

Oehl, F., Sieverding, E., Palenzuela, J., Ineichen, K., & da Silva, G. A. (2011b). Advances in Glomeromycota taxonomy and classification. *IMA Fungus, 2*(2), 191–199.

Oehl, F., Sánchez-Castro, I., Palenzuela, J., & Silva, G. A. (2015). Palaeospora spainii, a new arbuscular mycorrhizal fungus from Swiss agricultural soils. *Nova Hedwigia, 101*, 89–102.

Olsson, P. A., Hammer, E. C., Pallon, J., van Aarle, I. M., & Wallander, H. (2011). Elemental composition in vesicles of an arbuscular mycorrhizal fungus, as revealed by PIXE analysis. *Fungal Biology, 115*, 643–648. https://doi.org/10.1016/j.funbio.2011.03.008.

Ortas, I., Sari, N., Akpinar, C., & Yetisir, H. (2013). Selection of arbuscular mycorrhizal fungi species for tomato seedling growth, mycorrhizal dependency and nutrient uptake. *European Journal of Horticultural Science, 78*, 209–218.

Pawlowska, T. E. (2005). Genetic processes in arbuscular mycorrhizal fungi. *FEMS Microbiology Letters, 251*, 185–192. https://doi.org/10.1016/j.femsle.2005.08.007.

Pieterse, C. M. J., Van der Does, D., Zamioudis, C., Leon-Reyes, A., & Van Wees, S. C. M. (2012). Hormonal modulation of plant immunity. *Annual Review of Cell and Developmental Biology, 28*, 489–521.

Porcel, R., Aroca, R., Azcon, R., & Ruiz-Lozano, J. M. (2006). PIP aquaporin gene expression in arbuscular mycorrhizal Glycine max and Lactuca sativa plants in relation to drought stress tolerance. *Plant Molecular Biology, 60*, 389–404.

Pozo, M. J., Cardier, C., Dumas-Gaudot, E., Gianinazzi, S., Barea, J. M., & Azcan-Aguilar, C. (2002). Localized vs systemic effect of arbuscular mycorrhizal fungi on defence responses to Phytophthora infection in tomato plants. *Journal of Experimental Botany, 53*, 525–534.

Redecker, D., Kodner, R., & Graham, L. E. (2000). Glomalean fungi from the Ordovician. *Science, 289*, 1920–1921.

Rodriguez, A., & Sanders, I. R. (2015). The role of community and population ecology in applying mycorrhizal fungi for improved food security. *The ISME Journal, 9*, 1053–1061.

Ronco, M., Ruscitti, M., Arango, M., & Beltrano, J. (2008). Glyphosate and micorrización induce changes in plant growth and in root morphology and architecture in pepper plants (Capsicum annum L.). *Journal of Horticultural Science and Biotechnology, 83*, 497–505.

Ruiz-Lozano, J. M., Porcel, R., & Aroca, R. (2006). Does the enhanced tolerance of arbuscular mycorrhizal plants to water deficit involve modulation of drought. Induced plant genes? *The New Phytologist, 171*, 693–698.

Sadhana, B. (2014). Arbuscular mycorrhizal fungi (AMF) as a biofertilizer. A review. *International Journal of Current Microbiology and Applied Sciences, 3*, 384–400.

Sanchez-Romera, B., Ruiz-Lozano, J. M., Zamarreño, Á. M., García-Mina, J. M., & Aroca, R. (2016). Arbuscular mycorrhizal symbiosis and methyl jasmonate avoid the inhibition of root hydraulic conductivity caused by drought. *Mycorrhiza, 26*, 111–122. https://doi.org/10.1007/s00572-015-0650-7.

Santos-Gonzalez, J. C., Finlay, R. D., & Tehler, A. (2007). Seasonal dynamics of arbuscular mycorrhizal root colonization in a semi natural grassland. *Applied and Environmental Microbiology, 73*, 5613–5623.

Smith, S. E., & Read, D. J. (2008). *Mycorrhizal symbiosis* (3rd ed.). London: Academic.

Smith, S. E., & Smith, F. A. (2011). Roles of arbuscular mycorrhiza in plant nutrition and growth: New paradigms from cellular to ecosystem scales. *Annual Review of Plant Biology, 62*, 227–250.

Stahl, P. D., Williams, S. E., & Christensen, M. (1988). Efficacy of native vesicular arbuscular mycorrhizal fungi after severe soil disturbance. *The New Phytologist, 110*, 347–354. https://doi.org/10.1111/j.1469-8137.

Symanczik, S., Błaszkowski, J., Chwat, G., Boller, T., Wiemken, A., & Al-Yahyaei, M. N. (2014). Three new species of arbuscular mycorrhizal fungi discovered at one location in a desert of Oman: Diversispora omaniana, Septoglomus nakheelum and Rhizophagus arabicus. *Mycologia, 106*, 243–259.

Symanczik, G., Thonar, S., Van der Heijden, Kahmen, Boller, & Mäder. (2017). Application of mycorrhiza and soil from a permaculture system improved phosphorus acquisition in Naranjilla. *Frontiers in Plant Science, 8*, 1–12.

Tian, H., Drijber, R., Zhang, J., & Li, X. (2013). Impact of long-term nitrogen fertilization and rotation with soybean on the diversity and phosphorus metabolism of indigenous arbuscular mycorrhizal fungi within the roots of maize (*Zea mays* L.). *Agriculture, Ecosystems and Environment, 164*, 53–61.

Volpe, V., Giovannetti, M., Sun, X. G., Fiorilli, V., & Bonfante, P. (2015). The phosphate transporters LjPT4 and MtPT4 mediate early root responses to phosphate status in non mycorrhizal roots. *Plant, Cell & Environment, 39*, 660–670.

Vos, C., Claerhout, S., Mkandawire, R., Panis, B., Waele, D. D., & Elsen, A. (2012). Arbuscular mycorrhizal fungi reduce root-knot nematode penetration through altered root exudation of their host. *Plant and Soil, 354*, 335–345.

Vosátka, M., Látr, A., Gianinazzi, S., & Albrechtová, J. (2012). Development of arbuscular mycorrhizal biotechnology and industry: Current achievement sand bottlenecks. *Symbiosis, 58*, 29–37.

Walder, F., Brulé, D., Koegel, S., Wiemken, A., Boller, T., & Courty, P. E. (2015). Plant phosphorus acquisition in a common mycorrhizal network: Regulation of phosphate transporter genes of the Pht1 family in sorghum and flax. *The New Phytologist, 205*, 1632–1645.

Watts-Williams, S. J., Turney, T. W., Patti, A. F., & Cavagnaro, T. R. (2014). Uptake of zinc and phosphorus by plants is affected by zinc fertiliser material and arbuscular mycorrhizas. *Plant and Soil, 376*, 165–175.

Westphal, A., Snyder, N. L., Xing, L., & Camberato, J. J. (2008). Effects of inoculations with mycorrhizal fungi of soilless potting mixes during transplant production on watermelon growth and early fruit yield. *Horticultural Science, 43*, 354–360.

Xie, X., Huang, W., Liu, F., Tang, N., Liu, Y., & Lin, H. (2013). Functional analysis of the novel mycorrhiza-specific phosphate transporter AsPT1 and PHT1 family from Astragalus sinicus during the arbuscular mycorrhizal symbiosis. *The New Phytologist, 198*, 836–852.

Soil Metagenomics: Unculturable Microbial Diversity and Its Function

20

Deep Chandra Suyal, Divya Joshi, Prasen Debbarma, Ravindra Soni, Biplab Das, and Reeta Goel

1 Introduction

Soil is a dynamic and most vibrant ecosystem in nature. It definitely serves as a vast reservoir for both known and unknown microorganisms inhabiting in different niche within the particular soil ecosystem. Each soil fraction i.e. sand, silt, clay and organic matter offer limitless diverse microhabitats for different soil inhabiting microbes (Rui et al. 2017). The reason behind such rich habitat is due to the soil environment which is so variable ranging from μm to mm scale and harbor distinct microbial communities. However, the surroundings of distinct niche may vary with microbial diversity, their abundances, biotic and abiotic features. Therefore, distinct microhabitat is dwelled by those microorganisms which showed the ability to adapt the environment and established the colony to the particular niche.

The most important factors that influence the microbial composition in the soil ecosystem may include the properties of soil physico-chemical features, soil pH, moisture and temperature (Fierer et al. 2009; Eccles et al. 2018). However, microorganisms are inhabited in only about 1% of the total available soil surface area regardless of its great biomass present in the soil ecosystem. Furthermore, it is estimated that microbes which are dormant at a particular given time accounts for as far as >95% of the overall biomass pool of microorganisms (Blagodatskaya and Kuzyakov 2013). Therefore, comprehending the structure of indigenous microbial diversity and dynamicity at present scenario stands one of the most challenging tasks in current soil ecology.

D. C. Suyal (✉) · D. Joshi · P. Debbarma · R. Goel
Department of Microbiology, CBSH, G.B. Pant University of Agriculture and Technology, Pantnagar, Uttarakhand, India

R. Soni · B. Das
Department of Agricultural Microbiology, College of Agriculture, Indira Gandhi Krishi Vishwavidyalaya, Raipur, Chhatisgarh, India

© Springer Nature Singapore Pte Ltd. 2019

A. Varma, D. K. Choudhary (eds.), *Mycorrhizosphere and Pedogenesis*,
https://doi.org/10.1007/978-981-13-6480-8_20

2 Need of Microbial Identification & Characterization

It is now well accepted that the microorganisms have highest level of diversity and abundance on the Earth. Nevertheless, the distribution of the microbial diversity at global scales is still partially understood (Joshi et al. 2017). As discussed earlier, the composition structure and microbial diversity are greatly influenced by the environmental factors. Thus, indexing, cataloging and documentation of the microorganisms are prerequisite for their exploration. Microbial diversity in any habitat is more related to the huge number of species existed at a given time. For any species with a similar number, the diversity of the community is more which are evenly distributed than the unevenly distributed community (Eccles et al. 2018).

As the soil microbial community plays crucial roles in soil health management, agro-ecosystem, biogeochemical cycling, availability of plant nutrition and turnover processes of organic matter in soil, they are tremendously affected by both natural and anthropogenic activities. For instances, numerous microbes which are beneficial to the ecosystem services are presently threatened due to poor agricultural practices, climate changing pattern, soil and land degradation, etc.. In recent years, the use synthetic fertilizers, herbicides, fungicides and various other pesticides become so prevalent that the soil microflora and their diversity are facing unprecedented alterations in their diversity. Therefore, characterization and identification of the microbes with the changing environment perhaps will give the bigger picture on how the microbes are shifting their functional characteristics and thriving in the threatened ecosystem.

3 Soil Metagenomics

Conventionally, the soil microbes were identified and characterized by different isolation techniques and culturing the microbes on various growth media and conditions. The prime setback of these culture based techniques is that the diversity of the microbial community would not be fully comprehended the whole scenario in any given environmental sample (Lim et al. 2018). However, the introduction of the culture independent approaches removes the hurdles and barriers to study the environmental samples and thus, the field of metagenomics came into existence to understand the community structure and function of microbes. Metagenomics initially targeted the shotgun sequencing, but now days it is equally useful for the studies related to marker genes *viz* 16S rRNA by employing NGS technologies (Goel et al. 2018). Therefore, primary target for the metagenomics are: (a) marker genes (b) metabolic behavior and (c) novel enzymes.

Culture independent techniques are involved on the direct extraction of soil DNA and later, investigate the molecular chronometer i.e. the genes encoding rRNA. The exploration of next generation sequencing and analysis have achieved in revealing the undiscovered microbial structure in various soil ecosystem without the requirement for cultivation (Lim et al. 2018). With this advancement, the microorganisms present at unconfined spatial range of biogeographic area could be describe and documented. Thorough study of the soil metagenome provided the functional characterization of soil microorganisms related to the genes active in nutrient cycling (Howe et al. 2016). However, efforts are being made to explore the predictions of gene functions in terms of their actual role *in situ*, especially in soil, where metagenomes can be trapped within biofilms (Goel et al. 2018).

4 Essential Steps of Metagenomics

Metagenomics has made significant advances in microbial diversity, ecology and evolution. It was originally initiated with the aim of DNA cloning and their functional screening (Lam et al. 2018). These primary projects not only proved the principle of the metagenomics, but also unraveled immense gene diversity within the microbial world.

4.1 Sampling and Processing (Fractionation and DNA Extraction)

Sampling is the foremost and crucial step in the metagenomics. The extracted DNA must be of high-quality for metagenomic library construction and sequencing (Lim et al. 2018). Further, fractionation or selective lysis is suitable for those communities which are linked to the host (Tully et al. 2018). Fractionation should be analyzed for sufficient target enrichment with minimal contamination.

4.2 Sequencing Technology

Metagenomic sequencing greatly relies upon the technologies used. Now days, NGS techniques *viz*. Illumina/Solexa systems and 454/Roche sequencing are being continuously used for metagenomic projects.

The 454/Roche system employs emulsion polymerase chain reaction (ePCR) on microscopic beads for amplification. These beads are placed in the wells of a picotitre plate which get pyrosequenced separately. The light emission during pyrosequencing is detected by the CCD detector which converts it into the sequence (Deshpande et al. 2018). In Illumina/Solexa technology, solid surface PCR technique is used for the amplification which produces identical DNA clusters which are ultimately sequenced (Lahens et al. 2017). Applied Biosystems SOLiD and Ion Torrent are the other NGS techniques which are being used now days (Deshpande et al. 2018).

4.3 Assembly

Contigs are important to get the full length sequence. Therefore, assembly of short reads becomes crucial in metagenomics which can be done by co-assembly and *de novo* assembly strategies. Co assembly can be done by using softwares *viz.* AMOS Newbler (Roche) or MIRA (Peterson et al. 2017). On the other hand, *de novo* assembly needs complex computational tools *viz.* MetaVelvet, Meta-IDBA etc. (Hoang et al. 2018).

4.4 Binning

Bining indicates the process of sorting of DNA in several groups of individual genomes. In the first step, compositional binning explores the conserved nucleotide composition of genomes that is also indicated in fragments of the particular genome. In second step, the unknown DNA fragment is searched against a a reference database to bin the sequence. Compositional binning algorithms include TACAO, S-GSOM etc. and similarity-based binning programs include MG-RAST, IMG/M etc. (Nair and Raja 2017). Few binning algorithms use both composition and similarity, like MetaCluster and PhymmBL (Ziels et al. 2018).

4.5 Annotation

If the aim of the study is reconstructed genome and large contigs are produced by assembly, then, IMG or RAST programs can be preferred (Zhu et al. 2018). However, in this approach, minimal length of contigs should be 30,000 bp or longer (Krawczyk et al. 2018). Alternatively, the entire community and be annotated by using unassembled or short reads. Annotation involves two basic steps: feature prediction and, gene function prediction.

In feature prediction sequences labeling is done (Krawczyk et al. 2018) while functional annotation includes mapping with current database. The sequences which cannot be mapped with are named as ORFans (Barrientos-Somarribas et al. 2018). They may create endless genetic novelty in metagenomic samples. Several reference databases can be used to give functional annotation *viz.* TIGRFAM, KEGG, eggNOG, PFAM etc. (Korhonen et al 2016).

4.6 Statistical Analysis

Statistical analysis of the metagenomic data is very important for their exploration. However, it needs proper experimental designs with appropriate replications (Forbes et al. 2018).

4.7 Sharing and Storage of Data

Metagenomic data sharing requires a good computational framework and storage facility. Some of the centralized services include CAMERA, MG-RAST and IMG/M (Nair and Raja 2017). They have standard formats for recording and documenting experimental data.

5 Metagenomics for Sustainable Agricultural Practices

Now days, agriculture needs low-input and higher output technologies to meet the challenges for agricultural sustainability (Joshi et al. 2017). Several metagenomic efforts have been done in the field of agriculture, however, many of them looks repetitive which do not hold any promise to help the marginal farmers. Therefore, productive studies are needed which could double the farmers income and assist the agriculture. Recent advancements suggested that rhizosphere metagenomics emerges as an extraordinary field of investigation due to the role of associated microorganisms in plant growth and development. Moreover, restoration of microbial diversity was found to enhance grain yield and soil health.

Hazardous effect of xenobiotics on soil health and their bio-remediation is being studied in recent years. Metagenomics can predict the community structure and show the effect on microbial groups of associated niche. PhyloChip and GeoChip based studies (Nair and Raja 2017) reveals the technological advancement in the field of metagenomics.

Sustainable agricultural practices at hilly and mountain agro-ecosystems is of particular interest now days. These ecosystems are consists of various microhabitats with great environmental fluctuations and genetic biodiversity. Earlier reports confirmed that mountain agricultural soils have large microbial resource with a great biodegradation potential (Giri et al. 2017a, b; Debbarma et al. 2017) and plant growth promotion activities (Kumar et al. 2014, 2018; Tomer et al. 2017; Suyal et al. 2014a). Moreover, Suyal et al. (2015b) and Soni et al. (2016) revealed the existence of *nif* from the Himalayan high altitude soils (Suyal et al. 2015a, b). Recently, differential proteomic studies of several Himalayan diaztrophs were documented (Soni et al. 2015; Suyal et al. 2014a, b, 2017).

The above mentioned studies are showing the recent metagenomic advancement in agriculture. Surely, other "omics" techniques will assist the metagenomics to achieve agricultural sustainability.

6 Conclusion

The field of metagenomics has been a groundbreaking technology that has made it possible to explore microbial diversity with its full potential. In addition, beyond estimating microbial load, it also helps in getting an idea about the microbes and its habitat. The metagenomic efforts, over recent years, with special reference to extreme habitats, have given a priority to explore untouched native microflora and their products.

Acknowledgements This work is funded by the Science and Engineering Research Board (SERB) Grant No. YSS/2015/001214 to DCS. First author (DCS) acknowledge SERB fellowship under young scientist scheme during the course of this study.

References

Barrientos-Somarribas, M., Messina, D. N., Pou, C., Lysholm, F., Bjerkner, A., Allander, T., & Sonnhammer, E. L. L. (2018). Discovering viral genomes in human metagenomic data by predicting unknown protein families. *Scientific Reports, 8*(1), 1–12. https://doi.org/10.1038/s41598-017-18341-7.

Blagodatskaya, E., & Kuzyakov, Y. (2013). Active microorganisms in soil: Critical review of estimation criteria and approaches. *Soil Biology and Biochemistry, 67*, 192–211.

Debbarma, P., Raghuwanshi, S., Singh, J., Suyal, D. C., Zaidi, M. G. H., & Goel, R. (2017). Comparative in situ biodegradation studies of polyhydroxybutyrate film composites. *3Biotech, 7*, 178. https://doi.org/10.1007/s13205-017-0789-3.

Deshpande, A., Lang, W., McDowell, T., Sivakumar, S., Zhang, J., Wang, J., & Scheet, P. (2018). Strategies for identification of somatic variants using the Ion Torrent deep targeted sequencing platform. *BMC Bioinformatics, 19*(1), 1–10. https://doi.org/10.1186/s12859-017-1991-3.

Eccles, D., Chandler, J., Camberis, M., Henrissat, B., Koren, S., Le Gros, G., & Ewbank, J. J. (2018). De novo assembly of the complex genome of Nippostrongylus brasiliensis using MinION long reads. *BMC Biology, 16*(1), 1–18. https://doi.org/10.1186/s12915-017-0473-4.

Fierer, N., Strickland, M., Liptzin, D., Bradford, M., & Cleveland, C. (2009). Global patterns in belowground communities. *Ecology Letters, 12*, 1238–1249.

Forbes, J. D., Knox, N. C., Peterson, C. L., & Reimer, A. R. (2018). Highlighting clinical metagenomics for enhanced diagnostic decision-making: A step towards wider implementation. *Computational and Structural Biotechnology Journal, 16*, 108–120. https://doi.org/10.1016/j.csbj.2018.02.006.

Giri, K., Rai, J. P. N., Pandey, S., Mishra, G., Kumar, R., & Suyal, D. C. (2017a). Performance evaluation of isoproturon-degrading indigenous bacterial isolates in soil microcosm. *Chemistry and Ecology*. https://doi.org/10.1080/02757540.2017.1393535.

Giri, K., Suyal, D. C., Mishra, G., Pandey, S., Kumar, R., Meena, D. K., & Rai, J. P. N. (2017b). Biodegradation of Isoproturon by *Bacillus pumilus* K1 Isolated from Foothill Agroecosystem of North West Himalaya. *Proceedings of the National Academy of Sciences India Section B Biological Sciences, 87*(3), 839–848. https://doi.org/10.1007/s40011-015-0667-x.

Goel, R., Kumar, V., Suyal, D. C., Narayan, D. B., & Soni, R. (2018). Toward the unculturable microbes for sustainable agricultural production. In V. S. Meena (Ed.), *Role of Rhizospheric microbes in soil*. Singapore: Springer Nature. https://doi.org/10.1007/978-981-10-8402-7_4.

Hoang, N. V., Furtado, A., Thirugnanasambandam, P. P., Botha, F. C., & Henry, R. J. (2018). De novo assembly and characterizing of the culm-derived meta-transcriptome from the polyploid sugarcane genome based on coding transcripts. *Heliyon, 4*(3), e00583. https://doi.org/10.1016/j.heliyon.2018.e00583.

Howe, A., Williams, R. J., Yang, F., Meyer, F., & Hofmockel, K. S. (2016). Characterization of core carbon cycling genes in fertilized prairie soils. *PLoS One, 11*(e0166578), 1–14.

Joshi, D., Kumar, S., Suyal, D. C., & Goel, R. (2017). The microbiome of the Himalayan ecosystem. In V. Kalia, Y. Shouche, H. Purohit, & P. Rahi (Eds.), *Mining of microbial wealth and metagenomics* (pp. 101–116). Singapore: Springer. https://doi.org/10.1007/978-981-10-5708-3_6.

Korhonen, P. K., Young, N. D., & Gasser, R. B. (2016). Making sense of genomes of parasitic worms: Tackling bioinformatic challenges. *Biotechnology Advances, 34*(5), 663–686. https://doi.org/10.1016/j.biotechadv.2016.03.001.

Krawczyk, P. S., Lipinski, L., & Dziembowski, A. (2018). PlasFlow: Predicting plasmid sequences in metagenomic data using genome signatures. *Nucleic Acids Research, 46*(6). https://doi.org/10.1093/nar/gkx1321.

Kumar, S., Suyal, D. C., Dhauni, N., Bhoriyal, M., & Goel, R. (2014). Relative plant growth promoting potential of Himalayan psychrotolerant *Pseudomonas jesenii* Strain MP1 against Native *Cicer arietinum* L., *Vigna mungo* (L.) Hepper; *Vigna radiata* (L.) Wilczek., *Cajanus cajan* (L.) Millsp. and *Eleusine coracana* (L.) Gaertn. *African Journal of Microbiology Research, 8*(50), 3931–3943.

Kumar, S., Suyal, D. C., Bhoriyal, M., & Goel, R. (2018). Plant growth promoting potential of psychrotolerant *Dyadobacter* sp. for pulses and finger millet and impact of inoculation on soil chemical properties and diazotrophic abundance. *Journal of Plant Nutrition, 41*(8). https://doi.org/10.1080/01904167.2018.1433211.

Lahens, N. F., Ricciotti, E., Smirnova, O., Toorens, E., Kim, E. J., Baruzzo, G., & Grant, G. R. (2017). A comparison of Illumina and Ion Torrent sequencing platforms in the context of differential gene expression. *BMC Genomics, 18*(1), 602. https://doi.org/10.1186/s12864-017-4011-0.

Lam, K. N., Martens, E. C., & Charles, T. C. (2018). Developing a bacteroides system for function-based screening of DNA from the human gut microbiome. *MSystems, 3*(3), 1–13. https://doi.org/10.1128/mSystems.00195-17.

Lim, M. Y., Song, E. J., Kim, S. H., Lee, J., Nam, Y., & Do. (2018). Comparison of DNA extraction methods for human gut microbial community profiling. *Systematic and Applied Microbiology, 41*(2), 151–157. https://doi.org/10.1016/j.syapm.2017.11.008.

Nair, G. R., & Raja, S. S. S. (2017). Decoding complex soil microbial communities through new age "Omics". *Journal of Microbial & Biochemical Technology, 9*(6), 301–309. https://doi.org/10.4172/1948-5948.1000382.

Rui, Y., Gleeson, D. B., Murphy, D. V., & Hoyle, F. C. (2017). Response of microbial biomass and CO_2-C loss to wetting patterns are temperature dependent in a semi-arid soil. *Scientific Reports, 7*(1), 1–12. https://doi.org/10.1038/s41598-017-13094-9.

Soni, R., Suyal, D. C., Agrawal, K., Yadav, A., Souche, Y., & Goel, R. (2015). Differential proteomic analysis of Himalayan psychrotolerant diazotroph *Pseudomonas palleroniana* N26 strain under low temperature diazotrophic conditions. *CryoLetters, 36*(2), 74–82.

Soni, R., Suyal, D. C., Sai, S., & Goel, R. (2016). Exploration of *nif*H gene through soil metagenomes of the western Indian Himalayas. *3 Biotech, 6*(1), 1–4.

Suyal, D. C., Shukla, A., & Goel, R. (2014a). Growth promotory potential of the psychrophilic diazotroph *Pseudmonas migulae* S10724 against Native *Vigna radiata* (L.) Wilczek. *3Biotech, 4*, 665–668.

Suyal, D. C., Yadav, A., Shouche, Y., & Goel, R. (2014b). Differential proteomics in response to low temperature diazotrophy of Himalayan psychrophilic nitrogen fixing *Pseudomonas migulae* S10724 strain. *Current Microbiology, 68*, 543–550.

Suyal, D. C., Yadav, A., Shouche, Y., & Goel, R. (2015a). Bacterial diversity and community structure of Western Indian Himalayan red kidney bean (*Phaseolus vulgaris* L.) rhizosphere as revealed by 16S rRNA gene sequences. *Biologia, 70/3*, 305–313.

Suyal, D. C., Yadav, A., Shouche, Y., & Goel, R. (2015b). Diversified diazotrophs associated with the rhizosphere of Western Indian Himalayan native red kidney beans (*Phaseolus vulgaris* L.). *3Biotech, 5*, 433–441.

Suyal, D. C., Kumar, S., Yadav, A., Shouche, Y., & Goel, R. (2017). Cold stress and nitrogen deficiency affected protein expression of psychrotrophic *Dyadobacter psychrophilus* B2 and *Pseudomonas jessenii* MP1. *Frontiers in Microbiology, 8*(430), 1–6. https://doi.org/10.3389/fmicb.2017.00430.

Tomer, S., Suyal, D. C., Rajwar, J., Yadav, A., Shouche, Y., & Goel, R. (2017). Isolation and characterization of Phosphate solubilizing bacteria from Western Indian Himalayan soils. *3Biotech, 7*(2), 95. https://doi.org/10.1007/s13205-017-0738-1.

Tully, B. J., Graham, E. D., & Heidelberg, J. F. (2018). The reconstruction of 2,631 draft metagenome-assembled genomes from the global oceans. *Scientific Data, 5*, 1–8. https://doi.org/10.1038/sdata.2017.203.

Zhu, C., Miller, M., Marpaka, S., Vaysberg, P., Rühlemann, M. C., Wu, G., & Bromberg, Y. (2018). Functional sequencing read annotation for high precision microbiome analysis. *Nucleic Acids Research, 46*(4). https://doi.org/10.1093/nar/gkx1209.

Ziels, R. M., Sousa, D. Z., Stensel, H. D., & Beck, D. A. C. (2018). DNA-SIP based genome-centric metagenomics identifies key long-chain fatty acid-degrading populations in anaerobic digesters with different feeding frequencies. *ISME Journal, 12*(1), 112–123. https://doi.org/10.1038/ismej.2017.143.

Microbial Interactions in Soil Formation and Nutrient Cycling

21

R. Koshila Ravi, S. Anusuya, M. Balachandar, and T. Muthukumar

1 Introduction

The soil is the outer layer covering the earth surface called pedosphere. Soil serves as a medium for plant growth, alters the earth's atmosphere through the liberation of the volatile substances, stores, supply and purifies water and it is a niche for organisms that in turn alter the soil. Soil formation involves physical, chemical and biological processes through which the parent rock materials are weathered or broken down into smaller particles. The broken rock materials combine with the organic matters produced by living organisms. Thus soil consists of organic and inorganic compounds, mineral particles and weathered pieces of rocks (Dominati et al. 2010). The major factors influencing the soil formation include the nature of the parent material, living organisms, climate, topography and time (Paul and Clark 1996). The interactions between these factors result in soil formation.

All the components of soil biota live and function in their habitat comprised of soil pore network, physiochemical constituents of components, biotic communities and environmental factors such as temperature and moisture. Biota in soil structure plays a pivotal role in soil function emphasizing the interaction between organisms and the physical construction of their environment called soil architecture. A primary physiochemical factor governing soil community structure is the individuals with different pH optima (Fierer and Jackson 2006). Soil moisture optima for the belowground biota vary between organism types as they rely on moisture films for their transport through the soil matrix while for mycorrhiza, it is not constrained (Augé 2004).

Bioweathering is an important process involving the dissolution or break down of rocks and minerals by soil microorganisms and plants via physical and chemical

R. Koshila Ravi · S. Anusuya · M. Balachandar · T. Muthukumar (✉)
Root and Soil Biology Laboratory, Department of Botany, Bharathiar University, Coimbatore, Tamil Nadu, India

© Springer Nature Singapore Pte Ltd. 2019

A. Varma, D. K. Choudhary (eds.), *Mycorrhizosphere and Pedogenesis*,
https://doi.org/10.1007/978-981-13-6480-8_21

mechanisms (Gadd 2007). This process contributes to soil formation and enhances the plant growth in different habitats and climate (Gulati et al. 2008; Mapelli et al. 2011). The soil stability and their formation are directly related with clay mineralogy and dissolution process, the occurrence of binding sources such as root exudates and fungal hyphae (Denef and Six 2005; Rillig and Mummey 2006). Soil microorganisms play a critical role in the formation of soil as they are involved in the biological transformations and develop most of the stable nutrients pools like carbon (C), nitrogen (N) and other vital nutrients (Schulz et al. 2013). Microbial symbioses like the lichens are important as they are the initial colonizers of rocks and therefore initialize the process of bioweathering and involve in early stages of soil mineral formation (Gadd 2017).

Mineralization by soil micro-organisms plays an important role in the environment as it releases trapped mineral nutrients [phosphorus (P), potassium (K), magnesium (Mg), calcium (Ca), and iron (Fe)] required for plant growth. Some of the living organisms such as mosses, cyanobacteria, microfungi and lichen present in the uppermost or top of the soil constitute biological soil crusts or biocrusts (Garcia-Pichel et al. 2003). These biocrusts develop mostly in dry or bare land and serve as a reservoir for C and N (Belnap et al. 2001). They are also formed in the wide space in between the vascular plants. Besides their role in C and dinitrogen (N_2) fixation, biocrusts also improve the soil stability, prevents soil erosion, soil water relationship, seed germination and make available nutrients to the plants (Kuske et al. 2012).

The intimate contact between the plant roots and microorganisms associated with the soil constitutes the rhizosphere. The microbial interaction in the rhizosphere is critical as microbes tend to modify the physical and chemical process during soil formation (Gregory et al. 2007). In the rhizosphere region, microorganisms remain active with higher microbial activity and soil factors affect the microbes that in turn contribute to nutrient cycling (Lambers et al. 2009). Mycorrhizal and saprophytic fungi and bacteria are responsible for mineral weathering in rhizosphere through acidification. The weathering and acidification processes lead to diverse weathering characteristics in mineral grains and thus supplying mineral nutrients for mycorrhizal plants in the rhizosphere (Koele et al. 2014). Mycorrhiza denotes a symbiotic relationship between plant root and the soil fungi (Smith and Read 2008). Apart from its role in plant growth promotion; mycorrhizal fungi have a key role in soil aggregation and improvisation of soil structure (Rillig and Mummey 2006). Soil management practices could change the physical and chemical characteristics of soil and microbiota dynamics.

Soil micro-organism constitutes bacteria, actinomycetes, fungi, protozoa, yeast, algae, worms and insects. Certain bacteria and all fungi being heterotropic, depend on the organic matter and obtain nutrients and minerals by decomposing them. Therefore they have different roles in nutrient cycling that keeps the soil in good and healthy condition for plant growth. The substrate in the soil increase the bacterial populations that feed on them and recycles the nutrients important for both plants and other soil organisms (Kuske et al. 2012). The expansion of bacterial population supports protozoa that predate bacteria. The increasing protozoan population, in turn, triggers the activity of mites which feed on protozoans. The substrate arrival expands the fungal population and the competition among the fungal species. Nematodes are prompted to feed on fungi and other nematodes species. Some nematophagous fungi are also capable of trapping and feeding on nematodes. In general, fungi consume and store more nutrients than bacteria due to the different proportions of C and N and thereby maintain the soil health (Paul and Clark 1996). A gram of soil may consist of several kilometers of fungal hyphae (Young and Crawford 2007). However, plant roots host diverse microorganisms (bacteria and fungi) that can stimulate each other forming specific interface between soil and plants. The nutrients stored in the cells of the micro-organisms prevent nutrient loss by leaching. This act as agents of nutrient exchange and helps to maintain the soil structure.

2 Microbial Activities in Soil Formation

One of the mechanisms for involvement of biota in soil formation and its functioning is the influence of higher plants, micro and mesofauna and microflora on the mineral decomposition (Sokolova 2011). The importance of mineral dissolution through the activities of microorganisms can be compared to that of the absorption of CO_2 by plants and atmospheric N_2 fixation by soil microbial communities (Schulz et al. 2013). Soil biota plays an important role in the degradation and decomposition of organic matters and contributes to humus formation. Microorganisms decompose the organic remnants and substances in the soil surfaces such as senesced plant leaves and other litter. The organic matters are utilized by the microbes as an energy source and thus increase their population in the soil. These microbes degrade the digestible materials leaving those that are not decomposed easily. This results in the formation of humus that holds the primary soil particles (clay, silt and sand) and forms secondary aggregates. Soil biota and humus aids in the soil formation and development of soil horizons (Martin and Haider 1971).

Due to their widespread distribution, rapid growth, metabolic diversity and colonization and adaption to extreme conditions, microorganisms occupies a central role in soil evolution and formation (Zhu et al. 2014). The litter decomposition process is intimately linked with microbial activities that modify the chemical structure of litter and manage soil C and N dynamics (Berg and McClaugherty 2014). Microorganisms play a key role in plant litter decomposition and formation of soil through their enzymatic activities (Helfrich et al. 2015). The microbial growth and their resulting biomass and necromass change the chemical composition of soil organic matter, as perceived in nutrient immobilization (Wanek et al. 2010; Cotrufo et al. 2013). Moreover, the transformations of microbial necromass and biomass influence soil stability, formation and fertility (Six et al. 2006; Fontaine et al. 2011). The microbial growth and decomposition could also be measured by amino acids and amino sugars and biomolecules which are very rich in microorganisms than in plant litter (Tremblay and Benner 2006). In addition, amino acids and amino sugars are essential constituents for N immobilization and C sequestration in the soil (Liang et al. 2007).

Microbial weathering is a geological process occurring on the Earth's surface that fundamentally refers to the microbial growth and reproduction (Bin et al. 2008). The microbial metabolites promote the dissolution of some substances from the rock due to the influence of microbial enzymes on the natural degrading rate of minerals. Microbial weathering alters the composition of minerals and rocks and thus leads to the liberation of elements such as, silica (Si), Fe, manganese (Mn), aluminium (Al) from silicates, oxides, carbonates that changes the proportion and contents of soil minerals (Bin et al. 2008). Calcareous rocks are mainly subject to chemical weathering (acidic dissolution of the calcium carbonate) whereas; siliceous rocks are mainly fractured as a consequence of freezing-thawing cycles. Siliceous rocks contain a number of minerals which contain essential elements and hence support microbial life. In contrary, the weathering of calcareous rocks liberates only a few elements which stimulate the growth of microbes. Serpentinite rocks may even release toxic compounds such as, nickel (Ni) and cadmium (Cd) which prevents the establishment of plant life and could also hinder microbial activities (Bratteler et al. 2006).

Microorganisms contribute to mineral weathering both through direct and indirect ways (Sokolova 2011). The direct effect of soil microflora occurs when microbial cells are directly in contact with the mineral surface. The adhesive property of microbes on the mineral particles may be due to their fixation on earlier absorbed compounds. On the other hand, the indirect effect of soil microbes includes varying products of soil microorganism's functioning that consist of chelating agents, bases and acids (Sokolova 2011). Microbial activity accelerates the release of P and sulphur (S) elements from the bedrock to supply living organisms with P and S, whereas C and N are not part of the mineral composition and are scarce in the initial soils. Soil aggregation is most crucial in controlling the structure and function of microorganisms and plant life (Kobierska et al. 2011). Thus, it is clear that the initial processes of soil formation and input of nutrients rely on the activity of microorganisms. The foremost principles of soil biota are biological weathering of the bedrock material and the formation of interfaces for nutrient turnover at vegetation free site.

3 Plant-Microbe Interaction

Microbiological activity is greater in the rhizosphere than in soil away from the plant roots. Plant-microbe interactions contribute to several soil processes such as nutrient cycling, C sequestration and ecosystem services. Plants play an important role in soil development and display a nutrient hotspot at initial sites of soil formation in terms of C, as they provide up to 40% of the photosynthetically fixed C to the microorganisms (Miniaci et al. 2007; Towe et al. 2010; Duc et al. 2009). In exchange, soil microbes provide N, P and other essential nutrients to the plant and also defend them against herbivores or parasites (Butler et al. 2003).

Plant roots secrete important compounds, which have a major role in the physical, chemical and biological interaction between plant roots and the rhizosphere (Moore et al. 2015). Alterations in the secondary metabolites released by the plants could influence the soil microbial communities (Bressan et al. 2009). The root exudates influence the biological and chemical activities of the soil thereby improves soil fertility (Altieri 2004). The root exudates supply a huge amount of C which activates the soil aggregate formation. The mucilaginous root exudates adhere to the soil particles and lead to short-term soil aggregation (Morel et al. 1991). The rhizosphere region usually has increased CO_2 and lower pH level and oxygen. Depending on the nutrients taken up by the plant roots from the soil, exudates could make the rhizospheric soil more alkaline or acidic. Rhizosphere activity modifies mineral surfaces; attack mineral structures, and also take up the weatherable soil minerals. All these activities stimulates mineral weathering and directs the formation of soil which makes rhizosphere the most dynamic environment in the soil. This intensification of mineral weathering is dependent on pH in addition to exudate and microbial communities in the rhizosphere. The capacity of soil microbial communities in mineral weathering is well documented (Favero-Longo et al. 2005; Uroz et al. 2007; Calvaruso et al. 2006).

4 Lichens in Soil Formation

Lichens play a vital role in soil formation. Lichens that inhabit rocks are referred to as saxicolous that include different morphology namely, foliose, crustose and fruticose (de los Rios et al. 2002). Several studies have shown the significant role of lichens in mineral and rock weathering and in soil formation (Chen et al. 2000; Begonha 2009). The close contact of fungi with the substratum and presence of algae on outer layers of the lichen thallus suggest that weathering capacity is critically due to the mycobionts (Chen et al. 2000). The deteriorating ability of rocks by lichens varies according to their growth form (de los Rios et al. 2002). For example, foliose lichens exert pressure on the rocks via fixation structures, whereas the thallus of crustose lichen gets fully developed and incorporated within the lithic substrate. The alterations in the volume of the thallus brought about by the lichens due to the expansion and narrowing by drying or freezing results in the mechanical breakdown of rocks (Ascaso et al. 2002).

During the metabolic activity, lichens produce organic acids that have a significant impact on weathering and decomposing the parent rocks (Belnap and Lange 2003). They have the capacity to break down the rock into smaller particles at a faster rate. Lichen undergoes both physical and chemical process during weathering of rocks. The physical process involves a mechanical breakdown of rocks through penetration of hyphae, expansion and contraction of lichen thallus and swelling of inorganic and organic salts generated during lichen activity. In the chemical weathering, lichens secrete organic acids, specifically oxalic acid, which efficiently break down minerals and chelate the metallic cations (Chen et al. 2000). In addition, lichens stimulate secondary mineral formation through turgor pressure and production of exopolysaccharide. These secondary minerals react with the cations of rock and cause disintegration and flaking of outer rock surface (Ranalli et al. 2009). The minerals formed by weathering of rocks by lichens possess enhanced surface corrosion. *Lecidea atrobrunnea, Rhizocarpon geographicum* and *Sporastatia testudinea* are capable of weathering the serpentinized rocks in alpine environments (Favero-Longo et al. 2005).

5 Bacteria in Mineral Weathering

Bacteria are not only involved in the biological processes such as biogeochemical cycles, providing nutrients to the plants, enhancing plant growth and in controlling plant pathogens (Hayat et al. 2010); they also have a crucial role in the initial stage of soil formation via modification of parental rock and in soil structuring (Paul and Clark 1996). Unlike other organisms, bacteria can adapt or tolerate extreme environmental conditions. The mineral particle forms a microenvironment to protect bacteria from stress conditions. Thus, bacteria obtain inorganic phosphate and energy from the mineral matrix or through the activities of other microbes. The mechanisms involved in mineral weathering include oxidoreduction reactions and production of organic acids and chelating agents (Uroz et al. 2007). Several bacteria that are involved in weathering of rocks and minerals, release large amounts of beneficial minerals from rocks to the plants, organic acids and fix N_2 and condense the rock particles thus forming mineral soil (Puente et al. 2009). These bacteria either in combination with other microbes or alone could mineralize through formation of complex microbial communities that associate with mineral surface (Uroz et al. 2009). For example, the species belonging to the genera *Bradyrhizobium*, *Collimonas*, and *Anabaena* are capable of mineral weathering (Männistö and Häggblom 2006; Calvaruso et al. 2009; Collingnon et al. 2011). The bacteria inhabiting the rock surfaces are different from those residing in the surrounding soil (Certini et al. 2004). The mineral particles that are colonized by bacterial communities commonly include quartz, granite, limestone or apatite (Gleeson et al. 2005; Carson et al. 2009). The chief elements such as Al, Ca or Si present in the mineral particles influence the structure of the bacterial communities. This lead to a new concept called 'mineralosphere', where certain microbes are selected for their capacity to utilize the inorganic nutrients that are released by soil minerals. Bacteria isolated from the rhizosphere soil and mineralosphere could supply nutrients to the plants in the nutrient-poor soil (Uroz et al. 2009). For example, one of the mineral weathering bacterial strain *Burkholderia glathei* PML1 promoted the growth of pine tree under nutrient stress condition in the presence of biotite (Calvaruso et al. 2006).

The composition of bacterial communities gets modified in the presence or absence of arbuscular mycorrhizal (AM) fungal hyphae (Marschner and Baumann 2003; Rillig et al. 2006). Moreover, the potential effect of bacterial hyphal colonizers on AM fungi and the AM fungal symbiosis is high. Several types of interactions between bacteria and AM fungi have been described (see Bonfante and Anca 2009). Bacteria could also enhance the rate of mineral dissolution and the ability of bacteria involved in mineral weathering varies according to the habitat they occupy (Huang et al. 2014). Mostly mineral weathering bacteria are isolated from the rhizosphere of trees and the ectomycorrhizosphere that forms tree root-soil boundary where nutrient exchange takes place (Calvaruso et al. 2007). The bacterial isolates isolated from the mycorrhizosphere region of the ectomycorrhizal fungus *Scleroderma citrinum* had higher mineral weathering capacity when compared to those isolated from bulk soil. The fungi select a bacterial community with greater mineral weathering ability in the bulk soil through carbon metabolism (Uroz et al. 2007). Likewise, Collignon et al. (2011) also reported the presence of potential mineral weathering bacteria (*Pseudomonas, Rhizobium, Burkholderia, Bacillus*) in tree rhizosphere of *Fagus sylvatica* and *Picea abies* was higher than in the bulk soil and the mineral weathering efficacy changes depending upon the seasons (Collignon et al. 2011). A Gram-negative, aerobic and motile mineral weathering bacterium, *Rhizobium yantingense* isolated from the surface of weathered rock was shown to possess high mineralization activity (Chen et al. 2015).

6 Non Mycorrhizal Endophytes in Mineral Weathering

Non mycorrhizal endophytes include both bacteria and fungi that colonize the plant tissues without causing any adverse effect to the host (Wilson 1995). These endophytes have the capability to transfer complex compounds (Wang and Dai 2011). Phosphate solubilization, rock degradation and N_2 fixation contributes to efficient rock-weathering bacterial endophytes (Lopez et al. 2011). Endophytic bacteria (*Bacillus* sp.) isolated from cactus roots have been reported as an efficient rock weathering microbe. This bacterial endophyte helps in the weathering of igneous rocks in nutrient-poor regions upon colonization of the plant roots (Puente et al. 2009). The bacterial endophytes, *Azotobacter vinelandii*, *Bacillus megaterium* and *Pseudomonas putida* isolated from roots of *Mammillaria fraileana* are capable of N_2 fixation and weathering of rocks into smaller particles (Lopez et al. 2011).

7 Role of Fungi in the Weathering Process

Fungi are prominent geoactive agents that are involved in the transformation of metals and minerals that in turn modify the chemistry and surface structure of rocks and minerals (Gadd 2017). Fungal communities that are capable of mineral weathering and dissolution include saprophytic fungi, lichen-forming fungi and mycorrhizal fungi (Hoffland et al. 2004). Symbiotic fungi or free-living fungi inhabit the outer surface of rocks and are recognized as one of the potential deteriorates of rocks and minerals (Warscheid and Braams 2000). The microcolonial fungi (black melanized colonies) that inhabit the exposed rock surfaces are tolerant to environmental stresses and produce filamentous hyphae that may penetrate into the rocks. These interactions may give rise to different types of surface coatings and secrete polysaccharides forming micropits in the rock surfaces. They may also form mutualistic relationship with algae on rock substrate in order to acquire C (Gorbushina 2007).

The colonization of fungi could lead to physical and biochemical changes in rocks. Fungi enter the solid materials through physical and chemical methods. The fungal hyphae penetrate along the weak points or spots on the surface of the rocks. The hyphae thus form ridges and grooves as the result of surface contours. At these spots, fungi weather the minerals by physical and chemical process. The exudates diffusion into the soil is prevented by these processes and thus enhances weathering mediated by fungi and result in the tunnel formation (Hoffland et al. 2002). Chemical weathering by fungi include the production of proton and ligand-based weathering agent. The hyphal tip growth involves the production of carbonic acids that aids in the breakdown of weak spots of rock surface and comprises proton based agents. Production of siderophores, polyphenolic and polysaccharide acids and organic anions are ligand based weathering mediators (Hoffland et al. 2004).

8 Mycorrhiza: Mineral Weathering

Mycorrhizal fungi play a major role in mineral weathering and in the formation of soil. The mechanism involved in weathering of rocks by the mycorrhizal fungi is well documented (Wallander and Thelin 2008; Brantley et al. 2011; Thorley et al. 2015). Taylor et al. (2009) suggested mechanisms through which mycorrhizal fungi influence the mineral weathering processes. This includes (a) disintegration of minerals through secretions like H^+ and organic chelators of low molecular weight; (b) respiration of plant roots and mycorrhizal fungi enhance CO_2 presence in the soil solution; (c) organic matter decomposition elevates the concentrations of high molecular weight organic acids and organic chelators in the soil solution, that are further utilized by heterotrophs to fuel respiration and hence returning base cations to the soil solution from the biota; and (d) the transpiration increase the water flow consisting of nutrients and base cations to plants. The adsorbed soil particles on to the plant roots and mycorrhizal hyphae decrease the soil erosion thus contributing to the soil development continuously. As symbiotic plant partners, mycorrhizal fungi extend into soils and act as biosensors for nutrients that are taken up by them and supplied to their host plants (Bücking and Kafle 2015). The mycorrhizal hyphae force mechanically and chemically alter the minerals to obtain the nutrient elements (Bonneville et al. 2009). The organic acids produced by mycorrhizal hyphae acidify their environment that helps in the breakdown of minerals (Uroz et al. 2011).

The mycorrhizal group, both AM and ectomycorrhizal fungi are actively involved in weathering of rocks (Koele et al. 2014). Mycorrhizal fungi mostly target the minerals that consist of essential and needful plant nutrients during the weathering process (Remiszewski et al. 2016). Arbuscular mycorrhizal fungi may impact the mineral weathering through respiration and proton release and extraradical mycelium efficiently bind soil particles and influence soil aggregation (Bago et al. 1996; Smith and Read 2008). Ectomycorrhizal fungi forms a layer of fungal material around the root tip and the hyphae growing outside the layer penetrates into the soil acting as the nutrient scavenger (Landeweert et al. 2001). A positive correlation was reported between densities of root tip of ectomycorrhizal fungi and tunnel frequency suggesting that ectomycorrhizal fungi may participate in the formation of mineral tunnels (Hoffland et al. 2002). The ectomycorrhizal fungi produce low molecular organic compounds and proton that might enhance the mineral weathering even under P deficiency (Smits et al. 2012). Arbuscular mycorrhizal fungi are capable of weathering biotite through which they contain acquire mineral elements (Sanz-Montero and Rodríguez-Aranda 2012).

In the zone of mineral–microbial contact, mineral dissolution, precipitation and clay mineral formation is influenced by the presence of the microbial cells and low molecular weight organic compounds produced by them. The organic compounds that are generated by the fungi impact the weathering process beneath the contact zone (Banfield et al. 1999). The ectomycorrhizal fungi enable weathering activity depending upon the chemical activity and colonization of the mycelia in the soil (Wallander et al. 1997). Acidification by the fungal hyphae due to the liberation of respired CO_2, the release of organic acids, biomineral precipitation, the occurrence of extracellular acidic polymer substances and proton efflux constitute the chemical activity of mycelia (Gadd 2007; Burford et al. 2003). The capability of ectomycorrhizal fungi in weathering has been investigated to P, K and Mg in mineral form (Fomina et al. 2006; Rosling et al. 2009). However, the process of induced weathering in relation to P availability in ectomycorrhizal fungi is yet to be determined. In a study, Quirk et al. (2012) reported that ectomycorrhizal fungi associated with gymnosperm released two folds more Ca from the weathering of silicate surface when compared to AM associated angiosperms.

9 Mycorrhiza and Soil Structure

Mycorrhizal fungi are intimately associated with plant roots, colonizing the root cortex as well as the surrounding soil. Mycorrhizal fungi are dominant among the fungal community in mineral soils (Lindahl et al. 2007). They play essential roles in terrestrial ecosystems serving as a sink for nutrient and carbon cycles. It is estimated that around 80% of plant N and P are acquired through mycorrhizal fungi (van der Heijden et al. 2015). The extensive extraradical hyphae facilitate the fungi to colonize and utilize nutrient-rich substrates in the soil and to absorb and translocate nutrients like P and C in soils contributing to plant fitness and soil quality (Ritz 2006).

As mycorrhizal colonization usually influences the soil structure, AM fungi may also probably affect soil water relations and therefore, the water relations of the host plants (Rilling and Mummey 2006). The extramatrical mycelia of ectomycorrhizal fungi obtain C from the soil through enzymatic breakdown of organic matter and from tree photosynthates. This contributes to the association between weathering of minerals in the soil and photosynthetically-assimilated C acquired from trees. Mycorrhizal plants may be considered as efficient competitors as they decrease the mineralizing populations in a scarcity of N and P or change the quality of the decomposing litter. In addition, ectomycorrhizal fungi modify the environment through acidification, organic acid exudation by hyphae (Rosling et al. 2009) and siderophores (Winklemann 2007).

Besides several beneficial aspects, mycorrhizal fungi also contribute to soil structure through soil aggregation (Rilling and Mummey 2006). Mycorrhizal fungi, plant roots and organic matter are considered as important traits in the development of soil structure (Daynes et al. 2013). Aggregation helps to maintain the soil porosity, biogeochemical cycle and water infiltration (Diaz-Zorita et al. 2002). Arbuscular mycorrhizal fungal hyphae are regarded as the primary soil aggregators thus, exhibiting a positive correlation between the AM fungal hyphae and aggregate stability (Borie et al. 2008). Soil aggregates are broadly classified as microaggregates (<250 μm in diameter) and macroaggregates (>250 μm in diameter) (Rilling and Mummey 2006). Microaggregates are formed by fungal hyphae and plant roots bounded by polysaccharides whereas extraradical fungal hyphae of AM fungi and fine plant roots associate to constitute macroaggregates by releasing a large amount of polysaccharides (Snyder and Vázquez 2005). Fungi can either influence the soil aggregation directly by combining the extracellular compounds produced by the fungi or indirectly by maintaining the soil particles through the hyphal network (Borie et al. 2008). As AM fungi dominate the soil component through constitution of around 30% of the soil microbial biomass (Olsson et al. 1999), they provide much more C when compared to saprobic fungi due to longer existence in the soil even after the removal of host plant. Therefore, AM fungi tend to be crucial component in relating biotic influences on soil aggregation (Borie et al. 2008).

Apart from the role of AM fungal hyphae in soil aggregation, AM fungi are well known to produce a non-water soluble and highly persistent glycoprotein called glomalin (Wright and Upadhyaya 1996) that have an important role in maintaining the soil structure and fertility (Fokom et al. 2012). Glomalin is also known as glycosylated glycoprotein (Gillespie et al. 2011). This glycoprotein is produced in cell walls of mycorrhizal fungi and persists in soil even after the death of the fungal hyphae (Driver et al. 2005). Glomalin commonly occurs in soils rich in insoluble humus or mineral fractions (Wright and Upadhyaya 1996). Owing to its adhesive properties, glomalin produces soil aggregates by combining fine soil particles together that aids in soil aeration (Purin and Rillig 2007). A positive correlation has been reported between C sequestration, soil aggregation and AM fungal density in a field study (Wilson et al. 2009). The amount of C and N from glomalin contributes to respectively 3% and 5% of soil C and N pools (Lovelock et al. 2004). Polyphenolic compounds such as humic acid and soil tannins have also been extracted along with glomalin (Whiffen et al. 2007; Jonathan and Javier 2006). Glomalin protects the fungal hyphae during translocation of nutrients to the hyphal tip from plants and to the plant from soil (Pal and Pandey 2014). The capability of AM fungal isolate to influence glomalin content and to the formation of extensive mycelial networks in the soil could impact the stability of soil aggregates through hyphal entrapment of soil particles. This suggests the need for the selection of potential AM fungal isolates that could be used for the improvement of soil quality and restoration of degraded lands (Bedini et al. 2009). Wu et al. (2012) studied the spatial distribution and relationship of glomalin with soil aggregates and root mycorrhization in the rhizosphere of *Citrus unshiu* and reported a positive correlation between glomalin related soil protein and plant roots. The study also revealed the fact that the secretion of glomalin decreased with increasing soil depth. Rillig and Steinberg (2002) observed an increased production of glomalin by *Rhizophagus intraradices* (=*Glomus intraradices*) under unfavorable environmental condition by facilitating the soil structure in lacking adequate soil pores.

10 Nutrient Cycling

Soil mineral weathering contributes in providing input of plant nutrients to ecosystems, thus preventing nutrient limitations (Chadwick et al. 1999). Moreover, the cations produced through mineral weathering neutralize the soil acidification, thereby enhancing the nutrient availability to plants (van Breemen et al. 1983). Clay particles formed as a result of weathering product contribute to the cation exchange capacity of the soil, decreasing the leaching of nutrients like K and ammonium and also positively correlates with soil organic matter and water holding capacity (Sollins et al. 1996). Weathering of rocks composed of silicate minerals releases a large amount of Ca and Mg that play a vital role in the C cycle as they are locked up as carbonates (Hartmann and Moosdorf 2011). The interaction between the process of geological leaching of plant nutrients and biological cycle that includes bioaccumulation process usually results in pedogenesis. The nutrients released during weathering processes are utilized by plants for their growth. In biological cycling, plants uptake specific nutrients from parent materials, atmosphere and water, thus through photosynthesis prepare organic matter, which is returned in the form of leaf or root residues to the soil. The decomposed organic matters from plant litters constitute an essential part of soil humus that enhances soil fertility (Zhu et al. 2014).

Nutrient cycling is an important ecological function that involves a defined, typically bounded, compartment which nutrients enter and leave via a range of pathways, and within which they are transformed via a myriad of chemical and biochemical reactions. In soil systems, a large proportion of these transformations are mediated by biota. Microbial communities play a crucial role in the nutrient cycle as they degrade organic materials; liberate inorganic nutrients that are taken up by plants; influence plant growth, availability of nutrients through several processes such as, chelation, oxidation, solubilization and reduction; stores and releases nutrients from microbial biomass (Marschner 2007). Microbes involved in N or P cycling are N_2- fixers, AM fungi and P- mobilizers.

Nitrogen cycling involves three important processes, N_2 fixation, denitrification and nitrification. Microorganisms take part in these processes as N_2 fixers, denitrifiers and nitrifiers (Stein and Klotz 2016). The N_2- fixing bacteria and AM fungi represent the most significant beneficial symbionts associated with nutrient cycling (Dos Santos et al. 2012; Schüßler and Walker, 2011). The capability of microorganisms to convert atmospheric N to ammonia is limited to bacteria that contain nitrogenase enzyme which combines hydrogen and N to form ammonia (de Bruijn 2015). The N_2 fixers are either free-living or symbiotic bacteria are commonly known as diazotrophs (Dixon and Khan 2004). The rhizobia bacteria can fix N_2 with leguminous plant through mutual symbiotic association, whereas, some of the actinomycetes fix N_2 and form nodules on the roots, that is known as actinorrhizal plants (Olivares et al. 2013). Other N_2 fixing bacteria include *Azotobacter*, *Bacillus*, *Clostridium*, *Frankia*, etc. The ammonium is converted into nitrate through soil bacteria such as *Nitrosomonas*, *Nitrobacter* etc. The last process of N cycling involves detrification that are carried out by denitrifiers like, *Pseudomonas* and *Clostridium* through which nitrates are converted to N_2 (Hayatsu et al. 2008).

Phosphorus is stored in the soils, bedrock and sediments and is not available directly to organisms (Ruttenberg 2002). Microorganisms including bacteria, fungi and actinomycetes have the capacity to solubilize and mineralize P (Alori et al. 2017). The most prominent soil bacteria involved in P are the species of *Agrobacterium*, *Pseudomonas*, *Acetobacter* and *Bacillus* (Babalola and Glick 2012; Kumar et al. 2014; David et al. 2014). Acidification, chelating organic acids, and siderophore production is involved in solubilizing Fe, Ca and aluminium phosphate from the soil (Marschner 2008). Phosphate solubilizers enhance the amount of orthophosphate leaving the nutrient to be absorbed by the plant roots (Richardson et al. 2009). Phosphatase enzymes are utilized by fungi and bacteria in mineralizing P (Jorquera et al. 2008). Some of the efficient fungal P solubilizers include *Fusarium*, *Cladosporium*, *Rhizoctonia*, and *Alternaria* (Sharma et al. 2013). Some fungi decompose and degrade the wood thus producing large amounts of oxalic acids. These oxalic acids might have a secondary effect on the release of P from the soil (Dutton and Evans 1996; Fransson et al. 2004).

Arbuscular mycorrhizal fungal symbioses are significant for the continual nutrient cycling in the plant community and thus avoid nutrient sequestration. During root colonization, the fungus grows in the surrounding soil of roots to establish a network of hyphae called extraradical mycelium. The extraradical mycelium uptake the nutrient from the soil with help of branched absorbing structures and transfer to a long distance of about 25 cm (Jansa et al. 2003). Arbuscular mycorrhizal fungi rely on the host plant for C compounds for their growth and metabolism, in return for mineral nutrients (especially P) that the extraradical mycelium takes up from the soil and transport to the root. For the growth and metabolism, AM fungi rely on the host plant for C sources and in return, the extraradical mycelium provides mineral nutrient from the soil to plant roots. This symbiotic relation of extraradical mycelium and intraradical mycelium is important for translocation, distribution and movement of mineral nutrients in the plant-soil environment (Richardson et al. 2009). The AM fungal contribution to plant nutrition through P uptake by mycelium is considered as the extension of the root system. AM fungi influence the uptake of other nutrients such as K, Ca, Zn, Cu or Fe (Liu et al. 2000).

Arbuscular mycorrhizal fungi stimulate the inorganic phosphate transporters which are present in the periarbuscular membrane (Xie et al. 2013). These genes are considered as functional markers for the AM symbiosis (Harrison 2012). Therefore, mycorrhizal plants absorb P directly via root epidermis and through the AM fungal pathway) that delivers P to the root cortex (Smith and Smith 2011). It is proposed that mycorrhizal N uptake is similar to Pi uptake pathway. The C transfer to fungi from the plant is transferred to plant sink organs, distributed at the arbuscular interface and are further hydrolyzed by cell walls (Ferrol and Pérez-Tienda 2009). Mycorrhizae can also influence the uptake of a range of elements by plants, including S, B, K, Ca, Mg, Na, Zn, Cu, Mn, Fe, Al and Si (Clark and Zeto 2000). Bacteria may generate various metabolites necessary for mineral weathering with the available nutrients (Bennett et al. 2001). Photoautotroph organisms fix C and penetrate into the soil through root deposition, specifically through soluble exudates that originate from growing roots at a faster rate and slowly by cells and tissue depositions. The senesced plants parts are accumulated on the soil surface and these organic matter in the soil are transformed by soil organisms during the assimilation of energy for growth and reproduction. Through this, the compounds are additionally transformed and cycled between the compartments (Paul 2007). This process yields stable soil organic matter which contributes to structural development.

11 Conclusion

It is clear that soil biota's as the biological engine carry out the myriad of the process which underpins soil function via biochemical pathways. Soil biota is involved in many aspects of soil functioning and delivery of the full range of ecosystem goods and services that soils support. Nevertheless, the virtual role of biota in soil production varies between systems. Soil organisms exist in the presence of other populations and the diverse members of biomass. Thus, emerging interactions persuade the community structure within soils having strong impact on growth and functions of individual organisms. The process of mineral weathering via microorganisms and thereby providing nutrients to plants is well documented. Nevertheless, the microbial enzymes, genes in the mineral weathering process are yet to be investigated. The transfer of electron and genes involved in interaction between microbes and minerals is still obscure. The information on the distribution of microorganisms on the mineral surface and its chemistry could help in better understanding of mineral weathering processes. Compared to the microcosm experiments, mesh bag incubation studies could be performed to determine the dissolution rate of minerals. The interaction between plants and microbes and the microbes and mineral particles constitute a major role in soil formation.

References

Alori, E. T., Glick, B. R., & Babalola, O. O. (2017). Microbial phosphorus solubilization and its potential for use in sustainable agriculture. *Frontiers in Microbiology, 8*, 971. https://doi.org/10.3389/fmicb.2017.00971.

Altieri, M. A. (2004). Linking ecologists and traditional farmers in the search for sustainable agriculture. *Frontiers in Ecology and the Environment, 2*, 35–42.

Ascaso, C., Wierzchos, J., Souza-Egipsy, V., & De los RÃos A. (2002). In situ evaluation of the biodeteriorating action of microorganisms and the effects of biocides on carbonate rock of the Jeronimos Monastery (Lisbon). *International Biodeterioration Biodegradation, 49*, 1–12.

Augé, R. M. (2004). Arbuscular mycorrhizae and soil/plant water relations. *Canadian Journal of Soil Science, 84*, 373–381.

Babalola, O. O., & Glick, B. R. (2012). The use of microbial inoculants in African agriculture: Current practice and future prospects. *Journal of Food, Agriculture and Environment, 10*, 540–549.

Bago, B., Vierheilig, H., Piché, Y., & Azcón-Aguilar, C. (1996). Nitrate depletion and pH changes induced by the extraradical mycelium of the arbuscular mycorrhizal fungus *Glomus intraradices* grown in monoxenic cultures. *The New Phytologist, 133*, 273–280.

Banfield, J. F., Barker, W. W., Welch, S. A., & Taunton, A. (1999). Biological impact on mineral dissolution: Application of the lichen model to understanding mineral weathering in the rhizosphere. *Proceedings of the National Academy of Sciences of the United States of America, 96*, 3404–3411.

Bedini, S., Pellegrino, E., Avio, L., Pellegrini, S., Bazzoffi, P., Argese, E., & Giovannetti, M. (2009). Changes in soil aggregation and glomalin-related soil protein content as affected by the arbuscular mycorrhizal fungal species *Glomus mosseae* and *Glomus intraradices*. *Soil Biology and Biochemistry, 41*, 1491–1496.

Begonha, A. (2009). Mineralogical study of the deterioration of granite stones of two Portuguese churches and characterization of the salt solutions in the porous network by the presence of diatoms. *Materials Characterization, 60*, 621–635.

Belnap, J., & Lange, O. L. (2003). Biological soil crusts: Structure, function, and management. In I. T. Baldwin, M. M. Caldwell, G. Heldmaier, O. L. Lange, H. A. Mooney, E.-D. Schulze, & U. Sommer (Eds.), *Ecological studies series 150* (Vol. 150, pp. 1–503). Berlin: Springer.

Belnap, J., Büdel, B., & Lange, O. L. (2001). Biological soil crusts: Characteristics and distribution. In J. Belnap & O. L. Lange (Eds.), *Biological soil crusts: Structure, function, and management* (Ecological studies analysis and synthesis, Vol. 150, pp. 3–30). Berlin/Heidelberg: Springer.

Bennett, P. C., Rogers, J. R., Choi, W. J., & Hiebert, F. K. (2001). Silicates, silicate weathering, and microbial ecology. *Geomicrobiology Journal, 18*, 3–19.

Berg, B., & McClaugherty, C. (2014). Decomposition as a process: Some main features. In B. Berg & C. McClaugherty (Eds.), *Plant litter: decomposition, humus formation, carbon sequestration* (pp. 11–34). Berlin/Heidelberg: Springer.

Bin, L., Ye, C., Lijun, Z., & Ruidong, Y. (2008). Effect of microbial weathering on carbonate rocks. *Earth Science Frontiers, 15*, 90–99.

Bonfante, P., & Anca, I. A. (2009). Plants, mycorrhizal fungi, and bacteria: A network of interactions. *Annual Review of Microbiology, 63*, 363–383.

Bonneville, S., Smits, M. M., Brown, A., Harrington, J., Leake, J. R., Brydson, R., & Benning, L. G. (2009). Plant-driven fungal weathering: Early stages of mineral alteration at the nanometer scale. *Geology, 37*, 615–618.

Borie, F., Rubio, R., & Morales, A. (2008). Arbuscular mycorrhizal fungi and soil aggregation. *Journal of Soil Science and Plant Nutrition, 8*, 9–18.

Brantley, S. L., Megonigal, J. P., Scatena, F. N., Balogh-Brunstad, Z., Barnes, R. T., Bruns, M. A., & Yoo, K. (2011). Twelve testable hypotheses on the geobiology of weathering. *Geobiology, 9*, 140–165.

Bratteler, M., Lexer, C., & Widmer, A. (2006). Genetic architecture of traits associated with serpentine adaptation of *Silene vulgaris*. *Journal of Evolutionary Biology, 19*, 1149–1156.

Bressan, M., Roncato, M. A., Bellvert, F., Comte, G., Haichar, F. Z., Achouak, W., & Berge, O. (2009). Exogenous glucosinolate produced by *Arabidopsis thaliana* has an impact on microbes in the rhizosphere and plant roots. *The ISME Journal, 3*, 1243–1257.

Bücking, H., & Kafle, A. (2015). Role of arbuscular mycorrhizal fungi in the nitrogen uptake of plants: Current knowledge and research gaps. *Agronomy, 5*, 587–612.

Burford, E. P., Fomina, M., & Gadd, G. M. (2003). Fungal involvement in bioweathering and biotransformation of rocks and minerals. *Mineralogical Magazine, 67*, 1127–1155.

Butler, J. L., Williams, M. A., Bottomley, P. J., & Myrold, D. D. (2003). Microbial community dynamics associated with rhizosphere carbon flow. *Applied and Environmental Microbiology, 69*, 6793–6800.

Calvaruso, C., Turpault, M. P., & Frey-Klett, P. (2006). Root-associated bacteria contribute to mineral weathering and to mineral nutrition in trees: A budgeting analysis. *Applied and Environmental Microbiology, 72*, 1258–1266.

Calvaruso, C., Turpault, M. P., Leclerc, E., & Frey-Klett, P. (2007). Impact of ectomycorrhizosphere on the functional diversity of soil bacterial and fungal communities from a forest stand in relation to nutrient mobilization processes. *Microbial Ecology, 54*, 567–577.

Calvaruso, C., Mareschal, L., Turpault, M. P., & Leclerc, E. (2009). Rapid clay weathering in the rhizosphere of Norway spruce and oak in an acid forest ecosystem. *Soil Science Society of America Journal, 73*, 331–338.

Carson, J. K., Campbell, L., Rooney, D., Clipson, N., & Gleeson, D. B. (2009). Minerals in soil select distinct bacterial communities in their microhabitats. *FEMS Microbiology Ecology, 67*, 381–388.

Certini, G., Campbell, C. D., & Edwards, A. C. (2004). Rock fragments in soil support a different microbial community from the fine earth. *Soil Biology and Biochemistry, 36*, 1119–1128.

Chadwick, O., Derry, L., Vitousek, P., Huebert, B., & Hedin, L. (1999). Changing sources of nutrients during four million years of ecosystem development. *Nature, 397*, 491–497.

Chen, J., Blume, H.-P., & Beyer, L. (2000). Weathering of rocks induced by lichen colonization – A review. *Catena, 39*, 121–146.

Chen, W., Sheng, X.-F., He, L.-Y., & Huang, Z. (2015). *Rhizobium yantingense* sp. nov., a mineral-weathering bacterium. *International Journal of Systematic and Evolutionary Microbiology, 65*, 412–417.

Clark, R. B., & Zeto, S. K. (2000). Mineral acquisition by arbuscular mycorrhizal plants. *Journal of Plant Nutrition, 23*, 867–902.

Collignon, C., Uroz, S., Turpault, M. P., & Frey-Klett, P. (2011). Seasons differently impact the structure of mineral weathering bacterial communities in beech and spruce stands. *Soil Biology and Biochemistry, 43*, 2012–2022.

Cotrufo, M. F., Wallenstein, M. D., Boot, C. M., Denef, K., & Paul, E. (2013). The Microbial Efficiency-Matrix Stabilization (MEMS) framework integrates plant litter decomposition with soil organic matter stabilization: Do labile plant inputs form stable soil organic matter? *Global Change Biology, 19*, 988–995.

David, P., Raj, R. S., Linda, R., & Rhema, S. B. (2014). Molecular characterization of phosphate solubilizing bacteria (PSB) and plant growth promoting rhizobacteria (PGPR) from pristine soils. *International Journal of Innovative Science Engineering and Technology, 1*, 317–324.

Daynes, C. N., Field, D. J., Saleeba, J. A., Cole, M. A., & McGee, P. A. (2013). Development and stabilisation of soil structure via interactions between organic matter, arbuscular mycorrhizal fungi and plant roots. *Soil Biology and Biochemistry, 57*, 683–694.

de Bruijn, F. J. (2015). Biological nitrogen fixation. In B. Lugtenberg (Ed.), *Principles of plant-microbe interactions* (pp. 215–224). Cham: Springer International.

De los Ríos, A., Wierzchos, J., & Ascaso, C. (2002). Microhabitats and chemical microenvironments under saxicolous lichens growing on granite. *Microbial Ecology, 43*, 181–188.

Denef, K., & Six, J. (2005). Clay mineralogy determines the importance of biological versus abiotic processes for macroaggregate formation and stabilization. *European Journal of Soil Science, 56*, 469–479.

Díaz-Zorita, M., Perfect, E., & Grove, J. H. (2002). Disruptive methods for assessing soil structure. *Soil and Tillage Research, 64*, 3–22.

Dixon, R., & Kahn, D. (2004). Genetic regulation of biological nitrogen fixation. *Nature Reviews. Microbiology, 2*, 621–631.

Dominati, E. J., Patterson, M. G., & Mackay, A. D. (2010). A framework for classifying and quantifying the natural capital and ecosystem services of soils. *Ecological Economics, 69*, 1858–1868.

Dos Santos, P. C., Fang, Z., Mason, S. W., Setubal, J. C., & Dixon, R. (2012). Distribution of nitrogen fixation and nitrogenase-like sequences amongst microbial genomes. *BMC Genomics, 13*, 162–162.

Driver, J. D., Holben, W. E., & Rillig, M. C. (2005). Characterization of glomalin as a hyphal wall component of arbuscular mycorrhizal fungi. *Soil Biology and Biochemistry, 37*, 101–106.

Duc, L., Noll, M., Meier, B. E., Burgmann, H., & Zeyer, J. (2009). High diversity of diazotrophs in the forefield of a receding alpine glacier. *Microbial Ecology, 57*, 179–190.

Dutton, M. V., & Evans, C. S. (1996). Oxalate production by fungi: Its role in pathogenicity and ecology in the soil environment. *Canadian Journal of Microbiology, 42*, 881–895.

Favero-Longo, S. E., Turci, F., Tomatis, M., Castelli, D., Bonfante, P., Hochella, M. F., Piervittori, R., & Fubini, B. (2005). Chrysotile asbestos is progressively converted into a non-fibrous amorphous material by the chelating action of lichen metabolites. *Journal of Environmental Monitoring, 7*, 764–766.

Ferrol, N., & Pérez-Tienda, J. (2009). Coordinated nutrient exchange in arbuscular mycorrhiza interface. In C. Azcon-Aguilar, J. M. Barea, S. Gianinazzi, & V. Gianinazzi-Pearson (Eds.), *Mycorrhizas: Functional processes and ecological impact* (pp. 73–87). Berlin/Heidelberg: Springer.

Fierer, N., & Jackson, R. B. (2006). The diversity and biogeography of soil bacterial communities. *Proceedings of the National Academy of Sciences of the United States of America, 103*, 626–631.

Fokom, R., Adamou, S., Teugwa, M. C., Boyogueno, A. D. B., Nana, W. L., Ngonkeu, M. E. L., Tchamen, N. S., Nwaga, D., Tsala Ndzomo, G., & Amvam Zollo, P. H. (2012). Glomalin related soil protein, carbon, nitrogen and soil aggregate stability as affected by land use variation in the humid forest zone of south Cameroon. *Soil and Tillage Research, 120*, 69–75.

Fomina, M., Charnock, J. M., Hillier, S., Alexander, I. J., & Gadd, G. M. (2006). Zinc phosphate transformations by the *Paxillus involutus*/pine ectomycorrhizal association. *Microbial Ecology, 52*, 322–333.

Fontaine, S., Henault, C., Aamor, A., Bdioui, N., Bloor, J. M. G., Maire, V., Mary, B., Revaillot, S., & Maron, P. A. (2011). Fungi mediate long term sequestration of carbon and nitrogen in soil through their priming effect. *Soil Biology and Biochemistry, 43*, 86–96.

Fransson, A.-M., Valeur, I., & Wallander, I. (2004). The wood-decaying fungus *Hygrophoropsis aurantiaca* increases P availability in acid forest humus soil while N addition hampers this effect. *Soil Biology and Biochemistry, 36*, 1699–1705.

Gadd, G. M. (2007). Geomycology: Biogeochemical transformations of rocks, minerals, metals and radionuclides by fungi, bioweathering and bioremediation. *Mycological Research, 111*, 3–49.

Gadd, G. M. (2017). Fungi, rocks, and minerals. *Elements, 13*, 171–176.

Garcia-Pichel, F., Johnson, S. L., Youngkin, D., & Belnap, J. (2003). Small-scale vertical distribution of bacterial biomass and diversity in biological soil crusts from arid lands in the Colorado Plateau. *Microbial Ecology, 46*, 312–321.

Gillespie, A. W., Farrell, R. E., Walley, F. L., Ross, A. R. S., Leinweber, P., Eckhardt, K.-U., Regier, T., & Blyth, R. I. R. (2011). Glomalin-related soil protein contains non-mycorrhizal-related heat-stable proteins, lipids and humic materials. *Soil Biology and Biochemistry, 43*, 766–777.

Gleeson, D. B., Clipson, N. J. W., Melville, K., Gadd, G. M., & McDermott, F. P. (2005). Mineralogical control of fungal community structure in a weathered pegmatitic granite. *Microbial Ecology, 50*, 360–368.

Gorbushina, A. A. (2007). Life on the rocks. *Environmental Microbiology, 9*, 1613–1631.

Gregory, A. S., Watts, C. W., Whalley, W. R., Kuan, H. L., Griffiths, B. S., Hallett, P. D., & Whitmore, A. P. (2007). Physical resilience of soil to field compaction and the interactions with plant growth and microbial community structure. *European Journal of Soil Science, 58*, 1221–1232.

Gulati, A., Rahi, P., & Vyas, P. (2008). Characterization of phosphate-solubilizing fluorescent pseudomonads from the rhizosphere of seabuckthorn growing in the cold deserts of Himalayas. *Current Microbiology, 56*, 73–79.

Harrison, M. J. (2012). Cellular programs for arbuscular mycorrhizal symbiosis. *Current Opinion in Plant Biology, 15*, 691–698.

Hartmann, J., & Moosdorf, N. (2011). Chemical weathering rates of silicate dominated lithological classes and associated liberation rates of phosphorus on the Japanese Archipelago-implications for global scale analysis. *Chemical Geology, 287*, 125–157.

Hayat, R., Ali, S., Amara, U., Khalid, R., & Ahmed, I. (2010). Soil beneficial bacteria and their role in plant growth promotion: A review. *Annales de Microbiologie, 60*, 579–598.

Hayatsu, M., Tago, K., & Saito, M. (2008). Various players in the nitrogen cycle: Diversity and functions of the microorganisms involved in nitrification and denitrification. *Soil Science & Plant Nutrition, 54*, 33–45.

Helfrich, M., Ludwig, B., Thoms, C., Gleixner, G., & Flessa, H. (2015). The role of soil fungi and bacteria in plant litter decomposition and macroaggregate formation determined using phospholipid fatty acids. *Applied Soil Ecology, 96*, 261–264.

Hoffland, E., Giesler, R., Jongmans, T., & van Breeman, N. (2002). Increasing feldspar tunneling by fungi across a North Sweden podzol chronosequence. *Ecosystems, 5*, 11–22.

Hoffland, E., Kuyper, T. W., Wallander, H., Plassard, C., Gorbushina, A. A., Haselwandter, K., Holmström, S., Landeweert, R., Lundström, U. S., Rosling, A., Sen, R., Smits, M. M., van Hees, P., & van Breemen, N. (2004). The role of fungi in weathering. *Frontiers in Ecology and the Environment, 2*, 258–264.

Huang, J., Sheng, X. F., Xi, J., He, L. Y., Huang, Z., Wang, Q., & Zhang, Z. D. (2014). Depth-related changes in community structure of culturable mineral weathering bacteria and in weathering patterns caused by them along two contrasting soil profiles. *Applied and Environmental Microbiology, 80*, 29–42.

Jansa, J., Mozafar, A., & Frossard, E. (2003). Long-distance transport of P and Zn through the hyphae of an arbuscular mycorrhizal fungus in symbiosis with maize. *Agronomie, 23*, 481–488.

Jonathan, J. H., & Javier, M. G. (2006). Bradford reactive soil protein in Appalachian soils: Distribution and response to incubation, extraction reagent and tannins. *Plant and Soil, 286*, 339–356.

Jorquera, M. A., Hernández, M. T., Rengel, Z., Marschner, P., & Luz Mora, M. (2008). Isolation of culturable phosphobacteria with both phytate-mineralization and phosphate-solubilization activity from the rhizosphere of plants grown in a volcanic soil. *Biology and Fertility of Soils, 44*, 1025–1034.

Kobierska, F., Jonas, T., Magnusson, J., Zappa, M., Bavay, M., Bosshard, T., Paul, F., & Bernasconi, S. (2011). Climate change effects on snow melt and discharge of a partly glacierized watershed in Central Switzerland (Soil Trec Critical Zone Observatory). *Applied Geochemistry, 26*, 60–62.

Koele, N., Dickie, I. A., Blum, J. D., Gleason, J. D., & de Graaf, L. (2014). Ecological significance of mineral weathering in ectomycorrhizal and arbuscular mycorrhizal ecosystems from a field-based comparison. *Soil Biology and Biochemistry, 69*, 63–70.

Kumar, S., Bauddh, K., Barman, S. C., & Singh, R. P. (2014). Amendments of microbial bio fertilizers and organic substances reduces requirement of urea and DAP with enhanced nutrient availability and productivity of wheat (*Triticum aestivum* L.). *Ecological Engineering, 71*, 432–437.

Kuske, C. R., Yaeger, C. M., Johnson, S., Ticknor, O. L., & Belnap, J. (2012). Response and resilience of soil biocrust bacterial communities to chronic physical disturbance in arid shrublands. *The ISME Journal, 6*, 886–897.

Lambers, H., Mougel, C., Jaillard, B., & Hinsinger, P. (2009). Plant–microbe–soil interactions in the rhizosphere: An evolutionary perspective. *Plant and Soil, 321*, 83–115.

Landeweert, R., Hoffland, E., Finlay, R. D., Kuyper, T. W., & van Breemen, N. (2001). Linking plants to rocks: Ectomycorrhizal fungi mobilize nutrients from minerals. *Trends in Ecology & Evolution, 16*, 248–254.

Liang, C., Zhang, X., & Balser, T. C. (2007). Net microbial amino sugars accumulation process in soil as influenced by different plant material inputs. *Biology and Fertility of Soils, 44*, 1–7.

Lindahl, B. D., Ihrmark, K., Boberg, J., Trumbore, S. E., Hogberg, P., Stenlid, J., & Finlay, R. D. (2007). Spatial separation of litter decomposition and mycorrhizal nitrogen uptake in a boreal forest. *The New Phytologist, 173*, 611–620.

Liu, A., Hamel, C., Hamilton, R. I., Ma, B. L., & Smith, D. L. (2000). Acquisition of Cu, Zn, Mn and Fe by mycorrhizal maize (*Zea mays* L.) grown in soil at different P and micronutrient levels. *Mycorrhiza, 9*, 331–336.

Lopez, B. R., Bashan, Y., & Bacilio, M. (2011). Endophytic bacteria of *Mammillaria fraileana*, an endemic rock-colonizing cactus of the Southern Sonoran Desert. *Archives of Microbiology, 193*, 527–541.

Lovelock, C. E., Wright, S. E., & Nichols, K. A. (2004). Using glomalin as an indicator for arbuscular mycorrhizal hyphal growth: An example from a tropical rainforest soil. *Soil Biology and Biochemistry, 36*, 1009–1012.

Männistö, M. K., & Häggblom, M. M. (2006). Characterization of psychrotolerant heterotrophic bacteria from Finnish Lapland. *Systematic and Applied Microbiology, 29*, 229–243.

Mapelli, F., Marasco, R., Rizzi, A., Baldi, F., Ventura, S., Daffonchio, D., & Borin, S. (2011). Bacterial communities involved in soil formation and plant establishment triggered by pyrite bioweathering on Arctic moraines. *Microbial Ecology, 61*, 438–447.

Marschner, P. (2007). Plant microbe interactions in the rhizosphere and nutrient cycling. In P. Marschner & Z. Rengel (Eds.), *Nutrient cycling in terrestrial ecosystems* (Soil biology series, pp. 159–182). Heidelberg: Springer.

Marschner, P. (2008). The role of rhizosphere microorganisms in relation to P uptake by plants. In P. J. White & J. Hammond (Eds.), *The ecophysiology of plant-phosphorus interactions series: Plant ecophysiology* (Vol. 7, pp. 165–176). Dordrecht: Springer.

Marschner, P., & Baumann, K. (2003). Changes in bacterial community structure induced by mycorrhizal colonisation in split-root maize. *Plant and Soil, 251*, 279–289.

Martin, J. P., & Haider, K. (1971). Microbial activity in relation to soil humus formation. *Soil Science, 111*, 54–63.

Miniaci, C., Bunge, M., Duc, L., Edwards, I., Burgmann, H., & Zeyer, J. (2007). Effects of pioneering plants on microbial structures and functions in a glacier forefield. *Biology and Fertility of Soils, 44*, 289–297.

Moore, J. A. M., Jiang, J., Courtney, M., Patterson, C. M., Melanie, A., Mayes, M. A., Wang, G., & Classen, A. T. (2015). Interactions among roots, mycorrhizas and free-living microbial communities differentially impact soil carbon processes. *Journal of Ecology, 103*, 1442–1453.

Morel, J. L., Habib, L., Plantureux, S., & Guckert, A. (1991). Influence of maize root mucilage on soil aggregate stability. *Plant and Soil, 136*, 111–119.

Olivares, J., Bedmar, E. J., & Sanjuán, J. (2013). Biological nitrogen fixation in the context of global change. *Molecular Plant-Microbe Interactions, 26*, 486–494.

Olsson, P. A., Thingstrup, I., Jakobsen, I., & Baath, F. (1999). Estimation of the biomass of arbuscular mycorrhizal fungi in a linseed field. *Soil Biology and Biochemistry, 31*, 1879–1887.

Pal, A., & Pandey, S. (2014). Role of glomalin in improving soil fertility. *International Journal of Plant and Soil Science, 3*, 112–129.

Paul, E. A. (2007). Soil microbiology, ecology and biochemistry in perspective. In E. A. Paul (Ed.), *Soil microbiology ecology and biochemistry* (3rd ed., pp. 3–24). San Diego: Academic.

Paul, E. A., & Clark, F. E. (1996). *Soil microbiology and biochemistry* (2nd ed.). San Diego: Academic.

Puente, M. E., Li, C. Y., & Bashan, Y. (2009). Rock-degrading endophytic bacteria in cacti. *Environmental and Experimental Botany, 66*, 389–401.

Purin, S., & Rillig, M. C. (2007). The arbuscular mycorrhizal fungal protein glomalin: Limitations, progress, and a new hypothesis for its function. *Pedobiologia, 51*, 123–130.

Quirk, J., Beerling, D. J., Banwart, S. A., Kakonyi, G., Romero-Gonzalez, M. E., & Leake, J. R. (2012). Evolution of trees and mycorrhizal fungi intensifies silicate mineral weathering. *Biology Letters, 8*, 1006–1011.

Ranalli, G., Zanardini, E., & Sorlini, C. (2009). Biodeterioration including cultural heritage. In M. Schaechter (Ed.), *Encyclopedia of microbiology* (3rd ed., pp. 191–205). Oxford: Elsevier.

Remiszewskia, K. A., Brycea, J. G., Fahnestocka, M. F., Pettitta, E. A., Blichert-Toft, J., Vadeboncoeur, M. A., & Bailey, S. W. (2016). Elemental and isotopic perspectives on the impact of arbuscular mycorrhizal and ectomycorrhizal fungi on mineral weathering across imposed geologic gradients. *Chemical Geology, 445*, 164–171.

Richardson, A. E., Barea, J. M., McNeill, A. M., & Prigent-Combaret, C. (2009). Acquisition of phosphorus and nitrogen in the rhizosphere and plant growth promotion by microorganisms. *Plant and Soil, 321*, 305–339.

Rillig, M. C., & Mummey, D. L. (2006). Mycorrhizas and soil structure. *The New Phytologist, 171*, 41–53.

Rillig, M. C., & Steinberg, P. D. (2002). Glomalin production by an arbuscular mycorrhizal fungus: A mechanism of habitat modification? *Soil Biology and Biochemistry, 34*, 1371–1374.
Rillig, M. C., Mummey, D. L., Ramsey, P. W., Klironomos, J. N., & Gannon, J. E. (2006). Phylogeny of arbuscular mycorrhizal fungi predicts community composition of symbiosis-associated bacteria: AMF-associated bacterial communities. *FEMS Microbiology Ecology, 57*, 389–395.
Ritz, K. (2006). Fungal roles in transport processes in soils. In G. M. Gadd (Ed.), *Fungi in biogeochemical cycles* (pp. 51–73). Cambridge: Cambridge University Press.
Rosling, A., Roose, T., Herrmann, A. M., Davidson, F. A., Finlay, R. D., & Gadd, G. M. (2009). Approaches to modelling mineral weathering by fungi. *Fungal Biology Reviews, 23*, 1–7.
Ruttenberg, K. C. (2002). The global phosphorus cycle. In A. S. Goudie & D. J. Cuff (Eds.), *The encyclopedia of global change* (Vol. 2, pp. 241–245). Oxford: Oxford University Press.
Sanz-Montero, M. E., & Rodríguez-Aranda, J. P. (2012). Endomycorrhizae in Miocene paleosols: Implications in biotite weathering and accumulation of dolomite in plant roots (SW Madrid Basin, Spain). *Palaeogeography Palaeoclimatology Palaeoecology, 333–334*, 121–130.
Schulz, S., Brankatschk, R., Dumig, A., Kogel-Knabner, I., Schloter, M., & Zeyer, J. (2013). The role of microorganisms at different stages of ecosystem development for soil formation. *Biogeosciences, 10*, 3983–3996.
Schüßler, A., & Walker, C. (2011). Evolution of the 'plant-symbiotic' fungal phylum, Glomeromycota. In S. Pöggeler & J. Wöstemeyer (Eds.), *Evolution of fungi and fungal-like organisms* (pp. 163–185). Berlin: Springer.
Sharma, S. B., Sayyed, R. Z., Trivedi, M. H., & Gobi, T. A. (2013). Phosphate solubilizing microbes: Sustainable approach for managing phosphorus deficiency in agricultural soils. *Springer Plus, 2*, 587.
Six, J., Frey, S. D., Thiet, R. K., & Batten, K. M. (2006). Bacterial and fungal contributions to carbon sequestration in agroecosystems. *Soil Science Society of America Journal, 70*, 555–569.
Smith, S. E., & Read, D. J. (2008). *Mycorrhizal symbiosis* (3rd ed.). New York: Academic.
Smith, S. E., & Smith, F. A. (2011). Roles of arbuscular mycorrhizas in plant nutrition and growth: New paradigms from cellular to ecosystems scales. *Annual Review of Plant Biology, 63*, 227–250.
Smits, M. M., Bonneville, S., Benning, L. G., Banwart, S. A., & Leake, J. R. (2012). Plant-driven weathering of apatite-The role of an ectomycorrhizal fungus. *Geobiology, 10*, 445–456.
Snyder, V. A., & Vázquez, M. A. (2005). Structure. In D. Hillel, J. H. Hatfield, D. S. Powlson, C. Rosenzweig, K. M. Scow, M. J. Singer, & D. L. Sparks (Eds.), *Encyclopedia of soils in the environment* (pp. 54–68). Oxford: Elsevier/Academic.
Sokolova, T. A. (2011). The role of soil biota in the weathering of minerals: A review of literature. *Eurasian Soil Science, 44*, 56–72.
Sollins, P., Homann, P., & Caldwell, B. A. (1996). Stabilization and destabilization of soil organic matter: Mechanisms and controls. *Geoderma, 74*, 65–105.
Stein, L. Y., & Klotz, M. G. (2016). The nitrogen cycle. *Current Biology, 26*, 94–98.
Taylor, L. L., Leake, J. R., Quirk, J., Hardy, K., Banwart, S. A., & Beerling, D. J. (2009). Biological weathering and the long-term carbon cycle: Integrating mycorrhizal evolution and function into the current paradigm. *Geobiology, 7*, 171–191.
Thorley, R. M., Taylor, L. L., Banwart, S. A., Leake, J. R., & Beerling, D. J. (2015). The role of forest trees and their mycorrhizal fungi in carbonate rock weathering and its significance for global carbon cycling. *Plant, Cell & Environment, 38*, 1947–1961.
Towe, S., Albert, A., Kleineidam, K., Brankatschk, R., Dumig, A., Welzl, G., Munch, J. C., Zeyer, J., & Schloter, M. (2010). Abundance of microbes involved in nitrogen transformation in the rhizosphere of *Leucanthemopsis alpina* (L.) Heywood grown in soils from different sites of the Damma glacier forefield. *Microbial Ecology, 60*, 762–770.
Tremblay, L., & Benner, R. (2006). Microbial contributions to N-immobilization and organic matter preservation in decomposing plant detritus. *Geochimica et Cosmochimica Acta, 70*, 133–146.

Uroz, S., Calvaruso, C., Turpault, M. P., Pierrat, J. C., Mustin, C., & Frey-Klett, P. (2007). Effect of the mycorrhizosphere on the genotypic and metabolic diversity of the soil bacterial communities involved in mineral weathering in a forest soil. *Applied and Environmental Microbiology, 73*, 3019–3027.

Uroz, S., Calvaruso, C., Turpault, M. P., & Frey-Klett, P. (2009). Mineral weathering by bacteria: Ecology, actors and mechanisms. *Trends in Microbiology, 17*, 378–387.

Uroz, S., Turpault, M. P., van Scholl, L., Palin, B., & Frey-Klett, P. (2011). Long term impact of mineral amendment on the distribution of mineral weathering associated bacterial communities from the beech *Scleroderma citrinum* ectomycorrhizosphere. *Soil Biology and Biochemistry, 43*, 2275–2282.

van Breemen, N., Mulder, J., & Driscoll, C. T. (1983). Acidification and alkalinization of soils. *Plant and Soil, 75*, 283–308.

van der Heijden, M. G. A., Martin, F. M., Selosse, M.-A., & Sanders, I. R. (2015). Mycorrhizal ecology and evolution: The past, the present and the future. *The New Phytologist, 205*, 1406–1423.

Wallander, H., & Thelin, G. (2008). The stimulating effect of apatite on ectomycorrhizal growth diminishes after PK fertilization. *Soil Biology and Biochemistry, 40*, 2517–2522.

Wallander, H., Wickman, T., & Jacks, G. (1997). Apatite as a P source in mycorrhizal and non-mycorrhizal *Pinus sylvestris* seedlings. *Plant and Soil, 196*, 123–131.

Wanek, W., Mooshammer, M., Blöchl, A., Hanreich, A., & Richter, A. (2010). Determination of gross rates of amino acid production and immobilization in decomposing leaf litter by a novel ^{15}N isotope pool dilution technique. *Soil Biology and Biochemistry, 42*, 1293–1302.

Wang, Y., & Dai, C. C. (2011). Endophytes: A potential resource for biosynthesis, biotransformation, and biodegradation. *Annales de Microbiologie, 61*, 207–215.

Warscheid, T., & Braams, J. (2000). Biodeterioration of stone: A review. *International Biodeterioration and Biodegradation, 46*, 343–368.

Whiffen, L. K., Midgley, D. J., & McGee, P. A. (2007). Polyphenolic compounds interfere with quantification of protein in soil extracts using the Bradford method. *Soil Biology and Biochemistry, 39*, 691–694.

Wilson, D. (1995). Endophyte – the evolution of the term, a clarification of its use and definition. *Oikos, 73*, 274–276.

Wilson, G. W., Rice, C. W., Rillig, M. C., Springer, A., & Hartnett, D. C. (2009). Soil aggregation and carbon sequestration are tightly correlated with the abundance of arbuscular mycorrhizal fungi: Results from long-term field experiments. *Ecology Letters, 12*, 452–461.

Winkelmann, G. (2007). Ecology of siderophores with special reference to the fungi. *Biometals, 20*, 379–392.

Wright, S. F., & Upadhyaya, A. (1996). Extraction of an abundant and unusual protein from soil and comparison with hyphal protein of arbuscular mycorrhizal fungi. *Plant and Soil, 198*, 97–107.

Wu, Q. S., He, X. H., Zou, Y. N., He, K. P., Sun, Y. H., & Cao, M. Q. (2012). Spatial distribution of glomalin-related soil protein and its relationships with root mycorrhization, soil aggregates, carbohydrates, activity of protease and β-glucosidase in the rhizosphere of *Citrus unshiu*. *Soil Biology and Biochemistry, 45*, 181–183.

Xie, X., Huang, W., Liu, F., Tang, N., Liu, Y., Lin, H., & Zhao, B. (2013). Functional analysis of the novel mycorrhiza-specific phosphate transporter AsPT1 and PHT1 family from *Astragalus sinicus* during the arbuscular mycorrhizal symbiosis. *The New Phytologist, 198*, 836–852.

Young, I. M., & Crawford, J. W. (2007). Interactions and self organisation in the soil-microbe complex. *Science, 304*, 1634–1637.

Zhu, Y. G., Duan, G. L., Chen, B. D., Peng, X. H., Chen, Z., & Sun, G. X. (2014). Mineral weathering and element cycling in soil-microorganism-plant system. *Science China Earth Sciences, 57*, 888–896.

Soil Health: The Contribution of Microflora and Microfauna

22

Kalaivani Nadarajah

1 Introduction

While biodiversity above ground has been much better documented and studied, researches conducted on the biodiversity below ground leaves much to be explored (Balvanera et al. 2006; Zavaleta et al. 2010; Hector and Bagchi 2007). When we explore the area around the root of any plant we are likely to find thousands of taxa within the soil made up of bacteria, fungi, insect, nematodes, earthworms and various other living micro and macrofauna and flora (Roesch et al. 2007; Bardgett et al. 2005). While they may remain hidden underground but these organisms contribute not just towards the biodiversity but also contribute towards soil biomass (Fierer et al. 2009; Wardle et al. 2004). The microflora and microfauna found within the soil can be drastically affected through agricultural activities. The crops, the fertilizers, the herbicides and chemical biocontrol agents can significantly change the composition of soil organisms. This therefore has resulted in concerns among agriculturists and soil biologist that the intensified usage of land, chemicals and the lack of activity in a plot of land may result in reduced biodiversity (Mäder et al. 2002; de Vries and Shade 2013). Reduced biodiversity in soil could affect biological processes in the soil and above soil such as nutrient acquisition, and nutrient cycling (de Vries and Shade 2013; van der Heijden et al. 2008). Most of the data collected to date on such drastic changes in land usage has been focused on changes observed in the soil bacterial community, myccorhizae, fungi and soil fauna (Bonkowski and Roy 2005; Griffiths et al. 2000; Maherali and Klironomos 2007; Bradford et al. 2002). As the organisms in the soil live in an interactive, interconnected web, therefore any changes in the community consequently will result in the alteration of the diversity,

K. Nadarajah (✉)
School of Environmental and Natural Resource Sciences, Faculty of Science and Technology, Universiti Kebangsaan Malaysia, Bangi, Selangor, Malaysia
e-mail: vani@ukm.edu.my

© Springer Nature Singapore Pte Ltd. 2019

A. Varma, D. K. Choudhary (eds.), *Mycorrhizosphere and Pedogenesis*,
https://doi.org/10.1007/978-981-13-6480-8_22

abundance and the interactive dynamics between the populations (Hunt and Wall 2002; Duffy et al. 2007).

Therefore in this chapter we will look into the main microflora and microfauna within the soil system. We will address them individually in order to determine the role these organisms play not just in soil diversity, but the contribution of these organisms to the functionality of the ecosystem. However, whatever is known of the organisms within the soil, it is still unlikely that we will be able to chart the complete story of the ecosystem interaction and determine how these organisms contribute to soil health and sustainable agriculture. In the past years studies have shown that plant diversity has its influence on multiple ecosystems defined as the ecosystem multi-functionality. We are yet to determine however, if the ecosystem multi-functionality is influenced by soil biodiversity. This only adds to the complexity in the flora, fauna and plant interactions (Maestre et al. 2012).

1.1 The Role of Microflora and Microfauna in Soil Health

As described above the microorganisms that are found within the soil does not just affect the biomass but it does result in consequences that can affect agriculture. These interaction can either be beneficial or a non-beneficial. The beneficial attributes of microflora are: protect crops, increase yield, improve nutrient cycling, biological pest control, maintenance of soil structure and function, and degradation of agrochemicals and pollutants (Newton and Chantal 2010). The non-beneficial microbes are there to cause disease and also to inhibit the microflora within the soil. The microbial diversity within the soil would include organisms such as bacteria, fungi, archae and algae. These organisms have their niche contribution to the soil from nutrient cycling to structure and formation. In addition to these microflora, the soil contains microfauna that are able to decompose material within the soil. These microfauna include organisms such as protozoa, and nematodes. These organisms have also been implicated in grazing plant roots to induce the excretion of exudates into the soil which contributes towards soil enrichment.

1.2 Contribution of Microflora

1.2.1 Bacteria

Through the advent of molecular techniques, the microbial diversity and structure has been further elucidated where more and more phyla have been determined and logged into the microbial diversity (Agrawal et al. 2015). However the depth of diversity in the soil is largely dependent on the soil type and richness. Additionally in most soils that have been analyzed to date the most abundant species that were detected are those belonging to the bacterial group especially those belonging to the phyla Proteobacteria, Acidobacteria and Actinobacteria (Lee et al. 2008, Nemergut et al. 2011; Kielak et al. 2016). Through the newly available molecular techniques, more microbes are being identified and assigned to their respective phyla. However through the sequence data that is being derived from 16S libraries there remain bacterial isolates with unknown identity. This can at times be about 10% of the sequence data derived which only indicates that there is so much more that we do not know about the soil microbial community (Janssen 2006; Nacke et al. 2011; Rincon-Florez et al. 2013).

The largest group of soil microorganisms reported belong to the phyla Proteobacteria. From the various groups of known Proteobacteria, it has been reported that the α-, β-, γ-, and δ-Proteobacteria are the most prevalent within rich soil areas such as the rhizosphere (Fierer et al. 2007; Spain et al. 2009). The α-Proteobacteria are known for their ability in recycling nutrients and toxic compound in the environment. The members of this group of Proteobacteria are either heterotrophs or autotrophs. Members such as the Bradyrhizobium, Rhizobium, *Nitrospira* sp. and *Nitrobacter* sp. are involved in the nitrogen cycle where Rhizobium and Bradyrhizobiums are nitrogen fixers while Nitrospira and Nitrobacter are nitrite oxidizers. In addition to the heterotrophs and autotrophs in alpha Proteobacteria, the phototrophic Rhodospirillum and Rhodobacter have also been reported as players in nitrogen cycling. Other members such as the *Sphingomonas* spp. are able to degrade toxic waste found in the environment while *Methylobacter* sp. and *Methylophilus* sp. act as methane oxidizers (Goldfarb et al. 2011).

In any given soil type the β-Proteobacteria is one that responds to changes in nutrient levels. Most of these organisms that belong in this group are heterotrophs, autotrophs, and methanotrophs. The best-known β-Proteobacteria heterotrophs in soil belongs to the genera Burkholderia, Alcaligenes, and Acidovorax. The largest group of bacteria within this phylum is the *Burkholderia* sp. which fluctuates in density based on nutrient availability or scarcity (Goldfarb et al. 2011). Burkholderia are able to metabolize amino acids, methane, minerals, sugars and other compounds to turn around carbon. Both Burkholderia and Collimonas species weather minerals (Uroz et al. 2007). Other economically important members of *β Proteobacteria* are able to use nitrate as their electron acceptor and can be used industrially to remove nitrate from wastewater by denitrification. A number of β-Proteobacteria are diazotroph, meaning that they can fix molecular nitrogen from the air as their nitrogen source for growth – this is important to the farming industry as it is a primary means of ammonium levels rising in soil without the presence of leguminous plants. An example of a methanotroph belonging to the β-Proteobacteria is *Methylomonas* sp.

The γ-Proteobacteria in soil are the most diverse in their nutritional needs including heterotrophs, lithotrophs, and phototrophs. Among the best-known heterotrophs are Pseudomonas and Xanthamonas. The γ-Proteobacteria also includes the photolithotrophs which functions under anaerobic conditions in light and are able to utilize sulphides and elemental sulphur as electron donors and CO_2 as their carbon sources. *Chromatium,* a photosynthetic, hydrogen sulfide oxidizing microorganisms produces sulphur as a waste product. Some γ-Proteobacteria are methane oxidizers, and many are symbiotic with geothermic ocean vent-dwelling animals. The fourth Proteobacteria group, δ-Proteobacteria is able to utilize sulphide and iron for growth and nutrition. This group of organisms has a predominantly aerobic group with *Myxobacteria* which is able to flourish even in unfavorable environments. Another strictly anaerobic genera contains most of the sulphate (*Desulfovibrio*, *Desulfobacter*, *Desulfococcus*, *Desulfonema*), sulphur reducing(e.g. Desulfuromonas spp.) and iron reducing bacteria (*Geobacter* spp.). The anaerobic condition in the soil results in the production of CO_2, ethanol and lactate that has been utilized by the delta bacteria as carbon sources (Davis et al. 2011; Goldfarb et al. 2011).

The next abundant species found in soil are the Acidobacteria which is found to fluctuate and differ with changes in soil pH (Lauber et al. 2009). Members of the phylum Acidobacteria make up an average of 20% (range, 5–46%) of soil bacterial communities and is almost as diverse and large as the phylum Proteobacteria (Dojka et al. 2000; Hugenholtz et al. 1998) with three main subclasses: Actinobacteridae, Acidimicrobidae, and Rubrobacteridae, the largest being Actinobacteridae. This phyla is largely heterotrophic aerobes (Jones et al. 2009; Davis et al. 2011). Through genome analysis this phyla was assigned as mainly oligotrophic with the ability to utilize various carbon sources (Ward et al. 2009). Members of this phyla have shown ability to withstand adverse conditions such as lack of nutrients, moisture content, pH and various other conditions that affect directly their effectiveness in their given environment (Janssen 2006). The Verrucomicrobia makes up an average of 7% (range, 0–21%) of soil bacterial communities with the class Spartobacteria being one of the most dominant (janssen et al. 2002; Sangwan et al. 2004). The ecology niche of Verrucomicrobia however remains poorly understood where very little is known of its dominant members nutrient, and physiological requirements (Bergmann et al. 2011).

Another group of microbes that occupy a niche in the soil microbiology are the Gram-positive microorganisms. This group has many lineages of Actinobacteria and Acidimicrobidae (e.g., see references Gremion et al. 2003; Lüdemann and Conrad. 2000). The members of the lineages of the Actinobacteridae lack phylogenetic depth with high phenotypic diversity Dojka et al. 2000, Garrity and Holt 2001; Joseph et al. 2003). Generally the actinobacteroids are aerobic heterotrophs. However we cannot dismiss the possibility that the subclasses such as Rubrobacteridae and Acidimicrobidae may utilize other forms of metabolic systems (Janssen 2006). Actinobacteria belonging to the subclass Actinobacteridae and isolates from soil include Arthrobacter, Rhodococcus, Streptomyces, and Mycobacterium (Goldfarb et al. 2011). Streptomyces are known for their ability to produce antimicrobial compounds. The Rubrobacteridae include the genera Rubrobacter and Solirubrobacter. Both genera are not common in soil culture collections. Rubrobacter are especially prevalent in desert soils and may resist ionizing radiation (Holmes et al. 2000). To withstand extreme conditions some of these organisms produce endospores. Some of these endospore-forming organisms belong to the genus Bacillus and Clostridium. Some of these Bacilli are able to utilize various carbon sources and are involved in nitrogen fixation and denitrification (Goldfarb et al. 2011).

Through the sequencing of soil sample it has been reported that the phylum Bacteroidetes make up about 5% of soil bacterial communities depending on soil pH. The organisms within this phyla range from aerobe to anaerobe and therefore the species composition of the soil is largely dependent on the oxygen content. Main genera belonging to these phyla that are found predominantly in the soil are

Hymenobacter, Flavobacterium, Pedobacter, and Chitinophaga. It has been suggested that Bacteroidetes are copiotrophs, because their relative abundance in soil may increase following carbon-addition (Fierer et al. 2007; Eilers et al. 2010). The relative abundances of Bacteroidetes and Actinobacteria tends to increase with environmental changes (Lauber et al. 2009). DeBruyn et al. (2011) suggesting that they are adapted to low soil moisture conditions.

1.2.2 Archae Bacteria

Several researcher have reported on the archae communities found in the soil through 16S rRNA studies (Bates et al. 2011; Prosser and Nicol 2012; Mukhtar et al. 2017). These studies have identified that the most widespread archae in soil is from the phylum Crenarchaeota (Treusch et al. 2005). Ammonium-oxidizing crenarchaea have been isolated from garden soil and even from the paddy field in various parts of the world (Tourna et al. 2011). It was initially thought that the ammonia oxidizing bacteria (AOB) were the drivers of the process in the soil. However recent discovery has confirmed that ammonia-oxidizing archaea (AOA) (Könneke et al. 2005; Treusch et al. 2005; Venter et al. 2004) may be the prominent drivers of this process underground. The initial reports to highlight the importance of AOA came from soil microbial survey conducted in various soil types in Europe, aquatic and terrestrial environments where AOA were clearly dominant among ammonia oxidizers (Leininger et al. 2006; He et al. 2007; Dang et al. 2008). To further strengthen the outcome of these studies, molecular approaches were applied to identify AOA abundance, diversity and ecosystem functionality in any given soil environment (Tourna et al. 2008; Jia and Conrad 2009; Offre et al. 2009). Environmental factors such as low nutrient availability and pH affect AOA's niche in the process of nitrification (Erguder et al. 2009). In recent studies by Zhang et al. (2010, 2012) SIP experiments have confirmed that environmental factors control AOA function and distribution. In high nutrient environment i.e. post fertilizing, it has been reported that the AOB is higher in density that the AOA. However, the relative importance and contribution of AOB and AOA in ammonia oxidation remains obscure. It has however been indicated that their contribution to nitrification varies with soil conditions. The Euryarchaeota, which are methanogens, are prevalent in waterlogged soils, which is ideal for this anaerobic organism (Angel et al. 2012). These organisms utilize the available complex organic matter in the soil and converts it into methane and carbon dioxide gas. The dominant methanogens found in the soil are members from the genera Methanosarcina, Methanosaeta, and Methanocella. Methanosarcina and Methanosaeta that are able to reduce acetate to methane.

1.2.3 Actinomycetes

Though Actinomycetes are generally Gram-positive, aerobic, mycelial bacteria; we address them separately here due to the role they play in soil ecology. These organisms are known to play a role in producing bioactive compounds and have been exploited by pharma and industries in development of antibiotics, vitamins and enzymes (Terkina et al. 2006). In addition to the above actinomycetes such as *Streptomyces* sp., Micromonospora and Norcadia have countlessly been reported as

Actinomycetaceae that are involved in enhancing soil fertility and disease suppression (Aghighi et al. 2004). As a major component of soil ecosystem this organism has been reported to bind atmospheric N_2 to produce ammonium for use by forest plants and trees (Brady and Weil, 2008). In addition these organisms are also actively involved in the degradation of organic matter and are great choice of organism for composting. The bioactive components secreted by this group of organism such as antimicrobials and siderophores are able to enhance plant growth and at the same time inhibit detrimental soil borne pathogens (Aghighi et al. 2004; Franco-Correa et al. 2010). The multifunctional Actinomycetes contribute carbon source through their exudates, supply easily assimilated nitrates, control root pathogens, and in general maintain good soil health (Govaerts et al. 2007). Enzymes such as phosphatases secreted by Actinomycetes assist with the mineralization of P sources in the soil (Richardson et al. 2009). Actinomycete also can improve both shoots and roots biomass accumulation. Tarkka and Frey-Klett (2008) considered the Actinomycetes such as Streptomyces species as Mycorrhiza Helper Bacteria (MHBs) as they are involved in the colonization of roots with mycorrhiza (Schrey et al. 2005).

1.2.4 Algae

This group of microscopic plant like organisms includes cyanobacteria, mosses, ferns liverwort, lichen, and grasses. The blue-green algae like Anabaena, Nostoc, Aulosira, Calothrix, and *Plectonema* spp. have been extensively studied and are involved in nitrogen fixation and have especially been isolated in paddy fields (Sahu et al. 2012). Due to their ability to adapt easily to various environments, these organisms have been used as inoculums in soil to improve soil fertility and thence resulting in better soil structure and yield in the paddy fields (Dhar et al. 2007). Through their ability to degrade and add organic matter to the soil, the Cyanobacteria are able to provide O_2, improve salinity, improve pH regulation, increase water holding capacity, solubilize P and other organic compounds in the soil (Kaushik 2004; Roger and Reynaud, 1982). Eventually when these organisms die, they contribute towards increase in soil biomass, reduction in soil salinity and growth of weeds (Saadatnia and Riahi, 2009). Azolla and Anabaena are both associated with nitrogen fixation and have been used effectively in rice fields to fix nitrogen (Sahu et al. 2012). Pioneer organisms like lichens, liverworts and mosses, colonize the substrate, degrading the rocks and incorporating organic and organic compounds through chemical change and organic process (Thomas, 2013). Through this method, they stabilize and moderate the microenvironment, making conditions suitable for colonizers, leading to the institution of higher plants and invertebrate animals.

Plants directly influence the soil community by their root growth and plant cover. The majority of microorganisms found in the soil are associated with plants roots that provide them with carbon and other nutrients. An agroforestry technology termed "fertilizer tree" is a system by which leguminous trees or woody shrubs are grown to produce nutrients back to soil and thus enriching it. This is one way plants may contribute towards the health of the soil system.

1.2.5 Fungi

The fungal communities within the soil are just as diverse as the soil bacteria. To date about seven fungal phyla have been identified as dominant in most soil types. They are the Ascomycota, Basidiomycota, Blastocladiomycota, Neocallimastigomycota, Glomeromycota, Chytridiomycota, and Microsporidia (Liu et al. 2006). One of the most commonly encountered fungal phyla is the Ascomycota where more than 15,000 species of these organisms, have been reported to live symbiotically with algae, and cyanobacteria. The symbiotic relationship between the algae and ascomycetes (bipartite) and sometimes cyanobacteria (tripartite) too, results in the formation of lichens (Lutzoni et al. 2001). The fungi provides the holdfast and the protection from radiation and desiccation while the algae conducts the photosynthesis. In the event if cyanobacteria are present in the interaction, it is able to fix atmospheric nitrogen. The organic acids they secrete help to break down primary substrates, thereby helping a soil profile to develop and facilitating primary succession of plants onto these new soils.

Another ascomycetes interaction involves the formation of ectomycorrhizal and/or ectoendomycorrhizal with plants. In these interactions the mycorrhizal fungi forms a symbiotic association with the roots. While the mycorrhiza improves nutrient absorption from the soil the plant provides sugars to the fungus. Through the formation of this interaction with mycorrhizae, plants are able to inhabit more environments as the arbuscular (AM) and ectomycorrhizal (EM) fungi increases the efficiency of nutrient absorption, improves water acquisition and protects against soil based pathogens. AM are more dominant then the EM. They are extremely important in utilization of inorganic soil phosphorus. As a saprobe, the EM decomposes organic material through a wide range of enzymes like amylases, proteases, lipases, and phosphatases. As these fungi receive their nutrition from plants, their in exhaustive fuel enables them to provide more enzymes compared to other saprotrophs (Lindahl et al. 2007). The products from the decomposing effect of these fungi are beneficial not just to the fungal community but also to the other micro and macro organisms found within the soil. Thus, by making substrates available to other soil organisms, saprotrophic fungi increases the biomass and diversity of soils and plays a critical role in decomposition. EM are not selective on host and therefore have been reported to form mycorrhizal network that enables sharing of nutrients, water, defence molecules and others between plants. Any seed or plant that grows within a mycorrhizal network benefits greatly from the resources shared within this network (Teste et al. 2009) contributing towards improved growth and survival of plants in ecosystems through their complex adaptive systems (Simard et al. 2012).

Besides the functions stated above, ascomycetes are also effective in the control of soil based pathogens and parasites. However, perhaps the most remarkable lifestyle of a member of the Ascomycota in soil is that of predators. Members of the family Orbiliaceae are carnivorous fungi with hyphae that are specialized to trap prey. Some species hyphae are spring-loaded, ring-shaped traps that respond to the movement of prey, which include a variety of soil mesofauna including protists, nematodes, tardigrades, and arthropods. The enzymes secreted by this group of organisms are also effective in inhibiting pathogens found in the soil. The second most important phyla is the Basidiomycota. This group of fungi is further divided into Pucciniomycotina, Ustilaginomycotina, and the Agaricomycotina (Kirk et al. 2008; Hibbett et al. 2007). Among these three subphyla, members of the subphyla Agaricomycotina are particularly important in temperate forests and woodland where they form the majority of ectomycorrhizae (as well as prized edible mushrooms). A few species in the Agaricomycotina are lichenized fungi (e.g. Omphalina) and therefore are involved in symbiotic activities between the fungi, algae and cyanobacteria.

1.3 Contribution of Microfauna

Soil contains various types of fauna. These include organisms such as protozoa, nematodes, earthworms, snails and various types of insects (Maha, 2013; Battigelli and Berch, 2002). These organisms affect the chemical and physical structure of the soil through the interactions within the soil (Sugiyarto, 2009). The fauna within the soil are further divided into macro and microfauna and in this section our focus will be towards two members of the microfauna group, the nematodes and the protozoa. These organisms have their advantages of soil microflora as bioindicators of soil health due to their ability to integrate the physical, chemical, and biological properties related with their food resources. In addition these organisms do not have a reproduction rate which is as rapid as microbes where it is stable temporally as it does not change as rapidly as microbes in variable environmental settings (Nannipieri et al. 1990).

1.3.1 Protozoa

This particular group of microfauna are classified as single celled animals which feed on other organisms such as bacteria in the soil and have been classified into groups based on their shape. Protozoa are found in greatest abundance near the surface of the soil, particularly in the upper 15 cm (6 inches). These organisms require water and move within the water films and the water filled pores in the soil. The protozoan groups are: (1) flagelattes which are motile through the use of flagella and feed primarily on bacteria; (2) Cilliates which are the least numerous protozoans and move with cillia and they feed on amoeba, flagellates and bacteria. They consume large numbers of organisms a day; finally there are the amoebae which are large and move through the production of pseudopods. Amoebas reside in the rhizosphere and at the root surface where they graze on bacteria populations. The life cycle of many protozoa consists of an active or trophozoite phase where the animal feeds and multiplies and a resting or cyst stage where the cell produces a thick coating. In the cyst stage, many species can withstand harsh environmental conditions and persist for many years until environmental conditions improve.

Protozoa play an important role in mineralizing nutrients for use both by microbe and the plant. Protozoa do not need the concentrations of C and N that they obtain from the bacteria that they ingest. Therefore the excess is secreted into the soil for use by microbes and the plant. Protozoa help to maintain the ecology of the soil through feeding on the bacteria. This results in stimulation of bacterial population, decomposition and soil aggregation and can lead to changes in the bacterial community. Protozoa are additionally a vital nutrient source for other soil organisms and also assist with the suppression of disease through competition or feeding on pathogens. Food and moisture determines the presence of protozoa in the soil. The amount of moisture will determine which protozoa is dominant. When examining any soil sample, there can be a difference in protozoa population size where rich soil has high content. Like bacteria, protozoa are particularly active in the rhizosphere next to roots. Mastigophora or flagellates tend to dominate in drier soils while Ciliophora or ciliates are abundant only if the soil moisture level is high. In bacterial-dominated soils like cultivated soils, flagellates and amoebae predominate. In general, high clay-content soils contain a higher number of smaller protozoa (flagellates and naked amoebae), while coarser textured soils and undisturbed or no-till soils contain more large flagellates, testate amoebae, and ciliates. Protozoa and bacterial-feeding nematodes compete for their common food resource, which is bacteria. Some soils have high numbers of either nematodes or protozoa, but not both (Nielsen and Winding 2002; Hoorman 2011).

1.3.2 Nematode

Nematodes or roundworms are non-segmented worms with tapered ends typically 1/500 of an inch (50 μm) in diameter and 1/20 of an inch (1 mm) in length. They have a head, and a tail with a well developed central nervous and fertility system with a complete digestive system, so they are considered the most primitive animal. They are small enough to fit in most soil pores and soil aggregates. They are classified in the animal phylum Nemata and are best known for causing infectious disease in plants and animals, but they also play an important role in soil and crop ecology. Nematodes are aquatic organisms so they require adequate soil moisture to move in the soil. A few species are responsible for plant diseases but far less is known about the majority of the nematode community that plays beneficial roles in soil. Many beneficial nematodes serve as biological pest control agents in managed systems and others regulate the natural ecosystem and soil nutrient cycling. A variety of nematodes function at several trophic levels of the soil food web. Nematodes are most abundant in the surface soil horizon.

Like protozoa, nematodes are also responsible for mineralization especially that of nitrogen. These ecologically significant organism have highly diverse feeding habits that enables them to adapt to any environment and play key roles in the above and below ground nutrient web (Borgonie et al. 2011). Due to their abundance they are involved in processes such as soil organic matter (SOM) decomposition and plant health (Wardle et al. 2004; Griffiths et al. 2001). Plant grazing nematodes are known to induce the secretion of root exudates into the soil which stimulates the microbial population (Denton et al. 1999) that results in a ‘priming effect’ which increases SOM decomposition (Kuzyakov 2002).

Nematodes can increase nitrogen mineralization to between 8–19% (Beare et al. 1997; Ferris et al. 1998; Ingham et al. 1985). Through their ingestion of both bacteria and fungi, nematodes produce in excess organic and inorganic compound (such as amino acids, NH_{4+} and PO_{4-}) which is then returned to the soil and utilized by the plants (Bonkowski et al. 2000; Trofymow and Coleman, 1982; Seastedt et al. 1988; Sohlenius et al. 1988; Ingham et al. 1985). The abundance and activity of these microbivorous nematodes may, in turn, also be regulated by predatory nematodes and other fauna, further modulating nutrient availability (Neher 2001). Predatory nematodes also regulate nitrogen mineralization by feeding on microbial grazing nematodes, a conduit by which resources pass from bottom to top trophic levels (Wardle and Yeates 1993).

Nematodes can be divided into five broad groups based on their diet with the first four groups being free living:

1. Bacterial-feeders (Bacterivorous) consume bacteria through a stoma, a large open channel.
2. Fungal-feeders feed (fungivorous) by puncturing the cell wall of fungi using a small slender stylet to suck out the internal contents.
3. Omnivores eat anything from microflora to microfauna and plant root systems. They may have different nutritional requirements at different stages of their life.
4. Predatory nematodes eat all types of nematodes and protozoa using a stylet. They eat smaller microorganisms whole or attach themselves to the cuticle of larger nematodes, scraping away until the prey's internal body parts can be extracted.
5. Root-feeders are plant parasites feeding on roots, and thus are not free-living in the soil because they live either inside or outside the plant root, depending on the plant root for a food source.

Fungivorous, bacterivorous, and omnivorous nematodes feed on microbes and excrete waste into the soil.

Despite years of studies, it is still difficult to understand and elucidate the role played by microfauna in the setting of soil health (Schimel and Bennett 2004). This is primarily due to the lack of environmental analysis and quantitative data that will enable to solidly decipher the role played by this group of organisms (Osler and Sommerkorn 2007).

1.4 Optimizing the Use of Living Organisms in the Maintenance of Soil Fertility

In soils, microflora and microfauna are involved in the process of cycling nutrients for use by plants and soil communities. They play a pivotal role in the cycling of C and N which are essential in producing essential items such as amino acids, DNA and RNA. Mineralization by these organisms supplies the nutrients to the soil in order to maintain soil fertility. Figure 22.1 depicts the interaction between the

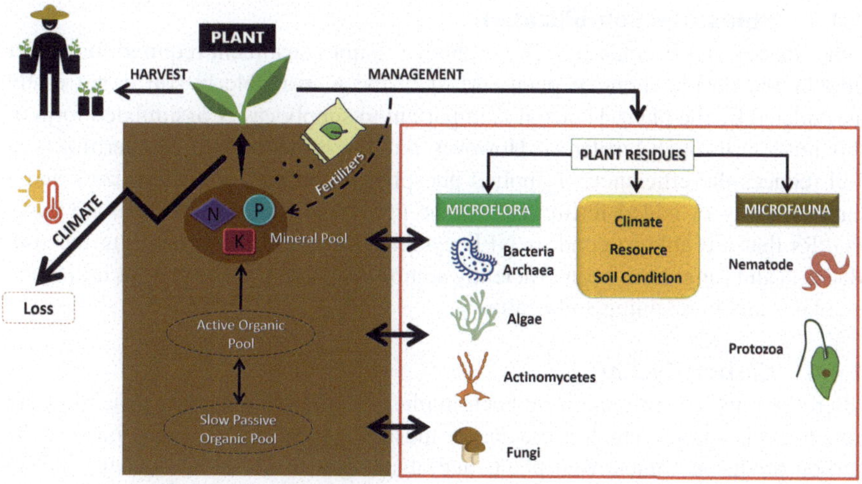

Fig. 22.1 The role played by microflora and microfauna in restoring soil health and contributing towards improved agricultural yields. The twin arrow shows the connection between the plant, microflora and microfauna residues and the source of nutrient pool. The nutrient pool enhances the growth and yield

microflora and microfauna and the role that they play in nutrient cycling and how this contributes to soil health and thus improved agricultural environment.

1.4.1 Nitrogen Cycling

All living organisms require N as a source for the synthesis of nucleic acids. For healthier and more productive crops, N is an essential nutrient where plants need inorganic nitrogen sources such as ammonium and nitrate (Schimel and Bennett 2004). Microbes play an important role in the nitrogen cycle where they are nitrogen fixers, nitrifiers, denitrifiers, anammox, and dissimilatory nitrate reduction to ammonia (DNRA). There are specific groups of organisms that conduct specific functions in nitrogen cycling. The type of microflora and microfauna in the soil will determine the efficiency of nitrogen cycling.

1.4.2 Phosphorus Cycling

Soil is known to contain phosphorus (P) in abundance. However this readily available P is precipitated in the presence of alkali pH. Soil microbiota are able to transform the soil based P in two ways: (1) mineralize organic P into inorganic phosphate enzymatically via fungi and bacteria. (2) transformation insoluble P into mobilized and solubilized form catalyzed enzymatically by specific group of organisms. The phosphorus produced is sufficient for their specific use and for secretion and use by other soil based organisms and plants (Plassard and Dell 2010; Wakelin et al. 2012).

1.4.3 Potassium Solubilization

Other than N and P, potassium (K) is another major constituent required for proper growth and development of plants. Most of the K available in soil is not easily assimilated by the plant. Hence it is important to supply easily assimilated form of this nutrient through fertilizers. However, fixation of added nutrients/fertilizers in soil reduces the efficiency of applied phosphorus and potassium fertilizers and a large quantity of added fertilizers become unavailable to plants. There are many PGPRs that are able to conduct NPK cycling in the soil thus enriching the soil. Bacteria and fungi have been efficiently solubilizing the K in the soil for utilization by plants and in retaining soil health.

1.4.4 Carbon Cycling

Microflora and microfauna have been implicated in carbon cycling. Microbes are able to cycle carbon which is crucial for most living organisms. While most of the carbon produced comes from plants and algae, cyanobacteria and lichen are also able to fix carbon in the ecosystem. Autotrophic soil microbes are able to fix carbon dioxide while the heterotrophic soil organisms, both fungi and bacteria are able to recycle organic material. The saprotrophs complete the carbon cycle by converting the primary producers organic material into CO_2 through the process of decomposition. Mineralization is also another process that occurs in environment where the most predominant organisms that contribute towards it are the fungi and protozoa. The organic compounds are completely mineralized to carbon dioxide, ammonia, and water (Eilers et al. 2010; Treseder et al. 2011).

1.5 Conclusions and Future Prospects

Food security is a global issue and it is major threat with the increasing world population, the declining land space and the degradation of soil quality due to over exploitation. Therefore it is important for us to identify ways and means by which we can increase yield per hectare by including soil communities into the equation. Soil organisms both microflora and microfauna play a crucial role in soil fertility and health. They have been reported to enhance the quality of degraded soil through nutrient cycles. Although there have been encouraging reports on the contribution of these organisms in improving nutrient availability and soil aggregation there still remains areas that require further study such as:

1. Studies have been conducted to test single versus co-inoculation and laboratory versus field-testing. Generally co-inoculation provides better results in soil health and better yield enhancement. However when tested in the greenhouse and later implemented in the field condition results may vary greatly. This therefore requires all isolates to be tested for efficacy at the field level.
2. Most often farmers choose to use fertilizers to enhance soil health and increase yield. This is most often due to very little attention being given to identifying microbial and microfauna communities that are indeed good for soil health.

3. Research on understanding the role of microflora and microfauna in nutrient cycling and to determine their efficiency in returning soil health in degraded soil is seriously lacking.
4. Determining the best combination of organisms for specific soil types.

The microflora and microfauna contribute towards soil health through the secretion and excretion of products. They achieve incorporation of nutrients to soil through the process of fixation, solubilization, chelation, mineralization, excretion and degradation. The nutrients and chemical components that are secreted play specific roles in changing soil chemistry and structure. Therefore for us to optimize the usage of microflora and microfauna in contributing towards soil health we need to first identify targets, understand their mechanism of function, understand their role and interaction in soil communities and finally their interaction and effect on plants and other soil microbiota. Hence we hope that with more research and a better understanding of these microbiota and their mechanism of action and contribution we are then able to use them more efficiently in agriculture as soil natural 'fertilizers' as opposed to the current excessive use of chemical fertilizers which may be a serious concern for the farmers as well as on the environment.

References

Aghighi, S., Shahidi Bonjar, G., & Saadoun, I. (2004). First report of antifungal properties of a new strain of Streptomyces plicatus (strain101) against four Iranian phytopathogenic isolates of Verticillium dahliae, a new horizon in biocontrol agents. *Biotechnology (Faisalabad), 3*, 90–97.

Agrawal, P. K., Agrawal, S., & Shrivastava, R. (2015). Modern molecular approaches for analyzing microbial diversity from mushroom compost ecosystem. *3Biotech, 5*, 853–866.

Angel, R., Claus, P., & Conrad, R. (2012). Methanogenic archaea are globally ubiquitous in aerated soils and become active under wet anoxic conditions. *The ISME Journal, 6*, 847.

Balvanera, P., Pfisterer, A. B., Buchmann, N., He, J. S., Nakashizuka, T., Raffaelli, D., & Schmid, B. (2006). Quantifying the evidence for biodiversity effects on ecosystem functioning and services. *Ecology Letters, 9*, 1146–1156.

Bardgett, R. D., Bowman, W. D., Kaufmann, R., & Schmidt, S. K. (2005). A temporal approach to linking aboveground and belowground ecology. *Trends in Ecology & Evolution, 20*, 634–641.

Bates, S. T., Berg-Lyons, D., Caporaso, J. G., Walters, W. A., Knight, R., & Fierer, N. (2011). Examining the global distribution of dominant archaeal populations in soil. *The ISME Journal, 5*, 908.

Battigelli, J. P., & Berch, S. (2002). *Soil Fauna in the sub-boreal spruce (sbs) installations of the long-term soil productivity (ltsp) study of Central British Columbia: One year results for soil Mesofauna and microfauna*. British Columbia: Ministry of Forests, Prince George, Prince Rupert and Caribou Forest Regions.

Beare, M., Reddy, M. V., Tian, G., & Srivastava, S. (1997). Agricultural intensification, soil biodiversity and agroecosystem function in the tropics: The role of decomposer biota. *Applied Soil Ecology, 6*, 87–108.

Bergmann, G. T., et al. (2011). The under-recognized dominance of Verrucomicrobia in soil bacterial communities. *Soil Biology and Biochemistry, 43*, 1450–1455.

Bonkowski, M., & Roy, J. (2005). Soil microbial diversity and soil functioning affect competition among grasses in experimental microcosms. *Oecologia, 143*, 232–240.

Bonkowski, M., Cheng, W., Griffiths, B. S., Alphei, J., & Scheu, S. (2000). Microbial-faunal interactions in the rhizosphere and effects on plant growth §. *European Journal of Soil Biology, 36*, 135–147.

Borgonie, G., et al. (2011). Nematoda from the terrestrial deep subsurface of South Africa. *Nature, 474*, 79.

Bradford, M., et al. (2002). Impacts of soil faunal community composition on model grassland ecosystems. *Science, 298*, 615–618.

Brady, N., & Weil, R. (2008). *Soil colloids: Seat of soil chemical and physical acidity*. Upper Saddle River: Pearson Education.

Dang, H., Zhang, X., Sun, J., Li, T., Zhang, Z., & Yang, G. (2008). Diversity and spatial distribution of sediment ammonia-oxidizing crenarchaeota in response to estuarine and environmental gradients in the Changjiang Estuary and East China Sea. *Microbiology, 154*, 2084–2095.

Davis, K. E., Sangwan, P., & Janssen, P. H. (2011). Acidobacteria, Rubrobacteridae and Chloroflexi are abundant among very slow-growing and mini-colony-forming soil bacteria. *Environmental Microbiology, 13*, 798–805.

De Vries, F. T., & Shade, A. (2013). Controls on soil microbial community stability under climate change. *Frontiers in Microbiology, 4*, 265.

DeBruyn JM, Nixon LT, Fawaz MN, Johnson AM, Radosevich M (2011) Global biogeography and quantitative seasonal dynamics of Gemmatimonadetes in soil. *Applied and Environmental Microbiology*. 05005-05011.

Denton, C. S., Bardgett, R. D., Cook, R., & Hobbs, P. J. (1999). Low amounts of root herbivory positively influence the rhizosphere microbial community in a temperate grassland soil. *Soil Biology and Biochemistry, 31*, 155–165.

Dhar, D. W., Prasanna, R., & Singh, B. (2007). Comparative performance of three carrier based blue green algal biofertilizers for sustainable rice cultivation. *Journal of Sustainable Agriculture, 30*, 41–50.

Dojka, M. A., Harris, J. K., & Pace, N. R. (2000). Expanding the known diversity and environmental distribution of an uncultured phylogenetic division of bacteria. *Applied and Environmental Microbiology, 66*, 1617–1621.

Duffy, J. E., Cardinale, B. J., France, K. E., McIntyre, P. B., Thébault, E., & Loreau, M. (2007). The functional role of biodiversity in ecosystems: Incorporating trophic complexity. *Ecology Letters, 10*, 522–538.

Eilers, K. G., Lauber, C. L., Knight, R., & Fierer, N. (2010). Shifts in bacterial community structure associated with inputs of low molecular weight carbon compounds to soil. *Soil Biology and Biochemistry, 42*, 896–903.

Erguder, T. H., Boon, N., Wittebolle, L., Marzorati, M., & Verstraete, W. (2009). Environmental factors shaping the ecological niches of ammonia-oxidizing archaea. *FEMS Microbiology Reviews, 33*, 855–869.

Ferris, H., Venette, R., Van Der Meulen, H., & Lau, S. (1998). Nitrogen mineralization by bacterial-feeding nematodes: Verification and measurement. *Plant and Soil, 203*, 159–171.

Fierer, N., Bradford, M. A., & Jackson, R. B. (2007). Toward an ecological classification of soil bacteria. *Ecology, 88*, 1354–1364.

Fierer, N., Strickland, M. S., Liptzin, D., Bradford, M. A., & Cleveland, C. C. (2009). Global patterns in belowground communities. *Ecology Letters, 12*, 1238–1249.

Franco-Correa, M., Quintana, A., Duque, C., Suarez, C., Rodríguez, M. X., & Barea, J.-M. (2010). Evaluation of actinomycete strains for key traits related with plant growth promotion and mycorrhiza helping activities. *Applied Soil Ecology, 45*, 209–217.

Garrity, G. M., & Holt, J. G. (2001). The road map to the manual. In D. R. Boone, R. W. Castenholz, & G. M. Garrity (Eds.), *Bergey's manual of systematic bacteriology* (Vol. 1, 2nd ed., pp. 119–166). Springer: New York.

Goldfarb, K. C., et al. (2011). Differential growth responses of soil bacterial taxa to carbon substrates of varying chemical recalcitrance. *Frontiers in Microbiology, 2*, 94.

Govaerts, B., et al. (2007). Influence of tillage, residue management, and crop rotation on soil microbial biomass and catabolic diversity. *Applied Soil Ecology, 37*, 18–30.

Gremion, F., Chatzinotas, A., & Harms, H. (2003). Actinobacteria might be a dominant part of the metabolically active bacteria in heavy-metal contaminated bulk and rhizosphere soil. *Environmental Microbiology, 5*, 896–907.

Griffiths, B. S., Ritz, K., Bardgett, R. D., Cook, R., Christensen, S., Ekelund, F., Sorensen, S. J., Baath, E., Bloem, J., de Ruiter, P. C., Dolfing, J., & Nicolardot, B. (2000). Ecosystem response of pasture soil communities to fumigation-induced microbial diversity reductions: An examination of the biodiversity-ecosystem function relationship. *Oikos, 90*, 279–294.

Griffiths, B. S., Ritz, K., Wheatley, R., Kuan, H. L., Boag, B., Christensen, S., Ekelund, F., Sorensen, S. J., Muller, S., & Bloem, J. (2001). An examination of the biodiversity-ecosystem function relationship in arable soil microbial communities. *Soil Biology and Biochemistry, 33*, 1713–1722.

He, J., Shen, J., Lm, Z., Zhu, Y., Zheng, Y., Xu, M., & Di, H. (2007). Quantitative analyses of the abundance and composition of ammonia-oxidizing bacteria and ammonia-oxidizing archaea of a Chinese upland red soil under long-term fertilization practices. *Environmental Microbiology, 9*, 2364–2374.

Hector, A., & Bagchi, R. (2007). Biodiversity and ecosystem multifunctionality. *Nature, 448*, 188.

Hibbett, D. S., et al. (2007). A higher-level phylogenetic classification of the Fungi. *Mycological Research, 111*, 509–547.

Holmes, A. J., Bowyer, J., Holley, M. P., O'donoghue, M., Montgomery, M., & Gillings, M. R. (2000). Diverse, yet-to-be-cultured members of the Rubrobacter subdivision of the Actinobacteria are widespread in Australian arid soils. *FEMS Microbiology Ecology, 33*, 111–120.

Hoorman, J. J. (2011). *The role of soil protozoa and nematodes Fact sheet: agriculture and natural resources* (pp. 1–5). Colombus: The Ohio State University Extension.

Hugenholtz, P., Goebel, B. M., & Pace, N. R. (1998). Impact of culture-independent studies on the emerging phylogenetic view of bacterial diversity. *Journal of Bacteriology, 180*, 4765–4774.

Hunt, H., & Wall, D. (2002). Modelling the effects of loss of soil biodiversity on ecosystem function. *Global Change Biology, 8*, 33–50.

Ingham, R. E., Trofymow, J., Ingham, E. R., & Coleman, D. C. (1985). Interactions of bacteria, fungi, and their nematode grazers: Effects on nutrient cycling and plant growth. *Ecological Monographs, 55*, 119–140.

Janssen, P. H. (2006). Identifying the dominant soil bacterial taxa in libraries of 16S rRNA and 16S rRNA genes. *Applied and Environmental Microbiology, 72*, 1719–1728.

Janssen, P. H., Yates, P. S., Grinton, B. E., Taylor, P. M., & Sait, M. (2002). Improved culturability of soil bacteria and isolation in pure culture of novel members of the divisions Acidobacteria, Actinobacteria, Proteobacteria, and Verrucomicrobia. *Applied and Environmental Microbiology, 68*, 2391–2396.

Jia, Z., & Conrad, R. (2009). Bacteria rather than Archaea dominate microbial ammonia oxidation in an agricultural soil. *Environmental Microbiology, 11*, 1658–1671.

Jones, R. T., Robeson, M. S., Lauber, C. L., Hamady, M., Knight, R., & Fierer, N. (2009). A comprehensive survey of soil acidobacterial diversity using pyrosequencing and clone library analyses. *The ISME Journal, 3*, 442.

Joseph, S. J., Hugenholtz, P., Sangwan, P., Osborne, C. A., & Janssen, P. H. (2003). Laboratory cultivation of widespread and previously uncultured soil bacteria. *Applied and Environmental Microbiology, 69*, 7210–7215.

Kaushik, B. (2004). Use of blue-green algae and Azolla biofertilizers in rice cultivation and their influence on soil properties. In *Microbiology and biotechnology for sustainable development* (pp. 166–184). New Delhi: CBS.

Kielak, A. M., Barreto, C. C., Kowalchuk, G. A., van Veen, J. A., & Kuramae, E. E. (2016). The ecology of Acidobacteria: Moving beyond genes and genomes. *Frontiers in Microbiology, 7*, 744.

Kirk, P., Cannon, P., Minter, D., & Stalpers, J. (2008). *Dictionary of the Fungi* (Vol. 396). Wallingford: CABI.

Könneke, M., Bernhard, A. E., José, R., Walker, C. B., Waterbury, J. B., & Stahl, D. A. (2005). Isolation of an autotrophic ammonia-oxidizing marine archaeon. *Nature, 437*, 543.

Kuzyakov, Y. (2002). Factors affecting rhizosphere priming effects. *Journal of Plant Nutrition and Soil Science, 165*, 382–396.

Lauber, C. L., Hamady, M., Knight, R., & Fierer, N. (2009). Pyrosequencing-based assessment of soil pH as a predictor of soil bacterial community structure at the continental scale. *Applied and Environmental Microbiology, 75*, 5111–5120.

Lee, S.-H., Ka, J.-O., & Cho, J.-C. (2008). Members of the phylum Acidobacteria are dominant and metabolically active in rhizosphere soil. *FEMS Microbiology Letters, 285*, 263–269.

Leininger, S., et al. (2006). Archaea predominate among ammonia-oxidizing prokaryotes in soils. *Nature, 442*, 806.

Lindahl, B. D., Ihrmark, K., Boberg, J., Trumbore, S. E., Högberg, P., Stenlid, J., & Finlay, R. D. (2007). Spatial separation of litter decomposition and mycorrhizal nitrogen uptake in a boreal forest. *New Phytologist, 173*, 611–620.

Liu, Y. J, Hodson, M. C., & Hall, B. D. (2006). Loss of the flagellum happened only once in the fungal lineage: Phylogenetic structure of kingdom Fungi inferred from RNA polymerase II subunit genes. *BMC Evolutionary Biology, 6*: 74. www.biomedcentral.com/1471-2148/6/74

Lüdemann, H., & Conrad, R. (2000). Molecular retrieval of large 16S rRNA gene fragments from an Italian rice paddy soil affiliated with the class Actinobacteria. *Systematic and Applied Microbiology, 23*, 582–584.

Lutzoni, F., Pagel, M., & Reeb, V. (2001). Major fungal lineages are derived from lichen symbiotic ancestors. *Nature, 411*, 937.

Mäder, P., Fliessbach, A., Dubois, D., Gunst, L., Fried, P., & Niggli, U. (2002). Soil fertility and biodiversity in organic farming. *Science, 296*, 1694–1697.

Maestre, F. T., et al. (2012). Plant species richness and ecosystem multifunctionality in global drylands. *Science, 335*, 214–218.

Maha, A. A. L. (2013). Ecological role of animal diversity in soil system (A Case Study at El-Rawakeeb Dry Land Research Station, Sudan) 1st Annual International Interdisciplinary Conference, AIIC 2013, 24–26 April, Azores, Portugal – Proceedings, pp. 345–350.

Maherali, H., & Klironomos, J. N. (2007). Influence of phylogeny on fungal community assembly and ecosystem functioning. *Science, 316*, 1746–1748.

Mukhtar, H., Lin, Y.-P., & Anthony, J. (2017). Ammonia oxidizing archaea and bacteria in east Asian paddy soils—A mini review. *Environments, 4*, 84.

Nacke, H., et al. (2011). Pyrosequencing-based assessment of bacterial community structure along different management types in German forest and grassland soils. *PLoS One, 6*, e17000.

Nannipieri, P., Grego, S., Ceccanti, B., Bollag, J., & Stotzky, G. (1990). Ecological significance of the biological activity in soil. *Soil Biochemistry*, 6, 293–355.

Neher, D. A. (2001). Role of nematodes in soil health and their use as indicators. *Journal of Nematology, 33*, 161.

Nemergut, D. R., et al. (2011). Global patterns in the biogeography of bacterial taxa. *Environmental Microbiology, 13*, 135–144.

Newton, L., & Chantal, H. (2010). Soil Biology of the Canadian Prairies. Agricultural Soils of the Prairies. PS&C. *Prairie Soils and Crops Journal, 3*, 16–24.

Nielsen, M., & Winding, A. (2002). Microorganisms as indicators of soil health. National Environmental Research Institute, Denmark Tech Rep:388.

Offre, P., Prosser, J. I., & Nicol, G. W. (2009). Growth of ammonia-oxidizing archaea in soil microcosms is inhibited by acetylene. *FEMS Microbiology Ecology, 70*, 99–108.

Osler, G. H., & Sommerkorn, M. (2007). Toward a complete soil C and N cycle: Incorporating the soil fauna. *Ecology, 88*, 1611–1621.

Plassard, C., & Dell, B. (2010). Phosphorus nutrition of mycorrhizal trees. *Tree Physiology, 30*, 1129–1139.

Prosser, J. I., & Nicol, G. W. (2012). Archaeal and bacterial ammonia-oxidisers in soil: The quest for niche specialisation and differentiation. *Trends in Microbiology, 20*, 523–531.

Richardson, A. E., Barea, J.-M., McNeill, A. M., & Prigent-Combaret, C. (2009). Acquisition of phosphorus and nitrogen in the rhizosphere and plant growth promotion by microorganisms. *Plant and Soil, 321*, 305–339.

Rincon-Florez, V. A., Carvalhais, L. C., & Schenk, P. M. (2013). Culture-independent molecular tools for soil and rhizosphere microbiology. *Diversity, 5*, 581–612.

Roesch, L. F., et al. (2007). Pyrosequencing enumerates and contrasts soil microbial diversity. *The ISME Journal, 1*, 283.

Roger, P.-A., & Reynaud, P.-A. (1982). Free—Living blue—Green algae in tropical soils. In *Microbiology of tropical soils and plant productivity* (pp. 147–168). Dordrecht: Springer.

Saadatnia, H., & Riahi, H. (2009). Cyanobacteria from paddy fields in Iran as a biofertilizer in rice plants. *Plant, Soil and Environment, 55*, 207–212.

Sahu, D., Priyadarshani, I., & Rath, B. (2012). Cyanobacteria–as potential biofertilizer. *CIBTech Journal of Microbiology, 1*, 20–26.

Sangwan, P., Chen, X., Hugenholtz, P., & Janssen, P. H. (2004). Chthoniobacter flavus gen. nov., sp. nov., the first pure-culture representative of subdivision two, Spartobacteria classis nov., of the phylum Verrucomicrobia. *Applied and Environmental Microbiology, 70*, 5875–5881.

Schimel, J. P., & Bennett, J. (2004). Nitrogen mineralization: Challenges of a changing paradigm. *Ecology, 85*, 591–602.

Schrey, S. D., Schellhammer, M., Ecke, M., Hampp, R., & Tarkka, M. T. (2005). Mycorrhiza helper bacterium Streptomyces AcH 505 induces differential gene expression in the ectomycorrhizal fungus Amanita muscaria. *New Phytologist, 168*, 205–216.

Seastedt, T., James, S., & Todd, T. (1988). Interactions among soil invertebrates, microbes and plant growth in the tallgrass prairie. *Agriculture, Ecosystems & Environment, 24*, 219–228.

Simard, S. W., Beiler, K. J., Bingham, M. A., Deslippe, J. R., Philip, L. J., & Teste, F. P. (2012). Mycorrhizal networks: Mechanisms, ecology and modelling. *Fungal Biology Reviews, 26*, 39–60.

Sohlenius, B., Boström, S., & Sandor, A. (1988). Carbon and nitrogen budgets of nematodes in arable soil. *Biology and Fertility of Soils, 6*, 1–8.

Spain, A. M., Krumholz, L. R., & Elshahed, M. S. (2009). Abundance, composition, diversity and novelty of soil Proteobacteria. *The ISME Journal, 3*, 992.

Sugiyarto, S. (2009). The effect of mulching technology to enhance the diversity of soil macroinvertebrates in Sengon-based agroforestry systems. *Biodiversitas Journal of Biological Diversity, 10*, 129–133.

Tarkka, M. T., & Frey-Klett, P. (2008). Mycorrhiza helper bacteria. In *Mycorrhiza – State of the art, genetics and molecular biology, eco-function, biotechnology, eco-physiology, structure and systematics* (pp. 113–132). Heidelberg: Springer.

Terkina, I., Parfenova, V., & Ahn, T. (2006). Antagonistic activity of actinomycetes of Lake Baikal. *Applied Biochemistry and Microbiology, 42*, 173–176.

Teste, F. P., Simard, S. W., & Durall, D. M. (2009). Role of mycorrhizal networks and tree proximity in ectomycorrhizal colonization of planted seedlings. *Fungal Ecology, 2*, 21–30.

Thomas, W. C. (2013). Role of arthropods in maintaining soil fertility. *Agriculture, 3*, 629–659. www.mdpi.com/journal/agriculture

Tourna, M., Freitag, T. E., Nicol, G. W., & Prosser, J. I. (2008). Growth, activity and temperature responses of ammonia-oxidizing archaea and bacteria in soil microcosms. *Environmental Microbiology, 10*, 1357–1364.

Tourna, M., et al. (2011). Nitrososphaera viennensis, an ammonia oxidizing archaeon from soil. *Proceedings of the National Academy of Sciences, 108*, 8420–8425.

Treseder, K. K., Kivlin, S. N., & Hawkes, C. V. (2011). Evolutionary trade-offs among decomposers determine responses to nitrogen enrichment. *Ecology Letters, 14*, 933–938.

Treusch, A. H., Leininger, S., Kletzin, A., Schuster, S. C., Klenk, H. P., & Schleper, C. (2005). Novel genes for nitrite reductase and Amo-related proteins indicate a role of uncultivated mesophilic crenarchaeota in nitrogen cycling. *Environmental Microbiology, 7*, 1985–1995.

Trofymow, L., & Coleman, D. (1982). The role of bacterivorous and fungivorous nematodes in cellulose and chitin decomposition. In *Nematodes in soil ecosystems* (pp. 111–138). Austin: University of Texas.

Uroz, S., Calvaruso, C., Turpault, M.-P., Pierrat, J.-C., Mustin, C., & Frey-Klett, P. (2007). Effect of the mycorrhizosphere on the genotypic and metabolic diversity of the bacterial communities involved in mineral weathering in a forest soil. *Applied and Environmental Microbiology, 73*, 3019–3027.

Van Der Heijden, M. G., Bardgett, R. D., & Van Straalen, N. M. (2008). The unseen majority: Soil microbes as drivers of plant diversity and productivity in terrestrial ecosystems. *Ecology Letters, 11*, 296–310.

Venter, J. C., Remington, K., Heidelberg, J. F., Halpern, A. L., Rusch, D., Eisen, J. A., et al. (2004). Environmental genome shotgun sequencing of the Sargasso Sea. *Science, 304*, 66–74. https://doi.org/10.1126/science.1093857.

Wakelin, S., et al. (2012). Response of soil microbial communities to contrasted histories of phosphorus fertilisation in pastures. *Applied Soil Ecology, 61*, 40–48.

Ward, N. L., et al. (2009). Three genomes from the phylum Acidobacteria provide insight into the lifestyles of these microorganisms in soils. *Applied and Environmental Microbiology, 75*, 2046–2056.

Wardle, D. A., & Yeates, G. W. (1993). The dual importance of competition and predation as regulatory forces in terrestrial ecosystems, evidence from decomposer food-webs. *Oecologia, 93*, 303–306.

Wardle, D. A., Bardgett, R. D., Klironomos, J. N., Setälä, H., Van Der Putten, W. H., & Wall, D. H. (2004). Ecological linkages between aboveground and belowground biota. *Science, 304*, 1629–1633.

Zavaleta, E. S., Pasari, J. R., Hulvey, K. B., & Tilman, G. D. (2010). Sustaining multiple ecosystem functions in grassland communities requires higher biodiversity. *Proceedings of the National Academy of Sciences, 107*, 1443–1446.

Zhang, L.-M., Offre, P. R., He, J.-Z., Verhamme, D. T., Nicol, G. W., & Prosser, J. I. (2010). Autotrophic ammonia oxidation by soil thaumarchaea. *Proceedings of the National Academy of Sciences, 107*, 17240–17245.

Zhang, L.-M., Hu, H.-W., Shen, J.-P., & He, J.-Z. (2012). Ammonia-oxidizing archaea have more important role than ammonia-oxidizing bacteria in ammonia oxidation of strongly acidic soils. *The ISME Journal, 6*, 1032.

Molecular Microbial Biodiversity Assessment in the Mycorrhizosphere

23

Kalaivani Nadarajah and Ilakiya Sharanee Kumar

1 Introduction

The microbial flora within the soil plays a crucial role in ensuring the plant's well-being and the richness of the soil. These organisms have niche activities that contribute either through nutrient uptake, nutrient cycles, suppression of disease, growth enhancement and many more processes (Jacobi et al. 2017; Muller et al. 2016). While studying the soil microbial structure, it has been noted that mycorrhizae also play a role in the root ecosystem. This therefore has resulted to the widening of the rhizosphere terminology to mycorrhizosphere, which includes the fungal component of this community (Sehgal and Sagar 2017). As mycorrhizae and the soil microorganisms contribute to the overall well-being and productivity of plants, the understanding of the interactions involved between the plant-microbe-soil is absolutely crucial. The understanding derived from these interactions is imperative in improving soil health and crop production.

A major group of fungi in the root system is the arbuscular mycorrhizal (AM) which is known to form symbiosis with the host root systems. Currently at least 160 taxa have been identified and a brief analysis via molecular techniques has indicated that these numbers are conservative. Research conducted on soil microbiology has shown that bacterial communities also interact with the AM fungi in the root. They affect the root-fungi interaction directly through (i) provision of energy, (ii) exudates that improve AM function such as germination, growth, receptivity and recognition, (iii) alteration of soil pH, and (iv) exudates that inhibit the detrimental organisms in the soil. Indirectly, these bacteria can affect the growth, yield, soil structure and root exudates in a mycorrhizae based interaction. The direct impact of the soil bacteria interaction with the root and the mycorrhizae has mostly been

K. Nadarajah (✉) · I. S. Kumar
School of Environmental and Natural Resource Sciences, Faculty of Science and Technology, Universiti Kebangsaan Malaysia, Bangi, Selangor, Malaysia
e-mail: vani@ukm.edu.my

© Springer Nature Singapore Pte Ltd. 2019

A. Varma, D. K. Choudhary (eds.), *Mycorrhizosphere and Pedogenesis*,
https://doi.org/10.1007/978-981-13-6480-8_23

positive in greenhouse trials (Ross 1980; Tommerup 1985; Wilson et al. 1988). Frequent reports have cited that AM improves plants nutrient uptake and improves disease resistance in their host. Other organisms such as N fixers and P solubilizer are known to work with AM in jointly improving plants growth and development (Puppi et al. 1994).

Now that we have accomplished the definition of the microbial composition within this area, we arrive now at a problem that is constantly faced by soil microbiologists which is the appropriate tools to study the community, diversity and structure. The initial techniques that were utilized by microbiologists such as general serial dilution, plating and the biochemical assays have all met with their limitations especially when addressing soil microorganisms that are tedious or difficult to culture. As group of non-culturable and difficult organisms make up a large portion of soil microbes, it is essential that these organisms are identified so that their role and function within the ecosystem is understood (Amann et al. 1995). The endosymbionts remain largely unexplored and require elucidation for better understanding of the microbial diversity in the ecosystems (Bianciotto et al. 1996, 2000). Therefore, to study the mycorrhizae population and the immense unculturable and culturable organisms within the soil, technologies that are high throughput and able to screen large quantities of material quickly and accurately is required. Through the advent of molecular biology, several molecular biology and omics platforms have been established which enable us to address the need to analyze large microbial samples, including unculturable organisms, at high accuracy, at improved costing and reduced time (Hugenholtz et al. 1998, 2001; Quince et al. 2009). The molecular assessment techniques have provided means to study various soil ecosystems (Elshahed et al. 2008; Finlay and Medzhitov 2007; Liu et al. 2007). This chapter endeavors to provide an overview of molecular assessment tools that are available and their applications and limitations in studying the mycorrhizosphere community with the overall aim of using this information to enhance plant well-being and positively contributing to sustainable agriculture.

2 Molecular Detection of Microorganisms in the Mychorrhizosphere

Compared to the morphological and biochemical methods that have been employed to date, the molecular approaches promise better opportunities to analyze the full diversity of the microbial community. The continuous advancement in technologies and platforms related to molecular studies allows for rapid profiling of communities to identify microbial groups present and thus making the information readily available for mutual benefit of scientists from various different fields (Fakruddin and Mannan 2013).

As mentioned above, all methods utilized for the analysis of the mycorrhizosphere has to be inclusive of mycorrhizae and other fungal and bacterial species found within the sphere. Therefore there is a continuous quest for methods that would provide precise coverage of microbial diversity at ideal cost and at a time

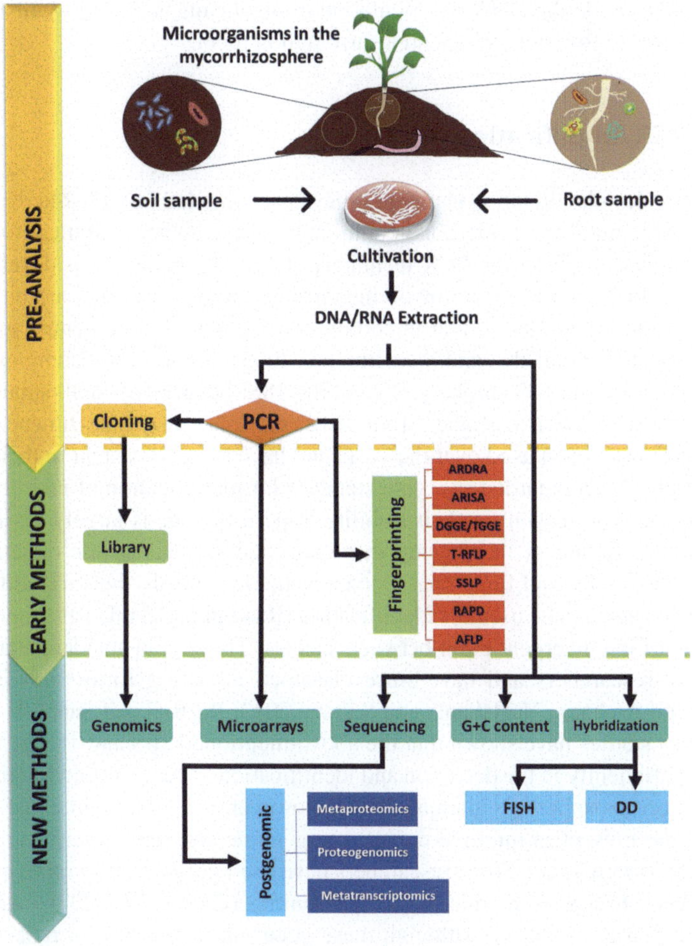

Fig. 23.1 Links most of the methods available to analyze soil sample from the mycorrhizosphere. This figure provides processes that encompass all the analysis starting from the early methods up to the current cutting-edge platforms available to analyze soil samples

effective manner. Figure 23.1 shows the diagrammatic representation of various methods that may be utilized to conduct microbial analysis on the mycorrhizosphere and the components within this zone.

2.1 Nucleic Acid Isolation

As in most molecular techniques especially those that require PCR, nucleic acid is a routine requirement that needs to be fulfilled. Appropriate soil and root samples are needed for successful isolation of nucleic acid. Samples are collected aseptically from the roots and the soil aggregates around the root within the mycorrhizosphere.

This would provide the DNA representation of organisms including symbionts that are found within the root and the area surrounding them.

2.2 PCR Amplification

Most molecular identification techniques have been divided into PCR and non-PCR based. A large number of microbial community and structure analytical tools have been developed utilizing the PCR technique. The PCR technique is developed on the basis of PCR amplification involving specific target genes that are either prokaryotic or eukaryotic based and in certain cases genes that are genus, species or function specific (González and Saiz-Jiménez 2005). Some of the commonly used marker genes are the 16S (prokaryote) and the 18S (eukaryote) small subunit ribosomal RNA (rRNA) (Dakal and Arora 2012; González and Saiz-Jiménez 2005). These genes have been used over the years and have been consistent in their results that they have been regarded as a gold standard for identification of microbes from environments. The reason for their stability in performance is largely attributed to the ubiquitous nature of these genes in prokaryotic and eukaryotic organisms, as they are structurally and functionally conserved. Henceforth, the variability in the conserved region can be used for identification (Rastogi and Sani, 2011) and for the estimation of the divergent point between species. These gold standards have been utilized by researchers and have since classified all living forms into Eukarya, Bacteria and Archaea (Neelakanta and Sultana 2013; Rastogi and Sani 2011).

Previous studies have shown that the above-mentioned 16S and 18S genes have been used efficiently in the detection and identification of bacteria and fungi present in the mycorrhizosphere (Nadarajah 2017). However, we need to note that the 16S gene may be present in multiple copies in a genome and thus it may be useful to have alternative markers. Some researchers have used genes such as *rpoS, gyrB* and *recA* in their studies of the microbial communities (Case et al. 2007; Tacão et al. 2005; Waleron et al. 2008). Although these genes show promise in reflecting the evolutionary history and diversity within a community (van Elsas et al. 2006), the limited availability of sequence databases for these genes in contrast to 16S and 18S hampers the extensive use of these candidates. However, it is hoped that the continuous submission of data on these alternative candidates to databases will eventually result in these genes being used routinely in soil microbial analysis. One definite indication for a need of new alternative marker genes comes from the difficulty in resolving pseudomonads through the utilization of 16S rRNA. This is due to the fact that pseudomonads have slightly distinct roles and these functions are supported by different sets of accessory genes. Costa et al. (2007) reported that the global regulator *gacA* gene was able to resolve the pseudomonads at a higher resolution compared to the universal 16S rRNA gene. Amplification of genes from DNA/RNA of microbial communities such as *amoA*, *nifH*, *nirK*, *nirS*, and *dsrA* facilitated studies on microbial processes such as nitrogen fixation, denitrification and sulfate reduction. Microbial catabolic diversity can be elucidated through advanced studies on enzymes-coding genes that are involved in carbon utilization.

In addition to the 18S rRNA, another commonly used molecular marker in the identification of fungi is the internal transcribed spacer region located between 18S and 28S rRNA which consists of internal non-coding regions ITS1, ITS2 and 5.8S rRNA gene. These regions are highly conserved and may be beneficial in studies that aim to show the similarities between evolutionarily distant organisms and sequences with high genetic variability (ITS regions) which will especially be useful in determining genera and species. Apart from that, the ITS regions are of particular importance in molecular diagnostics of molds, because they are present in all fungi in a large number of copies, which increases the sensitivity and specificity of the PCR reaction (Atkins and Clark 2004; Ciardo et al. 2007, 2010). Further, other than the ITS primers mentioned above (especially IT1 and ITS4 which is widely used), certain studies have also utilized universal eukaryotic primers such as NS31 (Simon et al. 1992) in combination with AM2 and AM3 (Santos-González et al. 2007) which produces the amplified 5.8S rRNA gene. The PCR products amplified from environmental DNA can be analyzed by (i) genetic fingerprinting, (ii) clone libraries, or (iii) by combination of these techniques or (iv) new next generation technologies.

Besides the standard PCR process, quantitative PCR or qPCR is being applied in analysis of DNA extracted from soil. The extracted DNA is subjected to qPCR to quantify the number of target genes of 16S or any other functional genes (*amoA, rpo,* or *nifH*). Though it has been successfully utilized in soil studies (Kolb et al. 2003), this method provides a bias picture of the number of targets and does not detect similar genes with slightly varied sequence or similar function. Nonetheless, this method is still quite efficient at portraying the effects of the environment on the gene and gene expression and thus is efficient in mapping the diversity of the microbial communities in various environmental conditions within the soil or the mycosphere.

2.3 Preparation of Library

Following PCR amplification is the construction of libraries that carry the amplified PCR product. The establishment of the cloned libraries provide a means to analyze PCR products obtained from 16S and 18S rRNA genes. A metagenomic analysis of any given microbial community involves the construction of libraries which involves the isolation of metagenome DNA, the fragmentation, cloning and transformation, followed by screening and bioinformatics analysis of the clones (Mocali and Benedetti 2010).

There are plasmids such as cosmids, fosmids and BACs that may be utilized in the construction of these libraries depending on the size of inserts involved. In most studies conducted to date the preferred host for cloning and expression studies of the metagenome is *Escherichia coli*. However, over the years there are new host that have been included into the repertoire such as *Pseudomonas aeruginosa*, *Rhizobium leguminosarum* and *Streptomyces lividans*. These host have been chosen for some specific application such as analysis and detection of bioactive compounds (Mocali and Benedetti 2010; Streit and Schmitz 2004).

These metagenomic libraries are then subjected to analysis based on the objective(s) of the study, which could be anything from determining the presence of a gene to the identification of clones with a desired function. Some of these activities may not require the sequencing of the libraries as it may be involved in identification of a specific gene, enzymes or metabolites. For instance, a study on the mycorrhizosphere indicates that the genes of interest may be directly involved in process of nitrogen fixation, nutrient acquisition, quorum sensing and others. However, in projects that require determination of community diversity and structure, there is a need for sequencing which would incur a higher cost into the projects. Hence it is quite common for projects like this to include a prescreening strategy such as fingerprinting to ensure smaller number of clones are subjected to the process of sequencing followed by analysis that adopts bioinformatics tools (Coutinho et al. 2013; Deja-Sikora et al. 2007; Gonzalez et al. 2003; Mocali and Benedetti 2010; McNamara et al., 2006). The following section will address the importance and application of fingerprinting techniques.

2.4 Fingerprinting Techniques

The genetic fingerprinting technique prompts to electrophoretically analyzing PCR based products that have been amplified from metagenomic DNA. There are several types of fingerprinting tools that have been developed over the years that may be utilized in the microbial fingerprinting of the mycorrhizosphere. These techniques include: ARDRA (amplified rDNA restriction analysis), ARISA (automated ribosomal intergenic spacer analysis), SSCP (single strand conformation polymorphism), T-RFLP (terminal restriction fragment length polymorphism) and DGGE/TGGE (denaturing/temperature gradient gel electrophoresis).

Fingerprinting techniques have been used in the detection of microbial cells and in visualizing the quantitative profiles of the composition within a given ecosystem. Conducting genetic fingerprinting has permitted the researchers to explore the diversity within a community especially for communities that involve non-culturable and difficult organisms. Although a composition of the community is provided, this method by no means provides a direct taxonomic identification of microorganisms (Dakal and Arora 2012; González and Saiz-Jiménez 2005). The basic procedure of this protocol is the isolation of a given sample DNA, which is followed by amplification of any specific genes mentioned above and visualization of the product on an electrophoretic gel. The banding profiles generated from these amplified products represents the data which will be analyzed (Muyzer 1999; Rastogi and Sani 2011).

2.4.1 Amplified Ribosomal RNA Restriction Analysis (ARDRA)

ARDRA is utilized in monitoring the communities within changing environments. In this particular technique, the rDNA is amplified via PCR and digested using restriction enzyme before visualization of the restricted fragments via gel electrophoresis. This technique allows the capture of microbial community structure information but unfortunately it does not give a picture on diversity and phylogeny (Cetecioglu et al. 2012; Rastogi and Sani 2011). ARDRA-ITS allows the inquiry of microorganisms without any information on the genome organization. The conserved domain within the amplified rDNA is interrupted by the non-coding variable of ITS1 and ITS 2, which allows for differentiation. This is useful to exhibit the differences at the species and subspecies levels. However one of the major limitation of the ARDRA technique is that it does not provide any details about the microbial population present in the sample (Gich et al. 2000).

2.4.2 Automated Ribosomal Intergenic Spacer Analysis (ARISA)

ARISA has been efficiently used to shed light on the richness and diversity of microbial communities. This culture independent method was developed towards the end of the twentieth century to differentiate between the size and nucleotide variation within the intergenic spacer region that exists between the 16S and 23S ribosomal subunits (Cardinale et al. 2004; Fisher and Triplett 1999; Popa et al. 2009). The variation within the intergenic spacer region is analyzed within an automated capillary laser detection system. This method of analysis utilizes universal primers that cause multiple peaks and limits the ability of the system. In addition, it is very difficult to interpret results for fingerprints obtained for uncultured microorganisms (Popa et al. 2009).

2.4.3 Denaturing or Temperature Gradient Gel Electrophoresis (DGGE/TGGE)

Denaturing or temperature gradient gel electrophoresis are molecular techniques based on PCR-amplified molecular markers (16S rRNA or 18S rRNA genes) separated by gradient polyacrylamide gels electrophoresis based on either chemical gradient (denaturing) or temperature gradient (Rastogi and Sani 2011). Both these techniques involve partial denaturation of DNA within domains that is largely dependent on the DNA sequences of these domains. Differences in nucleotide sequences will cause difference in temperature of melting for this particular domain and therefore result in variable migration rates through the polyacrylamide gel (Muyzer 1999; Muyzer et al. 1993; Muyzer and Smalla 1998; Więckowicz 2009).

These techniques allow for the detection of approximately 50% of differences in sequence of fragments which can go up to 500 bp. Besides providing the ability to determine the differences between these fragments, this technique also has the added advantage of excision of respective bands from the gel for amplification followed by sequencing. The sequence data obtained from these fragments may be utilized to generate phylogenetic correlations of the microbial diversity in a given sample. However, one limitation of this technique lies in the short fragments generated i.e. up to 500 bp. These short fragments make it a bit hard to separate the fragments effectively to make concrete interpretation of the results. However from literature review of past studies of microbial diversity and communities, the DGGE/TGGE techniques have been successfully used to interpret the microbial communities of bacteria (Gaylarde et al. 2012; Piñar et al., 2009, 2013), cyanobacteria (Gaylarde et al. 2012), archaea (Piñar et al. 2001a, b) and fungi (Giacomucci et al. 2011). As the mycorrhizosphere has all these groups of organisms, this technique remains a method of choice for microbial diversity studies.

2.4.4 Terminal Restriction Fragment Length Polymorphism (T-RFLP)

The T-RFLP method is a marriage between multiple techniques, which includes RFLP, PCR, nucleic acid electrophoresis, and comparative genomics. This fingerprinting technique is used as a supplement to the ARDRA method. The only difference between the ARDRA technique and T-RFLP is that one of the two primers used in this technique is fluorescent labeled (Liu et al. 1997; Więckowicz 2009). The amplified fragment is then restricted with enzymes and fractionated through polyacrylamide gel electrophoresis. As a consequence of the digestion, only the labeled fluorescent end is detected in the gel profiles and these detected bands greatly simplify the analysis of any microbial population in soil (Cetecioglu et al. 2012; Rastogi and Sani 2011). The variation in the number, size and peak height obtained from the analysis of these restriction fragments will provide the data on the biodiversity of the population. However for complete quantitative analysis of the polymorphisms of the restricted bands, the resulting banding profiles may be compared against configured databases to provide valuable comparative community analysis (Rastogi and Sani 2011). This method is applied in identifying the strains, comparative analysis on microbial communities and the estimation of phylogenetic divergence within the community. Community dissection at a higher level may be obtained by inclusion of primers that are specific to phylogenetic groups in the T-RFLP protocol.

2.4.5 Single Strand Conformation Polymorphism (SSCP)

This technique differentiates samples based on the migration mobility in polyacrylamide gel resulting from the variation in the protein structure. This variation is caused by differences in the secondary structure of folded DNA which is a result of sequence differences of single-stranded DNA (ssDNA). Therefore, any given population of fragments of the same size may separate with different mobilities in a non-denaturing PAGE due to the variable conformational change. All fragment lengths analyzed are of uniform size i.e. approximate range of 150–400 bp. Unlike the other gel techniques, this method does not require GC clamped primers nor does it require gel gradients (Cetecioglu et al. 2012; Rastogi and Sani 2011). The SSCP-PCR is ideal to detect polymorphisms that results from mutation in the DNA which contributes in conformational change (Orita et al. 1989). In some circumstances, this technique has been used as an alternative to the DGGE/TGGE. The disadvantages of this systems is that the fragments are between 150–400 bp and that these single stranded DNA fragments are able to form multiple conformations that may be represented as multiple bands (Cetecioglu et al. 2012; Rastogi and Sani 2011).

2.4.6 Random Amplification of Polymorphic DNA (RAPD)

RAPD is based on PCR of randomly chosen single primers that anneal to complimentary sequences in the DNA (Agrawal and Shrivastava 2013). Once these primers are annealed in inverted orientation to the template, several bands are amplified. The products are then fractioned through a gel and the presence or absence of the polymorphic bands in the profile allows for the polymorphism assay. RAPDs are able to distinguish isolates to their taxonomic level based on the primers used. However, while the RAPD method is quick and convenient, this technique has its glitches in reproducibility therefore requiring optimization in every fingerprinting exercise to ensure robustness of data. This technique has been used to elucidate the genetic difference and species diversity in many environments studied (Singh et al. 2005).

2.4.7 Amplification Fragment Length Polymorphism (AFLP)

AFLP is a more robust and stringent method with reproducibility and ability to provide quantifiable data. This method produces a more complex fingerprint compared to RAPD. To provide this quality of data, the technique requires good quality and quantity of DNA in addition to requiring reasonably good experimental skill set (Karp et al. 1996). While AFLP is suitable for determination of genetic distance, mapping and fingerprinting analysis, this method is not amenable for use in comparative genomics involving fast evolving microbes. AFLP is not suitable for use in homologous genomes analysis too (Karp et al. 1996).

2.4.8 Restriction Fragment Length Polymorphism (RFLP)

RFLP is a technique used where restriction endonucleases are used to digest DNA of organisms. Different organisms with different genome content are likely to be digested at different locations within the genome by the same endonuclease. The fragments generated will be different not just in size but in numbers too. The DNA fragments are generally digested with different endonucleases and the profiles are visualized via gel electrophoresis. Therefore, the restriction profiles visualized are able to distinguish the differences between species and also up to strain levels (Avise 1994). Compared to RAPD, the RFLP techniques provides several advantages as follows: (i) any DNA source may be utilized for the analysis; (ii) their codominance is independent of the environment, and (iii) markers mapped to a population are not stressed but rather the effect of phenotypic mutations.

2.5 DNA Sequencing

The Sanger's sequencing method has been used for more than a decade. This method has since been improved on for better efficacy, cheaper cost and rapid data generation (Mecler and Nawrot 2007; Rastogi and Sani 2011). Over the last few years, several new next generation sequencing techniques have been developed using primarily platforms such as 454-based/pyrosequencing and Illumina/Solexa's Genome Analyzer (Margulies et al. 2005). These high-throughput technologies have since become a method of choice for metagenomes and metatranscriptome sequencing projects. The pyrosequencing technique enables the sequencing of DNA or RNA samples from the soil (Lauber et al. 2009; Roesch et al. 2007; Urich et al. 2008). This technique leaves out library generation, template preparation and capillary sequencing (Rothberg and Leamon 2008). The multi-parallelism of the 453 system allows the generation of 450-bp reads of thousands to millions run at once. The

Solexa platform offers higher throughput compared to the 454 but at smaller read lengths. While sequencing is generally looked upon as an unbiased technique, it still is dependent on the quality and quantity of the DNA or RNA. The 454 platform can be used together with the Illumina/Solexa platform where the 454 can generate a longer read and the Illumina/Solexa can fill in the gaps in the data through its high throughput (Quince et al. 2009). Through sequencing it is possible to obtain the information on the most abundant of species to the most rare organisms in the biosphere giving novel insight into the soil microbial communities (Elshahed et al. 2008; Liu et al. 2007; Roesch et al. 2007). Some of the limitations of these methods are in the financial and analysis of large datasets generated through bioinformatics. This method still remains as the most detailed tool for study of microbial diversity, community structure and gene expression (metatranscriptomics) across diverse soils (Lauber et al. 2008, 2009; Urich et al. 2008). We assume that with time, this technology will improve in sensitivity and therefore supersede any other techniques such as the microarray (Lauber et al. 2008, 2009; Roesch et al. 2007; Urich et al. 2008). Programs such as MEtaGenome Analyzer are used to align and assemble the sequence obtained into a finished sequence. These sequences are made available in databases such as National Center for Biotechnology Information (NCBI) and Genomes Online Database (GOLD) for common use by the research community.

2.6 Bioinformatic Tools and Databases Used in Metagenomics

In order for us to make sense of the large amount of data that is generated from soil microbiology studies, the bioinformatics tools and databases are a crucial medium to support the analysis and information generation from these studies. Determination of a sequence homology between an investigated product and thousands of sequences collected in public (National Center for Biotechnology Information NCBI, GenBank), or commercial databases is possible by using suitable computer programs, such as BLAST which is among the most widely used ones (Mecler and Nawrot 2007). The BLAST algorithm is a heuristic program which performs “local” alignments, based on shortcuts, and its task is to conduct a quick search (Tatusova and Madden 1999). An advantage of the molecular tools such as metagenome analysis is the ability for this method to also elucidate the non-culturable and problematic organisms whether from soil or any environment.

2.7 Determination of the DNA Base Ratio (Mole Percent G+C)

A classical genotyping method used in determination of bacterial taxa is the mole percentage of cytosine plus guanosine where the G+C percentage has been reported to be between the range of 20–80% in the bacterial world (Vandamme et al. 1996). The G+C percentage can be determined through thermal denaturation method, HPLC and the buoyant density method (De Ley 1970; Mandel and Marmur 1968; Mesbah et al. 1989). It has been reported that microorganisms differ in the G+C content and related groups differ slightly in their G+C percentage (i.e. 3–5%) (Nüsslein and Tiedje 1999; Tiedje et al. 1999). Through density gradient centrifugation based on G + C content, the fractionation of the total community DNA is determined. The fractionated profile will then provide the information on the relative abundance of any genus or taxa. These profiles can be analyzed further using techniques such as DGGE/ARDRA to provide greater detail on the community diversity.

2.8 Fluorescence In Situ Hybridization (FISH)

The fluorescent hybridization probe technique is employed to detect the presence of rRNA at cellular level with the aid of an epifluorescence microscope. This technique enables correlations to be made with regards to cell metabolic state through the intensity of fluorescent signals in cell. Over the years this technique has advanced in the type of fluorescent dyes developed which have better sensitivity, and multiple fluorochromes. The signals have also been amplified through reporter enzymes, where the catalyzed reporter deposition FISH, with tyramide-labeled fluorochromes, allows enhanced signal emissions (Rogers et al. 2007). Further FISH has been used in combination with secondary-ion mass spectrometry (SIMS) where 16S rRNA probes are used to identify microbes by *in situ* NanoSIMS imaging (Li et al. 2008). This technique is suitable for the detection of microbial density and metabolic state in any given soil sample (Caracciolo et al. 2010).

2.9 DNA: DNA Hybridization (DDH)

This technique allows for the entire genome comparison between strains based on nucleotide level similarities/dissimilarities. In this technique, all the steps that comprise extraction, denaturation and incubation of the sample DNA are conducted in conditions that allows for hybridization and re-association. As comparisons are down to the nucleotide level, the DDH technique is able to differentiate to the species level the organisms within the soil sample. In conducting the DDH analysis a 70% standard was stipulated while a 97% delineation was recommended for the 16S rRNA gene sequence homology (Goris et al. 2007) for species level differentiation. However, this method is not suitable for differentiation at the genus level (Krieg and Holt 1984). In addition, there has also been some inquiry into the suitability of utilizing data obtained from short oligonucleotides and mispairing to extrapolate to whole genomes. Currently, the conversion of DNA-DNA hybridization to whole genome sequence similarities is rather unachievable (Vandamme et al. 1996). There are three forms of hybridization available: The Southern blotting which enables the identification of DNA molecules through DNA/RNA probes, Northern blotting involves RNA molecules analyzed with RNA/DNA probes and finally, Western blot whereby proteins are probes with specialized antibody probes.

2.10 Microarray

For the microarray method the soil DNA that is obtained in fluorescent labeled and brought in contact with the microarray. The array contains thousands upon thousands of oligo-probes that are either 16S based (Phylochip) or functional gene related (Geochip) which hybridizes to the soil DNA at homologous positions. Following hybridization, the signal output from the chips is digitally analyzed. Through the phylogeny relationship analysis (Phylochip) and the functional analysis of the population (Geochip), a high throughput picture is obtained of the heterogeneity of the microbial samples. In a highly diverse sample such as soil, distinguishing complexity may prove to be problematic. When highly abundant 16S rRNA genes fragments are available, cross hybridization becomes an issue due to shared sequence similarities to non-target probes which results in weak signals that are false positive. The currently available phyloarrays can be paired with various techniques, which include 16S cloning, and sequencing or the utilization of fingerprinting techniques such as PCR DGGE. Other than the phyloarray, the Geochip has been utilized successfully to studies the nutrient recycling processes in the soil sample from the Antarctic (Yergeau et al. 2007) where the association between the abundance of these functional genes corresponded to their respective abiotic factors. Functional gene array accompanied with quantitative PCR and enzyme assays has greatly facilitated in validating the microarray hybridization results and thus provides a reliable method on deriving information on the functional element of the microbe (He et al. 2007; Neelakanta and Sultana 2013; Yergeau et al. 2007). However, the lack of robustness and the inability to produce data on novel sequence types is a constraint to the application of the functional gene array. Hence, the information can only be accessed based on the existing breadth of known functions/genes (DeSantis et al. 2007; Yergeau et al. 2007). Despite such challenges, the microarray provides a quick glimpse at the functionality of soil and mycorrhizospheric microbial population (Van Elsas and Boersma 2011).

2.11 Reverse Sample Genome Probing (RSGP)

RSGP is a technique that has been employed to analyze microbiota and to determine dominancy within these species. In this method, the genomic DNA will be isolated from pure cultures and hybridized to determine fragment that underwent cross-hybridization less than 70% which is then followed by the preparation of genomic arrays and finally random labeling of total communities and internal standards. This method is useful only when low diversity is observed in the mixture of total community DNA and internal standard (Greene and Voordouw 2003).

2.12 Postgenomic Approaches

The *in situ* gene expression of microbe can't be deduced from DNA-based molecular approaches (Rastogi and Sani 2011). Therefore postgenomic approaches such as metatranscriptomics and metaproteomics are applied with the available comprehensive metagenomic databases to connect the genetic potential to the functionality in microbial communities (Rastogi and Sani 2011).

2.12.1 Metaproteomic

Metaproteomic is a study on proteins retrieved from environmental microorganisms at a certain point in a microbe's life cycle. (Keller and Hettich 2009; Wilmes and Bond 2006). It functions mainly by providing valuable insights into the interactions between proteins and data on the quantity of proteins. In doing so, there is an opportunity for the elucidation of physiological roles of microbial communities (Keller and Hettich 2009). For example from a soil sample, a few important proteins, enzymes, and chaperones associated in the biodegradation of chlorophenoxy acid were identified through proteomic analysis (Benndorf et al. 2007; Rastogi and Sani 2011). Metaproteomic study encompasses the extraction of proteome from a sample from environment followed by separation of the proteome through one and two-dimensional electrophoresis to produce a proteofingerprint of community and finally the digestion of protein spots that will be then identified through several analyzes (Rastogi and Sani 2011). The advancement in techniques such as chromatography and mass spectroscopy (MS-based proteomics) has enabled microbiologists to perform the profiling of the proteome of microbiota which are high-throughput (Rastogi and Sani 2011). Besides, services provided in the Web like ExPASy (Expert Protein Analysis System; http://www.expasy.org/) provides various tools to identify and characterize the protein mass fingerprinting data (Rastogi and Sani 2011).

2.12.2 Proteogenomics

Most of the protein sequences obtained through proteomic analysis could not be identified with certainty as proteins are poorly related to the available database sequences. As a consequence, protein sequences remain unidentified in terms of their functionality and phylogenetic characteristics (Rastogi and Sani 2011). To overcome this limitation, a new technique known as proteogenomics which integrates metaproteomic and metagenomic approaches has effectively increased the identification of the sequences of protein where the sample of which the proteins were extracted and subjected to metagenomic analysis (Banfield et al. 2005). This method was adopted in a study conducted on phyllosphere bacterial communities which results in an increased number of identified protein, suggesting that most of the microbial communities in phyllosphere were different genetically as compared to those readily available in databases (Delmotte et al. 2009).

2.12.3 Metatranscriptomics

Metatranscriptomics encompass random sequencing of mRNA transcripts obtained from microbiota at a given location and period (Moran 2009). While metagenomics provides information on the genes, this technique further examines the global transcription of genes to comprehend the activity and expression of microbial genes in their natural environments. This technique also surveys the differential expression of genes and their regulation in accordance to the changing environment (Rastogi and Sani 2011). Transcriptomic study can be done by isolating the RNAs in the microbe and selecting the mRNA by synthesizing the cDNA through the portrayal of poly-A tail. However, due to the lack of the poly-A tail in prokaryotic species, rRNA will have to be coextracted together with mRNA and this may lead to massive background sequences (Bashiardes et al. 2016; Rastogi and Sani 2011). Over the years, some improvements have been made to overcome this limitation whereby mRNAs are selectively enriched through subtractive hybridization of rRNA for gene transcript analysis. Besides, double-RNA method is also used in a study to analyze the community based on the total RNA pool which provides a means to study the structure and biochemical properties of microbes all in one go (Urich et al. 2008; Rastogi and Sani 2011). This study produced rRNA tags and mRNA-tags that facilitated understanding of the phylogenetic composition of soil microbial communities from sandy soil samples (Rastogi and Sani 2011). Another study successfully discovered transcripts associated to various biogeochemical processes. (Poretsky et al. 2005).

2.13 Concluding Remarks

In this chapter, we addressed the issue of assessing the microbial community within the mycorrhizosphere. As mentioned above, this zone has not been extensively studied mainly due to the inability at times to separate the organisms (endosymbionts) from the host and to also culture some of the bacteria and fungi in the lab. These have posed an obstacle on obtaining a clear picture of the soil ecosystem. Throughout the chapter we have provided a background on the various techniques that are now available for those who seek to decipher the mycorrhizosphere community. We begin with the basic DNA and RNA extraction to the library construction and the application of PCR techniques in fingerprinting of the samples. The more recent techniques however such as microarray and sequencing provide larger amount of information on the microbiota that is evident within the soil community.

Through the availability of the multiple techniques that have been outlined and the continuous advancements made in each technology we posit that with time we will be able to gather core information on the microbial structure and community within the mycorrhizosphere. However more importantly, we need clearly defined objectives and scope of research and use the techniques or combination of techniques to get a better overview of the ecosystem. In addition, while information of microbial diversity is useful, we need to focus on the functionality of these organisms. The molecular based post-genomic techniques such as metagenomics, proteogenomic and metatranscriptomics has provided a new level of understanding into the different and fascinating processes that occurs within the microbial communities. Through the utilization of the above tools, the interactions within the microcosm might be directly assessed.

However, given the overall nature of these methods, it is strongly recommended that to obtain a better overview of the ecosystem, studies should:

1. Directly analyse microorganisms based on molecular methods
2. Detect microbial activities through methods that enable *in situ* analysis
3. Isolate and question the contribution of these organisms in their given eco-physiological behavior and thence use this to predict their *in situ* behavior.

Through the information derived from the molecular studies conducted on the soil sample, we believe that various questions with regards to relationships, diversity, products and application may be answered. However as with any knowledge, the more we unravel, the more questions will arise. Figure 23.2 provides a diagrammatic representation of the outcome of molecular soil analysis.

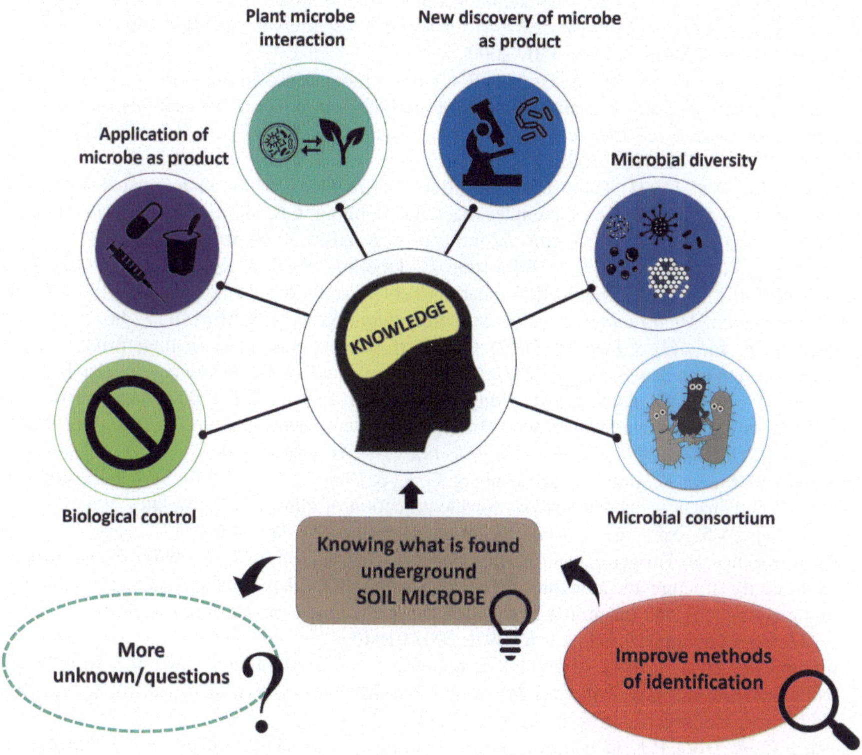

Fig. 23.2 Shows how the information derived from the mycorrhizosphere may be used to answer several questions with regards to soil health and the advancement of knowledge and techniques

References

Agrawal, P. K., & Shrivastava, R. (2013). Molecular markers. In J. Saxena, I. Ravi, & M. Bhaunthiyal (Eds.), *Advance in biotechnology* (pp. 25–39). New Delhi: Springer.

Amann, R. I., Ludwig, W., & Schleifer, K.-H. (1995). Phylogenetic identification and in situ detection of individual microbial cells without cultivation. *Microbiological Reviews, 59*, 143–169.

Atkins, S. D., & Clark, I. M. (2004). Fungal molecular diagnostics: A mini review. *Journal of Applied Genetics, 45*, 3–15.

Avise, J. C. (1994). Molecular tools. In *Molecular markers, natural history and evolution* (pp. 44–91). Boston: Springer. https://doi.org/10.1007/978-1-4615-2381-9.

Banfield, J. F., Verberkmoes, N. C., Hettich, R. L., & Thelen, M. P. (2005). Proteogenomic approaches for the molecular characterization of natural microbial communities. *OMICS: A Journal of Integrative Biology, 9*, 301–333. https://doi.org/10.1089/omi.2005.9.301.

Bashiardes, S., Zilberman-Schapira, G., & Elinav, E. (2016). Use of metatranscriptomics in microbiome research. *Bioinformatics and Biology Insights, 10*, BBI. S34610.

Benndorf, D., Balcke, G. U., Harms, H., & Von Bergen, M. (2007). Functional metaproteome analysis of protein extracts from contaminated soil and groundwater. *The ISME Journal, 1*, 224. https://doi.org/10.1038/ismej.2007.39.

Bianciotto, V., Minerdi, D., Perotto, S., & Bonfante, P. (1996). Cellular interactions between arbuscular mycorrhizal fungi and rhizosphere bacteria. *Protoplasma, 193*, 123–131. https://doi.org/10.1007/BF01276640.

Bianciotto, V., Lumini, E., Lanfranco, L., Minerdi, D., Bonfante, P., & Perotto, S. (2000). Detection and identification of bacterial endosymbionts in arbuscular mycorrhizal fungi belonging to the family Gigasporaceae. *Applied and Environmental Microbiology, 66*, 4503–4509. https://doi.org/10.1128/AEM.66.10.4503-4509.2000.

Caracciolo, A. B., Bottoni, P., & Grenni, P. (2010). Fluorescence in situ hybridization in soil and water ecosystems: A useful method for studying the effect of xenobiotics on bacterial community structure. *Toxicological & Environmental Chemistry, 92*, 567–579. https://doi.org/10.1080/02772241003620244.

Cardinale, M., et al. (2004). Comparison of different primer sets for use in automated ribosomal intergenic spacer analysis of complex bacterial communities. *Applied and Environmental Microbiology, 70*, 6147–6156. https://doi.org/10.1128/AEM.70.10.6147-6156.2004.

Case, R. J., Boucher, Y., Dahllöf, I., Holmström, C., Doolittle, W. F., & Kjelleberg, S. (2007). Use of 16S rRNA and *rpoB* genes as molecular markers for microbial ecology studies. *Applied and Environmental Microbiology, 73*, 278–288. https://doi.org/10.1128/AEM.01177-06.

Cetecioglu, Z., Ince, O., & Ince, B. (2012). Gel electrophoresis based genetic fingerprinting techniques on environmental ecology. In *Gel electrophoresis-advanced techniques*. Rijeka: InTech.

Ciardo, D. E., Schär, G., Altwegg, M., Böttger, E. C., & Bosshard, P. P. (2007). Identification of moulds in the diagnostic laboratory—An algorithm implementing molecular and phenotypic methods. *Diagnostic Microbiology and Infectious Disease, 59*, 49–60.

Ciardo, D. E., Lucke, K., Imhof, A., Bloemberg, G. V., & Böttger, E. C. (2010). Systematic internal transcribed spacer sequence analysis for identification of clinical mold isolates in diagnostic mycology: A 5-year study. *Journal of Clinical Microbiology, 48*, 2809–2813.

Costa, R., Gomes, N., Krögerrecklenfort, E., Opelt, K., Berg, G., & Smalla, K. (2007). *Pseudomonas* community structure and antagonistic potential in the rhizosphere: Insights gained by combining phylogenetic and functional gene-based analyzes. *Environmental Microbiology, 9*, 2260–2273. https://doi.org/10.1111/j.1462-2920.2007.01340.x.

Coutinho, M., et al. (2013). Microbial communities on deteriorated artistic tiles from Pena National Palace (Sintra, Portugal). *International Biodeterioration & Biodegradation, 84*, 322–332. https://doi.org/10.1016/j.ibiod.2012.05.028.

Dakal, T. C., & Arora, P. K. (2012). Evaluation of potential of molecular and physical techniques in studying biodeterioration. *Reviews in Environmental Science and Bio/Technology, 11*, 71–104. https://doi.org/10.1007/s11157-012-9264-0.

De Ley, J. (1970). Reexamination of the association between melting point, buoyant density, and chemical base composition of deoxyribonucleic acid. *Journal of Bacteriology, 101*, 738–754.

Deja-Sikora, E., Sikora, M., Golebiewski, M., & Tretyn, A. (2007). Metagenomic libraries as sources of genes useful for biotechnology. *Biotechnologia, 4*, 125–139.

Delmotte, N., et al. (2009). Community proteogenomics reveals insights into the physiology of phyllosphere bacteria. *Proceedings of the National Academy of Sciences, 106*, 16428–16433. https://doi.org/10.1073/pnas.0905240106.

DeSantis, T. Z., Brodie, E. L., Moberg, J. P., Zubieta, I. X., Piceno, Y. M., & Andersen, G. L. (2007). High-density universal 16S rRNA microarray analysis reveals broader diversity than typical clone library when sampling the environment. *Microbial Ecology, 53*, 371–383. https://doi.org/10.1007/s00248-006-9134-9.

Elshahed, M. S., et al. (2008). Novelty and uniqueness patterns of rare members of the soil biosphere. *Applied and Environmental Microbiology, 74*, 5422–5428. https://doi.org/10.1128/AEM.00410-08.

Fakruddin, M., & Mannan, K. (2013). Methods for analyzing diversity of microbial communities in natural environments. *Ceylon Journal of Science (Biological Sciences), 42*. https://doi.org/10.4038/cjsbs.v42i1.5896.

Finlay, B. B., & Medzhitov, R. (2007). Host-microbe interactions: Fulfilling a niche. *Cell Host & Microbe, 1*, 3–4.

Fisher, M. M., & Triplett, E. W. (1999). Automated approach for ribosomal intergenic spacer analysis of microbial diversity and its application to freshwater bacterial communities. *Applied and Environmental Microbiology, 65*, 4630–4636.

Gaylarde, C. C., Rodríguez, C. H., Navarro-Noya, Y. E., & Ortega-Morales, B. O. (2012). Microbial biofilms on the sandstone monuments of the Angkor Wat complex, Cambodia. *Current Microbiology, 64*, 85–92. https://doi.org/10.1007/s00284-011-0034-y.

Giacomucci, L., et al. (2011). Microbial deterioration of artistic tiles from the façade of the Grande Albergo Ausonia & Hungaria (Venice, Italy). *Microbial Ecology, 62*, 287–298. https://doi.org/10.1007/s00248-011-9812-0.

Gich, F. B., Amer, E., Figueras, J. B., Charles, A., Abella, C. A. A., Balaguer, M. D., & Poch, M. (2000). Assessment of microbial community structure changes by amplified ribosomal DNA restriction analysis (ARDRA). *International Microbiology, 3*, 103–106.

González, J. M., & Saiz-Jiménez, C. (2005). Application of molecular nucleic acid-based techniques for the study of microbial communities in monuments and artworks. *International Microbiology, 8*, 189.

Gonzalez, J., Ortiz-Martinez, A., Gonzalez-delValle, M., Laiz, L., & Saiz-Jimenez, C. (2003). An efficient strategy for screening large cloned libraries of amplified 16S rDNA sequences from complex environmental communities. *Journal of Microbiological Methods, 55*, 459–463.

Goris, J., Konstantinidis, K. T., Klappenbach, J. A., Coenye, T., Vandamme, P., & Tiedje, J. M. (2007). DNA–DNA hybridization values and their relationship to whole-genome sequence similarities. *International Journal of Systematic and Evolutionary Microbiology, 57*, 81–91. https://doi.org/10.1099/ijs.0.64483-0.

Greene, E. A., & Voordouw, G. (2003). Analysis of environmental microbial communities by reverse sample genome probing. *Journal of Microbiological Methods, 53*, 211–219.

He, L., et al. (2007). A microRNA component of the p53 tumour suppressor network. *Nature, 447*, 1130. https://doi.org/10.1038/nature05939.

Hugenholtz, P., Goebel, B. M., & Pace, N. R. (1998). Impact of culture-independent studies on the emerging phylogenetic view of bacterial diversity. *Journal of Bacteriology, 180*, 4765–4774.

Hugenholtz, P., Tyson, G. W., Webb, R. I., Wagner, A. M., & Blackall, L. L. (2001). Investigation of candidate division TM7, a recently recognized major lineage of the domain Bacteria with no known pure-culture representatives. *Applied and Environmental Microbiology, 67*, 411–419. https://doi.org/10.1128/AEM.67.1.411-419.2001.

Jacobi, S. K., Xi, L., Maltecca, C., Borst, L., Smith, A., & Odle, J. (2017). Dietary prebiotics and arachidonic acid (ARA) modulate intestinal injury and microbial taxa following acute dextran sodium sulfate induced colitis in formula-fed piglets. *The FASEB Journal, 31*, lb324–lb324.

Karp, A., Seberg, O., & Buiatti, M. (1996). Molecular techniques in the assessment of botanical diversity. *Annals of Botany, 78*, 143–149.
Keller, M., & Hettich, R. (2009). Environmental proteomics: A paradigm shift in characterizing microbial activities at the molecular level. *Microbiology and Molecular Biology Reviews, 73*, 62–70. https://doi.org/10.1128/MMBR.00028-08.
Kolb, S., Knief, C., Stubner, S., & Conrad, R. (2003). Quantitative detection of methanotrophs in soil by novel pmoA-targeted real-time PCR assays. *Applied and Environmental Microbiology, 69*, 2423–2429. https://doi.org/10.1128/AEM.69.5.2423-2429.2003.
Krieg, N., & Holt, J. (1984). *Bergey's manual of systemic bacteriology* (Vol. 1). The William & Wilkins Co: Baltimore. https://doi.org/10.1007/978-0-387-21609-6.
Lauber, C. L., Strickland, M. S., Bradford, M. A., & Fierer, N. (2008). The influence of soil properties on the structure of bacterial and fungal communities across land-use types. *Soil Biology and Biochemistry, 40*, 2407–2415.
Lauber, C. L., Hamady, M., Knight, R., & Fierer, N. (2009). Pyrosequencing-based assessment of soil pH as a predictor of soil bacterial community structure at the continental scale. *Applied and Environmental Microbiology, 75*, 5111–5120. https://doi.org/10.1128/AEM.00335-09.
Li, T., Wu, T. D., Mazéas, L., Toffin, L., Guerquin-Kern, J. L., Leblon, G., & Bouchez, T. (2008). Simultaneous analysis of microbial identity and function using NanoSIMS. *Environmental Microbiology, 10*, 580–588. https://doi.org/10.1111/j.1462-2920.2007.01478.x.
Liu, W.-T., Marsh, T. L., Cheng, H., & Forney, L. J. (1997). Characterization of microbial diversity by determining terminal restriction fragment length polymorphisms of genes encoding 16S rRNA. *Applied and Environmental Microbiology, 63*, 4516–4522.
Liu, Z., Lozupone, C., Hamady, M., Bushman, F. D., & Knight, R. (2007). Short pyrosequencing reads suffice for accurate microbial community analysis. *Nucleic Acids Research, 35*, e120. https://doi.org/10.1093/nar/gkm541.
Mandel, M., & Marmur, J. (1968). [109] Use of ultraviolet absorbance-temperature profile for determining the guanine plus cytosine content of DNA. *Methods in enzymology, 12*, 195–206. Elsevier. https://doi.org/10.1016/0076-6879(67)12133-2.
Margulies, M., et al. (2005). Genome sequencing in microfabricated high-density picolitre reactors. *Nature, 437*, 376. https://doi.org/10.1038/nature03959.
McNamara, C. J., Perry, T. D., Bearce, K. A., Hernandez-Duque, G., & Mitchell, R. (2006). Epilithic and endolithic bacterial communities in limestone from a Maya archaeological site. Microbiology. *Ecology, 51*, 51–64. https://doi.org/10.1007/s00248-005-0200-5.
Mecler, I., & Nawrot, U. (2007). Molecular methods applied in microbiological diagnosis. *Mikologia Lekarska, 14*, 280.
Mesbah, M., Premachandran, U., & Whitman, W. B. (1989). Precise measurement of the G+ C content of deoxyribonucleic acid by high-performance liquid chromatography. *International Journal of Systematic and Evolutionary Microbiology, 39*, 159–167. https://doi.org/10.1099/00207713-39-2-159.
Mocali, S., & Benedetti, A. (2010). Exploring research frontiers in microbiology: The challenge of metagenomics in soil microbiology. *Research in Microbiology, 161*, 497–505. https://doi.org/10.1016/j.resmic.2010.04.010.
Moran, M. A. (2009). Metatranscriptomics: Eavesdropping on Complex Microbial Communities-Large-scale sequencing of mRNAs retrieved from natural communities provides insights into microbial activities and how they are regulated. *Microbe, 4*, 329.
Muller, M. P., et al. (2016). Antimicrobial surfaces to prevent healthcare-associated infections: A systematic review. *Journal of Hospital Infection, 92*, 7–13.
Muyzer, G. (1999). DGGE/TGGE a method for identifying genes from natural ecosystems. *Current Opinion in Microbiology, 2*, 317–322. https://doi.org/10.1016/S1369-5274(99)80055-1.
Muyzer, G., & Smalla, K. (1998). Application of denaturing gradient gel electrophoresis (DGGE) and temperature gradient gel electrophoresis (TGGE) in microbial ecology. *Antonie van Leeuwenhoek, 73*, 127–141.

Muyzer, G., De Waal, E. C., & Uitterlinden, A. G. (1993). Profiling of complex microbial populations by denaturing gradient gel electrophoresis analysis of polymerase chain reaction-amplified genes coding for 16S rRNA. *Applied and Environmental Microbiology, 59*, 695–700.

Nadarajah, K. (2017). Rhizosphere interactions: Life below ground. In *Plant-microbe interaction: An approach to sustainable agriculture* (pp. 3–23). Singapore: Springer. https://doi.org/10.1007/978-981-10-2854-0_1.

Neelakanta, G., & Sultana, H. (2013). The use of metagenomic approaches to analyze changes in microbial communities. *Microbiology Insights, 6*, MBI. S10819. https://doi.org/10.4137/MBI.S10819.

Nüsslein, K., & Tiedje, J. M. (1999). Soil bacterial community shift correlated with change from forest to pasture vegetation in a tropical soil. *Applied and Environmental Microbiology, 65*, 3622–3626.

Orita, M., Suzuki, Y., Sekiya, T., & Hayashi, K. (1989). Rapid and sensitive detection of point mutations and DNA polymorphisms using the polymerase chain reaction. *Genomics, 5*, 874–879.

Piñar, G., Ramos, C., Rölleke, S., Schabereiter-Gurtner, C., Vybiral, D., Lubitz, W., & Denner, E. B. (2001a). Detection of indigenous *Halobacillus* populations in damaged ancient wall paintings and building materials: Molecular monitoring and cultivation. *Applied and Environmental Microbiology, 67*, 4891–4895.

Piñar, G., Saiz-Jimenez, C., Schabereiter-Gurtner, C., Blanco-Varela, M. T., Lubitz, W., & Rölleke, S. (2001b). Archaeal communities in two disparate deteriorated ancient wall paintings: Detection, identification and temporal monitoring by denaturing gradient gel electrophoresis. *FEMS Microbiology Ecology, 37*, 45–54. https://doi.org/10.1111/j.1574-6941.2001.tb00852.x.

Piñar, G., Ripka, K., Weber, J., & Sterflinger, K. (2009). The micro-biota of a sub-surface monument: The medieval chapel of St. Virgil (Vienna, Austria). *International Biodeterioration & Biodegradation, 63*, 851–859.

Piñar, G., Piombino-Mascali, D., Maixner, F., Zink, A., & Sterflinger, K. (2013). Microbial survey of the mummies from the Capuchin Catacombs of Palermo, Italy: Biodeterioration risk and contamination of the indoor air. *FEMS Microbiology Ecology, 86*, 341–356. https://doi.org/10.1111/1574-6941.12165.

Popa, R., Popa, R., Mashall, M. J., Nguyen, H., Tebo, B. M., & Brauer, S. (2009). Limitations and benefits of ARISA intra-genomic diversity fingerprinting. *Journal of Microbiological Methods, 78*, 111–118. https://doi.org/10.1016/j.mimet.2009.06.005.

Poretsky, R. S., et al. (2005). Analysis of microbial gene transcripts in environmental samples. *Applied and Environmental Microbiology, 71*, 4121–4126. https://doi.org/10.1128/AEM.71.7.4121-4126.2005.

Puppi, G., Azcón, R., & Höflich, G. (1994). Management of positive interactions of arbuscular mycorrhizal fungi with essential groups of soil microorganisms. In *Impact of arbuscular mycorrhizas on sustainable agriculture and natural ecosystems* (pp. 201–215). Springer. https://doi.org/10.1007/978-3-0348-8504-1_16.

Quince, C., et al. (2009). Accurate determination of microbial diversity from 454 pyrosequencing data. *Nature Methods, 6*, 639. https://doi.org/10.1038/nmeth.1361.

Rastogi, G., & Sani, R. K. (2011). Molecular techniques to assess microbial community structure, function, and dynamics in the environment. In *Microbes and microbial technology* (pp. 29–57). Springer.

Roesch, L. F., et al. (2007). Pyrosequencing enumerates and contrasts soil microbial diversity. *The ISME Journal, 1*, 283. https://doi.org/10.1038/ismej.2007.53.

Rogers, S. W., Moorman, T. B., & Ong, S. K. (2007). Fluorescent in situ hybridization and micro-autoradiography applied to ecophysiology in soil. *Soil Science Society of America Journal, 71*, 620–631. https://doi.org/10.2136/sssaj2006.0105.

Ross, J. (1980). Effect of nontreated field soil on sporulation of vesicular-arbuscular mycorrhizal fungi associated with soybean. *Phytopathology, 70*, 1200–1205.

Rothberg, J. M., & Leamon, J. H. (2008). The development and impact of 454 sequencing. *Nature Biotechnology, 26*, 1117. https://doi.org/10.1038/nbt1485.

Santos-González, J. C., Finlay, R. D., & Tehler, A. (2007). Seasonal dynamics of arbuscular mycorrhizal fungal communities in roots in a seminatural grassland. *Applied and Environmental Microbiology, 73*, 5613–5623. https://doi.org/10.1128/AEM.00262-07.

Sehgal, A. K., & Sagar, A. (2017). Ectomycorrhiza and fungal diversity in the mycorrhizosphere of *Pinus gerardiana. International Journal of Pure Applied Bioscience, 5*, 475–483.

Simon, L., Lalonde, M., & Bruns, T. (1992). Specific amplification of 18S fungal ribosomal genes from vesicular-arbuscular endomycorrhizal fungi colonizing roots. *Applied and Environmental Microbiology, 58*, 291–295.

Singh, S., Khawale, R., Singh, R., & Vimala, Y. (2005). Use of random amplified polymorphic DNA (RAPD) analysis to confirm genetic stability of *in vitro* regenerated grape plantlets. *Indian Journal of Horticulture, 62*, 12–15.

Streit, W. R., & Schmitz, R. A. (2004). Metagenomics–the key to the uncultured microbes. *Current Opinion in Microbiology, 7*, 492–498. https://doi.org/10.1016/j.mib.2004.08.002.

Tacão, M., Moura, A., Alves, A., Henriques, I., Saavedra, M. J., & Correia, A. (2005). Evaluation of 16S rDNA-and gyrB-DGGE for typing members of the genus *Aeromonas. FEMS Microbiology Letters, 246*, 11–18. https://doi.org/10.1016/j.femsle.2005.03.033.

Tatusova, T. A., & Madden, T. L. (1999). BLAST 2 sequences, a new tool for comparing protein and nucleotide sequences. *FEMS Microbiology Letters, 174*, 247–250.

Tiedje, J. M., Asuming-Brempong, S., Nüsslein, K., Marsh, T. L., & Flynn, S. J. (1999). Opening the black box of soil microbial diversity. *Applied Soil Ecology, 13*, 109–122. https://doi.org/10.1016/S0929-1393(99)00026-8.

Tommerup, I. (1985). Inhibition of spore germination of vesicular-arbuscular mycorrhizal fungi in soil. *Transactions of the British Mycological Society, 85*, 267–278. https://doi.org/10.1016/S0007-1536(85)80188-1.

Urich, T., Lanzén, A., Qi, J., Huson, D. H., Schleper, C., & Schuster, S. C. (2008). Simultaneous assessment of soil microbial community structure and function through analysis of the meta-transcriptome. *PLoS One, 3*, e2527. https://doi.org/10.1371/journal.pone.0002527.

Van Elsas, J., & Boersma, F. (2011). *A review of molecular methods to study the microbiota of soil and the mycosphere European Journal of Soil Biology, 47*, 77–87.

van Elsas, J. D., Trevors, J. T., Jansson, J. K., & Nannipieri, P. (2006). *Modern soil microbiology*. CRC Press.

Vandamme, P., Pot, B., Gillis, M., De Vos, P., Kersters, K., & Swings, J. (1996). Polyphasic taxonomy, a consensus approach to bacterial systematics. *Microbiological Reviews, 60*, 407–438.

Waleron, M., Waleron, K., Geider, K., & Lojkowska, E. (2008). Application of RFLP analysis of *recA*, *gyrA* and *rpoS* gene fragments for rapid differentiation of *Erwinia amylovora* from *Erwinia* strains isolated in Korea and Japan. *European Journal of Plant Pathology, 121*, 161–172. https://doi.org/10.1007/s10658-007-9260-3.

Więckowicz, M. (2009). Molekularne metody identyfikacji mikroorganizmów w złożonych ekosystemach. *Post Mikrobiol, 48*, 67–73. (in Polish).

Wilmes, P., & Bond, P. L. (2006). Metaproteomics: Studying functional gene expression in microbial ecosystems. *Trends in Microbiology, 14*, 92–97. https://doi.org/10.1016/j.tim.2005.12.006.

Wilson, G., Hetrick, B. D., & Kitt, D. G. (1988). Suppression of mycorrhizal growth response of big bluestem by non-sterile soil. *Mycologia*, 338–343. https://doi.org/10.2307/3807630.

Yergeau, E., Kang, S., He, Z., Zhou, J., & Kowalchuk, G. A. (2007). Functional microarray analysis of nitrogen and carbon cycling genes across an Antarctic latitudinal transect. *The ISME Journal, 1*, 163. https://doi.org/10.1038/ismej.2007.24.